Plants in Danger
What do we know?

INTERNATIONAL UNION FOR CONSERVATION
OF NATURE AND NATURAL RESOURCES

Plants in Danger
What do we know?

STEPHEN D. DAVIS, STEPHEN J.M. DROOP, PATRICK GREGERSON,
LOUISE HENSON, CHRISTINE J. LEON, JANE LAMLEIN
VILLA-LOBOS, HUGH SYNGE AND JANA ZANTOVSKA

Threatened Plants Unit,
IUCN Conservation Monitoring Centre
c/o The Royal Botanic Gardens, Kew, U.K.

Published by the International Union for Conservation of Nature and Natural Resources,
Gland, Switzerland, and Cambridge, U.K. 1986

IUCN

The International Union for Conservation of Nature and Natural Resources (IUCN) is a network of governments, non-governmental organizations (NGOs), scientists and other conservation experts, joined together to promote the protection and sustainable use of living resources.

Founded in 1948, IUCN has more than 500 member organizations from 116 countries, including 58 State Members. Its six Commissions consist of more than 2000 experts on threatened species, protected areas, ecology, environmental planning, environmental policy, law and administration, and environmental education. IUCN

- monitors the status of ecosystems and species throughout the world;
- plans conservation action, both at strategic level through the World Conservation Strategy and at the programme level through its programme of conservation for sustainable development;
- promotes such action by governments, inter-governmental bodies and non-governmental organizations;
- provides the assistance and advice necessary to achieve such action.

From 1984 IUCN and the World Wildlife Fund have been implementing a Plant Conservation Programme, designed "to assert the fundamental importance of plants in all conservation activities". *Plants in Danger: What do we know?* is a part of this programme.

The IUCN Conservation Monitoring Centre (CMC) is the division of IUCN that provides a data service to IUCN and to the conservation and development community. CMC's primary function is the continuous collection, analysis, interpretation and dissemination of data as a basis for conservation. CMC produces a wide variety of specialist outputs and analyses as well as major outputs such as the Red Data Books and Protected Areas Directories. CMC is based in the U.K. at Cambridge and Kew. Enquiries about the centre or book orders should be addressed to:

IUCN Conservation Monitoring Centre,
219(c) Huntingdon Road, Cambridge, CB3 0DL, U.K.

The designations of geographical entities in this book, and the presentation of the material, do not imply the expression of any opinion whatsoever on the part of IUCN concerning the legal status of any country, territory, or area, or of its authorities, or concerning the delimitation of its frontiers or boundaries.

Published by IUCN, Gland, Switzerland, and Cambridge, U.K. 1986

Prepared with financial support from the World Wildlife Fund, the Trust Fund for the United Nations Environment Stamp Conservation Fund, the United Nations Environment Programme and the Natural Environment Research Council (U.K.) on behalf of the European Research Councils through the European Science Foundation.
A contribution to GEMS — The Global Environment Monitoring System.

ISBN 2-88032-707-5

Printed by Unwin Brothers Ltd, The Gresham Press, Old Woking, Surrey, U.K.
Typeset by Parchment (Oxford) Ltd., 60 Hurst Street, Oxford OX4 1HD
Cover design by James Butler and Stephen Droop
Figures by Reginald Piggott
Book design by James Butler
Cover photograph by M.P. Price (Bruce Coleman Ltd.): Fire, Merritt Island, Florida, U.S.A.

Contents

Appendices

Geographical Index

Preface

Over the last ten years, a vast amount has been written and published on threatened plants, often in rather inaccessible places. Numerous countries have prepared Red Data Books of their threatened flora. Yet it is also clear that plant conservation is not succeeding in most parts of the world and is not yet fully accepted as a fundamental part of conservation as a whole. One reason may simply be that many conservationists do not know how much information on plants is already available. This would not be unduly surprising, as most efforts to list threatened plants have emerged from herbaria and botanic gardens, rather than from conservation groups. Botanists are concerned about the threats to the plants they study from day to day and anxious to provide at least an assessment of the problem. Yet, although individual botanists may be the best people to assess which species are in danger, conservation organizations, with successful track records in other fields of conservation, are surely in a far better position to turn that knowledge into effective action on the ground.

The purpose of this book is to provide these conservation organizations with a concise guide to information on threatened plants. Rather than providing information on each threatened plant, which would be impossible in one book, we show how to find that information. The entries are arranged alphabetically by country, so as to answer the questions, "Where can I find out about the flora of any country, which species in that flora are threatened, and who may be trying to save them?"

The book forms part of the IUCN/WWF Plant Conservation Programme. This is a set of around 90 activities, derived from the philosophy and principles of the World Conservation Strategy. Long overdue, its aim is two-fold: firstly, to provide a strategic basis for plant conservation, and secondly, by means of model projects, to show how this knowledge can be applied on the ground. As part of the first aim, IUCN is preparing about 10 books and major papers, of which this book is one. Others include an illustrated account for the layman of why plant conservation is important (*Green Inheritance* by Anthony Huxley, 1985) and a Conservation Strategy for Botanic Gardens (1985-6). At early stages of preparation are a book on the principles and practice of plant conservation, and a Red Data Book of plant sites where high numbers of plant species could be saved. Other activities cover education, training and institution-building. Special themes, in addition to threatened plants, are the issue of genetic resources, the status of economic plants and the role of botanic gardens in conservation.

The concepts developed in the strategic part of the programme are being applied in field projects in 16 selected countries. These include, for example, a rescue programme for the critically endangered Mauritian flora; land use surveys of threatened areas like the Usambaras and Ulugurus of Tanzania; support for large plant-rich national parks like La Amistad (Costa Rica), Taï (Ivory Coast) and Manu (Peru); support for planning networks of protected areas in Borneo and Irian Jaya; conservation of medicinal plants in Sri Lanka, of teosintes in Mexico and of multipurpose palm species in Latin America; and education about plant conservation in India.

As these activities show, research on threatened plants and rescue of their populations are only part of plant conservation. Yet it is on this aspect that most of the research and data-gathering has concentrated, at least until very recently. *Plants in Danger: What do we know?* charts the results of that work, but intentionally does not extend to other, more recent, topics in plant conservation. For instance, few references are given on the conservation of economic plants; in this case, and in others, the priority is not so much data synthesis as conceptual development and pilot projects which will show, for example, how the genetic variation of economic plants can best be conserved *in situ* as well as *ex situ*. As the spotlight widens to include topics such as the conservation of medicinal plants and the better use of traditional knowledge about plants useful to man, it seemed sensible to document the quite remarkable progress that has been made in the last decade or so in finding out which species are threatened.

It is our hope and intention that the knowledge outlined in this book will encourage action to save the threatened plants documented so assiduously by botanists all over the world. Although more research is needed, enough is known about the threats to plant life for action to be taken now: for instance, creation of national parks and biosphere reserves, better use of botanic gardens, and enactment of more effective laws to control plant collecting and plant trade. For of all the changes that man can make to the Earth, none is more permanent or more wasteful than the extinction of a species.

Acknowledgements

This book could not have been written without a great deal of help from many people. It is our pleasure to acknowledge and thank over 400 botanists who helped us and contributed information. Virtually all whom we approached offered their help. We are most grateful. The response we received, literally overwhelming at times, and the masses of additional data accumulated, are the main reason the book was delayed from its original publication date at the end of 1984.

We would like to thank especially those scientists who reviewed and commented on the drafts for whole regions and contributed so much to the book. Their help was vital in ensuring overall consistency and completeness. In some cases, reviewers most kindly spent many hours carefully checking manuscripts, finding obscure and difficult references for us and sharing their knowledge with us. Here we thank in particular C.D. Adams (Caribbean), P.S. Ashton (Asia), M.M.J. van Balgooy (Asia), F.R. Fosberg (Pacific), J.B. Gillett (Africa), B. MacBryde (New World), R. Polhill (Africa), G.T. Prance (New World Tropics), P.H. Raven and his colleagues at Missouri Botanical Garden (all the tropics) and V.M. Toledo (Latin America). We also thank L. McMahan and J. McKnight at WWF-U.S. for their help with the New World accounts and WWF-U.S. in general for their continued support to CMC. We thank especially those botanists who contributed country accounts for us; we want to mention here the contribution of R.A. DeFilipps, who not only wrote the account for the U.S.A. (with P. Gregerson), by far the longest in the book, but also gave extensive help with many other accounts.

We also warmly thank our colleagues in the Library and Herbarium at Kew. Preparing the book has drawn heavily on the splendid facilities of the Kew Library and we are most grateful to the staff for patiently coping with our many requests. Above all, we would like to thank the staff of the Herbarium, in particular the Keeper, G.Ll. Lucas, for their continued support. The Threatened Plants Unit of IUCN's Conservation Monitoring Centre developed within the Kew Herbarium and continues to benefit greatly from its presence there. IUCN is deeply grateful to the Royal Botanic Gardens, Kew, and to its Director, E.A. Bell, for their magnificent support that has now lasted over 10 years.

The sections for countries of Latin America were written by Patrick Gregerson and Jane Lamlein Villa-Lobos of the Smithsonian Institution, with whom IUCN has a co-operative arrangement for data-gathering on threatened plants in that region. IUCN is most grateful to the Smithsonian for their help and acknowledges with pleasure the contributions of their scientists.

In a sense the real authors of this book are the very many experts who spared time to comment, and in many cases rewrite, the accounts for the places on which they are the acknowledged experts. For their help and for sharing their knowledge, we thank E. Adjanohoun, J.M. Aguilar Cumes, J.R. Akeroyd, D.M. Al-Eisawi, A.H. Al-Khayat, A. Alnen, R.M. Alfaro, S.I. Ali, S. Andrews, G.W. Argus, E.O.A. Asibey, G.G. Aymonin, J.A. Bacone, P. Bamps, C. Barclay, W.T. Barker, T.M. Barkley, T. Baytop, H.E. Beaty, S. Beck, L.J. Beloussova, D. Benkert, G. Benl, R.W. Boden, P. Boniface, I. Bonnelly de Calventi, A. Borhidi, J. Bosser, D. Bramwell, F.J. Breteler, P. Broussalis, R.E. Brown, R.K. Brummitt, W. Burger, W. Burley, R. Burton, R. Bye, L.J.T. Cadet, J. Cerovsky, J.D. Chapman, A.O. Chater, M.N. Chaudhri, A. Cheke, S. Cheng-kui, M. Chilcott, G.L. Church, S. Cochrane, M. Cohen, N.H.A. Cole, J.B. Comber, P. Condy, M. Conrad, M.J.E. Coode, T.A. Cope, F. Corbetta, R.A. Countryman, P. Coyne, P.J. Cribb, J.R. Croft, B.S. Croxall, K. Curry-Lindahl, W. D'Arcy, J.-P. D'Huart, E. D'Souza, A.

Danin, B. De Winter, R.A. DeFilipps, H. Demirez, G. Dennis, G. Dihoru, M. Dillon, M.G. Dlamini, C.H. Dodson, D.D. Doone, L.E. Dorr, F. Dowsett-Lemaire, J. Dransfield, A.M. Dray, R.W. Dwyer, J. Dwyer, E. Einarsson, J.M. Engel, H. Ern, L. Escobar, R. Faden, P. Fairburn, L. Farrell, J.M. Fay, J. Feilberg, K. Ferguson, A.A. Ferrar, H. Fink, M.A. Fischer, J.J. Floret, E. Forero, L.L. Forman, B. Fredskild, J.D. Freeman, F. Friedmann, I. Friis, E. Gabrielian, Z.O. Gbile, C. Geerling, D. Geltman, A.H. Gentry, A. George, B. Gibbs-Russell, M.G. Gilbert, D.R. Given, L. Godicl, E.E. Gogina, P. Goldblatt, P. Gölz, L.D. Gómez P., C. Gómez-Campo, J.-J. de Granville, W. Greuter, C. Grey-Wilson, V.I. Grubov, C.V.S. Gunatilleke, M.N. el Hadidi, W. Hahn, A.V. Hall, N. Hallé, O. Hamann, H. Hamburger, L. Hämet-Ahti, A.C. Hamilton, A. Hansen, W.Z. Hao, R.M. Harley, I. Hedberg, I.C. Hedge, D. Henderson, A.J. Hepburn, F.N. Hepper, D. Herbst, V.H. Heywood, F.-C. Ho, K. Høiland, L. Holm-Nielsen, S. Holt, J. Holub, M. Houser, K.-S. Hsu, T.-C. Huang, O. Huber, C.J. Humphries, H.G. Hundley, D.R. Hunt, J. Hunziker, T. Ingelög, H. Jacques-Félix, P. Jaeger, S.K. Jain, H. Jasiewicz, C. Jeffrey, J. Jensen, J. Jérémie, R. Johns, M.C. Johnston, J.-C. Jolinon, L.D. Jornez, M.G. Karrer, K. Kartawinata, D.L. Kelly, H. Keng, R. Kiesling, R. Kiew, R.A. King, R.B. de Klee, E. Köhler, J. Kornas, R. Kral, B.A. Kuzmanov, R. Kwok, E. Landolt, E. Lanfranco, P. Lantz, S.E. Lauzon, C.C. Lay, J.-P. Lebrun, T.B. Lee, Y.N. Lee, J.H. Leigh, R. Letouzey, G.P. Lewis, R.W. Lichvar, J.C. Lindeman, H.P. Linder, A.H. Liogier, Phan Ke Loc, B. Løjtnant, D. Long, A.H. Lot, J. Lovett, S. Lyster, H.S. MacKee, D.A. Madulid, W. Marais, F. Markgraf, C. Martin, P.C. Martinelli, B. Mathew, S.J. Mayo, D. McClintock, B.R. McDonald, R.D. Meikle, J.E. Mendes Ferrão, J. Mennema, A.G. Miller, J. Miller, M.J. Mitchell, N. Mohner, D. Money, T. Monod, F. Monterroso, D.M. Moore, W.H. Moore, Ph. Morat, S.A. Mori, N. Morin, L. Morse, M. Muñoz Schick, T. Müller, D.F. Murray, C. Nelson S., F. Németh, E. Ni Lamha, D.H. Nicolson, H. Niklfeld, H. Nishida, C. Norquist, M. Numata, C. Ochoa, H. Ohba, J.C. Okafor, R. Olaczek, L. Olivier, P. Olwell, S. Orzell, R.T. Pace, J. Page, C. Pannell, F.H. Perring, D. Philcox, A. Phillipps, B.R. Phillips, D. Phitos, R.E.G. Pichi-Sermolli, J. Pickard, S. Pignatti, G.E. Pilz, E. Pingitore, A.R. Pinto da Silva, A. Pinzl, M. Plotkin, A.C. Podzorski, D.M. Porter, D.A. Powell, R. Press, S. Price, A. Radcliffe-Smith, T.P. Ramamoorthy, A.L. Rao, W. Rauh, L. Reichling, S.A. Renvoize, S.A. Robertson, W.A. Rodgers, J.A. Rodrigues de Paiva, M. Romeril, W. Rossi, J.H. Rumely, J. Rzedowski, M.-H. Sachet, Md. Salar Khan, M.J.S. Sands, C. Sargent, M. Scannell, J. van Scheepen, C. Scheepers, F.M. Schlegel, M. Schmid, J. Schwegman, J.W. Scott, K. Scriven, M. Segnestam, K.H. Sheikh, G. Sheppard, T. Shimizu, A. Shmida, S. Siwatibau, A.C. Smith, W.A. Smith, T. Smitinand, S. Snogerup, J.C. Solomon, G.V. Somner, B.A. Sorrie, M. Soto, R. Spichiger, J. Steyermark, A.L. Stoffers, W. Strahm, H.E. Strang, A. Strid, A.M. Studart da Fonseca Vaz, T.F. Stuessy, H.-J. Su, A. Sugden, H. Sukopp, J. Suominen, J.D. Supthut, D. Sutton, W.R. Sykes, A.L. Takhtajan, E. Tanner, C. Taylor, Y. Te-Tsun, A.D. Thompson, G. Thor, Dao Van Tien, C.C. Townsend, G. Traxler, G. Troupin, C. Tydeman, P. Uotila, K. Vollesen, S. Vuokko, M. Wadhwa, F.H. Wadsworth, S. Wahlberg, M. Walters, S.M. Walters, D.A. Webb, L. Webb, E. Weinert, O. Weiskirchner, D.W. Weller, T. Wendt, H. van der Werff, M. Werkhoven, A. Whistler, F. White, T.C. Whitmore, G.E. Wickens, S.R. Wilbur, R.T. Winterbottom, J.R.I. Wood, K. Woolliams, T. Wraber, A. Wünschmann, F. Yaltirik, T. Zanoni, E. Zardini, A. Zimmermann and E.M. van Zinderen Bakker, with apologies to anyone whom we may have forgotten.

We thank those in IUCN who have helped make this book possible, in particular M.F. Tillman, Director of the Conservation Monitoring Centre, J.A. McNeely, Director of the Programme and Policy Division, and O. Hamann, Plants Officer. We thank L. Wright,

IUCN Publications Officer, for seeing it through production and issuing it, and D.C. Mackinder, N.P. Phillips and S. Luckcock, in the Computer Services Unit, for help with the word-processing. His fellow authors would also like to thank Stephen Droop, now a professional publisher in his own right, for his meticulous work in compiling the appendices and in proof-reading the whole book.

Naturally we wish to give particular thanks to our financial sponsors, without whom none of the work could have been done. The preparation of the European accounts was done under a grant from the U.K. Natural Environment Research Council (NERC), on behalf of the European Research Councils, co-ordinated through the European Science Foundation. The CMC receives generous financial support from the United Nations Environment Programme (UNEP), under their Global Environment Monitoring System (GEMS), and from the World Wildlife Fund (WWF). In this case WWF have given an additional grant towards publication that will enable 500 copies of this book to be donated to botanical institutions and conservation organizations in those countries where funds for buying books are hard to obtain. We warmly thank our sponsors for all this support.

Outline of the book

In the pages that follow, we provide information about data sources on plants for each country and island group in the world. Most islands are given a separate account, whatever their political affiliations, because so often their flora is very different from that of the parent country. We have only placed the island account next to that of its parent country where both are close geographically, the island is not oceanic and the floras are similar; otherwise the islands are placed in the alphabetical sequence. For example, Corsica may be found after the account for France, but Guadeloupe and Martinique, French *départements*, are placed in the main sequence.

We have included most islands other than those inshore ones and those that have little or no flora. The main omissions are in the Arctic, where there are few, if any, endangered plants. We have had difficulty in finding the correct names for some of the islands, and have found it quite impossible to be wholly consistent in geographical names. The literature on small islands, although fascinating, is obscure and difficult to find and we are conscious that some of the accounts are far from complete. We would be glad to know of any errors.

The information in each account is arranged under the following headings, although where data are lacking or where the accounts are very short, some or all of the headings have been omitted for the sake of clarity.

Area In square kilometres, mostly taken from *The Times Atlas of the World, Comprehensive Edition* (Times Books, London, 1983 version).

Population Taken from the *UN World Population Chart*, 1984, prepared by the Population Division of the Department of International, Economic and Social Affairs, United Nations. The figures are estimates, to the nearest thousand, for the middle of the year. In a few cases, mostly small islands, different sources were used and these are indicated, with a date wherever possible.

Floristics Here we outline the size of the flora and its affinities, with, where relevant, notes on areas of high diversity and endemism.

In most cases we have tried to give two figures: the number of species of native vascular plants, and the number of endemic taxa. The first of these usually comes from the floristic literature, being either a tally of species recorded or an estimate of species predicted to occur in the country or island. It has been a pleasant surprise to find estimates and totals for so many countries. We are unable to present figures for only a handful of countries, principally Uruguay, the two Yemens, and the two Koreas. We should emphasize that the figures are not always strictly compatible from one country to another; taxonomic concepts vary, as does the extent of knowledge. But we do feel that this set of figures, never drawn together before as far as we can assess, provides a sharp comment on how the diversity of plants is spread over the Earth.

The second number we have tried to include is the number of endemics; by this we mean plants strictly confined to the island, island group or country concerned, rather than plants that are of an endemic nature, i.e. confined to small areas, whether in one country or not. These figures are usually taken from the IUCN database, as IUCN has been accumulating information on endemic plants for many years.

Vegetation Our aim has been to provide a succinct account of the principal vegetation types in each country and to outline the mosaic they form. This is no easy task, even for professional phytosociologists, and we have invariably found this the most difficult section to write. As botanists, with mostly a taxonomic and ecological rather than a phytosociological training, we have learned greatly from the process but are very aware of the deficiencies in what we have written. We hope, nevertheless, that the accounts will be of some use in providing a birds-eye picture of the natural vegetation that remains; the tremendous help that we have had from the numerous botanists who have reviewed the accounts should ensure, too, that they are not wholly inaccurate.

In writing these sections, we have deliberately not followed any one system of classifying vegetation, and have tried to follow a structural rather than a phytosociological approach. As White (1983) says, "The remark made long ago by Richards, Tansley and Watt (1939, 1940) in discussing Burtt Davy's (1938) classification of tropical woody vegetation, namely that existing knowledge is inadequate for the construction of a world-wide natural classification, still remains true." We have also tried to avoid the more baffling and complex terms used by some vegetation scientists.

The sections vary greatly from region to region, those for Europe, predominantly a man-made landscape, being the most difficult. For Africa, we have had the benefit of F. White's masterpiece on the vegetation – the AETFAT vegetation map and descriptive memoir (White, 1983). We have followed this closely and as a result the accounts of the vegetation for Africa are better, shorter and more consistent than those for other regions.

Where possible, especially in tropical forest countries, we have added figures on the extent of vegetation remaining, and of the rate of loss, although in no sense do we provide more than a brief introduction to the literature. Here, too, difficulties intrude for those who seek to summarize. We have, in fact, tended to quote from two very eminent but very different, indeed often contradictory, accounts. The first is the series of books by FAO/UNEP under the overall title *Tropical Forest Resources Assessment Project*, specifically *Forest Resources of Tropical Africa*, of *Tropical Asia* and of *Tropical America* (the latter in Spanish). These massive tomes were compiled by FAO from figures requested from governments. The second source is Norman Myers' *Conversion of Tropical Moist Forests* (Myers, 1980), a report prepared for the U.S. National Research Council and published by the National Academy of Sciences. As Myers himself (1984) points out, the discrepancy lies with the two sets of criteria used. He looked at significant conversion of primary forests, that is, destruction plus degradation, whereas the FAO/UNEP study focused instead on outright elimination of forests, that is, destruction alone. From the point of view of biological values, the Myers figures are therefore likely to be the more useful, because it is well known that modification of tropical forests tends to cause loss of plant diversity. When these differences are taken into account, Myers (1984) claimed that the figures for overall loss of tropical forests were quite similar: a deforestation rate in 1980 of 76,000 sq. km per year according to the FAO/UNEP study, and a figure for outright elimination from the Myers study of 92,000 sq. km per year. In both cases one should emphasize that the largest countries with tropical forest are often the least well documented so that the overall estimates are figures to be treated with caution.

Checklists and Floras This section is included to provide a taxonomic basis for the sections that follow on threatened species. The aim is to cite those works that conservationists would use, so we take a selective view of the botanical literature. Where a comprehensive Flora has just been completed, we have added none of the older works since these would only be required by the taxonomic specialist. But where a Flora has not been written, or is still incomplete, we have included those older works that will be needed to cover the gaps. Often, where modern Floras are still only just beginning, as in many South American countries, we have included references to monographs for the larger individual families. We have also included botanical bibliographies whenever we could find them. In European countries, and some others, we have included plant atlases and national botanical journals.

We should emphasize just how selective we have been, especially for countries with an extensive botanical literature. The bibliography of Mexican botany, for example, runs to 1015 pages (Langman, 1964). The second edition of *Taxonomic Literature*, TL2, in seven massive volumes, will list 15-16,000 titles, mostly published before 1939 and will not be complete, covering just the important works (M.R. Crosby and P.H. Raven, pers. comm.).

While in the final stages of producing this book, D.G. Frodin's *Guide to Standard Floras of the World* was published. This gives very detailed accounts of all the Floras published up to 1980, country by country, and is the result of many years of careful research. The Floras section of our book is fundamentally different as we list only selected works. Nevertheless, quick perusal of Frodin showed a high degree of consistency between the accounts. In only a few cases have we taken the liberty of adding a book or paper from Frodin's accounts and all these instances are cited (e.g. "from Frodin"). We salute Dr Frodin's *magnum opus* and commend it for those who require a more detailed and complete account.

> **Field-guides** Again our choice is selective, especially for those countries like Britain and the United States where very many field-guides have been published over the years. In numerous other countries, however, there is not even a simple guide to the common species.

> **Information on Threatened Plants** This is the core of the book. We have tried to include all lists of threatened plants and Plant Red Data Books, but have not listed papers on one or two threatened species only, unless they give valuable background on threatened species in the area concerned. Some of the major works have been reviewed in the *Threatened Plants Newsletter*, issued by the Threatened Plants Unit about twice a year and sent to those who contribute data to the CMC; these reviews are mentioned where they provide a useful summary of a work or give new information.

News of national databases on threatened species are also given, but this is a recent development in most countries. The maps (see below) summarize the coverage of Red Data Books for countries around the world and the conclusions from this are outlined in the following section.

Where known, we give figures for the number of species (and in some cases infraspecific or lower taxa) falling into each of the IUCN Red Data Book Categories, used as a measure of the degree of threat to wild populations of individual taxa. These categories are defined at the end of the introductory section and outlined with examples in a booklet available from the Threatened Plants Unit at Kew. Most of the figures are taken from the CMC database on plants. In some instances, we quote the number of plants in *The IUCN Plant Red Data Book* (Lucas and Synge, 1978), especially where these are the only readily available examples of threatened species from a particular country. It is important to remember, however, that *The IUCN Plant Red Data Book* contains only examples, chosen to show the types of threats, habitats and areas affected. The aim was to find a few examples for each country, so the accounts are not representative of the places where the most extinctions are happening.

> **Laws Protecting Plants** This section covers legislation specifically to protect plants. It includes details of the type of protection offered and the taxa covered. With the

exception of Europe, information on plants protected by law is still rudimentary; the very extensive database of the IUCN Environmental Law Centre in Bonn, West Germany, covers the individual species of fauna that receive legal protection, but not yet flora. The great size and complexity of that database, which depends on a standard list of animals, at least vertebrates, show how difficult it will be to compile similar records for plants.

Details on laws relating to protected area legislation are not given; for this the reader should consult the IUCN Directories of Protected Areas, of which the volume for the Neotropics is available (IUCN, 1982). Volumes for Africa, Asia and Oceania are in advanced stages of preparation.

Voluntary Organizations Here are listed those non-governmental organizations (NGOs), sometimes called citizen groups, that include plant conservation and botany in their remit. Many, but not all, are members of IUCN.

Botanic Gardens This section was included to reflect the very great importance that IUCN attaches to the role that Botanic Gardens can play in conservation. In the accounts of some countries, for reasons of space, only gardens subscribing to IUCN's Botanic Gardens Conservation Co-ordinating Body are included. More details of the Botanic Gardens of the world may be found in the *International Directory of Botanical Gardens IV* (4th Edition), compiled under the aegis of the International Association of Botanical Gardens (IABG) (Henderson, 1983). A survey of botanic gardens, undertaken by V.H. Heywood and P.S. Ashton for the preparation of an IUCN Botanic Gardens Conservation Strategy, has greatly increased the number of Botanic Gardens on which recent data is available; there are now over 1300 institutions recorded in the IUCN database as Botanic Gardens although not all may qualify in the scientific sense.

Useful Addresses These include, for example, the main conservation agency in the country and the CITES management and scientific authorities. For the most part, herbaria are not included, being very effectively covered by the very meticulous and accurate *Index Herbariorum* (Holmgren, Keuken and Schofield, 1981), which describes about 1400 herbaria.

Additional References This is a very selective section, including additional references cited in the text, as well as further books and articles on conservation and botany in the country concerned that are especially useful. We have made a special effort to include references to national vegetation maps here.

After the country and island accounts, we provide three appendices. The first gives the references that occur so often they are not repeated in full on each country or island account. The second provides a geographical index to the references in Appendix 1, with an indication of subject matter. It may be helpful in finding references for a region rather than for a country. The third is a table showing which countries have ratified or acceded to the three global conservation conventions that relate to plants – the Convention on International Trade in Endangered Species of Wild Fauna and Flora (CITES), the Convention concerning the Protection of the World Cultural and Natural Heritage (The World Heritage Convention), and the Convention on Wetlands of International Importance especially as Waterfowl Habitat, usually known as the Ramsar Convention.

The final part of the book is an index of countries, islands and island groups mentioned in the data sheets, and important old or alternative names (even if these are not mentioned in the text) given as synonyms, followed by the current name. The page number given is that at the beginning of the relevant data sheet, rather than the page number of every occurrence. Geographical entities such as mountains, rivers or regions are not included in the index.

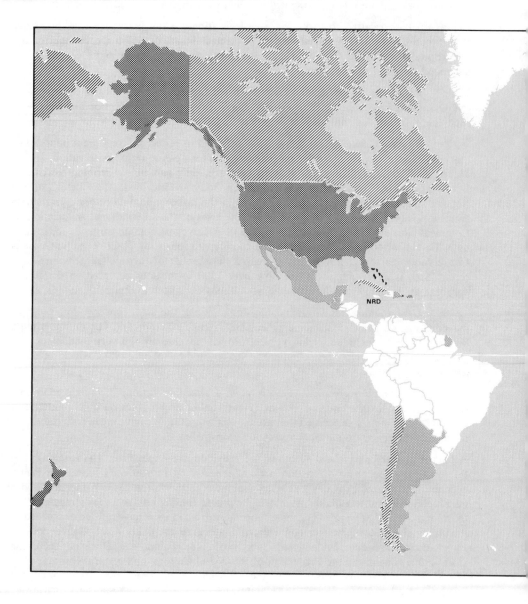

NRD

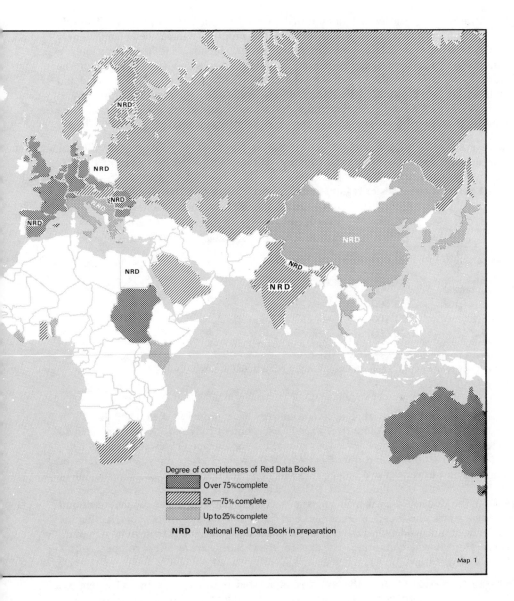

Degree of completeness of Red Data Books

Over 75% complete

25 — 75% complete

Up to 25% complete

NRD   National Red Data Book in preparation

Map 1

Plants in Danger: What we know so far

There is now a very substantial amount of knowledge on threatened plants. It is mostly very recent: for example, by 1970, only Belgium had produced a threatened plant list, only Ronald Melville was cataloguing threatened plants globally, and there was only a scattering of papers on plant conservation. Today, almost all the countries of the "North", as defined by the Brandt Report and so including Australia, New Zealand and South Africa, have produced Red Data Books listing their threatened plants. Several countries of the "South", notably India, have produced exemplary lists too, and are following them up with programmes to conserve the plants they have listed as threatened.

The coverage of Red Data Books is shown in Map No. 1. Some figures for numbers of threatened species in the regions and countries of the "North" are also given in Table 1.

Table 1 Selected countries or regions of the "North" with Red Data Books

Country/Region	Species	Rare & threatened taxa	Extinct taxa	Endangered taxa
Australia	25,000	1716	117	215
Europe[1]	11,300	1927	20	117
New Zealand	2000	186	4	42
South Africa	23,000	2122	39	107
U.S.S.R.	21,100	653	c. 20	c. 160
U.S.A.[2]	20,000	2050	90	?

Sources: Country accounts and CMC database
1. Excludes European U.S.S.R, Azores, Canary Islands and Madeira
2. Continental U.S.A.

In Europe, for example, all but five countries have produced Red Data Books or threatened plant lists, and those five, with the exception of Italy and Albania, are likely to produce reports soon. There is also a regional list for Europe (Threatened Plants Unit, IUCN Conservation Monitoring Centre, 1983), which covers only species rare or threatened on a European scale, and a rather incomplete list for the neighbouring region of North Africa and the Middle East (IUCN Threatened Plants Committee Secretariat, 1980). In the United States, there is a mass of lists and reports covering both the nation and individual states: the situation is complex and rather untypical of other countries, but the profligacy of independent initiatives and activities gives perhaps a glimpse of how the data may develop elsewhere in future. The data for the U.S.S.R. are also very complex, with a plethora of literature.

In North America, 10-11% of the taxa have been listed as rare or threatened. The figure is rather higher in Europe, probably because of the combination of extensive threats to vegetation in the northern, industrialized countries, and the high degree of plant endemism in the southern Mediterranean countries. In northern European countries, the number of world-threatened species tends to be low; this is a reflection of the poverty of the flora, mostly consisting of widespread species that have invaded since the last Ice Age. The lists of nationally threatened species, however, tend to be several times greater, typically of 200 taxa or more.

For the Southern Hemisphere, there is a list for South Africa (Hall *et al.*, 1980), although this is heavily weighted in favour of the Cape. A good list is available for Australia, now in its third version, though it is known to be incomplete for the fast disappearing Queensland rain forests and for the extraordinarily diverse flora of Western Australia. Botanists estimate that as many as 7000 plant species await discovery in Australia, mainly in the western region. In temperate Latin America, there is a list for Chile, but not yet for Argentina.

Of all countries, the problem of threatened plants has perhaps been best documented in New Zealand. First to appear, in 1976, was a register – or list – of 314 taxa under consideration for threatened status (Given, 1976). Then, in 1976-1978, sets of loose-leaf sheets were issued; each sheet covered an individual species, with emphasis on the exact localities and populations in each locality. This was not a public document, but was designed to provide the practising conservationist with the information needed on the most critically threatened plants. This was followed by a popular, illustrated book on conservation of the New Zealand flora (Given, 1981a), an official Red Data Book covering plants and animals (Williams and Given, 1981), and a paper describing the whole documentation process (Given, 1981b).

Within these regions, the highest percentages of rare and threatened species are from those areas with a mediterranean climate – the Mediterranean basin countries themselves, Western Australia, the Cape of South Africa and California. Raven (1976) estimates that these regions contain at least 25,000 plant species; a high percentage of them, maybe as many as half, are narrow endemics, and it is these plants, mostly in the IUCN Rare or Endangered categories, that dominate the threatened plant lists for U.S.A., Europe, Australia and South Africa. To give two examples, Californian endemics account for 669 of the 2050 threatened species in the U.S. and, according to Hall *et al.* (1984), the Cape Floristic Kingdom contains 1621 threatened plants, including 36 Extinct, 98 Endangered and 137 Vulnerable.

In addition, threatened plant lists and Plant Red Data Books have been prepared for many islands. For example, the Canary Islands are well covered by the list for Spain (Barreno *et al.*, 1984), a Red Data Book for Mauritius, sponsored by IUCN/WWF, is in preparation and several lists have been prepared for the species-rich islands of Hawaii. Emphasis, however, has been more on listing the endemics and assigning threatened categories to them rather than preparing comprehensive Red Data Books. Nevertheless these lists show convincingly the very high degree of species endangerment on islands, especially on tropical oceanic islands.

Most important for conservation of biological diversity are those islands with large endemic floras (endemic here means taxa confined to the island concerned). Those with over 1000 endemics are listed in Table 2. They are all very ancient land masses, unlike most oceanic islands which are of more recent geological origin. These islands contain remarkable floras that are very distinct, often isolated biologically and relicts of floras no longer seen today. This is demonstrated by the high degree of endemism among genera and even families. In all of them the vegetation is acutely threatened, but only for Cuba is there a comprehensive assessment of which species are at risk. A more detailed survey and assessment of the conservation status of these floras is an urgent world priority. For Cuba, Borhidi and Muñiz (1983) list 959 species as threatened or extinct, 832 of them endemics. For the Dominican Republic there is a partial list of 133 species (Jiménez, 1978) as well as extensive unpublished material.

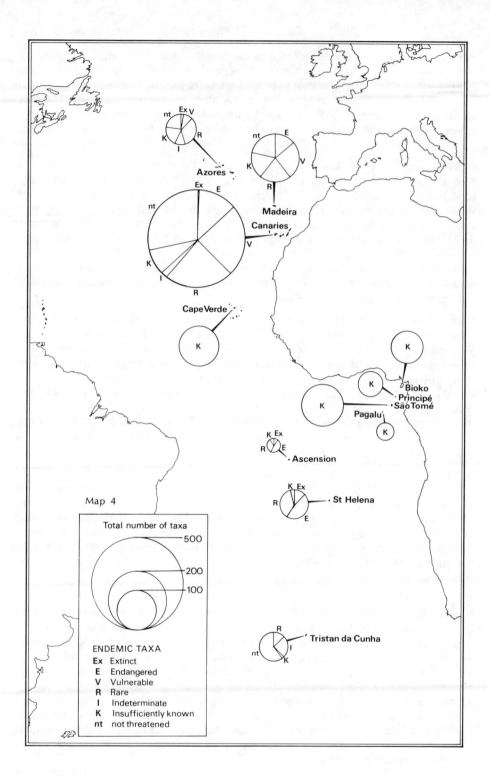

Azores

Madeira

Canaries

Cape Verde

Bioko
Principé
São Tomé
Pagalu

Ascension

St Helena

Tristan da Cunha

Map 4

Total number of taxa

500

200

100

ENDEMIC TAXA
Ex Extinct
E   Endangered
V   Vulnerable
R   Rare
I   Indeterminate
K   Insufficiently known
nt not threatened

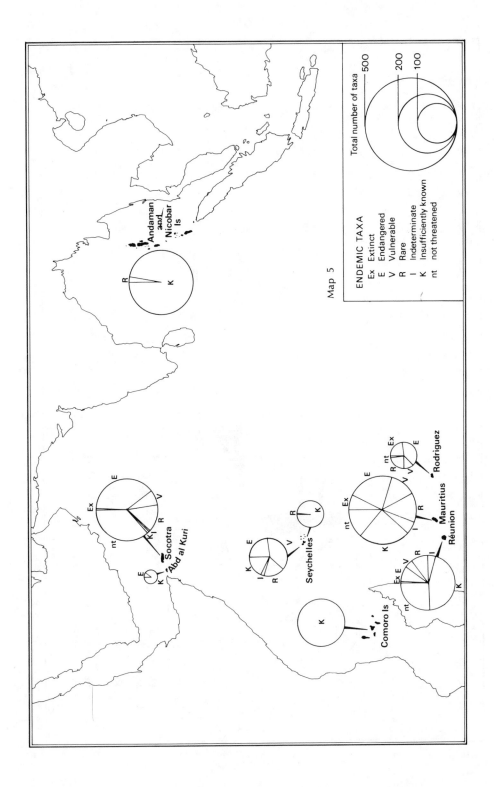

ENDEMIC TAXA

Ex	Extinct
E	Endangered
V	Vulnerable
R	Rare
I	Indeterminate
K	Insufficiently known
nt	not threatened

Map 5

Total number of taxa

Andaman and Nicobar Is

Socotra
Abd al Kuri

Seychelles

Comoro Is

Réunion
Mauritius
Rodriguez

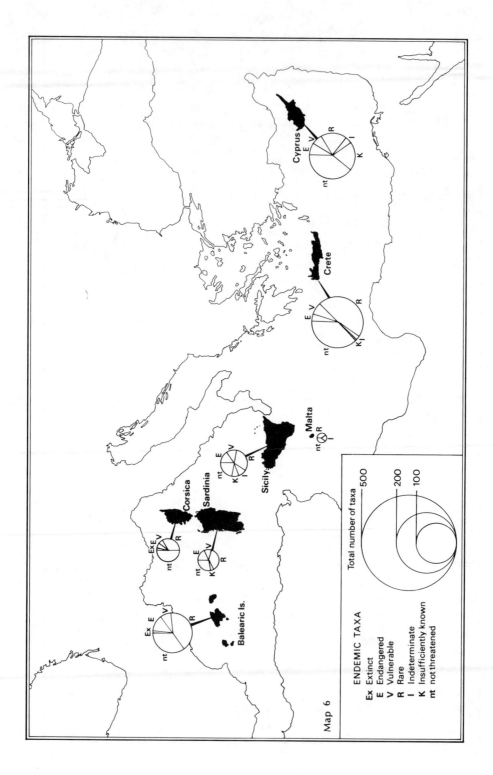

Map 6

ENDEMIC TAXA

Ex Extinct
E   Endangered
V   Vulnerable
R   Rare
I   Indeterminate
K   Insufficiently known
nt not threatened

Total number of taxa

500
200
100

Cyprus

Crete

Malta

Corsica

Sardinia

Sicily

Balearic Is.

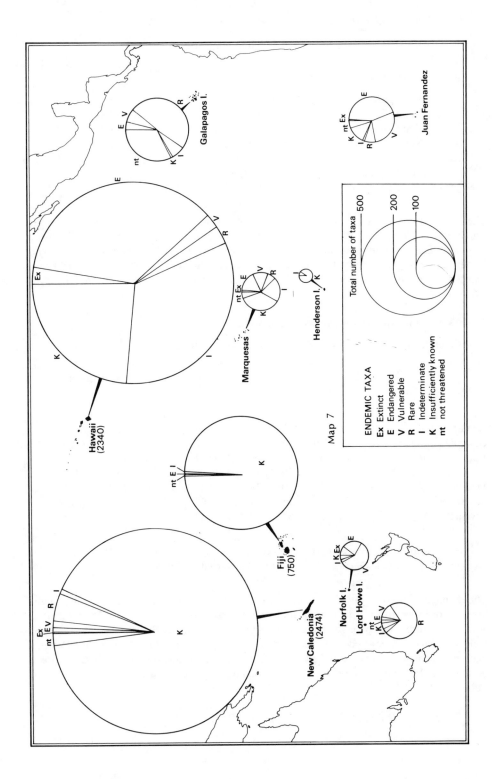

Galapagos I.

Juan Fernandez

Hawaii
(2340)

Marquesas

Henderson I.

Map 7

ENDEMIC TAXA
Ex Extinct
E   Endangered
V   Vulnerable
R   Rare
I   Indeterminate
K   Insufficiently known
nt not threatened

Total number of taxa
500
200
100

Fiji
(750)

New Caledonia
(2474)

Norfolk I.

Lord Howe I.

xxvii

Table 2 Oceanic islands with over 1000 endemic plant species

Country	Size of flora	Endemics
Madagascar	10,000-12,000 spp.	c. 80%
Cuba	6000-7000 spp.	3000-4000
Hispaniola²	5000 spp.	1800
New Caledonia	3250 spp.	2474

1. Excludes Australia, New Zealand, Borneo, New Guinea.
2. Hispaniola comprises the nations of Haiti and Dominican Republic.

There are, of course, many other islands with rich floras, although none except the four listed above have over 1000 endemics. Here, paradoxically, data are usually more complete. Some examples are given in Table 3, using the IUCN Red Data Book categories to define the degree of threat. The islands least well documented are in the Caribbean, especially the Lesser Antillean chain, where a new Flora is in preparation (Howard, 1974-).

Table 3 Endemic vascular plant taxa from selected oceanic islands

50-1000 endemics, over 75% of the flora assigned to categories

	Ex	E	V	R	I	K	nt	Total	Rare or threatened
Azores	1	–	5	18	6	11	14	55	30 (55%)
Canary Islands	1	126	119	132	5	26	160	569	383 (67%)
Galapagos	–	9	15	111	15	2	77	229	150 (66%)
Juan Fernandez	1	52	32	9	1	17	6	118	95 (81%)
Lord Howe Island	–	2	10	58	3	–	2	75	73 (97%)
Madeira	–	17	30	39	–	22	23	131	86 (66%)
Mauritius	19	65	35	39	14	69	39	280	172 (61%)
Seychelles	–	21	35	15	2	17	–	90	73 (81%)
Socotra	1	84	17	29	1	2	81	215	132 (61%)

Similar data exist for Puerto Rico (234 endemics) but are not yet converted to IUCN criteria. Figures for Seychelles are for the Granitic Islands only, and so exclude Aldabra.

The damage to the vegetation of smaller islands often gives an indication of what might happen to larger areas in future. In many cases destruction started centuries ago, often with the introduction of goats (in case of shipwreck). On the British dependency of St Helena, destruction started with the introduction of goats in 1513, which within 75 years had formed herds which stretched for nearly 2 km and devastated the flora before a botanist could even visit the island. Some examples of devastated floras are given in Table 4.

Table 4 Small oceanic islands with devastated floras

	Ex	E	V	R	I	K	nt	Total	Rare or threatened
Ascension Island	1	5	–	4	–	1	–	11	10 (91%)
Bermuda	3	4	1	6	–	?	?	?	14
Norfolk Island	5	11	29	–	1	2	–	48	46 (96%)
Rodrigues	10	20	8	8	–	–	2	48	46 (96%)
St Helena	7	23	–	17	–	2	–	49	47 (96%)

A number of tropical countries have prepared or are preparing threatened plant lists, despite the difficulties. Among the most prominent are:

India	Several lists prepared, covering in total c. 900 threatened species.
Nepal	National Plant Red Data Book, funded by WWF-US.
Pakistan	Identification of threatened plants is part of the WWF-Pakistan Plants Programme.
Peninsular Malaysia	Database on threatened plants being created by the Malayan Nature Society.

The programme in India is one of the most comprehensive, with a full-time team stationed around the country, and is described in the country account. In China, a basic list of 354 threatened plants has been prepared and a Red Data Book covering their status is in preparation, the English translation being due in 1985. Another good example is Egypt, where a threatened plant list is being prepared by the National Herbarium. There are lists of various kinds for El Salvador, French Guiana, Guatemala, Kenya, South Korea, Peninsular Malaysia, Mexico, Tanzania and Viet Nam, and IUCN hold additional data for countries including Ivory Coast, Malawi, Mozambique and Senegal.

In middle America (Mexico to Panama), an IUCN project has prepared a threatened plant list for the region, using where possible the floristic accounts prepared for the forthcoming *Flora Mesoamericana*. Out of many thousands of species screened, including 7622 single country endemics, 1620 have been found to be rare or threatened. This is undoubtedly a major under-estimate; because of the lack of botanical knowledge, it was not possible to assign categories to 5115 of the 7622 single-country endemics.

There is, then, a large and rapidly expanding literature of Red Data Books and national threatened plant lists but like much conservation information, it is mostly 'grey literature', that is reports, often a typescript, produced in low numbers and known only to a handful of people. Such reports are usually available on request but cannot be said to be published.

To make all these data more accessible and to enable comparisons to be made from country to country, the Threatened Plants Unit of IUCN's Conservation Monitoring Centre, based at Kew, has made a database drawing upon much of the information contained in the national Red Data Books and threatened plant lists. At present, this is principally a matrix of the name of the plant, its distribution by countries and the IUCN Red Data Book category for its degree of threat, applied at country and at world level, where known. A major development of the database is planned for 1986-7. This will enable the database to include plant distributions by localities, with presence in named protected areas, synonyms, life-form, bibliographic references and data sources. It will include continued research on effective coding schemes for threats and for habitats.

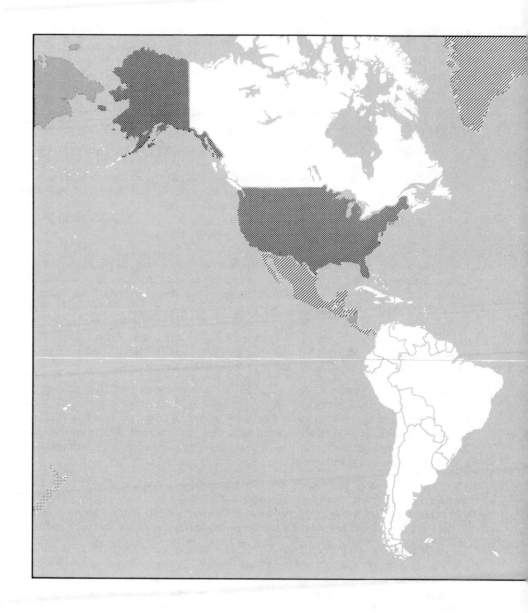

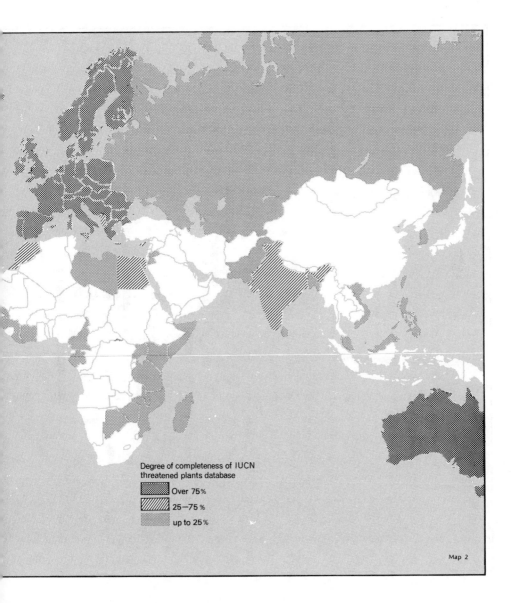

Degree of completeness of IUCN
threatened plants database

Over 75%

25—75 %

up to 25%

Map 2

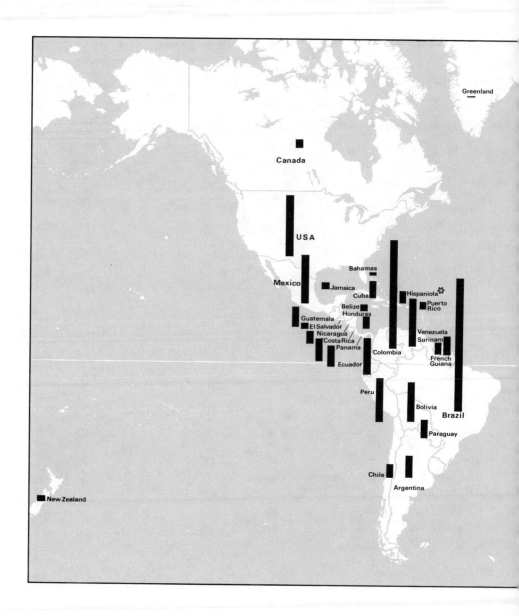

Greenland

Canada

USA

Mexico

Bahamas

Jamaica

Cuba

Belize
Honduras

Guatemala

El Salvador

Nicaragua

Costa Rica

Panama

Colombia

Ecuador

Peru

Hispaniola

Puerto
Rico

Venezuela

Surinam

French
Guiana

Bolivia

Brazil

Paraguay

Chile

Argentina

New Zealand

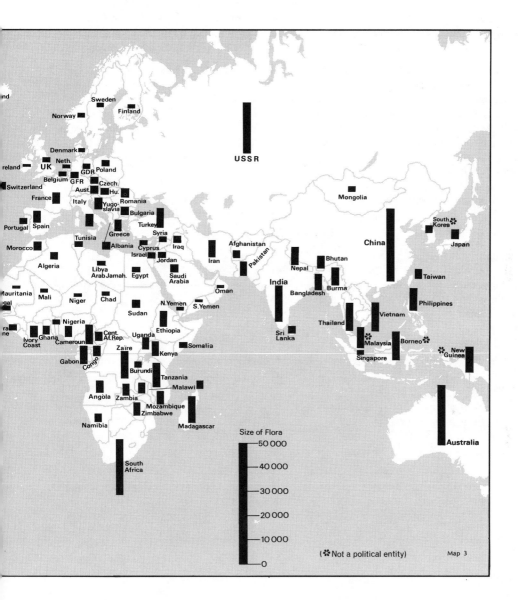

Sweden

Norway

Finland

USSR

Denmark

Neth

UK

Belgium GDR Poland

Switzerland GFR Czech.

France Aust. Hu.

Italy Romania

Yugoslavia Bulgaria

Portugal Spain

Turkey

Mongolia

South Korea ✿

Japan

Tunisia Greece Syria

Morocco Albania Cyprus Iraq

Israel Jordan

Algeria Iran

Libya
Arab Jamah. Egypt

Afghanistan

Pakistan

China

Taiwan

Mauritania Mali

Niger Chad

Saudi
Arabia

Oman

Nepal

Bhutan

Burma

Philippines

Sudan

N.Yemen S.Yemen

India

Bangladesh

gal

ra
ne

Nigeria

Ivory Ghana
Coast Cameroun

Cent.
Af.Rep.

Uganda

Ethiopia

Thailand

Vietnam

Gabon Congo

Zaire

Kenya

Somalia

Sri
Lanka

Malaysia ✿

Borneo ✿

Burundi

Tanzania

Singapore

New
Guinea ✿

Angola Zambia

Malawi

Mozambique
Zimbabwe

Namibia

Madagascar

Australia

South
Africa

Size of Flora

━ 50 000

━ 40 000

━ 30 000

━ 20 000

━ 10 000

━ 0

(✿ Not a political entity)

Map 3

So far the database contains records on 34,266 plant taxa, 15,870 of which are threatened; these comprise 42,569 plant-area records (18 September 1985). Detailed data-sheets, comprising one or more sheets of text, are held on c. 300 of the threatened plants, including those published as *The IUCN Plant Red Data Book* (Lucas and Synge, 1978). There are still a number of threatened plant lists not incorporated into the database. It is reasonable to assume, therefore, that the total number of known threatened plants will rise to perhaps as high as 20,000 taxa within the coming months as these data are incorporated.

Even so, this will barely cover many tropical countries, especially those where rain forest is the dominant vegetation. For some time IUCN's working estimate of the number of threatened plants has been 25,000-30,000; the terminology has, however, been unsatisfactory, as it is not possible to predict which species will be lost and when; the IUCN/WWF Plant Advisory Group in 1985 estimated that as many as 60,000 plant species could become extinct by 2050 if present trends continue – the greatest loss of plant species that has *ever* occurred during a short period of time. This estimate is entirely compatible with the figures outlined above.

Table 5 Families with most threatened species in the IUCN database

Name of Family	No. Threatened Species[1]	No. Species[2]
Compositae	1430	c. 25,000
Leguminosae	941	17,000
Orchidaceae	712	c. 18,000
Palmae	546	c. 2780
Rubiaceae	524	c. 7000
Liliaceae	495	c. 3500
Euphorbiaceae	487	Over 5000
Labiatae	477	c. 3000
Gramineae	460	c. 9000
Cruciferae	443	c. 3000

1. Source: IUCN database, 18 September 1985;
2. Source: Heywood (1978).

Constraints to the identification of threatened species

The concept of a species threatened with extinction is a simple one, yet, as the previous section shows, national threatened plant lists and Red Data Books so far cover only part of the world. Whereas most countries of the "North", with predominantly temperate vegetation, are well covered, there are few Red Data Books for the countries of the "South", where the vegetation is mostly tropical.

Yet as Map No. 3 shows, most of the world's plants grow in the tropics. Roughly two-thirds of the world's flora is tropical, half of it in Latin America and half shared between Africa and Asia. Comparison of the maps shows a sharp discrepancy between those regions with Red Data Books and those regions with most plants.

Although hardly surprising, this is obviously a matter of great concern. The richer the flora, obviously the more difficult it is to identify which species are threatened as the information on each species tends to be less. Indeed, for much of the world, the information on each species is so small that it is not possible to assess chances of survival at all. Numerous tropical plants are only known from a handful of herbarium specimens, often collected many years ago and frequently poorly documented. It is not known whether the plant is common, even dominant, over a large area, or extremely rare.

This problem is made more difficult by the distribution patterns of many tropical plants. The rare species in regions of mediterranean climate and on islands tend to be endemics, plants only known from one small place. Here the threat to a site can be equated to the threat to a species. Destroy the vegetation on the site and the species will disappear. But in the tropics, especially in tropical rain forests, plant species tend to have very scattered distributions. One small piece of forest may contain hundreds of different tree species, each one with only a few individuals per square kilometre. So the plants tend to be thinly scattered over a very extensive range. If part of the forest is to be cut down, it is usually not possible to say which species will become extinct and which will not. No one knows the critical point when species start to be lost.

But perhaps most serious of all is the great imbalance in resources for botanical research between the temperate and tropical regions. The flora of Britain has barely more species than the 1560 hectare island of Barro Colorado in the Panama Canal, whose luxuriant forest, although secondary, contains 1369 species (Croat, 1978). Britain, however, is probably the best botanized country in the world. Every plant is plotted on a 10 km square grid and thousands of amateurs regularly contribute plant records to the journals. There are probably more botanists competent on the British flora than species for them to identify! The country is covered by a voluminous literature with Floras for each county as well as for the nation itself. Yet in countries like Bolivia (15-18,000 species) and Colombia (estimated 45,000 species), a handful of botanists grapple with floras that are largely unknown. No expert can identify more than a small fraction of a tropical forest flora, at least without resource to an herbarium. The sad truth is that most botanists live and work in countries far away from most of the world's plants.

Also, of course, there are thousands, possibly tens of thousands, of plant species not yet discovered, the greatest proportion being from Latin America. This is from a generally accepted world total of around 250,000 species of vascular plants.

Conclusions for the future

There is, then, plenty of information on which plants are threatened. But most of the information is on the countries with least flora. There is very little information on plant conservation needs for those parts of the world where plant diversity is greatest and where threats to plant life may be most acute. Nor has specific action to save plants been particularly widespread or effective.

The IUCN/WWF Plant Advisory Group, meeting at Missouri Botanical Garden in December 1984, expressed the need for plant conservation in this way:

"Plants are a primary resource of fundamental importance for human life. Rapid population growth, together with the excessive and increasing demands that are placed on the world's resources by our societies, are threatening to destroy a major portion of our common heritage of plants. This threat is especially evident in the tropics and subtropics, where at least two-thirds of the plants of the world occur, and where the process of deforestation is proceeding at an alarming rate. Widespread poverty, famine, and political instability, for example in Africa, are manifestations of the same processes that are driving plants to extinction and, by doing so, seriously limiting our future options for developing sustainable relationships between man and his living resources.

All human beings depend upon plants, directly or indirectly, for their lives, as do most other forms of life: at least four million different kinds of organisms depend on about 250,000 kinds of plants. But unless we immediately begin to take drastic and innovative measures to preserve them, it is likely that tens of thousands of plant species will disappear forever during our lives or those of our children. Their loss would amount to a fundamental and permanent change in the character of life on Earth, a life whose wealth is characterized by great diversity.

Some 20 plants provide more than 85% of our food and only a few hundred species are cultivated widely. Most plant species have never been examined to see if they might have properties that would make them useful as food or for other purposes in our modern industrial age, and thousands of species have not even been given a name or described scientifically."

So, with a problem of such magnitude, what should be done? IUCN's response to such questions is the World Conservation Strategy, which provides a conceptual basis linking conservation with development. The task now is to work out precisely how the Strategy can be implemented. Applying the principles of the World Conservation Strategy to the problems of plant conservation:

1. *We need more botanists!* M.R. Crosby and P.H. Raven (pers. comm., 1985) estimate that there are about 3000 plant taxonomists in the world today. They estimate that six times more plant taxonomists are needed to study the world's flora to an adequate extent before it is too late.

This is a target to impress upon science research councils and other funding agencies. It is equally important to promote the correction of the imbalance between where the plants grow and where the botanists live. We should surely do all we can to encourage young botanists starting on their careers to work on tropical plants. The potential for discovering useful new plants is greater in the tropics than elsewhere, but many tropical plants are being lost before they are properly understood.

The goal should be to complete surveys of plant diversity and distribution in those areas, predominantly tropical, where they are lacking. The need is most acute in tropical Latin America, where there are an estimated 90,000 plant species, far more than for any other region on Earth; inventories have been prepared for only a few countries, e.g. Guatemala and Panama, and those are known to be far from complete. Without the basic knowledge of plant distributions, it is impossible to plan for the conservation of plant diversity. Inventories are the cornerstone of plant conservation.

2. *We need more Red Data Books!* This book shows that preparing Red Data Books is possible for many parts of the world. Yet there are still many gaps in the coverage.

Looking at the completed accounts, it is clearly not possible to make a quantitative assessment of priorities around the world. The data are too diffuse and the local knowledge of floras too variable for that. Yet it may be useful to have a more subjective assessment. On the basis of the evidence presented, and from our knowledge of compiling a Red Data Book, we would suggest that national plant Red Data Books are feasible and necessary in the following countries, where they should be treated as priorities:

Country	Approx. No. Species
Argentina	9000
Turkey	8000
Italy	5000
Yugoslavia	4800
Japan	4000
Morocco	3600
Saudi Arabia	3500
Canada	3200
Portugal	2500
Israel	2300
Jordan	2200
Cyprus	2000

Looking at the islands, we can be more objective. Clearly priority should be given to those islands with over 1000 endemics – Madagascar, Cuba, Hispaniola and New Caledonia, as outlined in Table 2. In each case the floras are not well known and far more work is urgently needed. The next priority is those oceanic islands whose floras have not been assessed for threatened species; all those with over 50 endemics are listed below:

Island	No. endemic taxa
Jamaica	c. 910
Taiwan	c. 900 [1]
Sri Lanka	c. 900 [1]
Fiji	c. 700
Caroline Islands	293 [2]
Trinidad and Tobago	215 [1]
Ogasawara-Gunto	151 [1,2]
Réunion	c. 150 [1]
Vanuatu	c. 150
Tubuai	140 or less
Comoro Islands	136 [4]
Bahamas	121 [1]
São Tomé	108
Marquesas Islands	103 [1]
Samoa	? 100 or more [3]
Cape Verde	92
Mariana Islands	81 [2]

Notes: 1. Some partial assessments of threatened species – see country accounts for details
2. Omits monocotyledons
3. Covered as American Samoa and Western Samoa.
4. From a checklist published in 1917.

The book shows that listing threatened plants is best done by a major botanical institution in each country, rather than by any international centre. This policy ensures botanical accuracy and provides the marriage of literature and herbarium groundwork with field knowledge. Work in the library and herbarium may show which species are very restricted in range or confined to vulnerable habitats, but only field knowledge, preferably accumulated over many years, can say which of them are threatened. This policy also helps to build a political and scientific climate within the country to go beyond the data-gathering and design a conservation programme to save the plants so listed.

The aim of a Red Data Book is simple: to provide such data as are required to help create a situation where action may be taken to prevent the plants from becoming extinct. IUCN believes that each country should develop its own approach and produce a book in its own style yet counsels two standards that will promote international collaboration and allow comparisons to be made from one country to another. The first is that countries use the IUCN Red Data Book categories as a measure of the *degree* of threat to individual taxa. Countries should by all means use other coded means of assessment of their own devising, numerical or subjective, but should also apply the IUCN categories to the species listed. The categories have been fixed for many years and used in virtually all the Red Data Books and lists that have appeared in recent years – the main exception is the United States, where categories of similar nomenclature but different meaning are used, following terminology in the U.S. Endangered Species Act of 1973.

The second standard is precision over the full range of the plants listed, in particular whether each is confined to the area covered by the Red Data Book or not. If it is, clearly it will be threatened on a world scale and should receive priority over species common in other countries. If the species does occur outside the country, it is very useful to give some indication, however brief, of its full range – is it a pantropical weed, for example, or does it just extend over the border into a neighbouring country where it may be equally threatened?

It may also be possible for Red Data Books to include information on the sites where most plants could be saved. This is an especially useful approach, particularly where data are lacking on individual species. Indeed, the best way to save most tropical plants is to preserve relatively large areas of intact vegetation, and it is certainly easier to assess where these sites should be rather than to identify threatened species. Using this approach, IUCN is preparing a *Plant Sites Red Data Book*; this will contain accounts of about 150 botanical sites indicative of those in greatest need of protection around the world, and where plant species diversity and/or endemism is particularly high. It is not intended to be a comprehensive account of all sites in danger, but rather an indication of those sites where most plants could be saved. Country accounts of savable plant sites will be even more useful.

3. *We need more detailed monitoring!* Identifying a species as threatened is only the very first step in its conservation. There are many other elements of information that are needed. The most critical of these are data on precisely where the plant occurs – its present localities and data on population biology – how many plants occur at each locality, what are the bottlenecks in the life-cycle which are critical to expansion of the population, and so on. The techniques to do this are fairly sophisticated, following for the most part J.L. Harper's work on population biology (Harper, 1977). Good examples of such studies are few, some being given in a conference partly devoted to this theme on the biological aspects of rare plant conservation (Synge, 1981); this shows rather clearly that the techniques available are as diverse as the number of experimenters!

Henefin and colleagues (1981) have designed guidelines for data-gatherers on the preparation of status reports on rare or endangered plant species. These were designed for the requirements of the U.S. Endangered Species Act of 1973. Their very structured approach provides a lengthy and useful checklist of factors to consider.

Equally important as studies on individual threatened species are studies on the plant communities in which the species occur, especially on the ecology of the vegetation. An understanding of succession, for example, can be critical in ensuring the survival of individual plant populations. Experience shows how important such knowledge can be before rescue attempts are undertaken. In a number of cases, the fate of the plant has been harmed by well-meaning but incorrect conservation action. This is especially true for small, very vulnerable sites, where mistakes can be fatal. A good example is the story of *Ranunculus ophioglossifolius* in Britain, outlined by Frost (1981). Early efforts to conserve the principal population, confined to a tiny site of 1/12 acre, eliminated the plants altogether! The tragedy is that for those regions where most plants are threatened, conservation of individual plants has never been attempted, so there are no stories of success or failure to recount. This is all the more serious because of the large number of economic species in the tropics.

Far more knowledge is still needed on the basic management of protected areas, especially in the tropics, to ensure that the species they contain will survive in future centuries. To our knowledge, there has also been little systematic assessment of which rare and threatened species are in existing protected areas. IUCN is keen to encourage inventories of protected areas as a basic first step in assessing what is protected so far. Few park managers have a list of the plants in the sites they manage and those lists that do exist are often difficult to obtain and unreliable. The IUCN/WWF Plant Advisory Group recommended surveys of the plants in Unesco Biosphere Reserves as a first step and as a way of uniting the biological and conservation communities. This is being taken up by Unesco.

Indeed, in the coming decades, as habitats continue to decline, the emphasis may move away from identifying threatened species towards cataloguing the occurrence of all species in protected areas. One can then ask the question, "Which species are *not* protected at all?", and give priority to them. This is a much less subjective approach. The difficulties, however, are formidable: to derive the list of species not protected, one needs first of all an agreed list or database of plants of the region concerned, followed by lists of species occurring in protected areas. The necessary agreement in taxonomy is still far off, but modern computer technology should act as a spur for regional and specialist plant databases, linked together in a network.

In contrast, the data are far better on which threatened species are conserved *ex situ* in Botanic Gardens, although it is generally accepted that this approach is a second-best solution and is unlikely to succeed for most tropical plants. IUCN's Botanic Gardens Conservation Co-ordinating Body links together about 250 Botanic Gardens into a world network for plant conservation. Surveys have shown that c. 4400 of the c. 16,000 known threatened species are recorded in cultivation. It is likely that this total will expand rapidly, double even, as exchange of electronic media becomes possible between Botanic Gardens and IUCN; the computerization of individual gardens record schemes and a limited measure of standardization, with the provision of an International Transfer Format, is the subject of an IUCN project this year.

4. *We need more conservation action!* Despite the impressive amount of information catalogued in this book, action to save plants in danger is scattered, often small in scale and rarely effective. Indeed, it is hard to find more than a handful of examples where a species, once threatened, has been rescued and is now conserved and safe for the future.

Despite this rather depressing fact, it would appear that success is possible in the predominantly temperate countries such as those of Europe and North America. Here, no more species should be lost. As Lucas and Synge (1978) outline, relatively few of the listed species are Endangered and most are confined to very small areas which can usually be protected without great difficulty. Indeed, relatively small protected areas may be adequate. Botanic Gardens, moreover, can not only cultivate the plants but also reintroduce them, maintain their habitats and even own and manage reserves for them. Once the individual facts on threats, habitats, sites and populations are known, successful conservation of most plant species is likely to prove far less difficult and costly than that of animals. The requirement is on the one hand for the political will to act and on the other for sufficient, energetic and skilful manpower to take protective action for the numerous species involved.

In much of the tropics, however, one has to recognize, and regretfully accept, that species losses are now virtually inevitable. The best answer is to build a network of protected areas – national parks and nature reserves – covering representative samples of the best habitat types. Clearly the priority is to find those areas with the most diversity and to protect them. Setting up one large tropical forest reserve can save hundreds if not thousands of species.

Sometimes a single species can act as a symbol and rallying point for a whole programme of habitat conservation. Project Tiger, an initiative of the Indian Government supported by the World Wildlife Fund, led not only to an increase in tigers from about 1800 to 3000, but, even more important, to a revitalizing of India's protected areas network, with benefits to the numerous plants and animals with which the tiger shared its habitat.

Yet, obviously, the creation of protected areas is principally a means of buying time. Protected areas cannot be effective in the middle of an over-populated and poverty-striken environment; the pressures and temptations are too great when protected areas become lush but forbidden pockets of vegetation surrounded by degraded land.

To counter this possibility, managers of protected areas are changing their tactics. In developing countries, meeting human needs for food, health and shelter has to be the primary goal. Rather than "set land aside", protected area managers want to protect it from gross outside disturbance so that the benefits continue to radiate out into the surrounding countryside; these include surplus animals for food and a continual supply of fresh water in the streams, to give two examples. A vital concept is the buffer zone, a broad and possibly undefined area between the park and the surrounding countryside. The buffer zone can continue to be used in traditional and sustainable ways, e.g. for gathering firewood and wild fruit, for grazing limited numbers of cattle and for gathering medicinal herbs.

Since the United States declared Yellowstone National Park in 1872, parks have been created all over the world. In the decade between 1972 and 1982, major protected areas, excluding the smallest, rose from 212 million hectares to around 386 million hectares – an impressive 55 percent increase. Yet this covers only a small proportion of the Earth's surface, at a time when vegetation is being destroyed faster than ever before. It is a race against time; most areas will have to be saved before the 1990s. The timescale for global conservation is desperately short.

The other main remedy, just as important, is to find better means of using land so that wild plants continue to grow there and the land remains productive. New ways to grow sustainable crops in tropical rain forest environments and to prevent desertification will not only contribute greatly to sustainable development but will save wild plants as well. To help achieve this, botanists should be included in land-use planning teams, particularly in tropical regions where the available knowledge is especially limited. Land-use specialists such as agronomists and foresters should be included in conservation-orientated discussions as a matter of routine. Conservationists must also work more actively with agriculturists and foresters, bearing in mind that conservation and sustainable development can succeed properly only if they go hand in hand. The subject is far too broad to go into here but is vital for the future of the plant kingdom.

5. *We need more education and training!* None of the activities outlined above will happen unless there is the trained and skilled manpower to implement them. Indeed an investment in training and institution-building can often be the most productive of all. There is a severe shortage of well-trained scientists and technicians with conservation skills, especially in the tropics. Specifically, the proper management of germplasm reserves differs greatly from park management in general, and is seldom based on adequate information.

To address this important problem, the IUCN/WWF Plant Advisory Group recommended that increased efforts should be made to provide training at all levels, to incorporate conservation principles as a normal part of biological and botanical training, and to encourage the preparation of outstanding textbooks and other curricular materials on plant conservation. The Group felt that the establishment of specific degrees in conservation might be extended to universities other than those where such degrees are offered at present.

Training needs to go hand in hand with general education and awareness-building on the need for plant conservation. Here we need good educational materials and a cadre of enthusiasts to put across the concepts and practice of plant conservation in the press, on radio and on television. Better use should be made of Botanic Gardens, which can provide the most important single point of information for the public on plant conservation issues.

This section draws extensively on the conclusions of the first meeting of the IUCN/WWF Plant Advisory Group, outlined in full in *Threatened Plants Newsletter*, No. 14: 4-7 (1985).

Definitions of the IUCN Red Data Categories

Extinct (Ex)
Taxa which are no longer known to exist in the wild after repeated searches of their type localities and other known or likely places.

Endangered (E)
Taxa in danger of extinction and whose survival is unlikely if the causal factors continue operating.

Included are taxa whose numbers have been reduced to a critical level or whose habitats have been so drastically reduced that they are deemed to be in immediate danger of extinction.

Vulnerable (V)
Taxa believed likely to move into the Endangered category in the near future if the causal factors continue operating.

Included are taxa of which most or all the populations are *decreasing* because of over-exploitation, extensive destruction of habitat or other environmental disturbance; taxa with populations that have been seriously *depleted* and whose ultimate security is not yet assured; and taxa with populations that are still abundant but are *under threat* from serious adverse factors throughout their range.

Rare (R)
Taxa with small world populations that are not at present Endangered or Vulnerable, but are at risk.

These taxa are usually localized within restricted geographical areas or habitats or are thinly scattered over a more extensive range.

Indeterminate (I)
Taxa *known* to be Extinct, Endangered, Vulnerable or Rare but where there is not enough information to say which of the four categories is appropriate.

Out of danger (O)
Taxa formerly included in one of the above categories, but which are now considered relatively secure because effective conservation measures have been taken or the previous threat to their survival has been removed.

In practice, Endangered and Vulnerable categories may include, temporarily, taxa whose populations are beginning to recover as a result of remedial action, but whose recovery is insufficient to justify their transfer to another category.

Insufficiently known (K)
Taxa that are suspected but not definitely known to belong to any of the above categories, because of the lack of information.

N.B. For species which are neither rare nor threatened, the symbol 'nt' is used.

References for introductory chapters

Barreno, E. *et al.* (Eds) (1984). Listado de Plantas Endemicas, Raras o Amenazadas de España. *Informacion Ambiental. Conservacionismo en España*. No. 3. 7 pp.

Borhidi, A. and Muñiz, O. (1983). *Catálogo de Plantas Cubanas Amenazadas o Extinguidas*. Edit. Academia. 85 pp.

Burtt Davy, J. (1938). The classification of tropical woody vegetation-types. *Inst. Pap. Imp. For. Inst.* 13: 1-85.

Croat, T. (1978). *Flora of Barro Colorado Island*. Stanford Univ. Press, California. 943 pp.

FAO/UNEP (1981). *Tropical Forest Resources Assessment Project (in the Framework of the Global Environment Monitoring System – GEMS)*. UN 32/6.1301-78-04. Technical Reports nos. 1-3, Food and Agriculture Organization of the United Nations, Rome. (Comprises 3 separate reports: *Los Recursos Forestales de la América Tropical*. 343 pp. (Forest Resources of Tropical America; in Spanish); 2 – *Forest Resources of Tropical Africa*. 108, 586 pp. (In English and French); 3 – *Forest Resources of Tropical Asia*. 475 pp. (In English and French).)

Frodin, D.G. (1984). *Guide to Standard Floras of the World*. Cambridge University Press, Cambridge. 619 pp.

Frost, L.C. (1981). The study of *Ranunculus ophioglossifolius* and its successful conservation at the Badgeworth Nature Reserve, Gloucestershire. In Synge, H. (Ed.), *The Biological Aspects of Rare Plant Conservation*. Wiley, Chichester. Pp. 481-489.

Given, D.R. (1976, 1977, 1978). *Threatened Plants of New Zealand: A Register of Rare and Endangered Plants of the New Zealand Botanical Region*. DSIR, Christchurch. (Loose-leaf.)

Given, D.R. (1976). A register of rare and endangered indigenous plants in New Zealand. *N.Z. J. Bot.* 14(2): 135-149.

Given, D.R. (1981a). *Rare and Endangered Plants of New Zealand*. Reed, Wellington. 154 pp.

Given, D.R. (1981b). Threatened plants of New Zealand: documentation in a series of islands. In Synge, H. (Ed.), *The Biological Aspects of Rare Plant Conservation*. Wiley, Chichester. Pp. 67-80.

Hall, A.V., de Winter, M. and B., van Oosterhout, S.A.M. (1980). *Threatened Plants of Southern Africa*. South African National Scientific Programmes Report No. 45, Pretoria. 244 pp.

Hall, A.V., de Winter, B., Fourie, S.P. and Arnold, T.H. (1984). Threatened plants in southern Africa. *Biol. Conserv.* 28(1): 5-20.

Harper, J.L. (1977). *Population Biology of Plants*. Academic Press, London.

Henderson, D.M. (1983). *International Directory of Botanical Gardens IV*, 4th Ed., (first published 1963 as *Regnum Vegetabile* vol. 28). Koeltz Scientific Books, D-6240 Koenigstein, W.-Germany. 288 pp.

Henifin, M.S. *et al.* (1981). Guidelines for the preparation of status reports on rare or endangered plant species. In Morse, L.E. and Henifin, M.S. (Eds), *Rare Plant Conservation: Geographical Data Organization*. New York Botanical Garden. Pp. 261-282.

Heywood, V.H. (Ed.) (1978). *Flowering Plants of the World*. Oxford Univ. Press. 336 pp.

Holmgren, P.K., Keuken, W. and Schofield, E.K. (1981). *Index Herbariorum: Part 1 The Herbaria of the world*, 7th Ed. Scheltema & Holkema, Utrecht and Antwerp. 452 pp.

Howard, R.A. (Ed.) (1974-). *Flora of the Lesser Antilles, Leeward and Windward Islands*. 3 vols so far. Arnold Arboretum, Mass.

Huxley, A. (1985). *Green Inheritance : The World Wildlife Fund Book of Plants*. Anchor Press/Doubleday, Garden City, New York. 193 pp.

IUCN Commission on National Parks and Protected Areas (CNPPA) (1982). *IUCN Directory of Neotropical Protected Areas*. Published for IUCN by Tycooly International Publishing Ltd, Dublin. 436 pp.

IUCN Threatened Plants Committee Secretariat (1980). First Preliminary Draft of the List of Rare, Threatened and Endemic Plants for the Countries of North Africa and the Middle East. Mimeo, IUCN, Kew. 170 pp.

Jiménez, J. de J. (1978). Lista tentativa de plantas de la República Dominicana que deben protegerse para evitar su extinción. *Coloquio Internacional sobre la practica de la conservación*, Santo Domingo.

Langman, I.K. (1964). *A Selected Guide to the Literature of the Flowering Plants of Mexico*. Univ. Pennsylvania Press, Philadelphia. 1015 pp.

Lucas, G. and Synge, H. (1978). *The IUCN Plant Red Data Book*. IUCN, Switzerland. 540 pp.

Myers, N. (1980). *Conversion of Tropical Moist Forests*. (A report prepared for the Committee on Research Priorities in Tropical Biology of the National Research Council.) National Academy of Sciences, Washington, D.C. 205 pp.

Myers, N. (1984). *The Primary Source: Tropical Forests and Our Future*. Norton, New York. 399 pp.

Raven, P.H. (1976). Ethics and attitudes. In Simmons, J.B. *et al.* (Eds), *Conservation of Threatened Plants*. Plenum Press, New York and London. Pp. 155-179.

Richards, P.W., Tansley, A.G. and Watt, A.S. (1939, 1940). The recording of structure, life-form and flora of tropical forest communities as a basis for their classification. *J. Ecol.* 28: 224-239 (1940). Also published as *Inst. Pap. Imp. For. Inst.*, No. 19 (1939).

Synge, H. (Ed.) (1981). *The Biological Aspects of Rare Plant Conservation*. Wiley, Chichester. 558 pp.

Threatened Plants Unit, IUCN Conservation Monitoring Centre (1983). *List of Rare, Threatened and Endemic Plants in Europe (1982 edition)*, 2nd Ed. Nature and Environment Series No. 27, Council of Europe, Strasbourg. 357 pp.

White, F. (1983). *The Vegetation of Africa. A Descriptive Memoir to Accompany the Unesco/AETFAT/UNSO Vegetation Map of Africa*. Natural Resources Research 20, Unesco, Paris. 356 pp.

Williams, G.R. and Given, D.R. (1981). *The Red Data Book of New Zealand: Rare and Endangered Species of Endemic Terrestrial Vertebrates and Vascular Plants*. Nature Conservation Council, Wellington. 175 pp.

Afghanistan

Area 636,267 sq. km

Population 14,292,000

Floristics About 3000 species (Kitamura, 1960-1966); estimated 25-30% species endemism (I.C. Hedge, 1984, *in litt.*). 23 endemic genera, most in the mountains (Hedge and Wendelbo, 1970). The flora includes Central Asiatic and Eastern elements (including many alpines found along the mountain chains of the Altai, Pamir Himalaya and south-west China); Himalayan elements in extreme east and north-east; Eurasiatic and Western elements (Stewart, 1982).

Vegetation In the south-west, mostly desert and semi-desert, with scant vegetation; in south and north-west, thorn scrub with many ephemerals, and grasses; in west and parts of south, open deciduous woodland with *Pistacia* and *Amygdalus*, together with mixed herb communities and steppe-like vegetation; *Artemisia* or *Haloxylon* where the soils are saline; much of the centre and east up to 3000 m, rising to 7000 m in the mountainous north-east; West Himalayan evergreen sclerophyllous forest, restricted to Nuristan and Safed Koh range (Stewart, 1982), with *Quercus* spp. up to 200 m, *Pinus gerardiana* (2100-2500 m), *Cedrus deodara* (2500-3100 m), *Picea smithiana* and *Abies wallichiana* at 2900-3300 m (Freitag, 1971); juniper woodland up to 3500 m; alpine vegetation mainly restricted to a few mountain ranges in east (Breckle, in Davis, Harper and Hedge, 1971).

Checklists and Floras Afghanistan is included in *Flora Iranica* (Rechinger, 1963-), cited in Appendix 1, and *Flore de L'Iran* (1943-1952), cited under Iran. Other relevant works:

Grey-Wilson, C. (1974). Some notes on the flora of Iran and Afghanistan. *Kew Bull.* 29(1): 19-81. (Annotated checklist of plants collected during 1971 expedition; notes on vegetation of Makran, Wakhan and Pamir regions of north-east Afghanistan.)

Hedge, I. and Wendelbo, P. (1964). *Studies in the Flora of Afghanistan*, 1. Norwegian Univ. Press, Oslo. 56 pp. (Annotated list of 7 ferns, 157 angiosperms collected on 1962 expedition; notes on vegetation.)

Kitamura, S. (1960-1966). *Flora of Afghanistan*, 3 vols. Kyoto University, Japan. (1,2 – Enumeration of plants collected during the Kyoto Univ. Scientific Expedition to Karakoram and Hindukush, 1955; details of distributions, Latin diagnoses of new species; 3 – additions and corrections.)

Information on Threatened Plants No national list available. *Ulmus wallichiana* was included in *The IUCN Plant Red Data Book* (1978).

Additional References

Breckle, S.-W., Frey, W. and Hedge, I.C. (1969, 1975). Botanical literature of Afghanistan. *Notes Roy. Bot. Gard. Edinburgh* 29: 357-371; 33: 503-521. (A useful bibliography of botanical literature and maps.)

Davis, P.H., Harper, P.C. and Hedge, I.C. (Eds) (1971). *Plant Life of South-West Asia*. Botany Society of Edinburgh. 335 pp. (See in particular S.W. Breckle on the vegetation in alpine regions of Afghanistan, pp. 107-116; H. Freitag on the natural vegetation of Afghanistan, pp. 89-106; P. Wendelbo on distributional patterns within the *Flora Iranica* area, pp. 29-41.)

Freitag, H. (1971). Die Natürliche Vegetation Afghanistans. Beiträge zur Flora und Vegetation Afghanistans, 1. *Vegetatio* 22: 285-344.

Frey, W. and Probst, W. (1978). *Vegetation und Flora des Zentralen Hindukus (Afghanistan)*. Reichart, Weisbaden. 126 pp.

Hedge, I.C. and Wendelbo, P. (1970). Some remarks on endemism in Afghanistan. *Israel J. Bot.* 19: 401-417.

Podlech, D. and Anders, O. (1977). Florula des Wakhan (Nordost-Afghanistan). *Mitt. Bot. München* 13: 361-502. (Includes annotated checklist, in German.)

Stewart, R.R. (1982). History and exploration of plants in Pakistan and adjoining areas. In Nasir, E. and Ali, S.I. (Eds), *Flora of Pakistan*. Pakistan Agricultural Research Council, Islamabad. 186 pp. (Published as a separate fascicle; see in particular pp. 155-174.)

Agalega Islands

Two small coralline islands c. 930 km north of Mauritius in the Indian Ocean, 10°20'S 56°20'E. The islands, c. 10 and c. 8 km long, are connected by a narrow sand bank. They are well wooded with coconut trees, casuarinas, and other trees; the cultivation of coconuts is the only industry on the islands. 91 species of plant were seen by the late J. Procter in 1972 (unpublished manuscript); 6 species recorded by Hemsley (1919), but 60 species more realistic (Procter, pers. comm. to S. Renvoize, reported in Renvoize, 1979). The islands are a dependency of Mauritius.

References

Hemsley, W.B. (1919). Flora of Aldabra: with notes on the flora of neighbouring islands. *Bull. Misc. Inf. Kew* 1919: 108-153. (Checklist, with descriptions of new species.)

Lincoln, G. (1893). Agalega Islands: a report to Sir H.E. Jerringham. Port Louis, Mauritius. Unpublished. 19 pp. (Illus.)

Renvoize, S.A. (1979). The origins of Indian Ocean island floras. In Bramwell, D. (Ed.), *Plants and Islands*. Academic Press, London. Pp. 107-129.

Albania

Area 28,748 sq. km

Population 2,985,000

Floristics 3100-3300 native vascular species, estimated by D.A. Webb (1978, cited in Appendix 1) from *Flora Europaea*; 24 national endemics (IUCN figures); c. 300 Balkan endemics. Elements: Central European, Mediterranean and alpine. Floristically diverse areas include serpentine and limestone rocks that support many Tertiary relict species.

Vegetation Little recent data on present extent and composition. According to Markgraf (1932) there are 4 natural vegetation zones stretching north-south: 1 – a narrow coastal belt, now largely agricultural with some maquis, phrygana and secondary steppe; 2 – a broad Mediterranean and transitional deciduous forest zone to the east; 3 – central European deciduous montane forests of beech dominating the eastern mountain belt, with scattered patches of Macedonian Pine (*Pinus peuce*); 4 – at highest elevations, mostly along Yugoslav border in the north and east, a subalpine and alpine zone.

Checklists and Floras One of the least known countries botanically in Europe, but covered by the completed *Flora Europaea* (Tutin *et al.*, 1964-1980) and the *Med-Checklist* (both cited in Appendix 1). No complete national Flora, but see Hayek (1924-1933, cited in Appendix 1), although the area delimited as Albania there does not exactly correspond to the limits of the modern state.

Most recent regional Floras:

Mitrushi, I. (1955). *Drurët dhe Shkurret e Shqipërisë.* Instituti i Shkencave, Tiranë. 604 pp. (Monocotyledons; dicotyledons; line drawings; maps.)

Mitrushi, I. (1966). *Dendroflora e Shqipërisë* (Tree Flora of Albania). Univ. Shtetëvor i Tiranes, Tiranë. 519 pp. (Partially supercedes above work; includes cultivated species; 617 line drawings.)

Paparisto, K., Qosja, X. and Demiri, M. (1962, 1965). *Flora e Tiranës, Ikonographia,* 2 vols. Univ. Shtetëvor i Tiranes, Tiranë. 520 pp., 515 pp. (Covers Tirana region only; habitats; vol. 2 contains 1300 line drawings.)

Checklists:

Alston, A.H.G. and Sandwith, N.Y. (1940). Results of two botanical expeditions to south Albania. *J. Bot.* 78: 119-126, 147-151, 167-174, 193-199, 219-224, 232-246. (Checklist for southern Albania.)

Bornmüller, J. (1933). Zur flora von Montenegro, Albanien und Mazedonien. *Magyar Bot. Lapok* 32(1/6): 109-142. (Angiosperm checklist.)

Jávorka, A. *et al.* (1926). Adatok Albania florajahoz. Additamenta ad floram Albaniae. 7. Anthophyta. *A Magyar Tud. Akad. Balk.-Kutat. tud. ered.* 3: 219-346. (Angiosperm checklist.)

Markgraf, F. (1931). Pflanzen aus Albanien 1928. *Denkschrift. Akad. Wiss. Wien Math.-naturw.* 102. 360 pp. (Checklist of vascular species compiled in 1928.)

Relevant botanical journal: *Buletin i Universitet Shtetëvor të Tiranes, Seria Shkencat Natyrore* (Bulletin of the State's University of Tirana, Series of Natural Sciences).

Information on Threatened Plants No national plant Red Data Book. Included in the European threatened plant list (Threatened Plants Unit, 1983, cited in Appendix 1) but for Albania this is based upon data from the 1920s; latest IUCN statistics, based upon this work: endemic taxa – E:1, V:2, R:11, I:6, K:2, nt:2; non-endemics rare or threatened worldwide – V:2, R:59, I:3 (world categories).

Useful Addresses
Botanical Institute of the University of Tirana, Tirana.

Additional References
Gölz, P. and Reinhard, H.R. (1984). Die Orchideenflora Albaniens. *Mitt. Bl. Arbeitskr. Heim. Orch. Baden-Württ.* 16(2): 193-394. (Comprehensive mapping register of the orchid flora of Albania; includes short history of floristic research.)

Hayek, A. von (1917, 1924). Beitrag zur Kenntnis der Flora des Albanisch-Montene-Grinischen Grenzgebietes. *Denkschrift. Akad. Wiss. Wien Math.-naturw.* 94 and 99. 224 pp. (Floristic knowledge about the flora of the Albanian-Montene-Grinischen border districts; illus.)

Markgraf, F. (1925). Botanische Reiseeindrücke aus Albanien. *Repert. Spec. Nov. Reg. Veget.* 36: 60-82. (Botanical journeys in Albania; descriptive account.)

Markgraf, F. (1932). Pflanzengeographie von Albanien. Ihre Bedeutung für Vegetation und Flora der Mittelmeerländer. *Bib. Bot.* 105. 132 pp. (Map; photographs.)

Markgraf, F. (1970). Die floristische Stellung und Gliederung Albaniens. *Feddes Repert.* 81(1-5): 215-222. (A descriptive account of the floristic composition and structure of the Albanian flora.)

Markgraf, F. (1974). Floristic report for Albania. *Mem. Soc. Brot.* 24(1): 5-7.

Ubrizsy, G. and Pénzes, A. (1960). Beiträge zur Kenntnis der Flora und der Vegetation Albaniens. *Acta Bot.* 6(1/2): 155-170.

Aleutian Islands

A continuous chain of about 12 large and 50 small islands, extending westwards for nearly 2000 km from the Alaska Peninsula to 172°W, close to the Commander Islands (Komendorskiye Ostrova – to U.S.S.R.). The Aleutians are a Territory of the U.S.A., cover 17,666 sq. km and have around 6700 people. Including the Commander Islands, the flora comprises 533 taxa of native and introduced vascular plants. "A few endemics." Floristic affinities to the Kamtchatka Peninsula of eastern U.S.S.R. rather than to the Arctic. Vegetation predominantly of heath, dominated by Ericaceae, with meadows in more sheltered places and fragments of alpine meadows in upland areas. The above taken from:

Hultén, E. (1960). *Flora of the Aleutian Islands*. Cramer, Codicote, Herts, U.K., and Hafner, New York. 376 pp., plus 533 distributional maps and 32 plates. (Includes westernmost Alaska Peninsula and with notes on the flora of the Commander Islands.)

For information on threatened plants, see:

Murray, D.F. (1980). *Threatened and Endangered Plants of Alaska*. U.S. Department of Agriculture, Forest Service and U.S. Department of the Interior, Bureau of Land Management. 59 pp. (42 species, dot maps, black ink drawings.)

Algeria

Area 2,381,745 sq. km

Population 21,272,000

Floristics 3139 species (Quézel and Santa, 1962-1963); 3150 species (Le Houérou, 1975). c. 250 endemic species (Quézel, 1964; 1978, cited in Appendix 1). The Ahaggar mountain massif in the south, and the north coast are especially rich.

Most of Algeria has a Saharan flora, but there is also a narrow coastal band with a Mediterranean flora, and a transition zone between the two. Mediterranean and African elements occur together on the Ahaggar massif.

Vegetation Mostly desert with little or no perennial vegetation, and semi-desert grassland and shrubland in the north. Coastal band of Mediterranean sclerophyllous forest. Saharomontane vegetation occurs on the Ahaggar massif, including tree, shrub and grassland communities. Mediterranean montane forests and altimontane shrubland occur on Grande Kabylie in the north.

For vegetation map see White (1983), cited in Appendix 1.

Checklists and Floras Algeria is included in the incomplete *Flore de l'Afrique du Nord*, the computerized *Atlas der Pflanzenwelt des Nordafrikanischen Trockenraumes* (Frankenberg and Klaus, 1980), *Flore du Sahara* (Ozenda, 1977), and is being covered in *Med-Checklist*; these are all cited in Appendix 1. See also:

Lapie, G. and Maige, A. (1915?). *Flore Forestière d'Algérie*. Orlhac, Paris. 359 pp. (Line drawings throughout. Also includes the more common woody plants of Tunisia, Morocco and southern France.)

Quézel, P. and Santa, S. (1962-1963). *Nouvelle Flore de l'Algérie et des Régions Désertiques Méridionales*, 2 vols. Centre National de la Recherche Scientifique, Paris. 1170 pp. (Descriptive keys, distributions; 20 black and white photographs in each volume.)

Information on Threatened Plants Algeria is included in the draft list for North Africa and the Middle East produced by IUCN Threatened Plants Committee Secretariat (1980), cited in Appendix 1.

Faurel, L. (1959). Plantes rares et menacées d'Algérie. In *Animaux et Végétaux Rares de la Région Méditerranéenne*. Proceedings of the IUCN 7th Technical Meeting, 11-19 September 1958, Athens, vol. 5. IUCN, Brussels. Pp. 140-155. (Includes lists of rare or threatened plants in different parts of Algeria.)

Mathez, J., Quézel, P. and Raynard, C. (1985). The Maghrib countries. In Gómez-Campo, C. (Ed.), *Plant Conservation in the Mediterranean Area*.

Latest IUCN statistics: endemic taxa – E:31, V:22, R:65, I:6, K:9, nt:38; non-endemic taxa rare or threatened worldwide – V:2, R:5, I:9 (world categories).

Botanic Gardens
Jardin d'Essais du Hamma, Rue de Lyon, Hamma.
University Botanic Garden, University d'Alger, Alger.

Useful Addresses
CITES Management Authority: Ministère de l'Hydraulique, de l'Environnement et des Fôrets, Ex Grand Seminaire, Kouba, Alger.
CITES Scientific Authority: Institut National de la Recherche Forestière, Arboretum de Baîenm, B.P. 37, Cheraga, Alger.

Additional References
Barry, J.P., and Faurel, L. (1973). Notice de la feuille de Ghardaia. Carte de la végétation de l'Algérie au 1:500,000. *Mém. Soc. Hist. Nat. Afr. N.*, n.s. 11: 1-125. (Map.)

Barry, J.P., Celles, J.C. and Faurel, L. (1974). Notice de la carte internationale du tapis végétal et des conditions écologiques. Feuille d'Alger au 1:1,000,000. Université d'Alger. 42 pp. (Map.)

Cannon, W.A. (1913). *Botanical Features of the Algerian Sahara*. Publication No. 178, Carnegie Institute, Washington. 81 pp. (84 black and white photographs.)

Guinet, P. (1958). Notice détaillée de la feuille de Beni-Abbès (coupure spéciale de la carte de la végétation de l'Algérie au 1:200,000). *Bull. Serv. Carte Phytogéogr., Sér. A., Carte de la végétation* 3: 21-96. CNRS, Paris.

Le Houérou, H.-N. (1975). Etude préliminaire sur la compatibilité des flores nord-africaine et palestinienne. In CNRS (1975), cited in Appendix 1. Pp. 345-350.

Quézel, P. (1964). L'endémisme dans la flore d'Algérie. *Compt. Rend. Somm. Séanc. Soc. Biogéogr.* 361: 137-149.

Quézel, P. and Bounaga, D. (1975). Aperçu sur la connaissance actuelle de la flore d'Algérie et de Tunisie. In CNRS (1975), cited in Appendix 1. Pp. 125-130.

American Samoa

The Samoan Archipelago is a chain of tropical, volcanic islands extending in a west-northwesterly direction in the South Pacific Ocean, 4200 km south-west of Hawaii and 1000 km north-east of Fiji. The archipelago is divided politically into American (or Eastern) Samoa and Western Samoa. American Samoa, an unincorporated territory of the United States, comprises 6 inhabited islands and about 20 small uninhabited islets. It includes Swains Island, which is geographically part of the Tokelau Islands. Western Samoa is covered separately.

Area 197 sq. km

Population 34,000

Floristics 489 vascular plant species, including naturalized introduced plants; 11 endemic species (Amersen *et al.*, 1982). Of the 140 fern species, 16 are endemic (Amersen *et al.*, 1982). Species endemism for the whole of the Samoan Archipelago is c. 25% (Whistler, 1980). The flora of American Samoa is closely allied to that of neighbouring Western Samoa, Fiji and Tonga.

Vegetation Lowland tropical evergreen rain forest, with *Diospyros, Dysoxylum, Pometia* and *Syzygium*, up to 300 m; montane forest, with *Dysoxylum*, at 300-700 m; *Syzygium samoense* cloud forest only found on Tau and Olosega at 500-930 m; small areas of montane scrub on Tutuila; mangroves and swamps near the coast. About two thirds of the native vegetation has been disturbed or cleared for settlements and agriculture. The area of disturbed forest (including *Rhus* secondary forest) was estimated to be c. 40 sq. km (U.S. Fish and Wildlife Service figures, quoted by Whistler, 1980).

Checklists and Floras

Amerson, A.B., Whistler, W.A. and Schwaner, T.D. (1982). *Wildlife and Wildlife Habitat of American Samoa*, 2 parts. U.S. Fish and Wildlife Service, Washington, D.C. (1 – Environment and ecology, with list of 15 "potentially threatened species"; 2 – flora and fauna, with checklist of 489 vascular plant species, most of which are native species; notes on distribution, endemics indicated.)

Christensen, C. (1943). A revision of the pteridophyta of Samoa. *Bull. Bernice P. Bishop Mus.* 177. 138 pp. (Covers both Western Samoa and American Samoa; revision of *Selaginella* by A.H.G. Alston.)

Christophersen, E. (1935, 1938). Flowering plants of Samoa. *Bull. Bernice P. Bishop Mus.* 128. 221 pp.; 154. 77 pp.

Parham, B.E.V. (1972). *Plants of Samoa*. DSIR Information Series no. 85, Govt Printer, Wellington, N.Z. 162 pp. (Short descriptions of plants from Western Samoa, arranged alphabetically by local names; many species also occur on American Samoa.)

Information on Threatened Plants The only available list is that of 15 "potentially threatened species", in Amerson, Whistler and Schwaner, cited above.

Additional References

Whistler, W.A. (1980). The vegetation of Eastern Samoa. *Allertonia* 2(2): 46-190.
Whistler, W.A. (1983). The flora and vegetation of Swains Island. *Atoll Res. Bull.* 262. 25 pp.

Andaman and Nicobar Islands

The Andaman and Nicobar Islands are island groups in the Bay of Bengal, the former of 204 large and small islands, and the latter of about 22 smaller islands. The islands are administered as a Union Territory of the Republic of India.

Area 8120 sq. km

Population 185,254 (1981 census, *Times Atlas*, 1983)

Floristics c. 2270 flowering plant species, of which 225 are endemic (Balakrishnan, 1977; Balakrishnan and Rao, 1984). The flora of the Andamans is related to that of Burma and north-east India, while that of the Nicobars is more closely related to that of Sumatra and Malaysia.

Vegetation The Andamans have tropical evergreen rain forest, rich in *Dipterocarpus* and *Pterocarpus*, tropical semi-evergreen rain forest and tropical moist deciduous forest. The Nicobars have tropical broadleaved evergreen rain forest, with *Terminalia*, *Mangifera*, *Calophyllum*, *Garcinia* and *Cyathea*. Remaining areas of rain forest are under severe pressures from logging and agriculture, particularly on the Andamans. Coastal areas of both the Andamans and Nicobars support mangrove forests, beach forests and littoral communities; scrub forest on the low flat islands of the northern Nicobars.

Checklists and Floras The Andaman and Nicobar Islands are included in the *Flora of British India* (Hooker, 1872-1897), cited in Appendix 1. For ferns see Beddome (1892), and the companion volume by Nayar and Kaur (1972), cited in Appendix 1. Rather dated accounts include:

Gamble, J.S. (1903). *A Preliminary List of the Plants of the Andaman Islands*. Chief Commissioner's Press, Port Blair. 51 pp.
Kurz, S. (1870). *Report on the Vegetation of the Andaman Islands*. Office of Govt Printing, Calcutta. 75 pp. (Includes enumeration of 660 phanerogams and 50 cryptogams; notes on distributions and main timber trees.)
Parkinson, C.E. (1923). *A Forest Flora of the Andaman Islands*. Govt Central Press, Simla. 325 pp. (Reprinted 1972 by Bishen Singh Mahendra Pal Singh, Dehra Dun. Keys, short descriptions of 540 native species.)

Information on Threatened Plants

Balakrishnan, N.P. (1977). Recent botanical studies in Andaman and Nicobar Islands. *Bull. Bot. Survey India* 19: 132-138. (Lists 136 'rare' and 'endangered' endemic species.)

Balakrishnan, N.P. and Rao, M.V.K. (1983). The dwindling plant species of Andaman and Nicobar Islands. In Jain, S.K. and Rao, R.R. (Eds), *An Assessment of Threatened Plants of India*. Botanical Survey of India, Howrah. Pp. 186-210. (Lists 110 threatened endemic taxa and 136 threatened non-endemics; notes on distribution.)

Botanical Survey of India (undated). Endangered flora of Andaman and Nicobar Islands. Mimeo, 5 pp. (Issued by the Botanical Survey of India, Andaman and Nicobar Circle, Port Blair; overview of vegetation and threats to species.)

Jain, S.K. and Sastry, A.R.K. (1980). *Threatened Plants of India – A State-of-the-Art Report*. Botanical Survey of India, Howrah. 48 pp. (Includes accounts of 11 threatened plants from the Andaman and Nicobar Islands.)

Thothathri, K. (1960). Studies on the flora of the Andaman Islands. *Bull. Bot. Survey India* 2: 357-373. (281 species listed, with notes on distribution and abundance on the islands.)

Useful Addresses

Botanical Survey of India, Andaman-Nicobar Circle, Regional Herbarium, Horticultural Road, Port Blair 744102, India.

Additional References

Kurz, S. (1876). A sketch of the vegetation of the Nicobar Islands. *J. Asiatic Soc. Bengal* 45(2): 105-164. (Includes notes on 624 vascular plant taxa.)

Melville, R. (1970). Endangered plants and conservation in the islands of the Indian Ocean. In IUCN, *11th Technical Meeting Papers and Proceedings, 2. Problems of Threatened Species*. IUCN New Series 18, Switzerland. Pp. 103-107.

Sahni, K.C. (1958). Mangrove forests in the Andamans and Nicobar Islands. *Indian Forester* 84: 554-562.

Thothathri, K. (1962). Contribution to the flora of Andaman and Nicobar Islands. *Bull. Bot. Survey India* 4: 281-296. (Floristic analysis; notes on vegetation.)

Andorra

The principality of Andorra is situated on the southern slopes of the Pyrenees, between Spain and France. It is surrounded by mountains, 2000-3500 m high, and nowhere falls below 900 m.

Area 465 sq. km

Population 34,000

Floristics and Vegetation Over 1000 native flowering plant species (Losa and Montserrat, 1950). The most floristically diverse areas occur on the alkaline rocks at Pic de Casamanyá, in the centre of the country, and in the north-west around Arinsal and Ordino. About one-third of the country is covered by forest of pine, fir, oak and birch, but a large proportion is plantation. Rich alpine meadows are widespread, although many mountain slopes have been developed for skiing, causing extensive damage.

Checklists and Floras No national Flora. See:

Losa, M. and Montserrat, P. (1950). *Aportación al Conocimiento de la Flora de Andorra*. Botánica 6. No. 53. 184 pp. Consejo Superior de Investigaciónes

Científicas, Zaragoza. (Without keys; an annotated checklist and floristic account including lower plants; black and white photographs; line drawings; maps.)

Stefenelli, S. (1979). *Guide des Fleurs de Montagne: Pyrenees – Massif-Central – Alpes – Apennins* (French adaptation). Duculot, Paris-Gembloux. 160 pp. (Colour photographs and ecological data for each species.)

The field-guides of Grey-Wilson (1979) and Polunin and Smythies (1973), both cited in Appendix 1, cover the flora.

Information on Threatened Plants None.

Angola

Area 1,246,700 sq. km

Population 8,540,000

Floristics (Excluding Cabinda) Estimates of size of flora include c. 5000 (Airy Shaw, 1947; J.-P. Lebrun, 1984, pers. comm.) and c. 4600 (calculated from figures quoted in Brenan, 1978, cited in Appendix 1). Endemism high; c. 1260 endemics, calculated from a sample of *Conspectus Florae Angolensis* (Exell and Gonçalves, 1973); this is second in Africa only to Zaïre. Districts with highest levels of endemism are Huilla, Benguela and Bie, in that order.

Flora predominantly Zambezian, but in northern third of country flora transitional between Zambezian and Guinea-Congolian. South-west coast with flora of Karoo-Namib and Kalahari-Highveld regions.

Vegetation Mostly rather uniform *Brachystegia-Julbernardia* (Miombo) woodland. Airy Shaw (1947) estimates that this type of woodland, together with other grassland and wooded grassland areas, occupies 90% of Angola. Only on the coastal belt and at the southern border do any major deviations from this type occur, and these include rain forest in the north, desert, montane forest, dry evergreen forest, Baobab associations, and various types of dry scrub. Zonation is well marked only in the south and south-west where desert and subdesert formations (containing the famed *Welwitschia mirabilis*), *Colophospermum mopane* (Mopane) bush and thorn scrub succeed one another as rainfall increases inland. Estimated rate of deforestation for closed broadleaved forest 440 sq. km/annum out of 29,000 sq. km (FAO/UNEP, 1981).

For vegetation map see White (1983), cited in Appendix 1.

Checklists and Floras

Carrisso, L. *et al.* (Eds) (1937-). *Conspectus Florae Angolensis*, 4 vols and 1 fascicle. Junta de Investigações do Ultramar, and later Junta de Investigações Científicas do Ultramar, Lisboa. (Fully annotated checklist with keys. Pteridophytes by E.A. Schelpe, 1977. Flora now produced in family fascicles; c. 45% published.)

Information on Threatened Plants No published lists of rare or threatened plants, but four examples of Vulnerable species are given by B.J. Huntley on p. 99 of Hedberg (1979), cited in Appendix 1.

IUCN has records of 808 species and infraspecific taxa believed to be endemic, including R:3, I:16, nt:8.

Botanic Gardens

Botanic Garden of Salazar and Floristic Reserve No. 1, Instituto de Investigação
Agronómica de Angola, C.P. 406, Huambo.

Additional References

Airy Shaw, J.K. (1947). The vegetation of Angola. *J. Ecol.* 35: 23-48.

Barbosa, L.A. Grandvaux (1970). *Carta Fitogeográfica de Angola*. Instituto de
Investigação Científica de Angola, Luanda. 323 pp. (With coloured vegetation map
1:2,500,000 and numerous black and white photographs.)

Exell, A.W. and Gonçalves, M.L. (1973). A statistical analysis of a sample of the flora
of Angola. *Garcia de Orta, Sér. Bot.* 1(1-2): 105-128.

Monteiro, R.F.R. (1970). *Estudo da Flora e da Vegetação das Florestas abertas do
Planalto do Bié*. Instituto de Investigação Científica de Angola, Luanda. 352 pp.
(With 35 black and white photographs and coloured vegetation map 1:500,000.)

Santos, R. Mendes Dos (1982). *Itinerários Florísticos e Carta da Vegetação do Cuando
Cubango*. Estudos, Ensaios e Documentos No. 137. Instituto de Investigação
Científica Tropical/Junta de Investigações Científicas do Ultramar, Lisboa. 266 pp.
(With coloured vegetation map 1:1,000,000.)

Teixeira, J. Brito (1968). Angola. In Hedberg, I. and O. (1968), cited in Appendix 1.
Pp. 193-197.

Werger, M.J.A. (1978), cited in Appendix 1. Citation includes list of relevant chapters.

Anguilla

A flat coralline island of 91 sq. km and 7000 inhabitants in the Leeward Islands of the
Eastern Caribbean, 113 km north-west of St Kitts. It is administered directly by the United
Kingdom as a Dependent Territory. The vegetation is mostly tropical evergreen bush and
low scrub. For botanical information, see the account on Antigua and Barbuda.
References specifically on Anguilla are:

Boldingh, I. (1909). A contribution to the knowledge of the flora of Anguilla, B.W.I.
Recueil des Travaux Botaniques Neerlandais 6: 1-36. (List of 50 vascular plants,
general ranges given.)

Box, H.E. (1940). Report upon collection of plants from Anguilla, B.W.I. *J. Bot.* 78:
14-16.

Antarctica

The continent of Antarctica covers 14 million sq. km. Almost the entire area is
permanently covered by ice. There is also a belt of pack ice, between 4 and 22 million sq.
km, surrounding the continent. In addition, there are a number of island groups extending
into the Southern Ocean and southern Indian Ocean (Crozet Islands, Kerguelen Islands,
New Amsterdam, Heard and Macdonald Islands) and South Atlantic Ocean (South
Orkney and South Shetland Islands).

The Crozet Islands, Kerguelen Islands, New Amsterdam, Heard and Macdonald Islands are rocky islets with mires in which the important peat-forming plants are bryophytes, tussock-forming grasses, cushion-forming flowering plants and other herbaceous communities. Much of the land is covered with snow throughout the year. Maritime Antarctica, the South Orkney and South Shetland Islands are even more barren and are within the limit of maximum pack ice extension.

For South Georgia and the South Sandwich Islands, see under the Falkland Islands (Islas Malvinas). Marion and Prince Edward Islands are covered separately.

Antarctic Continent 2 indigenous vascular plants (*Deschampsia antarctica* and *Colobanthus quitensis*), confined to the vicinity of the Antarctic Peninsula (Greene and Holtom, 1971).

Crozet Islands Area 505 sq. km; population of 30, permanent mission (1982); part of the French Southern and Antarctic Territory. 28 vascular plant species (Greene and Walton, 1975).

Heard and Macdonald Islands Area 412 sq. km; no permanent population; external territories of Australia. Heard Island has 8 vascular plant species, the Macdonald Islands have 3 (Greene and Walton, 1975).

Kerguelen Islands Area 7000 sq. km; population of 76, permanent mission (1982); part of the French Southern and Antarctic Territory. 29 vascular plant species, of which *Lyallia kerguelensis* is endemic and a further 7 species, including the famous Kerguelen Cabbage (*Pringlea antiscorbutica*), are confined to 2 or more sub-antarctic islands (Greene and Walton, 1975).

New Amsterdam Area 55 sq. km; population of 92 (1980), permanent mission (1982); part of the French Southern and Antarctic Territory. 55 vascular plant species (J. Jérémie, 1984, *in litt.*).

St Paul Area 7 sq. km; uninhabited; part of the French Southern and Antarctic Territory. Lowland slopes covered by *Poa novare* and *Spartina arundinacea*; wetter areas dominated by sedges, mainly *Scirpus nodosus*.

South Orkney Islands Area 620 sq. km; uninhabited; part of the British Antarctic Territory. 2 vascular plants, *Colobanthus quitensis* and *Deschampsia antarctica* (Brown, Wright and Darbishire, 1908).

South Shetland Islands Area 4700 sq. km; uninhabited; part of the British Antarctic Territory. 1 vascular plant (*Deschampsia antarctica*).

References

Brown, R.N.R., Wright, C.H. and Darbishire, O.V. (1908). The botany of the South Orkneys. Scottish National Antarctic Expedition. *Trans. Bot. Soc. Edinburgh* 23(1): 101-111. (Includes account of mosses and lichens.)

Chastain, A. (1958). La flore et la végétation des Iles de Kerguelen. *Mém. Mus. National Hist. Naturelle, Ser. B, Bot.* 11(1). 136 pp.

Clark, M.R. and Dingwall, P.R. (1985). Cited in Appendix 1.

Cour, P. (1959). Flore et végétation de l'Archipel de Kerguelen. *Terres Australes et Antarctiques Français* 8/9: 3-40.

Greene, S.W. and Holtom, A. (1971). Studies in *Colobanthus quitensis* (Kunth) Bartl. and *Deschampsia antarctica* Desv., 5. Distribution, ecology and performance on Signy Island. *Brit. Antarctic Survey Bull.* 28: 11-28.

Greene, S.W. and Walton, D.W.H. (1975). An annotated check list of the sub-antarctic and antarctic vascular flora. *Polar Record* 17(110): 473-484.

Hemsley, W.B. (1885). *Report on the Scientific Results of the Voyage of H.M.S. Challenger During the Years 1873-76.* Botany, vol. 1, part 2. London. (See in particular the section on the Crozets, including annotated checklist of 7 vascular plants, pp. 207-211; the Kerguelen Islands, including checklist of 21 vascular plants, pp. 211-243; the Macdonald Group, including checklist of lower plants and 5 vascular plants on Heard Island, pp. 245-258; New Amsterdam and St Paul, pp. 259-281.)

Hooker, J.D. (1844-1847). *The Botany of the Antarctic Voyage of H.M. Discovery Ships Erebus and Terror, in the Years 1839-1843*, 2 vols. London. (1 – Flora Antarctica, Lord Auckland's Group and Campbell's Island; 2 – Flora Antarctica, the Antarctic Region.)

IUCN (1984). *Conservation and Development of Antarctic Ecosystems.* IUCN, Switzerland. 36 pp.

Skottsberg, C. (1954). Antarctic flowering plants. *Bot. Tidsskr.* 51: 330-338.

Young, S.B. (1971). Vascular flora of the Kerguelen Islands. *Antarctic J. United States* 6(4): 110-111.

Antigua and Barbuda

Antigua One of the more northerly of the Leeward islands of the Lesser Antilles; low-lying, reaching only 415 m altitude; mostly under sugar cultivation. About 45 km WSW of Antigua is Redonda ("Round Island"), 1.3 sq. km, a fragment of a volcano and rising to 300 m; uninhabited apart from about 100 feral goats.

Barbuda 40 km north of Antigua; flat, only 30.5 m altitude; has a large lagoon and coastal sand-dunes.

Area Antigua: 279 sq. km; Barbuda: 160 sq. km

Population Antigua: 7300 (1979 estimate); Barbuda: 1500 (1979 estimate)

Floristics 724 angiosperms with 0.7% endemism (Box, 1938, see below, analysed by C.D. Adams). These figures are likely to change considerably as the *Flora of the Lesser Antilles* is published.

Vegetation
Antigua Mostly dry scrub woodland and man-made grassland; several types of seasonal forest, mostly low and secondary; in areas of low rainfall and limestone soils, several types of evergreen thicket and scrub; on the coast some mangrove and strand vegetation. Area of cultivation recorded as no more than 101 sq. km in 1960, decreasing, being replaced by secondary vegetation (Loveless, 1960). According to FAO (1974, cited in Appendix 1), 15.9% forested.

Barbuda Mostly natural bush, with trees in the higher terraces and more stunted bushland vegetation; grassy areas towards the windward coast; lower plains cultivated and grazed; some coastal mangrove and sand dunes.

Checklists and Floras Covered by the *Flora of the Lesser Antilles* (only monocotyledons and ferns published so far; Howard, 1974-), and by the family and generic monographs of *Flora Neotropica*. (Both are cited in Appendix 1.) See also:

Alston, A.H.G. and Box, H.E. (1935). Pteridophyta of Antigua. *J. Bot.* 73: 33-40.

Beard, J.S. (1944). Provisional list of trees and shrubs of the Lesser Antilles.
  Caribbean Forester 5(2): 48-67. (428 species in a table showing which are in the Leeward Is. but not which are on each island in the group.)

Howard, R.A. (1962). Botanical and other observations on Redonda, the West Indies.
  J. Arnold Arbor. 43: 51-66. (Includes account of vegetation and species list.)

Stehlé, H. and Stehlé, M. (1947). Liste complémentaire des arbres et arbustes des petites Antilles. *Caribbean Forester* 8: 91-123. (A further 328 species to Beard, 1944, in similar format.)

In 1938, H.E. Box prepared a check list, based on earlier records and collections and his own collections in Antigua and sight records in Barbuda. The taxonomy and nomenclature were revised by J.E. Dandy. Includes an historical introduction and an ecological description of the vegetation. Never published – copies at University of the West Indies Library, Mona, Jamaica; the Institute of Jamaica, Kingston; and the National Herbarium of Trinidad and Tobago, Trinidad. (C.D. Adams, 1984, pers. comm.).

Information on Threatened Plants None.

Voluntary Organizations

Antigua Archaeological Society, P.O. Box 103, St John's, Antigua. (Preparing a list of some of the plants of Antigua, Barbuda and Redonda, with some of their uses.)

Additional References

Harris, D.R. (1960). *The vegetation of Antigua and Barbuda, Leeward Islands, West Indies*. Prelim. Rep. Dep. Geog. Univ. Calif.

Harris, D.R. (1965). *Plants, Animals, and Man in the Outer Leeward Islands, West Indies. An ecological study of Antigua, Barbuda and Anguilla*. University of California Publications in Geography vol. 18. Univ. California Press, Berkeley. 164 pp. (With photographs and vegetation maps.)

Loveless, A.R. (1960). The vegetation of Antigua, West Indies. *J. Ecol.* 48(3): 495-527.

Wheeler, L.R. (1916). The Botany of Antigua. *J. Bot.* 54: 41-52.

Antipodes Islands

The Antipodes (21 sq. km) are an uninhabited, outlying island group of New Zealand, in the Pacific subantarctic, at 49°42'S, 178°50'E. The vegetation consists mainly of grassland and is relatively little disturbed. 62 vascular plant taxa (*Flora of New Zealand*, 1961, cited under New Zealand). One endemic, *Gentiana antipoda* (IUCN category: Rare). The islands were declared a Nature Reserve in 1961. For more information see Given, 1981a, cited under New Zealand.

Argentina

Area 2,777,815 sq. km

Population 30,094,000

Floristics Approximately 9000 species of vascular plants (J. Hunziker, 1984, pers. comm.), most in the tropical region; 25-30% endemic. Botanically the best known country in South America (Toledo, 1985, cited in Appendix 1). Areas of high endemism and diversity are: Provinces Patagonia, Puneña, Altoandina, del Monte and Paranaense (Hunziker, pers. comm.). The flora of the southern Andes has affinities to the flora of New Zealand.

Vegetation In the northeast, rain forest; in the northwest provinces of Jujuy and Salta subtropical semi-deciduous forest and subtropical evergreen seasonal submontane broadleaved forest (Unesco, 1981, cited in Appendix 1); in north central and central Argentina, the Gran Chaco, a mixture of xerophilous forest and savannas, with many halophytic and swamp associations. To the south the Pampa, a vast savanna and open prairie, without native trees, mostly grazed or cultivated; in Patagonia, the southern quarter of the country, mainly steppe and tundra, with coniferous forest in the west, low deciduous thicket in the northeast and subdesert deciduous shrubland and tundra in the south (Unesco, 1981). In the Andes, north to south, vegetation includes cloud forest and dry puna in the north, caespitose herbaceous communities all along and temperate forest in the south.

Checklists and Floras Recent floristic research in Argentina has focussed on the production of regional Floras, sponsored by the Instituto Nacional de Tecnología Agropecuaria (INTA):

Burkart, A. (1969-). *Flora Ilustrada de Entre Ríos (Argentina)*. Colección Científica del INTA, Buenos Aires. 6 vols planned, 3 completed: 2 – grasses (1969); 5 – Primulales to Plantaginales (1978); 6 – Rubiales to Campanulales (1974).
Cabrera, A.L. (1963-1970). *Flora de la Provincia de Buenos Aires*, 6 vols. INTA, Buenos Aires.
Cabrera, A.L. (1977-). *Flora de la Provincia de Jujuy, República Argentina*. INTA, Buenos Aires. 3 vols published out of 10; includes Pteridofitas (1977) and Compositae (1978). (To cover an estimated 3500 species.)
Cabrera, A.L. and Zardini, E.M. (1978). *Manual de la Flora de los Alrededores de Buenos Aires*, 2nd Ed. Acme, Buenos Aires. 755 pp.
Correa, M.N. (1969-). *Flora Patagónica*. INTA, Buenos Aires. 4 vols published, 8 projected.
Dimitri, M.J. (1962). La flora andino-patagónica. *Anal. Parques Nacionales* 9: 1-130.
Dimitri, M.J. (1974). *Pequeña Flora Ilustrada de los Parques Nacionales Andino-Patagónicos*. Publicación Técnica No. 46, Separada de los Anales de Parques Nacionales, Tomo 13. 122 pp.
Meyer, T. *et al.* (1977). *Flora Ilustrada de la Provincia de Tucumán*. Fundación Miguel Lillo, Tucumán. 305 pp.

Toledo (1985, cited in Appendix 1) refers to the following additional Floras as in progress: Centro de Argentina by A.T. Humziker (Museo Botánico de Córdoba), Provincia de Corrientes by A. Krapovickas (started in 1979), the Chaco by A. Digilio and the Pampa by G. Covas. A 1984 checklist of 1538 native genera is also referred to.

See also:

14

Boelcke, O., Moore, D.M. and Roig, F.A. (1985). *La Transecta Botánica de Patagonia Austral*. CONICET, Buenos Aires. (Vegetation, floristics, geology, human impact and climate for the Atlantic to Pacific Oceans between 51° and 52°S.; includes 2-sheet vegetation map; shorter English version being prepared for *Phil. Trans.* (London), 1985-6.)

Cabrera, A. and Ferrario, M. (1970). *Bibliografía Botánica de la Provincia de Buenos Aires, Plantas Vasculares*. Comisión de Investigaciones Científicas, Buenos Aires. 96 pp.

Descole, H.R. (1943-1956). *Genera et Species Plantarum Argentinarum*. Instituto Miguel Lillo. 5 vols, few families published.

Dimitri, M.J. (1972). *La Región de los Bosques Andino-Patagónicos*. Colección Científica del INTA, Buenos Aires.

Moore, D.M. (1983). *Flora of Tierra del Fuego*. Nelson, U.K., and Missouri Botanical Garden. 396 pp. (545 species, 3% endemic; illus., dot maps.)

Seckt, H. (1929-1930). *Flora Cordobensis*. Universidad Nacional, Córdoba. 632 pp.

Information on Threatened Plants There is no national Red Data Book. The following articles and papers contain information on threatened plants:

Endangered and Threatened Plants in the Republic of Argentina. Botanic Garden Journal of the Polish Academy of Sciences, Warsaw. (Not seen.)

Pingitore, E.J. (1976). The Republic of Argentina tree ferns. *Los Angeles Int. Fern Soc.* 3(10): 198-203; 3(11): 222-225; 3(12): 246-249. (Includes list of 8 Endangered and 2 Rare species.)

Pingitore, E.J. (1981). Especies vegetales en vías de extinción de la República Argentina. *Sociedad Horticola Argentina* 37: 10-13. (Tentative list of 69 threatened species.)

Pingitore, E.J. (1982). Especies interesantes de La Tierra del Fuego e Islas del Antarctico Sur. *Bol. Soc. Hort. Argentina* 38: 10-12. (Tentative list of 38 threatened species.)

Pingitore, E.J. (1983). Rare palms in Argentina. *Principes* 26(1): 9-18. (10 native palms, 7 listed as rare.)

Prance, G.T. and Elias, T.S. (Eds) (1977), cited in Appendix 1. See in particular A. Cabrera on endangered plants of Argentina, pp. 245-247; E. de la Sota on endangered plants and communities, pp. 240-244; J. Mickel on endangered pteridophytes, pp. 323-328; P. Ravenna on threatened bulbous plants, pp. 257-266.

Lists of threatened plants and plant communities, arranged by region, are given in Organización de los Estados Americanos (1967), cited in Appendix 1. 24 plants are listed in the Annex to the Convention on Nature Protection and Wildlife Preservation in the Western Hemisphere (1940).

Laws Protecting Plants No information. The U.S. Government has determined *Fitzroya cupressoides*, confined to Chile and Argentina, as 'Threatened' under the U.S. Endangered Species Act.

Voluntary Organizations

Associación Natura, 25 de Mayo 749, 1° Piso, Buenos Aires.

Centro de Ecología y Recursos Naturales Renovables, Universidad Nacional de Córdoba, C.C. 395, 5000 Córdoba.

Comité Argentino de Conservación de la Naturaleza, Avenida Santa Fe 1145, Buenos Aires.

Instituto de Investigaciones de las Zonas Aridas y Semiáridas, Parque Gral, San Martin, Mendoza.

Botanic Gardens

Departamento de Botánica Agrícola, Instituto Nacional de Tecnología Agropecuaria, 1712 Castelar, Provincia Buenos Aires.

Jardín Agrobotánico de Santa Catalina, Instituto Fototécnico de Santa Catalina, Llavallol, FNGR.

Jardín Botánico "Carlos Thays", Instituto Municipal de Botánica, Av. Santa Fe 3951, 1425 Buenos Aires.

Jardín Botánico de la Facultad de Agronomía y Veterinaria, Av. San Martin 4453, 1417 Buenos Aires.

An account of Argentinian botanic gardens is given in:

Sota, E. de la (1979). Argentina: the conservation of endemic and threatened plant species within botanic gardens. In Synge, H. and Townsend, H. (Eds), *Survival or Extinction*. Bentham-Moxon Trust, Kew. Pp. 95-99.

Useful Addresses

Fundación Vida Silvestre Argentina, Leandro N. Alem 968, 1001 Capital Federal, Buenos Aires.

CITES Management and Scientific Authorities: Dirección Nacional de Fauna Silvestre, Paseo Colon 922-2°, Piso Oficina 201, 1063 Buenos Aires; also (Scientific Authority only) Museo Argentino de Ciencias Naturales "Bernardino Rivadavia", Avenida Angel Gallardo 470, 1405 Buenos Aires.

Additional References

Cabrera, A.L. (1972). Estado actual del conocimiento de la Flora Argentina. *Mem. I Congreso Latinoamericano de Botánica*. Pp. 183-197. (Not seen.)

Cabrera, A.L. (Ed.) (1977). *Evolución de las Ciencias en la República Argentina. 1923-1972. Tomo VI. Botánica*. Sociedad Científica Argentina. (Not seen.)

Grassi, N. (Ed.) (1982). *Conservación Natural en la Rep. Argentina*. Simposio de las XVIII Jornadas Argentinas de Botánica. Tucumán. 130 pp.

La vegetación de la República Argentina (1951-1968). Various authors. 9 fascicles reported. INTA, Series Fitogeográfica. Buenos Aires.

Ascension Island

A barren volcanic island of 94 sq. km in the South Atlantic, c. 1300 km north-west of St Helena, 7°57'S 14°22'W. About 1050 residents, plus about 450 military personnel. An Island Dependency of St Helena, itself a Dependent Territory of the U.K. The highest point is the peak on the east/west ridge of Green Mountain (860 m). Flora of about 25 native vascular plants; these include 6 endemic fern species and 5 endemic flowering plant species; of these 1 is Extinct, 5 Endangered, 4 Rare and 1 Insufficiently Known. About 300 plants introduced deliberately or by accident; also goats, rabbits, donkeys, sheep. The status of the endemics is outlined in detail in:

Cronk, Q.C.B. (1980). Extinction and survival in the endemic vascular flora of Ascension Island. *Biol. Conserv.* 17(3): 207-219.

Other useful references:

Atkins, F.B., Baker, P.E., Bell, J.D. and Smith, D.G.W. (1964). Oxford Expedition to Ascension Island, 1964. *Nature* 204: 722-724.

Duffey, E. (1964). The terrestrial ecology of Ascension Island. *J. Appl. Ecol.* 1: 219-251. (Maps; includes outline of such vegetation as exists and assesses the impact of man.)

Packer, J.E. (1974). *Ascension Handbook: a concise guide to Ascension island, south Atlantic*, 2nd Ed. (1st Ed., 1968). Published privately, Georgetown. Unpaginated, but 1st Ed. 68 pp. (Includes a checklist of the flora, with line drawings.)

Rudmose Brown, R.N. (1906). Contributions towards the botany of Ascension. *Trans. Bot. Soc. Edinburgh.* 23: 199-204.

Auckland Islands

An outlying island group of New Zealand, comprising 7 uninhabited volcanic islands in the Pacific subantarctic. Total land area of 625 sq. km of which Auckland, the largest island, is 464 sq. km. 187 native flowering plant taxa, including 6 endemics. The vegetation, which has been modified by introduced goats, cattle, sheep, pigs and rabbits, includes coastal *Metrosideros* forest, scrub and grassland on higher ground and, above 500 m, exposed peatland. Adams Island was declared a Nature Reserve in 1910; the rest of the Auckland Islands were included in the reserve in 1934. There is a programme to reduce the numbers of introduced mammals (Clark and Dingwall, 1985, cited in Appendix 1).

The Auckland Islands are included in the *Flora of New Zealand* (1961, 1970, 1980), cited under New Zealand.

For information on threatened plants, see Given (1981a), cited under New Zealand. Latest IUCN statistics: endemic taxa – R:1; non-endemic taxa rare or threatened worldwide – V:1, R:4 (world categories).

Additional References

Godley, E.J. (1969). Additions and corrections to the flora of the Auckland and Campbell Islands. *N.Z. J. Bot.* 7: 336-348. (Covers 45 taxa.)

Johnson, P.N. and Campbell, D.J. (1975). Vascular plants of the Auckland Islands. *N.Z. J. Bot.* 13: 665-720. (Annotated checklist of 257 taxa including adventives.)

Australia

Area 7,682,300 sq. km

Population 15,519,000

Floristics c. 18,000 known native vascular plant species with an estimated 7000 yet to be named or recorded (*Flora of Australia*, 1981-). 80% species endemism; over 500 endemic genera. Species-rich areas include the Cape York Peninsula of northern Queensland, the South-Western Province and the Coolgardie region of Western Australia, the northern part of Northern Territory, the coastal regions of N.S.W., north-east Victoria and the Central Tablelands.

Vegetation Predominantly desert (receiving less than 250 mm mean annual rainfall) and semi-desert (250-500 mm rainfall). There are 2 extremely arid regions – the Nullarbor Plain in the south, and the Lake Eyre Basin/Simpson Desert in central Australia. *Acacia* and *Eucalyptus* shrublands cover 20% of Australia, mainly in centre and west; Mitchell Grass plains, dominated by *Astrebla*, cover vast areas of the north, extending into northern N.S.W.; Kangaroo Grass (*Themeda australis*) grassland in south-east, extensively modified for grazing; heathland in south, west and parts of Queensland and Tasmania, much has been cleared or drained (Leigh, Boden and Briggs, 1984); alpine communities in Tasmania, Victoria and N.S.W. (Beadle, 1981); open forests, dominated by *Eucalyptus*, *Callitris* and *Melaleuca*, cover large areas of inland Australia, from the Kimberleys in Western Australia, extending across the north to Queensland and west of the Great Dividing Range in N.S.W.; open forests of *Eucalyptus*, *Acacia* and *Casuarina*, in south-west Western Australia, Northern Territory to Queensland, Cape York to Victoria and Tasmania; cool temperate rain forest dominated by *Nothofagus* in Victoria, N.S.W. and Tasmania; subtropical and temperate rain forest mixtures in N.S.W. and outliers in north Queensland; subtropical rain forest in south Queensland and north New South Wales, in places reduced to small pockets; tropical rain forest and tropical monsoon forest in northern Australia.

c. 20,000 sq. km of all types of rain forest remain, out of an estimated 80,000 sq. km prior to European settlement. Clearing of forests continuing, mainly for agriculture, grazing and forest plantations; nearly all subtropical lowland forests destroyed and only a few thousand hectares of tropical lowland forest remain (Groves, 1981).

Checklists and Floras

Bentham, G. (1863-1878) *Flora Australiensis: A Description of the Plants of the Australian Territory*, 7 vols. Reeve, London. (Reprinted 1967 by Asher and Reeve, Amsterdam.)

Flora of Australia (1981-). 60 vols (including non-vascular plants) to be published over a 20-year period. Co-ordinated and edited by the Bureau of Flora and Fauna, Department of Arts, Heritage and Environment. Australian Government Publishing Service, Canberra. (5 vols published so far. 1 – Introduction, origin and evolution, keys to families; 4 – Phytolaccaceae to Chenopodiaceae, 5 families; 8 – Lecythidaceae to Bataceae, 19 families; 22 – Rhizophoraceae to Celastraceae, 17 families; 29 – Solanaceae.)

Checklists of large genera and families include:

Chippendale, G.M. and Wolf, L. (1981). *The Natural Distribution of Eucalyptus in Australia*. Australian National Parks and Wildlife Service Special Publication no. 6. 192 pp. (Checklist of 550 taxa, grid maps showing distributions.)

Clements, M.A. (1982). *Preliminary Checklist of Australian Orchidaceae*. National Botanic Gardens, Canberra. 216 pp. (List of over 600 accepted species names, with synonyms.)

Jones, D.L. and Clemensha, S.C. (1981). *Australian Ferns and Fern Allies*, 2nd Ed. Reed, Sydney.

There are many Floras at State and regional level; only a selection are cited here. For a comprehensive bibliography see Leigh, Boden and Briggs (1984) and the *Flora of Australia*, 1 (1981).

18

Bailey, F.M. (1899-1905). *The Queensland Flora with Plates Illustrating Some Rare Species*. Brisbane. (6 parts, General Index.)

Beadle, N.C.W., Evans, O.D., Carolin, R.C. and Tindale, M.D. (1982). *Flora of the Sydney Region*, 3rd Ed. Reed, Sydney. 724 pp. (Covers coastal N.S.W.; with line drawings and colour illus.)

Black, J.M. (1943-1957). *Flora of South Australia*, 2nd Ed., 4 parts. Govt Printer, Adelaide. (Part 1 – Lycopodiaceae to Orchidaceae has been revised and edited by J.P. Jessop, 1978, Woolman, Adelaide. A Supplement to the Flora by H. Eichler has been published by the Govt Printer, Adelaide, 1965.)

Burbidge, N.T. and Gray, M. (1970). *Flora of the Australian Capital Territory*. Australian National Univ. Press, Canberra. 447 pp. (Includes outline of vegetation of southern Tablelands; with line drawings.)

Curtis, W.M. (1956-1979). *Student's Flora of Tasmania*, parts 1-4, 4A. Govt Printer, Hobart.

Ewart, A.J. and Davies, O.B. (1917). *The Flora of the Northern Territory*. Govt Printer, Melbourne. 387 pp. (Annotated list with keys.)

Flora of New South Wales (1961-1978). National Herbarium of New South Wales. (Discontinued; covers ferns, gymnosperms and 16 flowering plant families, including grasses. Prior to 1971 published as a 'Flora Series' in *Contributions from the New South Wales National Herbarium*.)

Green, J.W. (1981). *Census of the Vascular Plants of Western Australia*. Western Australian Herbarium, South Perth. 113 pp. (Checklist of ferns, gymnosperms and angiosperms.)

Jacobs, S.W.L. and Pickard, J. (1981). *Plants of New South Wales: A Census of the Cycads, Conifers and Angiosperms*. Royal Botanic Gardens, Sydney. 226 pp. (Checklist of c. 6000 taxa, distributions indicated.)

Jessop, J.P. (Ed.) (1981). *Flora of Central Australia*. Reed, Sydney. (Includes c. 2000 species.)

Jessop, J.P. (Ed.) (1983). *A List of the Vascular Plants of South Australia*. Adelaide Botanic Gardens, State Herbarium and Dept of Environment and Planning. 87 pp. (Checklist of accepted names and synonyms.)

Stanley, T.D. and Ross, E.M. (1983-). *Flora of South-eastern Queensland*. Dept of Primary Industries, Brisbane. 545 pp. (3 vols projected; 1 – keys to dicotyledon families, treatments of 79 flowering plant families, 1983; 2 & 3 – in prep.)

Willis, J.H. (1962, 1972). *A Handbook to Plants in Victoria*, 2 vols. University Press, Melbourne. (1 – Ferns, conifers, monocotyledons; 2 – dicotyledons.)

Field-guides

Blombery, A.M. (1977). *Australian Native Plants*. Angus and Robertson, Sydney. 481 pp. (Keys to genera, line drawings and descriptions of selected plants.)

Francis, W.D. (1970). *Australian Rain-forest Trees*, 3rd Ed. Australian Govt Publ. Service, Canberra. 468 pp. (Keys, descriptions and field characters of mainly subtropical trees, covering mainly eastern Australia.)

Galbraith, J. (1977). *A Field Guide to the Wild Flowers of South-East Australia*. Collins, Sydney. 450 pp. (Includes temperate regions of N.S.W., Victoria, Tasmania, S. Australia and Queensland.)

Grieve, B.J. and Blackall, W.E. (1954-1975). *How to Know Western Australian Wildflowers: A Key to the Flora of the Temperate Regions of Western Australia*, 4 parts. Univ. of Western Australia Press, Nedlands.

Harris, T.Y. (1979). *Wild Flowers of Australia*, 8th Ed. Angus and Robertson, Sydney. 207 pp. (Keys to families, over 250 species illustrated in colour.)

Hodgson, M. and Paine, R. (1971). *A Field Guide to Australian Wildflowers*. Rigby, Adelaide. 251 pp.

Holliday, I. and Hill, R. (1974). *A Field Guide to Australian Trees*. Rigby, Adelaide. 229 pp. (Revised edition.)

Holliday, I. and Walton, G. (1975). *A Field Guide to Banksias*. Rigby, Adelaide. 141 pp.

Information on Threatened Plants The national list of threatened Australian plants has been revised twice; the first version was Specht *et al.* (1974), the second Hartley and Leigh (1979) and the third Leigh *et al.* (1981).

Hartley, W. and Leigh, J. (1979). *Plants at Risk in Australia*. Australian National Parks and Wildlife Service Occ. Paper no. 3. Canberra. (Provisional list of 2053 plants at risk.)

Leigh, J., Briggs J., and Hartley, W. (1981). *Rare or Threatened Australian Plants*. Australian National Parks and Wildlife Service Special Publication no. 7, Canberra. 178 pp. (2206 species listed as rare or threatened. Separate lists for Lord Howe, Macquarie, Norfolk, Philip and Christmas Islands; briefly reviewed in *Threatened Plants Committee Newsletter* No. 9: 18, 1982.)

Specht, R.L., Roe, E.M. and Boughton, V.H. (Eds) (1974). *Conservation of Major Plant Communities in Australia and Papua New Guinea*. Australian J. Bot. Supp. Series 7. 667 pp. (Detailed assessment of conservation status of all the major plant communities and species under threat in each State.)

Also relevant:

Good, R.B. and Leigh, J.H. (1983). The criteria for assessment of rare plant conservation. In Given, D.R. (Ed.), *Conservation of Plant Species and Habitats*. Nature Conservation Council, Wellington, N.Z. Pp. 5-28.

Leigh, J. and Boden, R. (1979). *Australian Flora in the Endangered Species Convention – CITES*. Australian National Parks and Wildlife Service Special Publication no. 3, Canberra. 93 pp. (Checklist of taxa covered then by CITES; list has since been revised; reviewed and outlined in *Threatened Plants Committee Newsletter* No. 7: 19-20, 1981.)

Leigh, J., Boden, R. and Briggs, J. (1984). *Extinct and Endangered Plants of Australia*. Macmillan, Melbourne. 369 pp. (Includes detailed case studies of 76 species presumed extinct and 203 which are endangered.)

Parsons, R.F., Scarlett, N.H. and Stuwe, J. (1981). A register of rare and endangered native plants in Victoria. *Threatened Plants Committee Newsletter* No. 7: 22-23. (Outline of a project to survey and document rare and threatened plants.)

Pryor, L.D. (1981). *Australian Endangered Species: Eucalypts*. Australian National Parks and Wildlife Service Special Publication no. 5, Canberra. 139 pp. (Data sheets, maps and photographs of 124 species at risk.)

A number of State lists of threatened plants have also been produced, including:

Rye, B.L. (1982). *Geographically Restricted Plants of Southern Western Australia*. Report no. 49. Dept of Fisheries and Wildlife, Perth. 63 pp.

Rye, B.L. and Hopper, S.D. (1981). *A Guide to the Gazetted Rare Flora of Western Australia*. Report no. 42. Dept of Fisheries and Wildlife, Perth. 211 pp.

A series of illustrated data sheets entitled *Rare Western Australian Plants* has been prepared by B.L. Rye for the Department of Fisheries and Wildlife, Perth, in 1982. (8 seen; notes on ecology, conservation measures; dot maps.)

Latest IUCN statistics: total rare and threatened endemic taxa – Ex:117, E:215, V:570, R:812, I:2, K:505, nt: not known; of these, statistics for State endemic taxa are – Ex:110, E:196, V:503, R:716, I:2, K:467, nt: not known.

Laws Protecting Plants There is legislation in each State and Territory for the protection of flora. Legislation is the most detailed in Western Australia, where 128 species are listed as 'Protected Flora' under the Wildlife Conservation Act Amendment Act 1979 of Western Australia. 65 of them are orchids. A further 100 taxa have been listed as 'Rare Flora' which are considered to be in danger of extinction, rare or otherwise in need of special protection; they can be taken from the wild only with the approval of the Minister for Fisheries and Wildlife. In Victoria the flora legislation is administered by the Forestry Commission while in all other States and Territories it is administered by the relevant nature conservation agency.

Voluntary Organizations
Australian Conservation Foundation, 672B Glenferrie Road, Hawthorn, Victoria 3122.
Australian Flora Foundation, c/o Botanic Gardens, Adelaide.
Society for Growing Australian Plants, c/o The Editor 'Australian Plants', 860 Henry Lawson Drive, Picnic Point, N.S.W. 2213.
Tropical Rainforest Society, Box 5918 CMC, Cairns 4870, Queensland.
WWF-Australia, Level 17, St Martins Tower, 31 Market Street, Sydney, N.S.W. 2000.

Botanic Gardens Many; for full list see Henderson (1983), cited in Appendix 1. See also:

Royal Australian Institute of Parks and Recreation (1984). *A Report on the Collection of Native Plants in Australian Botanic Gardens and Arboreta*. Canberra. 69 pp. (Lists 55 botanic gardens and arboreta growing native plants, with details of area, important plant groups in cultivation, and potential for extending collections.)

The principal botanic gardens include:

Adelaide Botanic Garden, North Terrace, Adelaide, S. Australia 5000.
Australian National Botanic Gardens, P.O. Box 158, Canberra, A.C.T. 2601.
Kings Park and Botanic Garden, Kings Park Road, West Perth 6005, W. Australia.
Royal Botanic Gardens, Mrs Macquaries Road, Sydney, New South Wales 2000.
Royal Botanic Gardens of Melbourne, Birdwood Avenue, South Yarra, Victoria 3141.
Royal Tasmanian Botanic Gardens, Queen's Domain, Hobart, Tasmania 7000.

Useful Addresses
Division of Plant Industry, CSIRO, P.O. Box 1600, Canberra City, A.C.T. 2601.
TRAFFIC Australia, P.O. Box 371, Manly 2095, N.S.W.
Western Australian Wildlife Authority, Department of Fisheries and Wildlife, 108 Adelaide Terrace, Perth, W. Australia 6000.
CITES Management and Scientific Authority: Australian National Parks and Wildlife Service, P.O. Box 636, Canberra, A.C.T. 2601.

The Council of Nature Conservation Ministers (CONCOM) Working Group on Endangered Flora provides a channel for enquiries from overseas. The CONCOM Secretariat is at the Department of Arts, Heritage and Environment, G.P.O. Box 1252, Canberra, A.C.T. 2601.

Additional References
Beadle, N.C.W. (1981). *The Vegetation of Australia*. Cambridge Univ. Press. 690 pp.
Groves, R.H. (Ed.) (1981). *Australian Vegetation*. Cambridge Univ. Press. 449 pp.

Morley, B.D. and Toelken, H.R. (Eds) (1983). *Flowering Plants in Australia*. Rigby, Adelaide. 416 pp. (Overview of more than 250 flowering plant families; keys to genera; distribution maps.)

Tracey, J.G. (1982). *The Vegetation of the Humid Tropical Region of North Queensland*. CSIRO, Melbourne. 124 pp.

For vegetation maps of Western Australia see:

Beard, J.S. *et al.* (1972-). *Vegetation Survey of Western Australia, 1:1,000,000 Vegetation Series*. Univ. of Western Australia Press, Nedlands. (1 – Kimberley; 2 – Great Sandy Desert; 3 – Great Victoria Desert; 4 – Nullarbor; 5 – Pilbara; 6 – Murchison; 7 – Swan area; each map with explanatory notes.)

Austria

Area 83,853 sq. km

Population 7,489,000

Floristics 2900-3100 native vascular species, estimated by D.A. Webb (1978, cited in Appendix 1) from *Flora Europaea*; 35 endemic taxa (IUCN figures). Elements: Central European (Pannonian), sub-Mediterranean and alpine. Areas of diversity: alpine grasslands and dry steppe regions bordering Hungary in the east.

Vegetation Most remaining semi-natural vegetation in west and central Alps and close to Hungarian border in far east. Central Alps: forest relicts of Arolla Pine (*Pinus cembra*) and European Larch (*Larix decidua*); eastern Alps: forests of beech and Norway Spruce (*Picea abies*) with relict stands of Black Pine (*Pinus nigra*), interspersed with meadows, pastures and arable land. In subalpine zone, Mountain Pine (*Pinus mugo*) and alder, with alpine heaths. On hills and lowlands north of Alps, patches of beech and hornbeam forests, amongst arable land and spruce plantations. Some riverine forests with poplars; those along Danube and March (Morava) rivers, recently threatened by construction of hydro-electric power stations. Eastern Austria mainly arable with vineyards, but with relicts of dry Pannonian steppe grassland and oak forests. Small sub-Mediterranean influence in south with *Ostrya carpinifolia* and *Fraxinus ornus* (M.A. Fischer, 1984, *in litt.*).

Total tree cover 39.1%; permanent pasture 26.7% (includes alpine grasslands, meadows and steppe); arable 20% (Poore and Gryn-Ambroes, 1980, cited in Appendix 1). For maps of vegetation and phytogeography see Wagner (1971).

Checklists and Floras Austria is covered by the 3 regional Floras, *Flora Europaea* (Tutin *et al.*, 1964-1980), *Illustrierte Flora von Mitteleuropa* (Hegi, 1935-), both cited in Appendix 1 and *Flora von Deutschland und seinen Angrenzenden Gebieten* (Schmeil and Fitschen, 1976, cited under F.R.G.). No modern national Flora, but see:

Fritsch, K. (1922). *Exkursionsflora für Österreich und die Ehemals Österreichischen Nachbargebiete*, 3rd Ed. C. Gerold, Wien. 824 pp. (Includes adjacent countries, but excludes the Province of Burgenland in eastern Austria; reprinted 1973 by Cramer, Liechtenstein.)

For a modern national checklist see:

Janchen, E. (1956-1967). *Catalogus Florae Austriae*, 1 vol. and 4 supplements. Springer-Verlag, Wien.

Janchen, E. (1977). *Flora von Wien, Niederösterreich und Nordburgenland*, 2nd Ed. Verein für Landeskunde von Niederösterreich und Wien, Wien. 757 pp.

See also:

Dalla Torre, K.W. and Sarnthein, L.G. von (1900-1913). *Flora von Tirol, Vorarlberg und Liechtenstein*, 6 vols. Wagner'schen Univ., Innsbruck.

Hayek, A. von. (1908-1956). *Flora von Steiermark*, 2 vols. Gebr. Borntraeger, Berlin and Naturwissenschaftlichen Verein für Steiermark, Graz.

For bibliographies see Hamann and Wagenitz (1977), cited in Appendix 1, and:

Ehrendorfer, F., Fürnkranz, D., Gutermann, W. and Niklfeld, H. (1974). Fortschritte der Gefässpflanzensystematik, Floristik und Vegetationskunde in Österreich, 1961-1971. *Verh. Zool.-Bot. Ges. Wien.* 114: 63-143.

Relevant journal: *Linzer Biologische Beiträge*, Linz. (Formerly *Mitt. Bot. Arbeits-gemeinschaft am Oberösterreichischen Landesmuseum*, Linz.)

Field-guides See Oberdorfer (1983), cited in Appendix 1, and:

Hegi, G., Merxmüller, H. and Reisigl, H. (1977). *Alpenflora. Die Wichtigeren Alpenpflanzen Bayerns, Österreichs und der Schweiz*. Parey, Berlin. 194 pp. (Introduction includes ecological descriptions of plant communities; lists protected plants; maps; illus.)

Höpflinger, F. and Schliefsteiner, H. (1981). *Naturführer Österreich*. Styria, Graz. 480 pp. (Flora and fauna; colour illus.)

Information on Threatened Plants National threatened plant list:

Niklfeld, H. and Karrer, G. (in prep.). *Rote Liste Gefährdeter Pflanzen Österreichs*. Bundesministerium für Gesundheit und Umweltschutz, Wien.

See also:

Kux, S., Kasperowski-Schmid, E. and Katzmann, W. (1981). *Naturschutz – Empfehlungen zur Umweltgestaltung und Umweltpflege II*. Österreichisches Bundesinstitut für Gesundheitswesen, Wien. 125 pp. (Includes principles and problems of nature conservation and countryside management; species protection; habitat protection; lists threatened animals, plants and protected areas; illus.)

There are threatened plant lists for 4 of the 9 Provinces – Burgenland, Kärnten, Salzburg and Steiermark:

Bach, H. (1978). *Kärntner Naturschutzhandbuch, Vol. 1*. Kärtner, Klagenfurt. 779 pp. (Includes threatened and protected plants, and threatened habitats in the Province of Kärnten; illus.)

Traxler, G. (1978). Verschollene und gefährdete Gefässpflanzen im Burgenland: Rote Liste bedrohter Gefässpflanzen (Extinct and endangered vascular plants in Burgenland: Red list of threatened vascular plants). *Natur und Umwelt im Burgenland* 1: 1-24. (Lists 619 regionally threatened flowering plants in Burgenland; conservation categories similar but not identical to those of IUCN.)

Traxler, G. (1980-1982, 1984). Zur Roten Liste der Gefässpflanzen des Burgenlandes. Nachträge, Ergänzungen und Berichtigungen (I)-(IV), (About the Red List of vascular plants in Burgenland. Additions, completions and corrections (I)-(IV).)

Natur und Umwelt im Burgenland 3(1): 9-14; 4(1): 22-25; 5(112): 3,4 and *Volk und Heimat* (1984) 3: 42-43.

Traxler, G. (1982). Liste der Gefässpflanzen des Burgenlandes (List of vascular plants in the Burgenland). *Veroffent. Internat. Clusius-Forschungsges. Güssing* 6: 1-32. (Checklist; includes conservation categories.)

Weiskirchner, O. (1979). *Rote Liste Bedrohter Farn- und Blütenpflanzen in Salzburg* (Red List of Threatened Ferns and Flowering plants in Salzburg). Amt d. Salzburger Landesregierung, Naturschutzreferat, Salzburg. 41 pp. (Lists c. 720 taxa.)

Zimmermann, A. and Kniely, G. (1980). Liste verschollener und gefährdeter Farn- und Blütenpflanzen für die Steiermark (List of missing and endangered ferns and flowering plants for Steiermark). *Mitt. Inst. Umweltwiss. Naturschutz* 3: 3-29. (Lists over 540 taxa including not threatened endemics.)

Zimmermann, A., Kniely, G., Maurer, W. and Melzer, H. (in prep.). *Atlas zur Liste Verschollener und Gefährdeter Farn- und Blütenpflanzen für die Steiermark*. Graz. (Distribution maps of species treated in Zimmermann and Kniely, 1980.)

Included in the European threatened plant list (Threatened Plants Unit, 1983, cited in Appendix 1); latest IUCN statistics, based upon this work: endemic taxa – V:1, R:7, I:1, K:6, nt:20; non-endemics rare or threatened worldwide – Ex:1, E:1, V:17, R:9, I:5 (world categories).

Laws Protecting Plants No federal legislation for plant species protection, but 150 taxa are protected by laws and ordinances issued by each of the 9 Provinces. Within each Province (Bundesland) there are 4 levels of protection; outlined in Kux *et al.* (1981) above. This supercedes the earlier publication:

Plank, S. (1975). *Gesetzlich Geschützte Pflanzen in Österreich*. Ludwig Boltzmann-Institut für Umweltwissenschaften und Naturschutz, Graz. 50 pp.

Voluntary Organizations
Österreichischer Naturschutzbund (ÖNB), Haus der Natur, 5010 Salzburg. (National Headquarters of the 9 Nature Protection Associations of the respective Provinces.)

WWF-Austria (Österreichischer Stiftverband für Naturschutz), Ottakringer Str. 120, Postfach 1, 1162 Wien.

Botanic Gardens
Alpengarten Franz Mayr-Melnhof, 8130 Frohnleiten.

Alpengarten im Oberen Belvedere (Verwaltung der Bundesgärten), Prinz-Eugen-Strasse 27, 1030 Wien 111.

Botanischer Garten des Landes Kärnten, Klinkstrasse 6, 9020 Klagenfurt.

Botanischer Garten der Universität für Bodenkultur, Gregor-Mendel-Strasse 33, 1180 Wien.

Botanischer Garten der Universität Graz, Holteigasse 6, 8010 Graz.

Botanischer Garten der Universität Innsbruck, Sternwartestrasse 15, 6020 Innsbruck.

Botanischer Garten der Universität Wien, Rennweg 14, 1030 Wien.

Botanischer Garten und Arboretum der Stadt Linz, Bancalariweg 41, 4020 Linz.

Schlosspark Schönbrunn, Verwaltung der Bundesgärten, Schönbrunn, 1130 Wien.

Useful Addresses
Institut für Botanik und Botanischer Garten der Universität Wien, Rennweg 14, 1030 Wien.

Institut für Umweltwissenschaften und Naturschutz, Österreichischen Akademie der Wissenschaften, Heinrichstrasse 5, 8010 Graz.

CITES Management Authority: Bundesministerium für Handel, Gewerbe und Industrie, Abteilung II/3, Landstrasser Hauptstrasse 55-57, 1031 Wien.

Additional References

Fischer, M.A. (1976). Österreichs Pflanzenwelt. *Naturgeschichte Österreichs*. 104 pp. (Vegetation descriptions; illus.)

Gutermann, W. and Niklfeld, H. (1974). Floristic report on Austria (1961-1971). *Mem. Soc. Brot.* 24: 9-23.

Maurer, W. (1981). *Die Pflanzenwelt der Steiermark*. Verlag für Sammler, Graz. 147 pp. (Includes geology, climate, floristics, vegetation and species case-studies in Steiermark Province; photographs; line drawings.)

Niklfeld, H. (1973). Über Grundzüge der Pflanzenverbreitung in Österreich und einigen Nachargebieten. *Verh. Zool.-Bot. Ges.* 113: 53-69.

Scharfetter, R. (1938). *Das Pflanzenleben der Ostalpen*. Wien. 419 pp. (Survey of vegetation of eastern Alps, covering most of Austria.)

Wagner, H. (1956). Die Pflanzengeographische Gliederung Österreichs. *Mitt. Geogr. Ges. Wien.* 98(1): 78-92.

Wagner, H. (1971). Natürliche Vegetation. In Bobek, H. (Ed.) *Atlas der Republik Österreich*. Map IV/3. Österr. Akad. d. Wissensch. Freytag-Berndt and Artaria. (Map of potential natural vegetation of Austria, 1: 100,000, with distribution maps for 90 taxa, including endemics, at 1: 3,000,000.)

Wolkinger, F. *et al.* (1981). *Die Natur- und Landschaftsschutz-gebiete Österreichs*. Österreichische Gesellschaft für Natur- und Umweltschutz, Wien.

Azores

A group of 9 volcanic islands (Flores, Corvo, Terceira, São Jorge, Pico, Faial, Graciosa, São Miguel and Santa Maria) in the Atlantic Ocean, about 1500 km from Lisbon and 1900 from Newfoundland.

Area 2235 sq. km

Population 259,800 (1979 estimate, *Times Atlas*, 1983)

Floristics About 600 native plants, 55 endemic; many introduced exotics, some harmful to the native flora (e.g. *Pittosporum undulatum* at low altitudes).

Vegetation Along the coast a cultivated zone, in which the shrub *Myrica faya* is characteristic. At 500-1350 m is a zone of scrub woodland, dominated by *Juniperus* and *Erica*, with *Laurus*, *Ilex* and other shrubs (Sjögren, 1973b). Laurel forest principally remains in the Pico da Vara area on eastern São Miguel, but also in small areas on Pico, Faial and São Jorge.

Checklists and Floras The Azores are covered by the completed *Flora Europaea* (Tutin *et al.*, 1964-1980) and the *Flora of Macaronesia* checklist, both cited in Appendix 1. Also relevant:

Fernandes, A. and R.B. (1980, 1983). *Iconographia Selecta Florae Azoricae*. 2 fascicles so far. Conimbriga. (Descriptions and line drawings; only pteridophytes and gymnosperms to date.)

Franco, J.A. (1971-). *Nova Flora de Portugal (Continente e Açores)*. Sociedade Astoria, Lisboa. 647 pp. (Incomplete, 1 vol. to date: Lycopodiaceae to Umbelliferae; covers mainland Portugal and the Azores.)

Hansen, A. (1970). *A Botanical Bibliography of the Azores*. Copenhagen. Mimeo. (Very comprehensive.)

Palhinha, R.T. (1966). *Catálogo das Plantas Vasculares dos Açores*. Sociedade de Estudos Açorianas Afonso Chaves, Lisboa. 186 pp. (Annotated checklist.)

Sjögren, E. (1973a). Vascular plants new to the Azores and to individual islands in the Archipelago. *Bol. Museu Municipal Funchal* 27: 94-120. (New records since Palhinha's 1966 catalogue.)

For a floristic study see:

Pinto da Silva, A.R. (1963). L'étude de la flore vasculaire du Portugal continental et des Açores les dernières années (1955-1961). *Webbia* 18: 397-412.

Information on Threatened Plants The only known list is that produced by the IUCN Threatened Plants Committee Secretariat (1980) for North Africa and the Middle East, cited in Appendix 1. Latest IUCN statistics, based on this work: endemic taxa – Ex:1, V:5, R:18, I:6, K:11, nt:14; non-endemics rare or threatened worldwide – V:1, R:2 (world categories).

Botanic Gardens IUCN/WWF have been asked by staff at the University of the Azores to fund the creation of a small botanic garden on São Miguel in which endangered plants would be propagated.

Additional References

Pinto da Silva, A.R. (1975). L'état actuel des connaissances floristiques et taxonomiques du Portugal, de Madère et des Açores, en ce qui concerne les plantes vasculaires. In CNRS, 1975, cited in Appendix 1. Pp. 19-28.

Sjögren, E. (1973b). Recent changes in the vascular flora and vegetation of the Azores Islands. *Mem. Soc. Brot.* 13. 453 pp. (Includes details on 414 taxa of vascular plants.)

Sjögren, E. (1973c). Conservation of natural plant communities on Madeira and in the Azores. In *Proc. 1 Intern. Congress pro Flora Macaronesica*. Pp. 148-153. (Not seen.)

Tutin, T.G. (1953). The vegetation of the Azores. *J. Ecol.* 41(1): 53-61.

Tutin, T.G. and Warburg, E.F. (1964). A vegetação dos Açores. *Açoreana* 6: 1-32.

Virville, A.D. de (1965). L'endémisme végétale dans les Iles Atlantides. *Rev. Gén. Bot.* 72 (857): 377-602.

Bahamas

A low-lying archipelago in a 1223 km long arc of the Atlantic Ocean, extending from the coast of Florida on the north-west almost to Haiti on the south-east; 30 major islands, 661 cays and nearly 2400 rocks.

Area 13,864 sq. km

Population 221,000

Floristics 1350 species of vascular plants; 121 taxa (8.83%) endemic to the archipelago (including Turks and Caicos islands) (Correll and Correll, 1982). Floristic relationships are with Florida, Cuba, Hispaniola and Yucatan.

Vegetation Some open pine forest on Grand Bahama, Abaco, New Providence; on the so-called Blackland soils, High and Low Coppice formations, the richest vegetation type in the islands, but now greatly modified for agriculture; on the coast, coppice on sand soils and stunted trees and shrubs on flat elevated rocks; some tidal flats and salt marshes; mangrove in protected locations of lee shores in all the larger islands and cays. Vegetation severely modified on the main islands (Correll and Correll, 1982). 23.2% forested (FAO, 1974, cited in Appendix 1).

Checklists and Floras The Flora is:

Correll, D.S. and Correll, H.B. (1982). *Flora of the Bahama Archipelago*. Cramer, FL-9490 Vaduz, Liechtenstein. 1692 pp. (715 illus. by Priscilla Fawcett; includes the Turks and Caicos Islands.)

Also relevant:

Britton, N. and Millspaugh, C.F. (1920). *The Bahama Flora*. Lancaster. New Era Printing Co., New York. 695 pp. (Reprinted 1962, by Hafner, New York.)
Patterson, J. and Stevenson, G. (1977). *Native trees of the Bahamas*. Privately published. 128 pp. (Colour illus., map.)

Information on Threatened Plants No national Red Data Book. The only known reference is:

Popenoe, J. (1984). Rare and threatened plants of the Bahamas. *Threatened Plants Newsletter* No. 13: 11. (Lists 21 species considered to be rare or threatened.)

Voluntary Organizations
The Bahamas National Trust, Nassau.

Useful Addresses
CITES Management Authority: Ministry of Agriculture, Fisheries and Local Government, P.O. Box N-3028, Nassau.

Additional References
Byrne, R. (1980). Man and the variable vulnerability of island life: a study of recent vegetation change in the Bahamas. *Atoll Res. Bull.* 240. 200 pp. (Illus., maps.)
Campbell, D.G. (1978). *The Ephemeral Islands: A Natural History of the Bahamas*. Macmillan Education Ltd., London. 151 pp.
Coker, W.C. (1905). Vegetation of the Bahama Islands. In Shattuck, G.B., *The Bahama Islands*. Geogr. Soc. Baltimore, John Hopkins Univ. Press. Pp. 185-270.
Gillis, W.T., Byrne, R. and Harrison, W. (1975). Bibliography of the natural history of the Bahama Islands. *Atoll Res. Bull.* 191: 1-123.
Howard, R.A. (1950). Vegetation of the Bimini Island group. Bahamas, B.W.I. *Ecol. Monogr.* 20(4): 317-349.
Taylor, N. (1921). Endemism in the Bahama flora. *Ann. Bot.* 35: 523-532.

Bahrain

A small island sheikdom of one island with several smaller satellite islands c. 30 km from the coast of Saudi Arabia about half way down the southern shore of the Persian Gulf, 26°N 50°30'E.

Area 661 sq. km

Population 414,000

Floristics Flora small, no endemics known; according to Good (1955), unlikely to be much over 175 species of vascular plants. Virgo (1980) quotes collecting lists of between 70 and 200 species. Affinities with the flora of Iraq.

Vegetation Mostly desert plant communities, with many sub-halophytic species. Two other localized communities: adventive flora of date gardens in cultivated northern part of island; halophytic vegetation of muddy shores (salt marsh and mangrove swamp).

Checklists and Floras

Bellamy, D.A. (1984). Additional flowering plants of Bahrain. In Hill, M. and Nightingale, T. (Eds), *Wildlife in Bahrain*. Third Biennial Report of the Bahrain Natural History Society. Pp. 90-96. (Additions to the checklist of Virgo, 1980; with 4 colour photographs.)

Good, R. (1955). The flora of Bahrain. In Dickson, V., *The Wild Flowers of Kuwait and Bahrain*. Allen and Unwin, London. Pp. 126-140. (Includes account of vegetation and checklist of vascular plants.)

Virgo, K.J. (1980). An introduction to the vegetation of Bahrain. In Hallam, T.J. (Ed.), *Wildlife in Bahrain*. Bahrain Natural History Society Annual Reports for 1978-1979, Bahrain Natural History Society. Pp. 65-109. (Includes an annotated and illustrated checklist of the flora.)

Most of the plants of Bahrain are included in the *Flora of Saudi Arabia* (Migahid, 1978, cited under Saudi Arabia). Descriptions of 86 plants recorded, mostly from the north, are given in Virgo (1980), see above.

Information on Threatened Plants None.

Voluntary Organizations

Bahrain Natural History Society, P.O. Box 20336, Manama.

Additional References

Vesey-Fitzgerald, D.F. (1957). The vegetation of central and eastern Arabia. *J. Ecol.* 45: 779-798. (With four black and white photographs and small-scale vegetation map.)

Zakis, M.M. (Ed.) (1978). *Comprehensive Study of Plant Ecology and Investigation into Possibility of Establishing a Botanic Garden in Bahrain*. Univ. Arab. States, Khartoum. (In Arabic; cited by Virgo, 1980.)

Bangladesh

Area 143,998 sq. km

Population 98,464,000

Floristics c. 5000 angiosperm species (Khan and Huq, 1972). The flora is mainly related to that of India; however, the flora of Chittagong and the Chittagong Hill Tracts is more related to that of Indo-China (S. Khan, 1984, *in litt.*).

Vegetation Mostly low-lying alluvial plains of the Ganges and Brahmaputra river systems with extensive marsh and sedge-land, much of the plains under rice and jute cultivation. Tropical semi-evergreen rain forest, on Chittagong hills in the south-east and in Sylhet; tropical moist semi-evergreen Sal (*Shorea robusta*) forest north of Dhaka, now mostly secondary. Extensive mangroves in the Sunderbans region at the mouth of Ganges, covering 6000 sq. km, the largest such tract in the world (Myers, 1980, cited in Appendix 1). Estimated rate of deforestation of closed broadleaved forests 80 sq. km/annum out of a total of 9270 sq. km (FAO/UNEP, 1981). However, Myers (1980, cited in Appendix 1) includes the tropical forests of Bangladesh as "undergoing broad-scale conversion at rapid rates" and predicts little forest could be left "by 1990 if not earlier".

Checklists and Floras

Datta, R.N. and Mitra, J.N. (1953). Common plants in and around Dacca. *Bull. Bot. Soc. Bengal* 7: 1-110. (Keys and descriptions of plants found in 16 km radius from Dhaka.)

Khan, S. and Huq, A.M. (Eds) (1972). *Flora of Bangladesh*. Bangladesh Agric. Res. Council, Dhaka. (27 fascicles to date covering 34 small families; no. 4 includes notes on vegetation types.)

Prain, D. (1903). *Bengal Plants*, 2 vols. Calcutta. (Reprinted by Botanical Survey of India, Calcutta, 1963.)

Prain, D. (1903). Flora of the Sundribuns. *Rec. Bot. Survey India* 2: 231-370.

Bangladesh is also covered by the *Flora of British India* (Hooker, 1872-1897), cited in Appendix 1. For ferns see Beddome (1892) and the companion volume by Nayar and Kaur (1972), both of which are cited in Appendix 1.

Information on Threatened Plants IUCN has a preliminary list of 35 threatened plants, prepared in 1984 by S. Khan, Bangladesh National Herbarium.

Botanic Gardens

Baldah Garden, Wari, Dhaka.
Mirpur Botanic Garden, Dhaka.

Useful Addresses

Bangladesh National Herbarium, 229 Green Road, Dhaka.
CITES Management Authority: The Chief Conservator of Forests, Government of Bangladesh, Bana Bhaban, Gulshan Road, Mohakhali, Dhaka-12.

Barbados

Barbados, 33.8 km long and 22.5 km broad, is the most easterly of the Caribbean islands. It is low-lying, coral and fertile, with a dense population and intensively cultivated for sugar cane.

Area 430 sq. km

Population 262,000

Floristics c. 700 native species, 6 endemic; over 10,000 introduced species and hybrids, many of which have become naturalized (National Conservation Commission, 1984, pers. comm.).

Vegetation Almost the entire island has been modified for cultivation, grazing and development; a few patches of coastal woodland remain as do a few isolated areas of mangrove swamp vegetation at Graeme Hall and St Lawrence; the greatest variety of plants on Barbados are in steep clefts in the upper coralline levels (the Gullies); sparse climbing xerophytic vegetation on rocky land and inland cliffs; dune vegetation of grass and sandy bushland of low shrub and trees nearly to the sea.

Checklists and Floras Covered by the *Flora of the Lesser Antilles, Leeward and Windward Islands* (only monocotyledons and ferns published so far; Howard, 1974-), and by the family and generic monographs of *Flora Neotropica*. (Both are cited in Appendix 1). The Island's Flora is:

Gooding, E.G.B., Loveless, A.R. and Proctor, G.R. (1965). *Flora of Barbados*. H.M.S.O., London. 486 pp.

Information on Threatened Plants None.

Voluntary Organizations
Caribbean Conservation Association, Savannah Lodge, The Garrison, St Michael.
National Conservation Commission, Codrington House, P.O. Box 807E, St Michael.
The Barbados National Trust, Ronald Tree House, No. 2, 10th Avenue, Belleville, St Michael.

Botanic Gardens
Andromeda Gardens, St Joseph.
Farley Hill National Park, St Peter.
Welchman Hall Gully, St Thomas.

Useful Addresses
The Bellairs Research Institute, McGill University, St James.

Additional References
Gooding, E.G.B. (1974). *The Plant Communities of Barbados*. Ministry of Education, Barbados. 243 pp.

Belgium

Area 30,519 sq. km

Population 9,877,000

Floristics 1600-1800 native vascular plant species, estimated by D.A. Webb (1978, cited in Appendix 1) from *Flora Europaea*; c. 1300 according to J.-P. d'Huart (*in litt.*, 1984). One extinct endemic (IUCN figure).

Vegetation Little natural vegetation. Relics of acid oakwoods and oak/beech-woods with birch in the north and east. In central Belgium original beechwoods now largely replaced by agriculture but with occasional patches of coppiced oak and hornbeam. Dry grassland drastically reduced; remaining pockets in south and east on sandy and calcareous soils. Some extensive areas of raised bog and moor survive in the east. Salt-marshes and dunes, once extensive along north coast, have almost completely been destroyed.

Checklists and Floras Covered by the completed *Flora Europaea* (Tutin *et al.*, 1964-1980, cited in Appendix 1). Selected national and regional Floras:

De Langhe, J.-E. *et al.* (1983). *Nouvelle Flore de la Belgique, du Grand-Duché de Luxembourg, du Nord de la France et des Régions Voisines*, 3rd Ed. Jardin Botanique National de Belgique, Meise. 1100 pp. (Ferns and flowering plants.)
Robyns, W. (Ed.) (1950-). *Flore Générale de Belgique*, several parts. Ministère de l'Agriculture, Jardin Botanique de L'Etat, Bruxelles. (Incomplete; ferns, gymnosperms, angiosperms to Thymelaeaceae, by A. Lawalrée; maps; illus.)

Atlas:

Rompaey, E. van and Delvosalle, L. (1978-1979). *Atlas de la flore Belge et Luxembourgeoise, Ptéridophtyes et Spermatophytes*, 2nd Ed., 2 vols. Jardin Botanique National de Belgique, Bruxelles. 116 pp; 293 pp; 1542 maps. (Distribution maps of majority of Belgian vascular plants, except the most widespread; 4 sq. km grid and explanatory text.)

National botanical journal: *Bulletin de la Société Royale de Botanique de Belgique*, Brussels.

Field-guides
De Sloover, J. and Goossens, M. (1981). *Guide des Herbes Sauvages*. Duculot, Gembloux. 217 pp.
Tercafs, R. and Thiernesse, E. (1978). *Guide Nature de l'Ardenne*. Duculot, Gembloux. 400 pp.

See also: Fitter, Fitter and Blamey (1974), cited in Appendix 1.

Information on Threatened Plants One of the first countries to publish a national plant Red Data Book:

Delvosalle, L., Demaret, F., Lambinon, J. and Lawalrée, A. (1969). *Plantes Rares, Disparues ou Menacées de Disparition en Belgique: L'Appauvrissement de la Flore Indigène*. Ministère de l'Agriculture, Service des Réserves Naturelles domaniales et de la Conservation de la Nature, No. 4. 129 pp. (Lists over 300 extinct and threatened vascular plants, and 148 threatened bryophytes; describes threats to the flora; maps.)

Other references:

D'Hose, R. and De Langhe, J.E. (1974-). Nieuwe Groeiplaatsen van zeldzame Planten in België (New locations of rare plants in Belgium). *Bull. Soc. Roy. Bot.* Belg. 107(1): 107-114. (Numerous papers in Dutch, starting with that given.)

Delvosalle, L. and Vanhecke, L. (1982). Essai du notation quantitative de la raréfaction d'espèces aquatiques et palustres en Belgique entre 1960 et 1980. In Symoens, J.J., Hooper, S.S. and Compère, P. (Eds), *Studies on Aquatic Vascular Plants*, Proceedings of the International Colloquium on Aquatic Vascular Plants, 23-25 January 1981, Brussels. Société Royale de Botanique de Belgique, Brussels. Pp. 403-409. (Quantifies the decline of aquatic and marsh plants using floristical data gathered by the Institut Floristique Belgo-Luxembourgeois.)

Lawalrée, A. (1971). L'appauvrissement de la flore belge. *Bull. Jard. Bot. Nat. Belg.* 41: 167-171.

Petit, J. (1979). Chromique de la Montagne Saint-Pierre: 2. Un liste rouge de plantes menacées. *Rev. Vervietoise Hist. Nat.* 36(7-9): 54-57.

Included in the European threatened plant list (Threatened Plants Unit, 1983, cited in Appendix 1); latest IUCN statistics, based upon this work: endemic taxa – Ex:1; non-endemics rare or threatened worldwide – E:2, V:5, R:2, I:1 (world categories).

In 1982 IUCN, under contract to the EEC through the U.K. Nature Conservancy Council, prepared a report (unpublished), *Threatened Plants, Amphibians and Reptiles, and Mammals (excluding Marine Species and Bats) of the European Economic Community*, which includes data sheets on 4 Endangered plants in Belgium.

In spring 1984 WWF-Belgium launched a national Plants Campaign in the Jardin Botanique National de Belgique, Meise, as part of their contribution to the International IUCN/WWF Plants Programme 1984-85. Further details available from WWF-Belgium and the Garden (addresses below).

Laws Protecting Plants National legislation in 1976 (Arrêté royal du 16 février 1976 relatif aux mesures de protection en faveur de certaines espèces végétales croissant à l'état sauvage) provides complete protection for 45 plant taxa and all Lycopodiaceae. Partial protection is given to a further 22 species and selected genera and families. For details see:

Lawalrée, A. (1981). *Plantes sauvages protegées en Belgique*. Jardin Botanique National de Belgique, Meise. 32 pp. (Describes habitats and threats of 64 protected species; colour photographs.)

Voluntary Organizations

Société Royale de Botanique de Belgique, Domaine de Bouchout, 1860 Meise.
WWF-Belgium, Chaussée de Waterloo 608, 1060 Brussels.

Botanic Gardens

Arboretum Geographique de Tervuren, Administration de la Donation Royale, Avenue du Derby 57, 1050 Bruxelles.
Arboretum Kalmthout, Weidestraat 60, 2600 Berchem.
Jardin Botanique National de Belgique, Domaine de Bouchout, 1860 Meise.
Jardin Botanique de l'Université de Liège, Sart Tilman, 4000 Liège.
Jardin Experimental Jean Massart, Université Libre de Bruxelles, Chaussée de Wavre 1850, 1160 Bruxelles.
Plantentuin der Rijksuniversiteit, K.L. Ledeganckstraat 35, 9000 Gent.
Station de Recherches des Eaux et Forêts, 1990 Groenendaal-Hoeilaart.

Useful Addresses

Centre d'Education pour la Protection de la Nature, Rue de la Paix 83, 6168 Chapelle-lez Herlaimont.

Institut Royal des Sciences Naturelles de Belgique, Rue Vautier 31, 1040 Bruxelles.

Ministère de l'Agriculture, Service de la Protection des Végétaux, Manhattan Centre, 21 Avenue du Boulevard, 1000 Brussels.

TRAFFIC-Belgium, WWF-Belgium, see above.

Additional References

Lawalrée, A. (1963). Aperçu sur l'étude de la flore vasculaire de la Belgique depuis 1945. *Webbia* 18: 107-127.

Lawalrée, A. (1978). *Introduction à la Flore de la Belgique*. Jardin Botanique National de Belgique, Meise. 67 pp. (Descriptive account; black and white photographs.)

Noirfalise, A. (1971). La conservation des biocoénoses en Belgique. *Bull. Jard. Bot. Nat. Belg.* 41: 219-230.

Tanghe, M. (1975). *Atlas de Belgique: Phytogéographie (Commentaire)*. Vaillant-Carmanne, Liège. 75 pp. (Detailed vegetation account with line drawings.)

Vanden Berghen, C. (1982). *Initiation à l'étude de la végétation*, 3rd Ed. Jardin Botanique National de Belgique, Meise. 263 pp. 134 figs.

Vanhecke, L. and Charlier, G. (1982). The regression of aquatic and marsh vegetation and habitats in the north of Belgium between 1904 and 1980: some photographic evidence. In Symoens, J.J., Hooper, S.S. and Compère, P. (Eds), *Studies on Aquatic Vascular Plants*, Proceedings of the International Colloquium on Aquatic Vascular Plants, 23-25 January 1981, Brussels. Société Royale de Botanique de Belgique, Brussels. Pp. 410-411.

Belize

Area 22,963 sq. km

Population 156,000

Floristics Toledo (1985, cited in Appendix 1), from published checklists, quotes 3240 species of vascular plants. (Gentry, 1978, cited in Appendix 1, quoting D.L. Spellman, pers. comm., had estimated 2500-3000 species.) 150 endemic species (IUCN figures). Flora is similar to that of the Yucatan peninsula in Mexico and of Petén in Guatemala.

Vegetation Over most of the country, broadleaved rain forest; in the northern half, where rainfall is lower, sometimes called semi- or quasi-rain forest; on river banks and in lowlands forest of Cohune palm (*Orbignya cohune*) associated with mahogany *Swietenia macrophylla*, which has been exploited almost to extinction; most of the rain forest is secondary due to effect of Mayan and present civilizations (D'Arcy, 1977); on the poor soils of the coastal plain and interior up to 1000 m, savannas and pine forests, mainly of *Pinus caribaea*; on the coast wet savannas and mangrove. Estimated rate of deforestation for closed broadleaved forest 90 sq. km/annum out of 12,570 sq. km (FAO/UNEP, 1981); this is similar to the NAS estimate that "some 11,000 sq. km may still support good-quality forest, albeit subject to some disruption through light-impact timber harvesting" (Myers, 1980, cited in Appendix 1).

Checklists and Floras Belize is covered by the *Flora Mesoamericana* Project, described in Appendix 1, and by the completed *Flora of Guatemala* and related articles in *Fieldiana* (cited under Guatemala), as well as by the family and generic monographs of *Flora Neotropica* (cited in Appendix 1). Country Floras and checklists are:

Dwyer, J.D. and Spellman, D.L. (1981). A list of the Dicotyledoneae of Belize. *Rhodora* 83: 161-236.

Spellman, D.L., Dwyer, J.D. and Davidse, G. (1975). A list of the Monocotyledoneae of Belize including a historical introduction to plant collection in Belize. *Rhodora* 77(809): 105-140. (Collections since 1959 with annotations for new country records.)

Standley, P.C. and Record, S.J. (1936). The forests and flora of British Honduras. *Field Mus. Nat. Hist., Bot. Ser.* 12: 1-432. (Description of forest types and annotated species list; 1981 angiosperms and gymnosperms, 134 pteridophytes.)

Williams, L.O. (1956). An enumeration of the Orchidaceae of Central America, British Honduras, and Panama. *Ceiba* 5: 1-256. (List of 97 species from Belize.)

See also:

Carnegie Institute of Washington (1936-1940). *Botany of the Maya Region: Miscellaneous Papers* 1-21. Washington, D.C. 2 vols. 802 pp.

Fosberg, F.R., Stoddart, D.R., Sachet, M.-H. and Spellman, D.L. (1982). Plants of the Belize Cays. *Atoll Research Bull.* 258. 77 pp. (Annotated checklist of 182 species of vascular plants.)

Information on Threatened Plants There is no national Red Data Book. IUCN is preparing a threatened plant list for release in a forthcoming report *The list of rare, threatened and endemic plants of Middle America*. Latest IUCN statistics, based upon this work: endemic taxa – R:6, I:1, K:141, nt:2; non-endemics rare or threatened worldwide – E:1, V:5, R:6, I:6 (world categories).

Threatened plants are mentioned in several papers in:

Prance, G.T. and Elias, T.S. (Eds) (1977), cited in Appendix 1. See in particular W.G. D'Arcy on endangered landscapes in the region (pp. 89-104) and J.T. Mickel on rare and endangered ferns (pp. 323-328).

Voluntary Organizations
Belize Audubon Society, P.O. Box 101, Belmopan. (Membership includes knowledgeable botanists.)

Useful Addresses
CITES implementation: Chief Forest Officer, Department of Forestry, Ministry of Natural Resources, Belmopan. (Note: Belize adheres to CITES, but is not considered a Party because it has not separately ratified the Convention since independence from the U.K. in 1981.)

Additional References
Hartshorn, G. *et al.* (1984). *Belize: Country Environmental Profile*. R. Nicolait & Assoc., Belize City. 2 parts – Executive Summary (8 pp.) and Field Study (151 pp.). (Latter contains list of tree species by G. Hartshorn (pp. 146-151) derived from works cited under Floras and Checklists, above, augmented by personal observations.)

Lundell, C.L. (1945). The vegetation and natural resources of British Honduras. In Verdoorn, F. (Ed.) (1945), cited in Appendix 1. Pp. 270-273. (Includes vegetation map.)

Romney, D.H. (Ed.) (1959). *Land in British Honduras*. Colonial Research Publications
no. 24, HMSO, London. 327 pp.

Benin

Area 112,622 sq. km

Population 3,890,000

Floristics c. 2000 species (H. Ern, 1984, *in litt.*); 11 endemic (Brenan, 1978, cited
in Appendix 1).

Floristic affinities predominantly Sudanian; in southernmost third of country affinities
Sudanian and Guinea-Congolian.

Vegetation Mostly Sudanian woodland with *Isoberlinia*, with a small area of
Sudanian woodland without characteristic dominants in extreme north, and, in the south,
lowland rain forest interspersed with secondary grassland and cultivation. In eastern Benin
there is semi-deciduous rain forest, but this is now represented only by some very small
reserves. Estimated rate of deforestation for closed broadleaved forest 12 sq. km/annum
out of 470 sq. km (FAO/UNEP, 1981).

For vegetation map see White (1983), cited in Appendix 1.

Checklists and Floras Benin is included in the *Flora of West Tropical Africa*,
cited in Appendix 1.

Information on Threatened Plants
Hedberg, I. (Ed.) (1979), cited in Appendix 1. (List for Benin, pp. 91-92, by E.J.
Adjanohoun, contains 48 species threatened in Benin: E:10, V:20, R:18.)

IUCN has records of 13 species and infraspecific taxa believed to be endemic; no
categories given.

Botanic Gardens
University Botanic Garden, Abomey-Calavi, near Cotonou.

Useful Addresses
Ministère du Développement Rural et de l'Environnement, Cotonou.
Université Nationale du Bénin, Herbier National du Bénin, B.P. 526, Cotonou.
CITES Management and Scientific Authorities: Direction des Eaux, Fôrets et Chasse,
Ministère des Fermes d'Etat, d'Elevage et de la Pêche, B.P. 393, Cotonou.

Additional References
Adjakidje, V. (1984). Contribution à l'étude botanique des savanes guinéennes de la
République Populaire du Bénin. Unpublished thesis, University of Bordeaux.
Adjanohoun, E. (1968). Le Dahomey. In Hedberg, I. and O. (1968), cited in Appendix
1. Pp. 86-91.
Akoegninou, A. (1984). Contribution à l'étude botanique des ilots de forêts denses
humides semi-décidues en République Populaire du Bénin. Unpublished thesis,
University of Bordeaux.
Aubréville, A. (1937). Les forêts du Dahomey et du Togo. *Bull. Com. Etud. Hist.
Scient. Afr. Occid. Fr.* 20. 112 pp. (With 18 black and white photographs.)

Paradis, G. (1983). A phytogeographic survey of southern Benin. In Killick, D.J.B. (1983), cited in Appendix 1. Pp. 579-585.

Bermuda

The Bermudas or Somers islands comprise 100 small limestone islands, c. 20 of them inhabited, in the west of the Atlantic Ocean, 917 km east of the coast of North Carolina, U.S.A. They are a self-governing dependent territory of the United Kingdom.

Area 54 sq. km

Population 54,670

Floristics 146 native species of flowering plants and 19 species of ferns, with 8.7% endemism (Britton, 1918). 17 endemic species recorded by B. Phillips, see below. Affinities with both the Old World Tropics and the Neotropics.

Vegetation Most of the vegetation has been modified; only small areas of natural vegetation remain, e.g. Paget and Devonshire marsh and the upland hills of Castle Harbour and Walsingham. Originally the endemic Bermuda Cedar (*Juniperus bermudiana*) was dominant, but 96% of its population was devastated by an introduced scale insect in 1942. A few mature trees survived and pockets of young Bermuda cedars are re-emerging in protected areas. To compensate, many exotic trees and shrubs were introduced in the 1950s and 1960s. Areas of protected mangroves exist in tidal inlets and around some sheltered bays. (B. Phillips, 1984, *in litt.*)

Checklists and Floras
Britton, N. (1918). *Flora of Bermuda*. Scribners, New York. 585 pp. (Illustrations and list of endemic species.)

Field-guides
Curtis, E.W. (1978). *Bermuda – a floral sampler*. Privately published. 54 pp. (Includes note on conservation, drawn illus., photographs.)

Information on Threatened Plants *The IUCN Plant Red Data Book* has one data sheet for Bermuda, on *Juniperus bermudiana*. In 1981, B. Phillips, of the Bermuda Department of Agriculture, prepared a set of data sheets on 30 Bermudan plants, 17 endemic, 2 of them mosses.

Laws Protecting Plants Tree Preservation Orders and Woodland Preservation Orders are used to protect areas of natural beauty or specimen trees. All remaining mangroves are so protected.

Voluntary Organizations
Bermuda Aquarium, Natural History Museum and Zoo (BAMZ), Conservation Volunteers, P.O. Box FL 145, Flatts, Smith's 3.
Bermuda National Trust, P.O. Box 61, Hamilton 5.
Walsingham Trust, Hamilton Paris.

Botanic Gardens
The Bermuda Botanical Gardens, Point Finger Road, Paget East.

Useful Addresses

Conservation Officer, Department of Agriculture and Fisheries, P.O. Box 834, Hamilton 5.

Additional References

Hayward, S.J., Gomez, V.H. and Sterrer, W. (Eds) (1981). *Bermuda's delicate balance: People and environment*. The Bermuda National Trust. 402 pp.

Bhutan

Area 46,620 sq. km

Population 1,388,000

Floristics Provisional estimate of 5000 vascular plant species (D. Long, 1984, *in litt.*). Country endemism very low, but 10-15% endemic to Eastern Himalayas. The subtropical flora has affinities with that of S.E. Asia, the temperate flora with that of China and Japan; Tibetan and Euro-Siberian species are also present (Grierson and Long, 1983).

Vegetation Tropical semi-evergreen forests in lowlands, temperate forests and scrub at high altitudes. Subtropical and tropical moist deciduous forests predominantly of Sal (*Shorea robusta*) on southern foothills of Himalayas at 200-1000 m, almost totally destroyed at low altitudes; warm temperate broadleaved forest at 1000-2000 m (some cleared for agriculture and timber); xerophytic Chir Pine (*Pinus roxburghii*) forest in deep dry valleys at 900-1800 m; cool temperate broadleaved forest at 2000-2900 m with evergreen *Quercus* and *Castanopsis* in drier areas, replaced by mixed forest in wetter areas; evergreen oak forest in central Bhutan, especially around Tongsa and on the hills above Mongar, between 1800-2600 m; various types of coniferous forests to 3800 m; juniper/rhododendron scrub and dry alpine scrub up to 4600 m (Grierson and Long, 1983). Estimated rate of deforestation of closed broadleaved forests 10 sq. km/annum out of a total of 14,900 sq. km (FAO/UNEP, 1981). Most clearance has taken place in the rich subtropical belt (Long, *in litt.*).

Checklists and Floras Bhutan is included in the *Flora of Eastern Himalaya* (1966-) and the *Flora of British India* (Hooker, 1872-1879), both of which are cited in Appendix 1.

Grierson, A.J.C. and Long, D.G. (1980). *A Provisional Checklist of the Trees and Major Shrubs (Excluding Woody Climbers) of Bhutan and Sikkim*. Royal Botanic Garden, Edinburgh. 51 pp.

Grierson, A.J.C. and Long, D.G. (1983-). *Flora of Bhutan: Including a Record of Plants from Sikkim*. Royal Botanic Garden, Edinburgh. (2 parts so far. Vol. 1(1) – vegetation, phytogeography, botanical bibliography of Bhutan and Sikkim; taxonomic treatments of all gymnosperms and 16 angiosperm families from Myricaceae to Polygonaceae; 1(2) – Phytolaccaceae-Moringaceae, 40 families.)

Subramanyam, K. (Ed.) (1983). Materials for the Flora of Bhutan. *Records Bot. Survey India* 22(2). 278 pp. (Enumeration of c. 200 vascular plants; notes on distribution, uses.)

There is additional information on the Bhutan flora, with newly described species, in the series 'Notes relating to the flora of Bhutan' in *Notes Royal Botanic Garden, Edinburgh*. Published parts – 36: 139-150 (1978); 37: 341-354 (1979); 38: 297-310 (1980); 38: 311-314 (1980); 40: 115-138 (1982).

Information on Threatened Plants None, except for:

Sahni, K.C. (1979). Endemic, relict, primitive and spectacular taxa in eastern Himalayan flora and strategies for their conservation. *Indian J. Forestry* 2(2): 181-190. (Mentions 30 taxa rare or threatened in the Himalayan region, including Bhutan; notes on vegetation.)

Additional References For useful background information on the Himalayan region see Lall and Moddie (1981), cited in Appendix 1.

Bismarck Archipelago

The Bismarck Archipelago, politically a part of Papua New Guinea, is situated east of the island of New Guinea, in the Bismarck Sea, south-west Pacific Ocean. The Bismarcks comprise volcanic islands, raised coral islands and low coral reefs. Area 49,658 sq. km (of which New Britain, the largest island, is 36,500 sq. km).

The vegetation consists of lowland tropical rain forest, extensive on New Britain; the lower limit of montane rain forest is 900 m (Whitmore, 1984, cited in Appendix 1); *Nothofagus* abundant between 1500-2800 m, in parts of central New Britain and eastern New Ireland; swamp forests with *Campnosperma* and *Terminalia* on coastal north-central New Britain; mangroves in north New Britain, New Ireland and New Hanover. Atoll/beach forest and large areas of coastal grasslands on New Britain. Bamboo and cloud forest probably present (Dahl, 1980, cited in Appendix 1).

The Bismarck Archipelago is included on the *Vegetation Map of Malaysia* (van Steenis, 1958) and on the vegetation map of Malesia (Whitmore, 1984), both covering the *Flora Malesiana* region at scale 1:5,000,000 and cited in Appendix 1.

No recent figure for size of flora. No information on threatened plants.

References

Peekel, G. E. (1947). Illustrierte Flora des Bismarck-Archipels für Naturfreunde. (Unpublished ms, Lae.)

Schumann, K. and Lauterbach, K. (1901, 1905). *Die Flora der Deutschen Schutzgebiete in der Südsee*, 2 vols. Leipzig. (Also covers north-east New Guinea; in German.)

Wagner, W.H., Jr. and Grether, D.F. (1948). The pteridophytes of the Admiralty Islands. *Univ. Calif. Publ. Bot.* 23(2): 17-110. (Keys, annotated enumeration, mainly covering Manus Island, with notes on localities, habitats, frequency.)

Bolivia

Based upon material by J.C. Solomon

Area 1,098,575 sq. km

Population 6,200,000

Floristics Estimated at 15,000 to 18,000 species, of which about 9000 recorded so far, reflecting the great diversity of vegetation in Bolivia (J.C. Solomon, 1984, pers. comm.). Probably the least collected country in South America (Prance, 1977). Floristic affinities with neighbouring countries: the upland Central Andean flora with Peru and Chile, the north-east flora with Brazilian Amazonia, the Pampus with Argentina, and the Chaco with Paraguay. Endemism uncertain but likely to be highest in the eastern Andean slopes (Yungas) and interior valleys (Solomon, pers. comm.).

Vegetation The Andes, stretching down western Bolivia, fall into three regions: the western Cordilleras (adjoining the Atacama Desert of Chile and Peru) with high alpine vegetation; the eastern Cordilleras, similar alpine vegetation but interspersed with temperate valleys; between them, at 3400-4300 m, cold semi-arid steppe (the Altiplano) dominated by low puna grassland with low shrubs, the northern part mostly cultivated. On the eastern flanks of the Eastern Cordilleras are very steep valleys with montane moist to pluvial forest and cloud forest (the Yungas); further south subtropical evergreen forest (the Tucumano-Boliviana forest); both these vegetation types lead into the evergreen seasonal lowland forest of the north-east, abutting Brazilian Amazonia; this extends 650,000 sq. km (9.1% of the total Amazon forest) (Unesco, 1981, cited in Appendix 1). In Santa Cruz (south-central Bolivia) are Pampas; in the south-east corner is the impenetrable thorn scrub and swamp of the Chaco Boreal, the northernmost part of the Gran Chaco of Argentina and Paraguay. In the extreme east this abuts the Pantanal of Brazil and Paraguay.

Estimated rate of deforestation for closed broadleaved forest 870 sq. km/annum out of 440,100 sq. km (FAO/UNEP, 1981).

Checklists and Floras Bolivia is covered by the family and generic monographs of *Flora Neotropica*, described in Appendix 1. Country accounts include:

Adolfo, H. (1962, 1966). *Plantas del valle de Cochabamba*. Editorial Canelas, Cochabamba. 2 fascicles.

Foster, R.C. (1958). A catalogue of the ferns and flowering plants of Bolivia. *Contr. Gray Herb.* 184: 1-223. (196 families listed.)

Foster, R.C. (1966). Studies in the Flora of Bolivia – IV. Gramineae. *Rhodora* 68: 97-120, 223-358.

Hitchcock, A.S. (1927). The grasses of Ecuador, Peru and Bolivia. *Contr. U.S. Nat. Herb.* 24(8): 291-556.

Kempff, N. (1976). *Flora Amazonica Boliviana*. Academia Nacional de Ciencias de Bolivia, La Paz. 71 pp.

Standley, P.C. (1931). The Rubiaceae of Bolivia. *Field Mus. Nat. Hist., Bot. Ser.* 7(3): 255-339.

Vasquez, R. and Dodson, C. (1982). Orchids of Bolivia. *Icones Plantarum Tropicarum* 6: 501-600. (Descriptions, illustrations, dot maps.)

The Missouri Botanical Garden and the Bolivian Academy of Sciences, through the Bolivian National Museum of Natural History, began a long-term floristic study of Bolivia in 1981. The first phases are a 3-year survey of two valleys above 1000 m in the Yungas near La Paz and an inventory of the Tariquia *Podocarpus* forest.

Information on Threatened Plants Four species are listed as threatened in Organización de los Estados Americanos (1967), cited in Appendix 1, whereas a further 9 are listed in the annex to the Convention on Nature Protection and Wildlife Preservation in the Western Hemisphere (1940). Also relevant:

Ravenna, P. (1977). Neotropical species threatened and endangered by human activity in Iridaceae, Amaryllidaceae and allied bulbous families. In Prance, G.T. and Elias, T.S. (Eds) (1977), cited in Appendix 1. Pp. 257-266.

Laws Protecting Plants Ley General Forestal de la Nación (Decreto 22686 of 13 August 1974), which covers the management and exploitation of forest resources and provided for the creation of the Centro Desarrollo Forestal (CDF) to administer Bolivian forestry, contains provisions that relate to forest inventories as well as to the creation of protected areas (Solomon, pers. comm.).

Voluntary Organizations
Asociación Boliviana Pro-Defensa de la Naturaleza (PRODENA), Casilla 989, La Paz.

Botanic Gardens
Jardín Botánico de Santa Cruz de la Sierra, Casilla 123, Santa Cruz.
Jardín Botánico "Martín Cárdenas", Casilla 538, Cochabamba.

Useful Addresses
Herbario Nacional de Bolivia, Cajón Postal 20127, La Paz.
Museo Nacional de Historia Natural, Casilla 5829, La Paz.
CITES Management Authority: Ministerio de Asuntos Campesinos y Agropecuarios, Centro de Desarrollo Forestal, Jefatura Nacional de Vida Silvestre, Parques Nacionales, Caza y Pesca, Av. Camacho 1471 6° Piso, Casilla de Correa No. 1862, La Paz.

Additional References
Aliaga de Vizcarra, I. (1978). *Bibliografía Boliviana de Recursos Vegetales*. Academia Nacional de Ciencias de Bolivia, La Paz. 14 pp.
Beck, S. (1982). Inventario y estudio de la flora Boliviana. *Ecología en Bolivia* 1: 14-21. (New journal; back cover contains simplified map of 'ecoregions' of Bolivia.)
Cárdenas, M. (1969). *Manual de Plantas Económicas de Bolivia*. Imprenta Ichtus, Cochabamba. 421 pp. (Not seen.)
Freeman, P.H., Cross, B., Flannery, R.D., Harcharik, D.A., Hartshorn, G.S., Simmonds, G. and Williams, J.D. (1980). *Bolivia: State of the environment and natural resources, a field study*. US-AID contract PDC-C-Q247. (Unpaged.)
Herzog, T. (1923). Die Pflanzenwelt der bolivischen Anden und ihres östlichen Vorlandes. In Engler, A. and Drude, O. (Eds), *Die Vegetation der Erde*, 15. Leipzig. 258 pp.
Prance, G. (1977). Floristic inventory of the tropics: Where do we stand? *Ann. Missouri Bot. Gard.* 64(4): 659-684.
Tosi, J., Unzueta, O., Holdridge, L. and Gonzalez, A. (1975). *Mapa Ecológico de Bolivia*. Ministerio de Asuntos Campesinos y Agropecuarios, La Paz.
Unzueta, O. (1975). *Memoria Explicativa: Mapa Ecológico de Bolivia*. Ministerio de Asuntos Campesinos y Agropecuarios, La Paz. 312 pp.

J.C. Solomon, at Missouri Botanical Garden, has compiled an extensive bibliography on the botany of Bolivia.

Botswana

Area 575,000 sq. km

Population 1,042,000

Floristics Number of species unknown. Brenan (1978, cited in Appendix 1) estimates 17 endemic species, from a sample of *Flora Zambesiaca*.

Split between Zambezian (north-eastern third of country), and Kalahari-Highveld regions.

Vegetation Mostly Kalahari *Acacia* wooded grassland and deciduous bushland (south-west), and Zambezian woodland without characteristic dominants (north-east), with a wide transition band between the two. In extreme south-west an area of sand-dunes with sparse grassland or wooded grassland. The Okavango delta in the north is occupied by herbaceous swamp and aquatic vegetation, while the Makarikari depression is surrounded by halophytic vegetation.

For vegetation maps see Wild and Barbosa (1967, 1968), and White (1983), both cited in Appendix 1.

Checklists and Floras Botswana is included in the incomplete *Flora Zambesiaca*, cited in Appendix 1.

Miller, O.B. (1952). *The Woody Plants of the Bechuanaland Protectorate*. Reprinted from the *J. S. Afr. Bot.* 18. National Botanic Gardens of South Africa, Kirstenbosch. 100 pp. (*Corrigenda* in *J. S. Afr. Bot.* 19:177-182.) (Short descriptions, specimen citations.)

Information on Threatened Plants
Hall, A.V. *et al.* (1980), cited in Appendix 1. (List for Botswana on p. 79 contains 15 non-endemic species and infraspecific taxa – V:1 (regional category), R:6, K:8.)

Useful Addresses
CITES Management and Scientific Authority: Ministry of Agriculture (Parks and Nature Conservation), Private Bag 003, Gaborone.

Additional References
Simpson, C.D. (1975). A detailed vegetation study on the Chobe River in north-east Botswana. *Kirkia* 10: 185-227.

Weare, P.R. and Yalala, A. (1971). Provisional vegetation map of Botswana. *Botswana Notes Rec.* 3: 131-147. (With vegetation map in colour.)

Werger, M.J.A. (1978), cited in Appendix 1. Citation includes list of relevant chapters.

Wild, H. (1968). Bechuanaland Protectorate. In Hedberg, I. and O. (1968), cited in Appendix 1. Pp. 198-202.

Bougainville

Bougainville is an island group, politically part of Papua New Guinea, situated north-west of the Solomon Islands in the south-west Pacific Ocean. Area 10,619 sq. km; population 77,880 (1970 census, *Times Atlas*, 1983). Bougainville, the largest island, reaches 2743 m at Mt Balbis. Large areas in the south have freshwater swamp forests. Bougainville also has lowland ridge forest, mixed lowland rain forest, mangroves, coastal forests (with

Calophyllum, Casuarina and *Terminalia*), secondary scrub and grasslands (Foreman, 1971). No figure for size of flora. No information on threatened plants.

References
Foreman, D.B. (1971). *A Check List of the Vascular Plants of Bougainville with Descriptions of Some Common Forest Trees*. Botany Bull. no. 5. Dept of Forests, Lae. 194 pp. (List of herbarium specimens; 58 trees described with line drawings.)

Heyligers, P.C. (1967). Vegetation and ecology of Bougainville and Buka Islands. *CSIRO Land Resources Series* 20: 121-145.

Thorne, A. and Cribb, P. (1984). Orchids of the Solomon Islands and Bougainville: a preliminary checklist. Royal Botanic Gardens, Kew. 33 pp. (Compiled from herbarium and literature records at Kew.)

Bounty Islands

The Bounty Islands (1.3 sq. km) are an outlying island group of New Zealand, consisting of about 13 rocky islets, stacks and wave-lashed rocks in the Pacific subantarctic, at 47°40'S, 179°10'E. No human interference. No vascular species (D. Given, 1984, *in litt.*).

Brazil

Area 8,511,965 sq. km

Population 132,648,000

Floristics Prance (1979) estimates over 55,000 species of flowering plants, considerably more than any other country in the world; of these, 25,000 to 30,000 occur only in Amazonia (G. Prance, pers. comm., quoted in Gentry, 1977, cited under Ecuador); in Bahia alone there are 129 genera and 850-950 species of Leguminosae (G. Lewis, 1984, pers. comm.).

Vegetation The main vegetation types of this vast country are Amazon rain forests, Caatinga, Cerrado, Pantanal and Atlantic coastal forests.

The Brazilian part of the Amazon forest covers 5,057,490 sq. km (Prance, 1979), 63% of the total Amazon forest and nearly 60% of Brazil; the largest extent of primary tropical rain forest in the world and botanically the least known part of Brazil; species composition very varied; besides the forests on high, non-flooded ground ("terra firme"), which occupies 90% of the area, are "savannas, Amazonian campinas on white sand, campina ... forests of the upper Rio Negro, swamp forest, transition forest, and montane forest" (Prance, 1977).

In the northeast is the caatinga, a semi-arid region dominated by succulents, drought-resistant deciduous thorny trees and shrubs. Central Brazil is mainly cerrado, which varies from dense evergreen lowland forest to medium-tall grassland with broadleaved evergreen trees; on the mountain chain up east central Brazil, above 900 m, is the floristically rich Campo Rupeste, mainly herbaceous vegetation on outcropping rocks and on sites of restricted drainage. Between the Amazon and the Chaco, on the border of Bolivia and reaching south to Paraguay and Argentina, is the Pantanal, a large swampland of c. 100,000 sq. km drained by the Río Paraguay; it is a mixture of open swamp, flooded and deciduous forest, Cerrado and Chaco; little known botanically (Prance and Schaller, 1982). Along the Atlantic coast from north of Pôrto Alegre south to Bahia is a strip of species-rich rain forests, reduced to relicts covering only 2-4% (S.J. Mayo, 1984, pers. comm.) of original extent; perhaps the most endangered tropical rain forests in the world.

Estimated rate of deforestation for closed broadleaved forest 13,600 sq. km/annum, out of a total of 3,562,800 sq. km (FAO/UNEP, 1981). Myers (1980, cited in Appendix 1) gives an analysis of the complex figures for deforestation in Brazil.

Checklists and Floras The part of Brazil north of the Tropic of Capricorn is covered by the family and generic monographs of *Flora Neotropica*, described in Appendix 1. The only published country-wide Flora is:

Martius, K.F.P., Eichler, A.W, and Urban, I. (1840-1906). *Flora Brasiliensis*. Facsimile reprint by Cramer, New York (1965).

More recent works are:

Angely, J. (1965). *Flora Analítica do Paraná*. Edições Phyton, Curitiba, Paraná. 728 pp. (Annotated list of 5287 species.)

Angely, J. (1969-1970). *Flora Analítica e Fitogeográphica do Estado de São Paulo*, 6 vols. Edições Phyton, São Paulo. (7251 species listed with dot maps.)

Flora Ecológica de Restingas do Sudeste do Brasil (1965-1978) (Various authors). Museu Nacional, Rio de Janeiro. (23 fascicles covering 24 families so far.)

Harley, R.M. and Mayo, S.J. (1980). *Towards a Checklist of the Flora of Bahia*. Royal Botanic Gardens, Kew. 250 pp. (A progress report on the Kew-CEPEC expeditions to Bahia in 1974 and 1977; systematic list of 1596 species; predicts "a total of 10,000 species for Bahia seems a conservative estimate".)

Luis, I.T. (1960). *Flora Analítica do Pôrto Alegre*. Instituto Geobiológico "La Salle". (Not seen.)

Pabst, G.F.J. and Dungs, F. (1975-1977). *Orchidaceae Brasilienses*, 2 vols. Brücke-Verlag K. Schmersow, Hildesheim, Germany. 926 pp. In Portuguese, German and English. (Watercolours of selected species.)

Reitz, P.R. (Ed.) (1965-). *Flora Ilustrada Catarinense*. Herbário "Barbosa Rodrigues", Itajaí, Santa Catarina. (Includes dot maps; by 1983 had covered 2759 species in 109 families (117 fascicles), including Bromeliaceae, 1983.)

Rizzo, J. (1981-). *Flora do Estado de Goiás, Coleção Rizzo*. Universidade Federal de Goiás, Goiânia. 4 vols so far – Plan of Collection; Meliaceae by L. Graça Amaral; Araliaceae by A.B. Peixoto; Myristicaceae by W. Rodrigues. (Dot maps.) Author estimates 9605 species (1978, quoted in Toledo, 1985, cited in Appendix 1.)

Schultz, A.R.H. and Homrich, M.H. (1955-1977). *Flora Ilustrada do Rio Grande do Sul*. Universidade Federal do Rio Grande do Sul, Pôrto Alegre. 12 vols. Complete. (Dot maps.)

Sobrinho, R.J. and Bresolin, A. (Eds) (1970-1977). *Flórula da Ilha de Santa Catarina*. Universidade Federal de Santa Catarina, Santa Catarina. 18 fascicles so far.

Teodoro Luis, I. (1960). *Flora Analítica de Pôrto Alegre*. Instituto Geobiologica "La Salle". Canoas. (Unpaged.)

In 1976 Brazil started *Programa Flora*, an inventory of vegetation and a computerized label data bank of Brazilian herbaria. The Programme is divided into regional projects: *Projeto Flora Amazônica* begun in 1977 and some 20 expeditions between then and 1983 have collected over 45,000 numbers; *Projeto Flora Nordeste* has also begun. The data bank for the Amazonian herbaria is now functional and enquiries about Amazonian plants can be made through the Conselho Nacional de Desenvolvimento Científico e Technológico in Brasília.

Toledo (1985, cited in Appendix 1) reports as "in progress" a *Flora of Minas Gerais*, by J. Angely, for a reported 11,156 species.

Field-guides

Centro de Pesquisas Florestais e Conservação da Natureza (1960, 1965). *Flores da Restinga* (54 pp.); *Arboreto carioca* (4 vols). CPFCN, Rio de Janeiro.

Ferri, M. Guimarães (1969). *Plantas do Brasil: Espécies do Cerrado*. Edgard Blücher, São Paulo. 239 pp. (Illus.)

Information on Threatened Plants There is no national Red Data Book; in 1985 FBCN (see below) start work on preparing a threatened plant list, grant-aided by IUCN/WWF (Project 3310). 8 plants are listed as threatened, with explanatory notes, in Organización de los Estados Americanos (1967), cited in Appendix 1, whereas 45 plant species are listed in the annex to the Convention on Nature Protection and Wildlife Preservation in the Western Hemisphere (1940). Threatened plants are mentioned in several papers in:

Prance, G.T. and Elias, T.S. (Eds) (1977), cited in Appendix 1. See in particular D. de Andrade-Lima on preservation of the flora of north-eastern Brazil (pp. 234-239), J.T. Mickel on rare and endangered ferns (pp. 323-328), H.E. Moore Jr. on endangerment in palms (pp. 267-282), P. Ravenna on endangered bulbous species (pp. 257-266).

Other references:

Carvalho, J.C.M. (1968). Lista das espécies de animais e plantas ameaçadas de extinção no Brasil. *Fundação Brasil. Conserv. Natureza, Bol. Inform.* 3: 11-16. (13 species listed.)

Casari, M.B. *et al.* (1980). Nove espécies ameaçadas ou em perigo de desaparecimento no Brasil. *Resumos do 31 Congresso Nacional de Botânica*. Sociedade Botânica do Brasil, Ilhéus. p.123.

Cavalcanti, D.F. (1981). Plantas em extinção no Brasil. *Fundação Brasil. Conserv. Natureza, Bol. Inform.* 16: 115-119.

Liddell, R. (1980). Collections and conservation of Brazilian orchids. In Sukshom, M.R. (Ed.), *Proceedings of the 9th World Orchid Conference*. Amarin Press, Thailand. Pp. 283-285.

Mori, S., Boom, B. and Prance, G. (1981). Distribution patterns and conservation of eastern Brazilian coastal forest tree species. *Brittonia* 33(2): 233-245.

Many individual case studies on endangered plants have been published in *Cadernos FEEMA Ser. Trab. Techn.*, the Bulletin of the Centro de Botânica do Rio de Janeiro, e.g. *Scaevola plumieri* in 18: 7-11 (1982); *Bumelia obtusifolia* in 18: 1-9 (1982); *Ficus lanuginosa* in 18: 3-35 (1982); *Dorstenia* in 1: 29-65 (1982).

Laws Protecting Plants Portaria No. 303 of 29 May 1968 is a regulation to implement the principal wildlife law in force (Lei No. 5197 of 3 January 1967); all trade, transport or export of 13 listed plants is prohibited, with the exception of scientific collection, for which a license is required from IBDF. The principal forestry law (Lei No. 4771 of 15 September 1965) covers trade in live plants and plant products; it is administered by IBDF (Fuller and Swift, 1984, cited in Appendix 1; lists the 13 species).

Voluntary Organizations

Associação de Defesa do Meio Ambiente (ADEMA), Rua Pedroso Alvarenga 1245-4°, and São Paulo, SP 04.531.

Associação Gaúcha de Proteção ao Ambiente Natural (AGAPAN), Caixa Postal 1996, Pôrto Alegre, RS 90.000.

Associação de Preservação da Flora e da Fauna (APREFFA), Caixa Postal 1176, Curitiba, PR 80.000.

Centro de Conservação da Natureza de Brasília, Edifício Antônio Venâncio da Silva, sala 512, Brasília, D.F.

Centro para Conservação da Natureza de Minas Gerais, Caixa Postal 2475, Belo Horizonte, MG 30.000.

Fundação Brasileira para a Conservação da Natureza (FBCN), Rua Miranda Valverde 103, CEP 22281, Rio de Janeiro.

União dos Defensores da Terra (OIKOS), Caixa Postal 51.570, São Paulo, SP 01.000.

Botanic Gardens

Horto Botânico, Divisão de Botânico do Museu Nacional, Quinta da Bôa Vista, Rio de Janeiro, Guanabara.

Jardim Botânico da Fundação Zoobotânica do Rio Grande do Sul (FZM), Caixa Postal 1188, Pôrto Alegre, RS 90.000.

Jardim Botânico, Instituto Básico de Biológia Médica e Agrícola (IBBMA), Caixa Postal 526, 18.610 Botucatu, São Paulo.

Jardim Botânico do Rio de Janeiro, Rua Jardim Botânico 1.008, 22.460 Rio de Janeiro.

Jardim Botânico de São Paulo, Instituto de Botânica, Caixa Postal 4005, 01000 São Paulo.

Museu de Historia Natural, Rua Gustavo da Silveira, 1035 Horto, Belo Horizonte, Minas Gerais.

Museu Paraense "Emílio Goeldi", Av. Magalhaes Barata 376, Caixa Postal 399, 66.000 Belém, Pará.

Parque Botânico do Morro Baú, Av. Marcos Ronder 800, 88.300 Itajaí, Santa Catarina.

Reserva Ecológia de IBGE, Edificio Venâncio II, 2° Andar, 70.302 Brasília, D.F.

Useful Addresses

FEEMA-DECAM, Herbário A. Castellanos, Estrada da Vista Chinesa 741, Alto da Bôa Vista, 20531 Rio de Janeiro.

Instituto Brasileiro de Desenvolvimento Florestal (IBDF), Esplanada dos Ministerios, Brasília 70.000.

Instituto Nacional de Pesquisas Amazonia (INPA), CP 478, Manaus, Amazonas, 69.000.

Museo Nacional, Quinta da Bôa Vista, Rio de Janeiro, RJ CEP 20940.

SEMA, Ministerio do Interior, Esplanada dos Ministerios, Brasília, D.F. 70054.

CITES Management Authority: Departamento de Parques Nacionais e Reservas Equivalentes, IBDF, Sain-Av. L4 Norte, Brasília, D.F.

CITES Scientific Authority (for flora): Jardim Botânica do Rio de Janeiro, see above.

Additional References

Ducke, A. and Black, G.A. (1953). Phytogeographical notes on the Brazilian Amazon. *An. Acad. Brasil. Ciências* 25(1): 1-46.

Eiten, G. (1972). The cerrado vegetation of Brazil. *Bot. Rev.* 38: 301-341.

Gentry, A. (1979). Extinction and conservation of plant species in Tropical America: a phytogeographical perspective. In Hedberg, I. (Ed.), *Systematic Botany, Plant Utilization and Biosphere Conservation*. Almqvist & Wiksell International, Stockholm, Sweden. Pp. 110-126. (Includes map of principal vegetation types.)

Pires, J.M. (1973). Tipos de vegetação de Amazônia. *Publ. Avulsas Museu Goeldi. Belém* 20: 179-202.

Pires, J.M. (1978). The forest ecosystems of the Brazilian Amazon: description, functioning and research needs. In Unesco/UNEP/FAO, *Tropical Forest Ecosystems*. Unesco, Paris. Pp. 607-627. (Substantial bibliography.)

Prance, G.T. (1977). The phytogeographic subdivisions of Amazonia and their influence on the selection of biological reserves. In Prance, G.T. and Elias, T.S. (Eds) (1977), cited in Appendix 1. Pp. 195-213.

Prance, G.T. (1979). The present state of botanical exploration: South America. In Hedberg, I. (Ed.), *Systematic Botany, Plant Utilization and Biosphere Conservation*. Almqvist & Wiksell International, Stockholm, Sweden. Pp. 55-70.

Prance, G.T. and Schaller, G.B. (1982). Preliminary study of some vegetation types of the Pantanal, Mato Grosso, Brazil. *Brittonia* 34: 228-251.

Rizzini, C.T. (1976, 1979). *Tratado de fitogeografia do Brasil*. São Paulo. HUCITEP/USP. 2 vols.

Veloso, H.P. (1966). *Atlas Florestal do Brasil*. Ministério da Agricultura, Rio de Janeiro. 82 pp.

British Indian Ocean Territory (Chagos Archipelago)

The British Indian Ocean Territory is situated to the south of the Maldive Islands between latitudes 5-10°S and longitudes 70-75°E. It includes the coral islands of the Chagos Archipelago (60 sq. km) of which Diego Garcia (47 sq. km) is the largest. Population 2000. Approximately 150 species of vascular plants (Fosberg and Bullock, 1971), of which about 100 are indigenous, mostly with pantropical or Indo-Pacific distributions. The vegetation consists of *Casuarina* woodland, mixed coconut woodland ("Cocos Bon-Dieu"), *Scaevola* scrub, marshland and relict broadleaved woodland with *Ficus*, *Morinda*, and *Terminalia*; some areas cleared for coconut plantations.

Three checklists of the flora are:

Fosberg, F.R. and Bullock, A.A. (1971). List of Diego Garcia vascular plants. In Stoddart, D.R. and Taylor, J.D. (Eds), Geography and ecology of Diego Garcia Atoll, Chagos Archipelago. *Atoll Res. Bull.* 149. 143-160. (Annotated list of 142 taxa from Diego Garcia.)

Willis, J.C. and Gardiner, J.S. (1901). The botany of the Maldive Islands. *Annals Royal Botanic Gardens Peradeniya* 1: 45-164. (Includes annotated list of 359 species recorded from Chagos Archipelago, Laccadives and Maldives.)

Willis, J.C. and Gardiner, J.S. (1931). Flora of the Chagos Archipelago. *Trans. Linn. Soc. Zoology* 19: 301-306. (Annotated checklist.)

British Virgin Islands

A Dependent Territory of the U.K., comprising 30 small islands, mostly uninhabited. The largest is the mountainous island of Tortola, 19 km long by 5.6 km wide. Other principal islands are Virgin Gorda, Jost Van Dyke and Anegada. The islands are hilly and volcanic, except for Anegada which is flat and formed of limestone and sand.

Area 153 sq. km

Population 13,000

Floristics No estimate for number of plant species. The Smith manuscript (see below) includes an analysis of the endemic taxa. Anegada has floristic affinities with Barbuda and Anguilla.

Vegetation Severely modified by man to mostly dry scrub woodland; scrub, principally of *Croton* spp. and thorny bushes are dominant where there is heavy grazing of feral goats and cattle; on higher ground 'xerophytic rain forest', a reduced type of evergreen forest; on Gorda Peak a better developed forest than anywhere else on the islands; vegetation of Anegada reduced to sandy scrub in the west and limestone scrub in the east (C. Pannell, 1976, *in litt.*); 6.7% forested (FAO, 1974, cited in Appendix 1).

Checklists and Floras

Britton, N. and Wilson, P. (1923-1930). Botany of Porto Rico and the Virgin Islands. *Scientific survey of Porto Rico and the Virgin Islands*, 5 (626 pp.) and 6 (663 pp.). New York Academy of Sciences, New York. (Keys, descriptions, general ranges and distributions by island.)

J. Smith, of Treasure Island Botanic Garden, Tortola, has prepared a manuscript entitled Native and naturalised flowering plants of the British Virgin Islands. It includes an outline of the vegetation, descriptions of endemic plants and summary of recorded species.

See also:

D'Arcy, W.G. (1967). Annotated checklist of the dicotyledons of Tortola, Virgin Islands. *Rhodora* 69: 385-450.

D'Arcy, W.G. (1975). Anegada Island: Vegetation and Flora. *Atoll. Res. Bull.* 188. 40 pp. (Illus. and maps.)

Liogier, A.H. (1965). Nomenclatural changes and additions to Britton and Wilson's "Botany of Porto Rico and the Virgin Islands". *Rhodora* 67(772): 315-361.

Liogier, A.H. (1967). Further changes and additions to the flora of Porto Rico and the Virgin Islands. *Rhodora* 69 (779): 372-376.

Little, E.L., Jr. and Wadsworth, F.H. (1964). *Common trees of Puerto Rico and the Virgin Islands*. Agriculture Handbook No. 249, U.S.D.A. Forest Service, Washington, D.C. 548 pp. (Keys, mainly to families; descriptions, illus., distributions.) Spanish edition by authors and J. Marrero, Editorial UPR, Puerto Rico, 1967.

Little, E.L., Jr. et al. (1974). *Trees of Puerto Rico and the Virgin Islands, Second volume*. Agriculture Handbook No. 449, U.S.D.A. Forest Service, Washington, D.C. 1024 pp. (2nd vol. to Little and Wadsworth, 1964, above; includes endemic, rare and endangered tree species.)

Little, E.L., Jr., Woodbury, R.O. and Wadsworth, F.H. (1976). *Flora of Virgin Gorda (British Virgin Islands)*. U.S. Forest Service Research Paper 21. Institute of Tropical Forestry, Río Piedras, Puerto Rico. 36 pp. (Illus. and map.)

Little, E.L., Jr. (1969). *Trees of the Jost Van Dyke (British Virgin Islands)*. U.S. Forest Service Research Paper 9. Institute of Tropical Forestry, Río Piedras, Puerto Rico. 12 pp. (Illus., checklist of 69 native and 18 introduced tree species, with notes on vegetation.)

Information on Threatened Plants

Ayensu, E.S. and DeFilipps, R.A. (1978). *Endangered and Threatened Plants of the United States*. Smithsonian Institution and WWF-U.S., Washington, D.C. Pp. 225-232 (Lists 102 'Endangered' and 'Threatened' taxa from Puerto Rico and the Virgin Islands, both U.S. and British, with a useful bibliography; 9 of them from the British Virgin Is.)

Little, E.L., Jr. and Woodbury, R.O. (1980). *Rare and Endemic Trees of Puerto Rico and the Virgin Islands*. Conservation Research Report No. 27, U.S.D.A. Forest Service, Washington, D.C. 26 pp.

Voluntary Organizations

The Virgin Islands Conservation Society (address not known).

Additional References

Anon (1960). Forestry. Extract from Report: British Virgin Islands, H.M.S.O., London 1957/58, (24). (A brief account of the preservation and conservation of the few existing fragments of forest.)

Fraser, H. (1958). Forest conservation in the British Virgin Islands. In Willan, R.L., *Forestry Development in the British Virgin Islands*. FAO, Rome. 26 pp.

Pannell, C. (1976). Section on vegetation. In *Report of the Cambridge Ornithological expedition to the British Virgin Islands 1976*. Cambridge University. Pp. 26-38.

Brunei

Area 5765 sq. km

Population 269,000

Floristics No overall figure for size of flora, but an estimated 2000 tree species (M. Jacobs, quoted in Unesco, 1974, cited in Appendix 1).

Vegetation Tropical evergreen rain forest, rich in dipterocarps, up to 1300 m; tropical montane rain forest to 1800 m; heath (kerangas) forest usually on sandy alluvial soils and high-altitude sandstone ridges (Brünig, 1974; Whitmore, 1975b, cited in Appendix 1); mangrove and peat swamp forest (with *Shorea albida*) occupy almost the entire coastline. Estimated rate of deforestation of closed broadleaved forest 50 sq. km out of a total of 3230 sq. km (FAO/UNEP, 1981). Myers (1980, cited in Appendix 1) estimates c. 4300 sq. km are still covered by relatively undisturbed primary forest, while secondary forests cover a further 1170 sq. km.

Brunei is included on the *Vegetation Map of Malaysia* (van Steenis, 1958), and on the vegetation map of Malesia (Whitmore, 1984), both covering the *Flora Malesiana* region at scale 1:5,000,000 and cited in Appendix 1.

Checklists and Floras Brunei is included in the incomplete, but very detailed *Flora Malesiana* (1948-), cited in Appendix 1. National accounts include:

Ashton, P. (1965). *Manual of the Dipterocarp Trees of Brunei State*. Oxford Univ. Press. 242 pp. (Keys, descriptions, notes on distribution.)

Browne, F.G. (1955). *Forest Trees of Sarawak and Brunei and Their Products*. Govt Printer, Kuching, Sarawak. 369 pp. (Descriptions of timber trees with notes on distribution and wood properties.)

Pukul, H.B. and Ashton, P.S. (1966). *A Checklist of Brunei Trees*. Govt of Brunei State. 132 pp. (List of trees, not including dipterocarps, arranged alphabetically by vernacular name; botanical names and notes on distribution within Brunei.)

Information on Threatened Plants None.

Additional References

Anderson, J.A.R. (1963). The flora of the peat-swamp forests of Sarawak and Brunei, including a catalogue of all recorded species of flowering plants, ferns and fern allies. *Gard. Bull. Singapore* 20: 131-228. (Enumeration of 33 pteridophytes and 395 flowering plant species; short descriptions and notes on distribution.)

Ashton, P.S. (1964). *Ecological Studies in the Mixed Dipterocarp Forests of Brunei State*. Oxford Forestry Memoirs 25. Clarendon Press, Oxford. 75 pp.

Brünig, E.F. (1974). *Ecological Studies in the Kerangas Forest of Sarawak and Brunei*. Borneo Literature Bureau, Kuching, Sarawak. 237 pp.

Bulgaria

Area 110,912 sq. km

Population 9,182,000

Floristics 3500-3650 native vascular species estimated by D.A. Webb (1978, cited in Appendix 1) from *Flora Europaea*; 53 endemics (IUCN figures). Elements: Atlantic, Central European and alpine, with Mediterranean and sub-Mediterranean influence in the south.

Areas of high endemism: Mt Slavjanka; Mt Pirin; Rhodope mountains; Stara Planina; north-eastern Bulgaria; Thracian Plain; Black Sea coast; Strandja Mts; Tundza hill region; and Mt Rila (Polunin, 1980, cited in Appendix 1). Many Tertiary relicts (e.g. *Haberlea rhodopensis*), especially in Rhodope Mts and Strandja and Slavyanka Mts (Stefanov, 1936).

Vegetation To the north of the Stara Planina (mountains running east-west across Bulgaria), Central European vegetation with steppe elements (*Stipa*, *Astragalus*, *Phlomis* spp.). On the Stara Planina, coniferous forest to 2000-2300 m with juniper and pine-scrub at higher altitudes and alpine flora (*Dryas*, *Empetrum* and *Salix* spp.). Deciduous oak and beech forests extend from the north-west with conifer forests of *Pinus heldreichii* and *P. peuce* in the south and south-west. Forests of *P. peuce* particularly well-developed in Bulgaria, forming pure stands above 1700 m in the Rila, Pirin and western Rhodope Mts. They cover 11,600 ha, about 3% of the country's conifer forests (Polunin and Walters, 1985, cited in Appendix 1). To the south, in the plain of Thrace, sub-Mediterranean maquis of *Quercus coccifera*, *Phillyrea*, *Cistus*.

Checklists and Floras Bulgaria is covered by the completed *Flora Europaea* (Tutin *et al.*, 1964-1980) and the *Med-Checklist* (both cited in Appendix 1). National Floras:

Jordanov, D. *et al.* (Ed.) (1963-). *Flora Reipublicae Popularis Bulgaricae*, 8 vols. Bulgarskata Akad., Sofiya. (Incomplete, a further 2-3 vols planned; vol. 1 contains an extensive historical account of Bulgarian floristic research; introductory text also in English; habitat and ecology details; line drawings.)

Stojanov, N. and Stefanov, B. (1966-1967). *Flora na Balgariya*, 4th Ed. by B. Kitanov. 2 vols. Nauka i Izkustvo, Sofia. (Includes habitat and ecological details; illus.)

For a bibliography see:

Kitanov, B. (1975). *Literature about the Flora and Plant Geography of Bulgaria. 1959-1968.* Bulgarische Akademie der Wissenschaften, Sofia. 270 pp. (In Bulgarian.)

National botanical journal: *Izvestiya na Botanicheskiya Institut* (Bulletin of the Institute of Botany), Bulgarian Academy of Sciences, Sofia.

Field-guides

Delipavlov, D. *et al.* (1983). *Opredelobal na Rastenijaba v Balgeorija.* Zemisdat, Sofia. 431 pp.

Gramatikov, D. (1974). *Identification of Wild and Cultivated Trees and Shrubs in Bulgaria.* Sofia. (In Bulgarian.)

Stojanov, N. and Kitanov, B. (1966). *Plants of the High Mountains in Bulgaria.* Nauka, Izkustvo, Sofia. 149 pp. (In Bulgarian; illus.)

Valev, S., Gancev, I. and Velcev, V. (1960). *Ekskurzionna Flora na Balgarija.* Narodna Prosveta, Sofia. 736 pp. (Native, naturalized and commonly cultivated plants; covers c. 2250 species.)

Also see Polunin (1980), cited in Appendix 1.

Information on Threatened Plants The national plant Red Data Book is:

Velchev, V., Kozuharov, S., Bondev, I., Kuzmanov, B. and Markova, M. (1984). *Red Data Book of the People's Republic of Bulgaria. Volume 1. Plants.* Bulgarian Academy of Sciences, Sofia. 447 pp. (Describes 763 threatened species; includes data about distribution, habitats, ecology; maps; line drawings.)

See also:

Kuzmanov, B. (1978). About the "Red Book of Rare Bulgarian Plants". *Phytology* (Bulgarian Academy of Sciences) 9: 17-32. (In Bulgarian, English summary; lists 150 rare Bulgarian plants.)

Other relevant publications include:

Dimitrov, D. (1977). Rare plant species of the Bulgarian Black Sea Coast. *Priroda* 26(3): 95-96.

Kruscheva, R. and Pirbanov, R. (1978). *Album of Protected and Rare Plants.* (In Bulgarian.)

Kuzmanov, B. (1981a). Mapping and protection of the threatened plants in the Bulgarian flora. In Velcev, V.I. and Kozuharov, S.I. (1981), *Mapping the Flora of the Balkan Peninsula.* 247 pp. (Not seen.)

Kuzmanov, B. (1981b). Balkan endemism and the problem of species conservation, with particular reference to the Bulgarian flora. *Bot. Jahrb. Syst.* 102(1-4): 255-270. (Lists Bulgarian and Balkan endemic vascular species; maps; illus.)

Stanev, S. (1975). *The Stars are Becoming Extinct in the Mountains: Stories about our Rare Plants*. Zemizdat, Sofia. 129 pp. (In Bulgarian; stories describing searches for rare plants.)

Stefanov, B. and Bânkov, M. (1978). Plants that are very rare in Bulgaria or that have recently disappeared and the cause of their decline. *Gorskostoponska Nauka* 15(6): 3-10.

Veltchev, V. and Stoeva, M. (1985). Population approach to the investigation of the threatened and rare species in the Bulgarian flora in connection with their conservation. In: MAB, Conservation of Natural Areas and the Genetic Material they Contain, International Symposium under Project 8 – MAB, 23-28 September 1985, Sofia. (In Bulgarian; English summary.)

Included in the European threatened plant list (Threatened Plants Unit, 1983, cited in Appendix 1); latest IUCN statistics, based upon this work: endemic taxa – Ex:1, E:4, V:10, R:18, I:8, K:2, nt:10; doubtfully endemic taxa – R:3, nt:3; non-endemics rare or threatened worldwide – V:16, R:23, I:10 (world categories).

Laws Protecting Plants The 1967 Law on Nature Protection prohibits the picking, damage, sale of, destruction to or digging up of 67 listed plant species.

Voluntary Organizations

Bulgarian Botanical Society, Institute of Botany, Acad. G. Bonchev Str., 1113 Sofia.

Botanic Gardens

Botanic Garden, University of Sofia, ul. Moskowska 49, P.O. Box 157, 1090 Sofia.

Hortus Botanicus Academia Scientiarum Bulgaricae, Str. "Akad. G. Bontshev," Clou I, 1113 Sofia.

Useful Addresses

Committee for Environmental Protection, Council of Ministers of the People's Republic of Bulgaria, 1000 Sofia.

Committee for Protection of Nature and Environment, State Council of Bulgaria, Trijadita 2, Sofia.

Concept for the Protection of the Natural Flora and Vegetation, Institute of Botany, Bulgarian Academy of Sciences, 13 Sofia.

Ministry of Forests and Protection of the Natural Environment, 17 Antim I Street, 4000 Sofia.

National Council for Nature Protection, Vitoshastz 18, 1000 Sofia.

Research Co-ordinating Centre for Conservation and Reproduction of the Environment, 2 Gagarin Street, 1113 Sofia.

Additional References

Kozuharov, S. (1975). On the endemism in the Bulgarian flora. In Jordanov, D. *et al.* (Eds), *Problems of Balkan Flora and Vegetation*. Proceedings of the 1st International Symposium on Balkan Flora and Vegetation, Varna, June 7-14 1973. Bulgarian Academy of Sciences, Sofia. Pp. 162-168.

Stefanov, B. (1936). Remarks upon the causes determining the relict distribution of plants. *Spis. Bulg. Acad. Sci.* 53: 133-179.

Stefanoff, B. and Jordanoff, D. (1931). Topographische Flora von Bulgarien. *Bot. Jahrb.* 64(5): 388-536.

Stoilov, D. *et al.* (1981). *Protected Natural Sites in the People's Republic of Bulgaria*. Committee on Environmental Protection, Council of Ministers of the People's Republic of Bulgaria, Sofia. 31 pp. (Translated from Bulgarian by I. Saraouleva.)

Stojanov, N. (1965). Phytogeographic elements in the flora of Bulgaria. *Rev. Roum. Biol. (Sér. Bot.)* 10(1-2): 69-70.

Velcev, V., Bondev, I. and Kozuharov, S. (1975). The problem of protection of the natural flora and vegetation in Bulgaria. In Jordanov, D. *et al.* (Eds), *Problems of Balkan Flora and Vegetation*. Proceedings of the 1st International Symposium on Balkan Flora and Vegetation, Varna, June 7-14 1973. Bulgarian Academy of Sciences, Sofia. Pp. 431-435.

Burkina Faso

Area 274,122 sq. km

Population 6,768,000

Floristics 1096 species from 618 genera in National Herbarium (it is assumed that all of these occur in Burkina Faso); degree of endemism unknown.

Floristic affinities predominantly Sudanian, but also Sahelian in extreme north.

Vegetation *Acacia* woodland in north; Sudanian woodland with *Isoberlinia* in south-west; large areas of more densely populated region in centre have been transformed into park-like savanna woodlands dominated by *Parkia biglobosa*, *Butyrospermum parkii* and *Acacia albida*; other dominants also occur.

For vegetation map see White (1983), cited in Appendix 1.

Checklists and Floras Burkina Faso is included in the *Flora of West Tropical Africa*, cited in Appendix 1.

Aubréville, A. (1959), cited under Ivory Coast. Although not actually including Burkina Faso, includes many of the same species.

IRBET (1983). *Inventaire de l'Herbier du CNRST de la Haute Volta.* CNRST, Ouagadougou.

Field-guides

Maydell, H.J. von (1983). *Arbres et Arbustes du Sahel.* Gesellschaft für Technische Zusammenarbeit, Eschborn 4, F.R.G.

Information on Threatened Plants None.

Laws Protecting Plants 15-20 species of economically important woody plants are given special protection.

Botanic Gardens

Centre National de la Recherche Scientifique et Technologique (CNRST), B.P. 7047, Ouagadougou.

Useful Addresses

Centre National de Semances Forestières (CNSF), B.P. 2682, Ouagadougou.

Equipe Ecologie et Forêts, Comite Permanent Interetats de Lutte contre la Secheresse dans la Sahel (CILSS), B.P. 7049, Ouagadougou.

Institute de Recherche en Biologie et Ecologie Tropicale (IRBET)/CNRST, B.P. 7047, Ouagadougou.

Ministère de l'Environnement et Tourisme, Ouagadougou.

Additional References

Terrible, M. (1976 or 1978). Végétation de la Haute Volta au millionème: carte et notices provisoires. In *Contribution à la Connaissance de la Haute Volta*. Bobo-Dioulasso.

Terrible, M. (1984). *Essai sur l'Ecologie et la Sociologie d'Arbres et Arbustes de Haute Volta*. Librairie de la Savane, Bobo-Dioulasso.

Burma

Area 678,031 sq. km

Population 38,513,000

Floristics About 7000 flowering plant species, including about 200 exotic species (Hundley and Chit Ko Ko, 1961). 1071 endemic vascular plant species (D. Chatterjee, 1939, quoted by Legris, 1974).

Vegetation Tropical lowland evergreen rain forest, mainly in south (Myers, 1980, cited in Appendix 1); tropical hill evergreen rain forest and temperate evergreen rain forest above 900 m in east, north and west; semi-evergreen rain forest in a narrow belt bordering arid central plain; mixed deciduous forest with teak (*Tectona grandis*) and dry dipterocarp forest in central Burma, under increasing pressure especially in the lowlands; coniferous forests in Shan and Chin States, with *Pinus khasya* between 1200-2500 m on dry slopes; oak and rhododendron forests on wetter slopes; 90,000 sq. km of bamboo forests throughout (Nao, 1974); dry forest and scrub formations where rainfall below 1000 mm, including 'than-dahat forest' with *Terminalia* and *Tectona*, thorn scrub with *Acacia* and *Ziziphus*, and 'indaing scrub forest' on lateritic soils, with *Pentacme siamensis* and *Shorea oblongifolia*.

According to government publications, forests cover 57% of Burma; however, analysis of recent satellite and air photographs by the FAO National Forest Inventory Project shows forest cover reduced to 42% by 1980 (Blower, 1985). This figure includes "degraded forests". According to Hundley (1984, *in litt.*), evergreen forests comprise 40% of the total forest cover; mixed deciduous forest 39%. There are 3650 sq. km of tropical lowland evergreen rain forests. Estimated rate of deforestation of closed broadleaved forest 1015 sq. km/annum out of a total of 311,930 sq. km (FAO/UNEP, 1981). Myers (1980, quoting Forest Department figures) states that about 1420 sq. km/annum of primary forest are modified, if not transformed, by shifting cultivation.

For vegetation map see:

Stamp, L. (1924). Notes on the vegetation of Burma. *Geographical J.* 64(3): 272. (Includes vegetation map, scale 1:8,000,000.)

Checklists and Floras Southern Burma is covered by the *Flora of British India* (Hooker, 1872-1897), cited in Appendix 1. National accounts include:

Hundley, H.G. and Chit Ko Ko, U. (1961). *List of Trees, Shrubs, Herbs and Principal Climbers, etc. Recorded from Burma with Vernacular Names*, 3rd Ed. Govt Printing Press, Rangoon. 532 pp.

Kurz, S. (1874-1877). Contributions towards a knowledge of the Burmese flora. *J. Asiatic Soc. Bengal* 43(2): 39-141; 44(2): 128-190; 45(2): 204-310; 46: 49-258. (Incomplete enumeration with notes on habitats and localities.)

Kurz, S. (1877). *Forest Flora of British Burma*, 2 vols. Govt Printer, Calcutta. (c. 2000 woody species and 2500 herbaceous species described; introductory chapter on vegetation. Reprinted by Bishen Singh Mahendra Pal Singh, Dehra Dun, 1974.)

Information on Threatened Plants A preliminary list of plants under consideration for threatened plant status includes 12 species, mainly trees. All orchids, *Dioscorea* and *Panax* are also under consideration (Hundley, *in litt.*). See also:

Blower, J. (1985). Conservation priorities in Burma. *Oryx* 19(2): 79-85. (Deals mainly with deforestation, protected areas and fauna; refers to 2 threatened trees.)

Laws Protecting Plants The Burma Forest Act, 1902 as amended to date, protects habitats of 22 species, as well as all smooth-barked *Dipterocarpus* in Kanyin, Lower Burma.

Botanic Gardens

Agri-Horticultural Society of Burma (Kandawgalay), Rangoon.
Government Botanical Gardens, Maymo.

Useful Addresses

Botany Department, Rangoon Arts and Sciences University, Rangoon.
Burma Forest School, Maymo.
Director General Forests of Burma, No. 62 Randeria Building, Rangoon.
Forest Research Institute, Yezin.

Additional References

Chatterjee, D. (1939). Studies on the endemic flora of India and Burma. *J. Royal Asiatic Soc. Bengal Sci.* 5: 19-67.

Legris, P. (1974). Vegetation and floristic composition of humid tropical continental Asia. In Unesco, *Natural Resources of Humid Tropical Asia*. Natural Resources Research 12. Paris. Pp. 217-238. (Includes bibliography of literature and vegetation maps.)

Nao, T.V. (1974). Forest resources of humid tropical Asia. In Unesco, *Natural Resources of Humid Tropical Asia*. Natural Resources Research 12. Paris. Pp. 197-215.

Rao, A.S. (1974). The vegetation and phytogeography of Assam-Burma. In Mani, M.S. (Ed.), *Ecology and Biogeography of India*. Junk, The Hague. Pp. 204-246.

Burundi

Area 27,834 sq. km

Population 4,503,000

Floristics 2500 species (quoted in Lebrun, 1976, cited in Appendix 1). Levels of endemism unknown, but unlikely to be high. Brenan (1978, cited in Appendix 1) gives a figure of 26 species endemic to Rwanda and Burundi, out of a c. 39% sample of the *Flore du Congo Belge et du Ruanda-Urundi*.

Floristic affinities with Lake Victoria and Afromontane regions.

Vegetation Mostly mosaic of East-African evergreen bushland and secondary *Acacia* wooded grassland. Large areas of Afromontane communities in the west. *Brachystegia-Julbernardia* (Miombo) woodland along south-east border. Small patches of transitional rain forest in north-west. Estimated rate of deforestation for closed broadleaved forest 4 sq. km/annum out of 140 sq. km (FAO/UNEP, 1981).

For vegetation map see White (1983), cited in Appendix 1.

Checklists and Floras Burundi is included in the incomplete *Flore du Congo Belge et du Ruanda-Urundi* (cited in Appendix 1), continued since 1972 as *Flore d'Afrique Centrale (Zaïre – Rwanda – Burundi)*. Burundi's plants of high altitudes are listed in *Afroalpine Vascular Plants* (Hedberg, 1957), cited in Appendix 1.

Lewalle, J. (1970). Liste floristique et répartition altitudinale de la flore du Burundi occidental. Université Officielle de Bujumbura. Cyclostyled. 84 pp. (c. 1700 species and infraspecific taxa listed.)

Information on Threatened Plants No published lists of rare or threatened plants; IUCN has records of 54 species and infraspecific taxa believed to be endemic; no categories assigned.

Additional References
Devred, R. (1958). La végétation forestière du Congo belge et du Ruanda-Urundi. *Bull. Soc. R. For. Belg.* 65: 409-468. (With vegetation map.)
Lebrun, J. (1956). La végétation et les territoires botaniques du Ruanda-Urundi. *Natural. Belges* 37: 230-256.
Lewalle, J. (1968). Burundi. In Hedberg, I. and O. (1968), cited in Appendix 1. Pp. 127-130.
Lewalle, J. (1972). Les étages de végétation du Burundi occidental. *Bull. Jard. Bot. Nat. Belg.* 42: 1-247. (With ten black and white photographs.)
Reekmans, M. (1980a). La flore vasculaire de l'Imbo (Burundi) et sa phénologie. *Lejeunia*, n.s. 100: 1-53.
Reekmans, M. (1980b). La végétation de la plaine de la Basse Rusizi (Burundi). *Bull. Jard. Bot. Nat. Belg.* 50: 401-444.

There is a series of vegetation and soil maps covering Zaïre, Rwanda and Burundi in c. 25 parts, published between 1954 and c. 1970 by the Institut National pour l'Etude Agronomique du Congo (INEAC); each is accompanied by a descriptive memoir, and several of the maps are to different scales. The series is called: *Carte des Sols et de la Végétation du Congo Belge et du Ruanda-Urundi*, or, more recently: ... *du Congo, du Rwanda et du Burundi*.

Cameroon

Area 475,500 sq. km

Population 9,467,000

Floristics c. 8000 species (Lebrun, 1976, cited in Appendix 1; certainly between 8000 and 10,000 (R. Letouzey, 1984, *in litt.*); 156 endemic species (but see below), with

c. 45 on Mt Cameroun (Brenan, 1978, cited in Appendix 1). This makes Cameroon one of the richest countries floristically in Africa.

Floristic affinities Sudanian in north, and Guinea-Congolian in south. Mt Cameroun and several other upland areas north-east from it hold Afromontane species. The lowland forests of south-west Cameroon are especially rich in endemics, with a number of diverse, species-rich communities.

Vegetation Extensive lowland rain forest interspersed with secondary grassland and cultivation, but considerable area of Sudanian woodland in northern part of country and sub-sahelian wooded grassland in extreme north. Also mangrove forest along coast. Inland, in a band more or less SW-NE, extensive Afromontane communities, including montane forest and grassland.

Estimated rate of deforestation for closed broadleaved forest 800 sq. km/annum out of 179,200 sq. km (FAO/UNEP, 1981). However, Myers (1980, cited in Appendix 1) quotes the following estimates for the amount of primary forest remaining: 175,000 sq. km (Dept of Forestry); 130,000 sq. km (Unesco, 1978). A further 60,000 sq. km have been given out as timber concessions.

For vegetation map see White (1983), cited in Appendix 1.

Checklists and Floras Mt Cameroun is included in *Hochgebirgsflora* (Engler, 1892), cited in Appendix 1.

Aubréville, A. *et al.* (Eds) (1963-). *Flore du Cameroun*. Ministère de l'Enseignement Supérieur et de la Recherche Scientifique, Yaoundé; Muséum National d'Histoire Naturelle, Paris. (27 fascicles so far; Flora less than half published.)

Letouzey, R. *et al.* (1978-1979). *Flore du Cameroun: Documents Phytogéographiques*, Nos. 1 & 2. Centre National de la Recherche Scientifique, Muséum National d'Histoire Naturelle, Paris. (2 portfolios: introduction, maps 1:5,000,000, and information on tree species of which the generic name begins with the letters "A" and "B".)

Field-guides Letouzey (1969-1972), cited in Appendix 1, contains information about the forests of Cameroon.

Information on Threatened Plants No published lists of rare or threatened plants; IUCN has records of 389 (see above) species and infraspecific taxa believed to be endemic: V:22, R:17, I:34, K:237, nt:79.

Laws Protecting Plants There is legislation forbidding the removal of trees less than a certain diameter, and the collection of some rare plants.

Botanic Gardens
Victoria Botanic Gardens, Limbe.

Useful Addresses
CITES Management Authority: Direction of Wildlife and National Parks, General Delegation for Tourism, Yaoundé.
CITES Scientific Authority: Wildlife College, B.P. 271, Garoua; and: Delegate General for Scientific Technological Research, B.P. 1457, Yaoundé.

Additional References
Letouzey, R. (1968a). *Etude Phytogéographique du Cameroun*. Lechevalier, Paris. 511 pp. (With 60 black and white photographs and several small-scale maps.)

Letouzey, R. (1968b). Cameroun. In Hedberg, I. and O. (1968), cited in Appendix 1.
    Pp. 115-121.

Campbell Islands

A group consisting of Campbell Island, of area 113 sq. km, and a number of offlying islets
and rocks, c. 700 km south of New Zealand, in the South Pacific Ocean. The islands are
remnants of a dissected volcanic dome; the highest point is Mount Honey (567 m). The
islands were declared a Reserve for Preservation of Fauna and Flora in 1954 and are
administered by the Department of Lands and Survey, New Zealand.

Area 114 sq. km

Population No permanent residents; 10-12 staff of the meteorological station on
Campbell Island (Clark and Dingwall, 1985, cited in Appendix 1).

Floristics 223 native vascular plant taxa, and 85 introduced taxa (Meurk and
Given, in prep.). 3 endemics (*Flora of New Zealand*, 1961, cited under New Zealand).

Vegetation Tussock grassland on steep coastal slopes; *Dracophyllum* and
Coprosma scrub found in sheltered gullies to 180 m; above 300 m, *Bulbinella* and rush
communities dominate an underturf of grasses, bryophytes and lichens. Virtually the
whole island is covered by thick peat deposits, often over 1 m deep; in wetter areas,
sphagnum bog and peat moors. The offshore islets have *Poa foliosa* grassland and
herbaceous communities. On Campbell Island, introduced sheep, and the burning of scrub
for pasture, has modified the vegetation, and has led to the erosion of peatlands. Cattle
have been completely removed and there is a programme to reduce the number of sheep
(Clark and Dingwall, in prep., cited in Appendix 1).

Checklists and Floras The Campbell Islands are included in the *Flora of New
Zealand* (1961, 1970, 1980), cited under New Zealand.

Information on Threatened Plants See Given (1981a), cited under New Zealand.
Latest IUCN statistics: world threatened non-endemic taxa – R:1 (world category).

Laws Protecting Plants It is illegal to collect or introduce plants without a permit.

Useful Addresses
Department of Lands and Survey, Private Bag, Wellington, New Zealand.

Additional References
Godley, E.J. (1969). Additions and corrections to the flora of the Auckland and
    Campbell Islands. *N.Z. J. Bot.* 7: 336-348. (Covers 45 taxa.)
Meurk, C.D. (1975). Contributions to the flora and plant ecology of Campbell Island.
    N.Z. J. Bot. 13: 721-742. (62 new plant records.)
Meurk, C.D. and Given, D.R. (in prep.). The vascular flora and plant communities of
    Campbell Island.
Sorenson, J.H. (1951). Botanical investigations on Campbell Island, 2: an annotated
    list of the vascular plants. *N.Z. DSIR, Cape Exped. Ser. Bull.* 7: 25-38.

Canada

Based upon material by G.W. Argus

Area 9,922,387 sq. km

Population 25,302,000

Floristics About 3220 native species of vascular plants and about 880 introduced species (Scoggan, 1978-1979). Most of the flora has recently reoccupied a landscape that was covered by ice sheets. There are, however, Pleistocene refugia on northern Ellesmere Island, central and northern Yukon, the mountains of Labrador and the Gaspe Peninsula, Quebec, the eastern coastal plain (now inundated), and the Queen Charlotte Islands, British Columbia. The most floristically diverse regions are southern British Columbia and southwestern Ontario.

Vegetation North of the tree-line, arctic tundra; on western mountains above the tree-line (which is at 900-2500 m, depending on latitude) alpine tundra; over about three-quarters of Canada coniferous forest, dominated by White Spruce (*Picea glauca*) and Black Spruce (*P. mariana*) extending from Newfoundland to Alaska; in British Columbia a complex assemblage of subalpine, montane and coastal coniferous forests; in a narrow band across central and western Canada, just north of the U.S. border, grassland – this includes fescue grassland, tall grass prairie (largely destroyed by agriculture and now confined to Manitoba), mixed grass and short grass prairie (southern Saskatchewan and Alberta), and Palouse Prairie (dry interior valleys of British Columbia); between the prairie and coniferous forest, in Central Canada, a transition zone characterized by Trembling Aspen (*Populus tremuloides*); between the coniferous forest and the tundra, transitional Taiga, characterized by open spruce woodlands with lichen ground cover; in eastern Canada, around the Great Lakes region, mainly deciduous forest, e.g. of maple, oak and other hardwood trees, but predominantly of conifers in some areas. (Partly from Skoggan, 1978, who outlines other plant communities).

Checklists and Floras The national Flora is:

Scoggan, H.J. (1978-1979). *The Flora of Canada*, 4 vols. National Museum of Natural Sciences, Ottawa. Publications in Botany 7. (Complete.)

North American checklists that include Canada are cited under the United States, which does not have a National Flora. Regional and provincial Floras and checklists include:

Boivin, B. (1967-1979). Flora of the Prairie Provinces: a handbook to the flora of the provinces of Manitoba, Saskatchewan and Alberta. *Phytologia* 15: 121-159, 329-446; 16: 1-47, 219-261, 265-339; 17: 57-112; 18: 281-293; 22: 315-398; 23: 1-140; 42: 1-24, 385-414; 43: 1-106, 223-251.

Calder, J.A. and Taylor, R.L. (1968). *Flora of the Queen Charlotte Islands*. Part 1, Systematics of the vascular plants. Canada Dept Agriculture, Research Branch, Monogr. 4(1). 659 pp.

Gleason, H.A. and Cronquist, A. (1963). Manual of Vascular Plants of Northeastern United States and Adjacent Canada. Van Nostrand, Princeton, New Jersey. 810 pp. (Covers New Brunswick, Prince Edward Island, and parts of Ontario and Quebec south of the 47th parallel.)

Hultén, E. (1968). *Flora of Alaska and Neighboring Territories: a manual of the vascular plants*. Stanford Univ. Press, Stanford, Calif. 1008 pp. (Covers Yukon and northwestern British Columbia.)

Marie-Victorin, E.C. (1964). *Flora Laurentienne*. Les Presses de l'Université de
  Montréal, Montréal, Québec. 925 pp.

Moss, E.H. (1983). *Flora of Alberta*, 2nd Ed. revised by J.G. Packer. Univ. Toronto
  Press. 687 pp. (Includes dot maps.)

Porsild, A.E. and Cody, W.J. (1980). *Vascular Plants of Continental Northwest
  Territories, Canada*. National Museums of Canada, Ottawa. 667 pp. (Includes line
  drawings and dot maps; area covered is between 60th parallel and Arctic Ocean, and
  from Yukon-Mackenzie border to west coast of Hudson Bay.)

Roland, A.E. (1947). The flora of Nova Scotia. *Proc. Nova Scotia Inst. Sci.* 21:
  94-642. (2nd Ed. in 2 parts by A.E. Roland and E.C. Smith, 1966, 1969.)

Scoggan, H.J. (1957). *Flora of Manitoba*. National Museum of Canada, Bulletin No.
  140. 619 pp.

Taylor, R.L. and MacBryde, B. (1977). *Vascular Plants of British Columbia: A
  descriptive resource inventory*. Botanical Garden, Univ. of British Columbia,
  Vancouver, Tech. Bull. No. 4. 754 pp.

Welsh, S.L. (1974). *Anderson's Flora of Alaska and Adjacent Parts of Canada*.
  Brigham Young Univ. Press, Provo, Utah. 724 pp. (Covers Yukon and northwestern
  British Columbia.)

Information on Threatened Plants The Rare and Endangered Plants Project
(address below) is publishing rare plant lists for the Canadian provinces and territories.
These are annotated lists, in English and French, of taxa with notes on habitat and
distribution but with only limited indications of degree of threat. Dot maps are included
except for the Ontario and Alberta lists. For further details of the programme see Argus
(1977).

Argus, G. W. and White, D. J. (1977). The rare vascular plants of Ontario. *Syllogeus* 14.

Argus, G.W. and White, D.J. (1978). The rare vascular plants of Alberta.
  Syllogeus 17.

Argus, G.W. *et al.* (Eds) (1982-). *Atlas of the Rare Vascular Plants of Ontario*. First
  two parts edited by G.W. Argus and D.J. White (1982, 1983), 3rd part by
  G.W. Argus and C.J. Keddy (1984), 4th and final part by G.W. Argus and
  K. Pryer (in prep.). National Museum of Natural Sciences, Ottawa, Ontario.
  (Supercedes Argus and White, 1977.)

Bouchard, A.D., Barabé, D., Dumais, M. and Hay, S. (1983). The rare vascular plants
  of Quebec. *Syllogeus* 48.

Douglas, G.W., Argus, G.W., Dickson, H.L. and Brunton, D.F. (1981). The rare
  vascular plants of the Yukon. *Syllogeus* 28.

Hinds, H. (1983). The rare vascular plants of New Brunswick. *Syllogeus* 50.

Maher, R.V., Argus, G.W., Harms, V.L. and Hudson, J.H. (1979). The rare vascular
  plants of Saskatchewan. *Syllogeus* 20. (Reviewed in *Threatened Plants Committee -
  Newsletter*, No. 5: 11, 1980.)

Maher, R.V., White, D.J., Argus, G.W. and Keddy, P.A. (1978). The rare vascular
  plants of Nova Scotia. *Syllogeus* 18.

Taylor, R.L., Douglas, G.W. and Straley, G. (in press). The rare vascular plants of
  British Columbia. *Syllogeus*.

White, D.J. and Johnson, K.L. (1980). The rare vascular plants of Manitoba.
  Syllogeus 27.

A computerized list of the rare and endangered vascular plants in Canada was compiled at
the University of Waterloo and last updated in 1978. Abbreviated version published as:

Kershaw, L.J. and Morton, J.K. (1976). Rare and potentially endangered species in the Canadian flora – A preliminary list of vascular plants. *Can. Bot. Assoc. Bull.* 9(2): 26-30.

The complete list was included as an appendix in:

Kershaw, L.J. (1976). A Phytogeographical Survey of Rare, Endangered and Extinct Plants in the Canadian Flora. M.Sc. Thesis, Univ. of Waterloo, Ontario.

Also relevant:

Guppy, G.A. (1977). Endangered plants in British Columbia. *Davidsonia* 8: 24-30.
Isnor, W. (1981). *Provisional Notes on the Rare and Endangered Plants and Animals of Nova Scotia.* Curatorial Report No. 46, Nova Scotia Museum, 1747 Summer Str., Halifax, Nova Scotia B3H 3A6. (Notes on identification, distribution, habitat and vulnerability for 82 vascular plants, with dot maps.)

The IUCN Plant Red Data Book includes 7 species for Canada. No IUCN statistics; there are c. 500 species that are rare throughout Canada.

The Committee on the Status of Endangered Wildlife in Canada (COSEWIC), a committee established in 1977 of the Canadian Federal-Provincial Wildlife Conference, is charged with preparing status reports and assigning status to Canadian species in jeopardy (Haber, 1983). This has been done for 19 plant species, including 7 'endangered', 8 'threatened' and 4 'rare'. The status reports are available at cost from Canadian Nature Federation (see below).

Laws Protecting Plants Complex and numerous; mostly at provincial level; for details see Argus (1977).

Voluntary Organizations
Canadian Nature Federation, 75 Albert Street, Ottawa, Ontario K1L 8B9.
Nature Conservancy of Canada, 22 Hillside Drive S., Toronto, Ontario M4K 2M2.
WWF-Canada, 60 St Clair Ave. E., Suite 201, Toronto, Ontario M4T IN5.

Botanic Gardens The following Canadian botanic gardens subscribe to the IUCN Botanic Gardens Conservation Co-ordinating Body.

Devonian Botanic Garden, University of Alberta, Room B-414, Biological Sciences Centre, Edmonton, Alberta T6G 2E9.
Jardin Botanique de la ville de Montréal, 4101 rue Sherbrooke est, Montréal, Québec H1X 2B2.
Oxen Pond Botanic Park, Memorial University of Newfoundland, St John's, Newfoundland A1C 5S7.
Royal Botanic Gardens, P.O. Box 3990, Hamilton, Ontario L8N 3H8.
University of British Columbia Botanical Garden, 6501 Northwest Marine Drive, Vancouver, B.C. V6T 1W5.
University of Guelph Arboretum, Guelph, Ontario N1G 2W1.

Useful Addresses
Committee on the Status of Endangered Wildlife in Canada (COSEWIC), Canadian Wildlife Service, Dept of the Environment, Ottawa, Ontario, K1A 0E7.
National Museum of Natural Sciences, National Museums of Canada, Ottawa, Ontario K1A 0M8.
Rare and Endangered Plants Project, Botany Division, National Museum of Natural Sciences, Ottawa, Ontario K1A 0M8.

CITES Management Authority: The Administrator, CITES, Canadian Wildlife Service, Dept of Environment, Ottawa, Ontario K1A 0E7.

Additional References

Argus, G.W. (1977). Canada. In Prance, G.T. and Elias, T.S. (Eds), cited in Appendix 1. Pp. 17-29.

Argus, G.W. and McNeill, J. (1974). Conservation of evolutionary centres in Canada. In Maini, J.S. and Carlisle, A., *Conservation in Canada: A Conspectus*. Dept of Environment, Canadian Forest Service Publication 1340. Pp. 131-141.

Haber, E. (1983). A report on the work of COSEWIC. *The Plant Press* 1(3): 45-47.

Morton, J.K. (Ed.) (1976). *Proceedings of the Symposium: Man's Impact on the Canadian Flora*. Canadian Botanical Association Bulletin, Suppl. to Vol. 9, No. 1.

Scudder, G.C.E. (1979). Present patterns in the fauna and flora of Canada. In Danks, H.V., Canada and its insect fauna. *Mem. Entomol. Soc. Can.* 108: 87-179.

Soper, J.H. (1979). Nature conservation in Canada. In Hedberg, I. (Ed.), cited in Appendix 1. Pp. 143-146.

Canary Islands

An archipelago in the Atlantic Ocean off the north-west coast of Africa, between Madeira and the Cape Verde Islands and belonging to Spain. Comprises the 2 Spanish metropolitan provinces of Las Palmas de Gran Canaria and Santa Cruz de Tenerife. Las Palmas province includes Gran Canaria, Fuerteventura and Lanzarote together with 3 islets – Alegranza, Graciosa, Lobos – and several uninhabited rocks. Santa Cruz de Tenerife province comprises the islands of Tenerife, La Palma, Gomera and Hierro.

Area 7273 sq. km

Population 1,394,288 (local figures, 1979)

Floristics About 2000 species of native and introduced vascular plants (D. Bramwell, 1985, pers. comm.), mostly of native Mediterranean species and introduced weeds and aliens. This includes a remarkable endemic flora of over 500 taxa (IUCN figures), with 19 endemic genera (Bramwell, 1976).

Generally considered to be a relict flora, with affinities to Tertiary Mediterranean flora; many endemics have their nearest relatives in South and East Africa, and even South America, being considered to be relicts of an African Tertiary 'Rand' flora (Bramwell, 1974, 1976 and 1985 pers. comm.).

Vegetation In the western and central islands extensive woods; in the eastern islands mostly xerophytic scrub, reflecting the more arid climate of North Africa. Bramwell (1974) lists 6 vegetation types, which show striking altitudinal zonation: semi-desert succulent scrub (0-700 m); juniper scrub (south slopes, 400-600 m); tree heath and evergreen forest, the former of *Erica arborea*, the latter of Lauraceae, forming the famous and species-rich laurel forests, of which only small areas remain (400-1300 m); savanna of *Pinus canariensis* (800-1900 m); montane scrub (1900-2500 m); and subalpine scrub (only on Pico de Teide, Tenerife, c. 2600 m). In Gran Canaria the laurel forest is now less than 1% of its original extent; on Tenerife about 10%.

Checklists and Floras The Canaries are covered by the *Flora of Macaronesia* checklist (Hansen and Sunding, 1979, cited in Appendix 1). There is no Canarian Flora, but see:

Kunkel, G. (1974-). *Flora de Gran Canaria*. 10 vols projected, 4 completed. Excmo. Cabildo Insular de Gran Canaria, Las Palmas. (1 - arboles y arbustos arbóreos; 2 - enredaderas, trepadoras y rastreras; 3 - las plantas suculentas; 4 - los subarbustos; illus.)

Lid, J. (1967). *Contributions to the Flora of the Canary Islands*. Universitetsforlaget, Oslo. 212 pp. (Annotated list of species with keys.)

Santos Guerra, A. (1983). *Vegetación y Flora de La Palma*. Editorial Interinsular Canaria, Tirso de Molina 8, Santa Cruz de Tenerife. 348 pp. (Includes catalogue of flora with distribution maps for most of the Macaronesian and Canarian endemics; maps of actual and potential vegetation; colour illus.)

The Jardín Botánico "Viera y Clavijo" is creating a computer database on the Canarian flora, developed from the *Flora of Macaronesia* checklist.

Three Canarian journals contain numerous articles on the flora: *Botanica Macaronésica*, published by Jardín Botánico "Viera y Clavijo", *Cuadernos de Botánica Canaria*, published privately, now discontinued, and *Vieraea*, published by Museo Insular de Ciencias Naturales, Tenerife.

Field-guides

Bramwell, D. and Z.I. (1974). *Wild Flowers of the Canary Islands.* Stanley Thornes, London. 261 pp. (Keys, descriptions, illus., mostly of the endemics; also describes areas of botanical interest.) Spanish edition as *Flores Silvestres de las Islas Canarias*, 2nd Ed., 1983, Editorial Rueda, Porto Cristo 13, Alcorcon, Madrid; German edition as *Kanarische Flora: Illustrierter Führer*, 1983, Editorial Rueda (without the keys and descriptions).

Kunkel, G. (1981). *Arboles y Arbustos de las Islas Canarias: Guia de Campo*. Coleccion Botánica Canaria, Vol. 1. 138 pp. (Line drawings.)

The Caja Insular de Ahorras de Gran Canaria, with the Jardín Botánico "Viera y Clavijo", have prepared a set of data cards with colour illustrations of Canarian plants, mostly endemic and threatened.

Information on Threatened Plants A national threatened plant list has recently been published:

Barreno, E. *et al.* (Eds) (1984). Listado de Plantas Endemicas, Raras o Amenazadas de España. *Informacion Ambiental. Conservacionismo en España*. No. 3. (Includes separate lists for peninsula Spain, Balearic Islands and Canary Islands; for the latter 578 threatened endemic taxa are listed; compiled with the agreement of numerous authoritative Spanish botanists, it is now the definitive list.)

Included in the European threatened plant list (Threatened Plants Unit, 1983, cited in Appendix 1. Latest IUCN statistics based upon Barreno (1984): endemic taxa - Ex:1, E:126, V:119, R:132, I:5, K:26, nt:160; non-endemics rare or threatened worldwide - E:1, V:17, R:2, I:1 (world categories). See also:

Bramwell, D. and Perez, J.P. (1982). Prioridades para la conservación de la diversidad genética en la flora de las Islas Canarias. *Botánica Macaronesica* 10: 3-17. (Classifies the species from the 1980 IUCN list in terms of the priorities of the World Conservation Strategy.)

Kunkel, G. (Ed.) (1975). *Inventario de los Recursos Naturales Renovables de la Provincia de las Palmas*. Excmo. Cabildo Insular, Las Palmas de Gran Canaria. 156 pp., maps. (Results of IUCN/WWF Project 817, undertaken by Asociación Canaria para la Defensa de la Naturaleza.)

The Environment Department of the Autonomous Government of the Canary Islands is preparing a protected area programme for the Canarian flora.

Voluntary Organizations Several local ecology groups, the most important being:

Asociación Canaria para la Defensa de la Naturaleza (ASCAN), c/o Presidente Alvear 50, Las Palmas de Gran Canaria.

Botanic Gardens

Jardín Botánico "Viera y Clavijo", Apto de Correos 14 de Tafira Alta, 35017 Las Palmas de Gran Canaria.
Jardín de Aclimatación de la Orotava, Puerto de la Cruz, Tenerife.

Useful Addresses

Gobierno de Canarias, Consejeria de Obras Públicas, Ordenación de Teritorio y Medio Ambiente, Edificio Hamilton, Santa Cruz de Tenerife.

Additional References

Bramwell, D. (1976). The endemic flora of the Canary Islands; distribution, relationships and phytogeography. In Kunkel, G. (Ed.), see below. Pp. 207-240.

Ceballos, L.C. and Ortuño, F. (1976). *Estudio sobre la Vegetación y Flora Forestal de las Canarias Occidentales*, 2nd Ed. Excmo. Cabildo Insular, Sta Cruz de Tenerife. 433 pp. (Covers Gomera, Hierro, La Palma, Tenerife; illus., vegetation maps.)

Hernández, P.H. (1979). *Natura y Cultura de las Islas Canarias*, 3rd Ed. La Cultura, Apto de Correos, 1012 Las Palmas de Gran Canaria.

Kunkel, G. (Ed.) (1976). *Biogeography and Ecology in the Canary Islands*. Junk, The Hague. 511 pp. (Includes essays on the Hierro laurisilva by E. Schmid (pp. 241-248), the introduced elements in the flora by G. Kunkel (pp. 249-266), the influence of man on Hierro vegetation by F. Kämmer (pp. 327-346) and on conservation by M. Sutton (pp. 479-483).)

Sunding, P. (1973). *A Botanical Bibliography of the Canary Islands*, 2nd Ed. Botanical Garden, Univ. of Oslo. 46 pp.

Canton and Enderbury Islands

Canton (9 sq. km) and Enderbury (6.5 sq. km) are low coral atolls 2620 km south-west of the Hawaiian islands and north of the Phoenix Islands in the Pacific Ocean. The islands are jointly administered by the United States and United Kingdom. There are no permanent inhabitants.

Canton (2°50'S, 171°40'W) has 14 native species and over 150 introduced weeds (Hatheway, 1955). Most of the flora consists of wide-ranging Indo-Pacific strand plants. The vegetation consists mainly of *Scaevola* and *Tournefortia* scrub, *Portulaca* herbaceous communities and a few *Cordia* trees and coconuts. Hatheway (1955) reported that 23% of the land surface of Canton had little or no natural vegetation, and a further 40% consisted of disturbed ground.

References

Degener, O. and I. (1959). Canton Island, South Pacific (Resurvey of 1958). *Atoll Res. Bull.* 64. 24 pp. (Includes notes on flora.)

Degener, O. and Gillaspy, E. (1955). Canton Island, South Pacific. *Atoll Res. Bull.* 41. 51 pp. (Checklist of introductions and notes on 68 species on Canton.)

Hatheway, W.H. (1955). The natural vegetation of Canton Island, an equatorial Pacific atoll. *Atoll Res. Bull.* 43. 9 pp.

Luomala, K. (1951). Plants of Canton Island, Phoenix Islands. *Occ. Papers Bernice P. Bishop Mus.* 20(11): 157-174. (59 taxa listed.)

Cape Verde

The Cape Verde Islands, 445 km off the west coast of Africa, consist of two groups of volcanic islands: Windward (Santo Antão, São Vicente, Santa Luzia, São Nicolau, Sal and Boa Vista) and Leeward (Maio, São Tiago, Fogo and Brava). They occupy 14°48'-17°12'N, 22°44'-25°22'W. The highest point is 2829 m on Fogo.

Area 4033 sq. km

Population 317,000

Floristics c. 659 species of vascular plants including introductions (Sunding, 1973, 1974); 92 endemics (Humphries, 1979).

Lowland species with tropical affinities; mountain species with Macaronesian or Mediterranean affinities.

Vegetation Original vegetation almost totally destroyed and potential vegetation impossible to assess. Mostly now lowland arid pastures with large numbers of goats, and agricultural crops and plantations on fertile slopes. More arid pastures above c. 1400 m, and more or less bare rocky summits at the highest altitudes on Fogo and Santo Antão.

For vegetation map see White (1983), cited in Appendix 1.

Checklists and Floras Cape Verde is included in the *Flora of Macaronesia* checklist (Hansen and Sunding, 1979), cited in Appendix 1.

Chevalier, A. (1935a). Les îles du Cap Vert. Géographie, biogéographie, agriculture. Flore de l'Archipel. *Rev. Bot. Appl. Agric. Trop.* 15: 733-1090. (Includes annotated checklist, pp. 867-1074.)

Chevalier, A. (1946). Additions à la flore des Iles du Cap Vert. In *Contribution à l'Etude du Peuplement des Iles Atlantides*, Mém. Soc. Biogéogr. 8: 349-356.

Sunding, P. (1973). *Check-list of the Vascular Plants of the Cape Verde Islands.* Botanical Garden, Univ. of Oslo, Oslo. 36 pp. (Includes distributions.)

Sunding, P. (1974). Additions to the vascular flora of the Cape Verde Islands. *Garcia de Orta, Sér. Bot.* 2(1): 5-30.

Information on Threatened Plants None.

Useful Addresses

Ministério de Desenvolvimento Rural, C.P. 50, Praia, S. Tiago.

Additional References

Barbosa, L.A. Grandvaux (1968a). L'archipel du Cap-Vert. In Hedberg, I. and O. (1968), cited in Appendix 1. Pp. 94-97.

Barbosa, L.A. Grandvaux (1968b). Vegetation. In Bannerman, D.A. and W.M. (Eds), *History of the Birds of the Cape Verde Islands*. Oliver and Boyd, Edinburgh. Pp. 58-61. (Birds of the Atlantic Islands, vol. 4.)

Chevalier, A. (1935b). Aperçu sur la végétation des îles de Cap Vert. *Compt. Rend. Somm. Séanc. Soc. Biogéogr.* 99: 21-24.

Humphries, C.J. (1979). Endemism and evolution in Macaronesia. In Bramwell, D. (Ed.), *Plants and Islands*. Academic Press, London. Pp. 171-199.

Sunding, P. (1977). A botanical bibliography of the Cape Verde Islands. *Bol. Mus. Munic. Funchal* 31: 100-109.

Sunding, P. (1979). Origins of the Macaronesian flora. In Bramwell, D. (Ed.), *Plants and Islands*. Academic Press, London. Pp. 13-40.

Teixeira, A.J. da Silva and Barbosa, L.A. Grandvaux (1958). A agricultura do Arquipélago de Cabo Verde. *Mem. Junta Invest. Ultram., Sér. 2 2*, and Mem. Trab. No. 26, Ministério do Ultramar, Lisboa. 178 pp. (With 10 maps in colour, 1:50,000-1:100,000; 77 plates of photographs.)

Cargados Carajos

A group of 22 coralline islands c. 350 km NNE of Mauritius in the Indian Ocean, 16°20'S, 59°20'E; made up of sand banks, shoals and islets. Total land area probably c. 4 sq. km. Also called St Brandon, after the name of the principal reef complex. Total 41 species, including 11 cultivated species, 13 weeds and 17 indigenous pantropical species. No endemics. (Staub and Guého, 1968; Renvoize, 1979.) The vegetation consists mostly of littoral scrub and herb mat; trees more or less absent except for a few stunted individuals.

References

Renvoize, S.A. (1979). The origins of Indian Ocean island floras. In Bramwell, D. (Ed.), *Plants and Islands*. Academic Press, London. Pp. 107-129.

Staub, F. and Guého, J. (1968). The Cargados Carajos shoals or St Brandon: resources, avifauna and vegetation. *Proc. Roy. Soc. Arts Sci. Mauritius* 3(1): 7-46. (Includes annotated checklist of plants.)

Caroline Islands

An archipelago of 70 islands in the west Pacific Ocean, to the east of the Philippines, and extending for over 2500 km, between latitudes 5°-10°N and longitudes 130°-165°E. In the west, the Palau Islands comprise volcanic islands, raised limestone islands and low coral atolls, including several hundred islets within a single reef system. The Yap Islands, northeast of the Palau Islands, are mainly metamorphic and old volcanic islands surrounded by broad fringing reefs. Further east are the Truk and Ponape Islands which include high volcanic islands surrounded by barrier reefs. The highest point is 791 m, on the island of Ponape. The Caroline Islands form part of the United Nations Trust Territory of the Pacific Islands administered by the United States.

Area 1170 sq. km

Population 85,910 (1980 estimate)

Floristics No overall figure for the size of the flora, but 992 taxa are dicotyledons, of which 609 are native, including 267 endemics (Fosberg, Sachet and Oliver, 1979, cited in Appendix 1). 201 native fern taxa, of which 26 are endemic; one native non-endemic gymnosperm (*Cycas circinalis*) (Fosberg, Sachet and Oliver, 1982, cited in Appendix 1). Most of the flora of the Carolines is related to that of Indo-Malesia and Melanesia-New Guinea.

Vegetation Evergreen rain forest and savanna woodlands on the Yap Islands; lowland rain forest, with *Campnosperma, Manilkara, Calophyllum, Eugenia, Ficus*, and tree ferns on the Palau, Truk and Ponape Islands; mixed forests on limestone on Fais, in the Yap group, and in southern Palau; montane rain forest on Ponape and Kusaie, in the Ponape group, and on the summit of Mt Winibot (480 m) on Tol, in the Truk group; mangrove forest on south-west and south-east coasts of Ponape, and south and north-west coasts of Kusaie.

Much of the natural vegetation has been cleared for coconut plantations (e.g. on Yap and Puluwat, in the Yap group) or disturbed by phosphate mining (e.g. on the raised coral island of Fais). Few areas of native vegetation remain on the Truk Islands, except on the high volcanic islands of Moen, Dublon, Uman, Fefan, Udot and Tol. Although the lowland forests on the Ponape Islands have been much disturbed, both Kusaie and the island of Ponape retain upland forests. See Fosberg (1973, cited in Appendix 1) for description of forests and conservation problems.

Checklists and Floras The Carolines are included in *Flora Micronesica* (Kanehira, 1933), the regional checklists of Fosberg, Sachet and Oliver (1979, 1982), and will be covered in the *Flora of Micronesia* (1975-), all cited in Appendix 1. Separate accounts for individual islands include:

Alkire, W.H. (1974). Native classification of flora on Woleai Atoll. *Micronesica* 10(1): 1-5. (Lists 84 species with vernacular names.)

Fosberg, F.R. (1969). Plants of Satawal Island, Caroline Islands. *Atoll Res. Bull.* 132. 13 pp. (Includes annotated checklist of 6 native fern species; 97 angiosperm taxa, of which 46 introduced.)

Fosberg, F.R. and Evans, M. (1969). A collection of plants from Fais, Caroline Islands. *Atoll Res. Bull.* 133. 15 pp. (Includes annotated checklist of 3 native fern species; 117 angiosperm taxa, of which 59 introduced.)

Fosberg, F.R., Otobed, D., Sachet, M.-H., Oliver, R.L., Powell, D.A. and Canfield, J.E. (1980). *Vascular Plants of Palau with Vernacular Names*. Smithsonian Institution, Washington, D.C. 43 pp. (Checklists; exotics indicated.)

Glassman, S.F. (1952). The flora of Ponape. *Bull. Bernice P. Bishop Mus.* 209. 152 pp. (Ponape has 249 indigenous angiosperms, 8 endemic.)

Glassman, S.F. (1957). The vascular flora of Ponape and its phytogeographical affinities. In *Proc. 8th. Pacific Science Congress – Botany*. Pp. 201-213.

Marshall, M. (1975). The natural history of Namoluk Atoll, eastern Caroline Islands. *Atoll Res. Bull.* 189. 53 pp. (Includes annotated list of 113 taxa; notes on vegetation.)

St John, H. (1948). Report on the flora of Pingelap Atoll, Caroline Islands, Micronesia, and observations on the vocabulary of the native inhabitants. Pacific Plant Studies 7. *Pacific Science* 2: 96-113. (Annotated checklist of 57 taxa, 32 indigenous.)

Stone, B.C. (1959). Flora of Namonuito and the Hall Islands. *Pacific Science* 13: 88-104. (Annotated checklist of 94 species, 52 indigenous.)

Stone, B.C. (1960). Corrections and additions to the Flora of the Hall Islands and to the Flora of Ponape. *Pacific Science* 14: 408-410.

Information on Threatened Plants 4 vascular plant species are listed as 'Endangered' in *Territorial Register* 2(1), 4 December 1976. (Adopted Regulations Title 45 – Fish, Shellfish and Game.)

Cayman Islands

A Dependent Territory of the U.K. comprising three islands – Grand Cayman (the largest – 35.5 km by 13 km at its widest), Cayman Brac and Little Cayman. They lie 240 km north-west of Jamaica and 772 km south of Miami. They are relatively flat and low-lying except for Cayman Brac, which is bordered by cliffs and reaches 43 m above sea-level.

Area 259 sq. km

Population 18,000

Floristics Just over 600 species of vascular plants, of which 102 are either cultivated or naturalized from cultivation (Proctor, 1984); 18 endemic species and 3 endemic varieties; affinities with other Antillean islands rather than Central America (Proctor, 1984).

Vegetation Little true woodland remains on Grand Cayman; a few isolated patches of dry evergreen forest (in the east of Grand Cayman and on Cayman Brac); in the uplands dry evergreen thicket, often reduced to pasture; littoral thicket on northern and eastern shores, grading inland to dry evergreen bush (much of western end of Grand Cayman); mangrove (mainly Grand Cayman); seasonal grassland swamp (West Bay area of Grand Cayman).

Checklists and Floras

Proctor, G.R. (1980). Checklist of the plants of Little Cayman. *Atoll Res. Bull.* 241: 71-80.

Proctor, G.R. (1984). *Flora of the Cayman Islands*. Kew Bulletin Additional Series XI. 834 pp. (Includes section on environment and plant associations by M.A. Brunt.)

Cayman Island records are also included in the Jamaican Flora by Adams (1972) and Proctor (1982), cited under Jamaica.

Information on Threatened Plants None.

Additional References
Sauer, J.D. (1982). *Cayman Islands Seashore Vegetation: A Study in Comparative Biogeography*. University of California Publications Geography Vol. 25. Univ. California Press, Berkeley. 161 pp.

Central African Republic

Area 624,977 sq. km

Population 2,508,000

Floristics Flora very poorly known. 3600 species (Sillans, 1958); almost certainly too low. Of these, c. 1000 occur in the rain forest with c. 10 endemic, and 2600 in the savanna with c. 90 endemic (Brenan, 1978, cited in Appendix 1). Endemics concentrated on mountain range in north-east.

Floristic affinities predominantly Sudanian, but also Guinea-Congolian in south.

Vegetation Mostly Sudanian woodland with *Isoberlinia, Terminalia* and *Combretum*, but also Sahelian woodland with *Acacia* in extreme north, and lowland rain forest interspersed with secondary grassland and cultivation in southern quarter of country. Very large area of unexploited moist forest round Bangassou south-east of centre of country. Estimated rate of deforestation for closed broadleaved forest 50 sq. km/annum out of 35,900 sq. km (FAO/UNEP, 1981).

For vegetation map see White (1983), cited in Appendix 1.

Checklists and Floras
Boulvert, Y. (1980?). *Catalogue de la Flore de l'Empire Centrafricaine*, 2 vols. ORSTOM, 20 rue Monsieur, Paris.
Guigonis, G. (1970). Liste des arbres et arbustes vivant dans la forêt dense et les galeries de la Republique Centrafricaine. Cyclostyled. 30 pp. (Lists 645 species.)
Tisserant, C. (1950). Catalogue de la flore de l'Oubangui-Chari. *Mem. Inst. Etud. Centrafricaines* 2. 165 pp. Imprimerie Julia, Toulouse.

Information on Threatened Plants No published lists of rare or threatened plants; IUCN has records of 117 species and infraspecific taxa believed to be endemic; no categories assigned.

Useful Addresses
CITES Management Authority: Direction des chasses, B.P. 830, Bangui.

Additional References
Aubréville, A. (1964). La forêt dense de la Lobaye. *Cah. Maboké* 2(1): 5-9.
Boulvert, Y. (1980). Végétation forestière des savanes Centrafricaines. *Bois Forêts Trop*. 191: 21-45. (With several maps and black and white photographs.)
Guigonis, G. (1968). République Centrafricaine. In Hedberg, I. and O. (1968), cited in Appendix 1. Pp. 107-111.
Lanly, J.P. (1966). La forêt dense centrafricaine. *Bois Forêts Trop*. 108: 43-55.
Sillans, R. (1958). *Les Savanes de l'Afrique Centrale Française*. Lechevalier, Paris. 423 pp. (Numerous illustrations throughout.)

Chad

Area 1,284,000 sq. km

Population 4,901,000

Floristics 1600 species, 1516 of which occur south of 16°N (Lebrun, 1976, cited in Appendix 1). The Tibesti Mountains in the extreme north are estimated to have 450 species (Lebrun, 1960, cited in Appendix 1; Maire and Monod, 1950). Level of endemism not known.

Flora with Saharan (north), Sahelian and Sudanian (south) affinities. The Tibesti Mountains have Mediterranean, Saharan, Sahelian and Afromontane elements.

Vegetation Northern part of country desert with little or no permanent vegetation. To the south, in the Sahelian zone, which has a short wet season, semi-desert grassland gradually replaced by dry wooded grassland with *Acacia* species. Further south the higher rainfall in the Sudanian Region supports woodland without characteristic dominants. The Tibesti mountains in the north of the country support a distinct form of montane vegetation, floristically rich and unrelated to the surrounding lowlands; it consists of grassland, woodland and shrubland, with communities of *Erica arborea* confined to narrow fissures on the higher peaks.

For vegetation map see White (1983), cited in Appendix 1.

Checklists and Floras The northern part of Chad, north of c. 16°N, is included in *Flore du Sahara* (Ozenda, 1977), and in the computerized *Atlas der Pflanzenwelt des Nordafrikanischen Trockenraumes* (Frankenberg and Klaus, 1980); both of these are cited in Appendix 1. See also:

Carvalho, G. and Gillet, H. (1960). Catalogue raisonné et commenté des plantes de l'Ennedi (Tchad Septentrional). *J. Agric. Trop. Bot. Appl.* 7: 49-96, 193-240, 317-378. (With 12 black and white photographs.)
Lebrun, J.-P., Audru, J., Gaston, A. and Mosnier, M. (1972). *Catalogue des Plantes Vasculaires du Tchad Méridional*. Etude Botanique No. 1, Institut d'Elevage et de Médecine Vétérinaire des Pays Tropicaux, Maisons-Alfort. 289 pp. (Annotated checklist covering only the tropical southern part of Chad, but including a useful botanical bibliography.)
Lebrun, J.-P. and Gaston, A. (1976). Premier supplément au "Catalogue des Plantes Vasculaires du Tchad Méridional". *Adansonia*, Sér. 2, 15(3): 381-390.
Lebrun, J.-P. and Gaston, A. (1977). Second supplément au "Catalogue des Plantes Vasculaires du Tchad Méridional". *Publ. Cairo Univ. Herb.* 7-8: 109-114.

Information on Threatened Plants No published lists of rare or threatened plants; IUCN has records of 49 species and infraspecific taxa believed to be endemic, including R:10, nt:4; no information for the rest.

Additional References
Gaston, A. (1980). *La Végétation du Tchad (Nord-Est et Sud-Est du Lac Tchad): Evolutions Recentes sous des Influences Climatiques et Humaines*. Institut d'Elevage et de Médecine Vétérinaire des Pays Tropicaux, Maisons-Alfort. (Colour map 1:1,000,000 covering about a quarter of Chad with unpublished descriptive thesis of 333 pp.)
Gillet, H. (1968a). Le peuplement végétal du massif de l'Ennedi (Tchad). *Mém. Mus. Nat. Hist. Nat. Paris, n.s. B*, 17. 206 pp. (With 66 black and white photographs.)

Gillet, H. (1968b). Tchad et Sahel Tchadien. In Hedberg, I. and O. (1968), cited in Appendix 1. Pp. 54-58.

Lebrun, J.-P. (1983). La flore des massifs Sahariens: espèces illusoires et endémiques vraies. In Killick, D.J.B. (1983), cited in Appendix 1. Pp. 511-515.

Maire, R. and Monod, Th. (1950). Etudes sur la flore et la végétation du Tibesti. *Mém. IFAN* 8. 140 pp., plus appendix.

Pias, J. (1970). *La Végétation du Tchad: Ses Rapports avec les Sols; Variations Paléobotaniques au Quaternaire*. Trav. Doc. ORSTOM 6. 47 pp. (With coloured vegetation map 1:1,500,000.)

Quézel, P. (1958). Mission botanique au Tibesti. *Mém. Inst. Rech. Sahariennes* 4. 357 pp. (Notes on the distribution of over 500 species, description of vegetation; 30 black and white photographs.)

Chatham Islands

An isolated group of islands c. 800 km east of South Island, New Zealand. The main islands (Chatham and Pitt) are surrounded by numerous islets, some of which are no more than precipitous rocks. Chatham (963 sq. km) is mostly low lying but reaches about 270 m in the south. It is geologically heterogenous, with schists, sandstones, limestones, and basaltic tuffs in the south. Pitt Island is more rugged and mainly basaltic. Blanket peat covers much of both Chatham and Pitt; several peat domes on Chatham. Some of the outlying islets, and parts of Pitt Island, are nature reserves. The Chatham Islands are administered by the New Zealand Department of Maori and Island Affairs.

Area 1235 sq. km

Population 751 (1981)

Floristics c. 300 vascular plant species (Devine, 1982); 35-40 endemic taxa (Given, 1984, *in litt.*); 2 endemic genera (*Embergeria* and *Myosotidium*).

Vegetation The original vegetation was probably a mosaic of Karaka (*Corynocarpus*) forest, swamp forest, and Tarahinau (*Dracophyllum*) forest in the lowlands, with mixed broadleaved forests in the uplands. *Sporadanthus* moorland and bogs were also extensive. Relatively intact natural vegetation occurs on the Southern Tablelands of the main island, but elsewhere vegetation has mostly been cleared for agriculture, or modified by draining, grazing and fires. By the end of the 1960s some 57 sq. km of Tarahinau forest remained, mainly on Chatham Island, together with 22 sq. km of *Sporadanthus* bog (Devine, 1982).

Checklists and Floras The Chatham Islands are included in the *Flora of New Zealand* (1961, 1970, 1980), cited under New Zealand. See also:

Mueller, F. (1864). *The Vegetation of the Chatham-Islands*. Melbourne. 86 pp. (Includes descriptive accounts of 87 vascular plant species.)

Information on Threatened Plants Given (1976, 1977, 1978), cited under New Zealand, includes 8 data sheets on threatened species from the Chatham Islands. *The Red Data Book of New Zealand* (Williams and Given, 1981), cited under New Zealand, includes data sheets on 9 species. *Myosotidium hortensia* is included in *The IUCN Plant Red Data Book* (1978). See also:

Given, D.R. (1983). Monitoring and science – the next stage in threatened plant conservation in New Zealand. In Given, D.R. (Ed.), *Conservation of Plant Species and Habitats*. Nature Conservation Council, Wellington. Pp. 83-101. (Lists 7 'endangered' and 8 'vulnerable' Chatham Island taxa, including non-endemics; population sizes indicated.)

Latest IUCN statistics: endemic taxa – E:6, V:4, R:6; non-endemic taxa rare or threatened worldwide – V:4, R:1, I:1 (world categories).

Additional References

Cockayne, L. (1902). A short account of the plant-covering of Chatham Island. *Trans. N.Z. Inst.* 34: 243-325.

Devine, W.T. (1982). Nature conservation and land-use history of the Chatham Islands, New Zealand. *Biol. Conserv.* 23: 127-140.

Several surveys have been undertaken since 1970, and are the subject of a number of unpublished reports by the Botany Division, DSIR, Christchurch, including:

Given, D.R. and Williams, P.A. (1985). *Conservation of Chatham Island Flora and Vegetation*. DSIR, Christchurch, 123 pp.

Kelly, G.C. (in prep.). Distribution and ranking of remaining areas of indigenous vegetation in the Chatham Islands. Department of Land and Surveys Resource Inventory. (Includes map with extended legend; not seen.)

Chile

Area 751,626 sq. km

Population 11,878,000

Floristics Over 5500 species of vascular plants (M. Muñoz, 1984, pers. comm.). Endemism over 50% at specific level (Fuenzalida, 1984), and 16% at generic level (Muñoz, pers. comm.). (Toledo, 1985, cited in Appendix 1, quoting from Gajardo-Michell, 1983, reports 4758 recorded species, 2698 of them endemic.) Floristic affinities with California, New Zealand, Tasmania and New Caledonia (Muñoz, pers. comm.).

Vegetation Very diverse due to Chile's extreme north-south length and high altitudes. In the north, a very dry region, which includes the Atacama Desert; vegetation varies from none on the northern coast to the Loma Formation (see under Peru) and deciduous scrub on the western side of the Andes; at high altitudes and on high plateaux, very dry puna and salt marsh communities. In Central Chile a Mediterranean climate permits growth of broadleaved evergreen shrubland in the south, and lowland and submontane forest on the Andean slopes. Much of central Chile is cultivated. In the southern third of the country, the only temperate rain forest in South America, the Valdivian forest, which is dense and rich in epiphytes (Unesco, 1981, cited in Appendix 1). In the extreme south, including Tierra del Fuego, temperate and subpolar evergreen moist forest and high Andean meadows. Native forests, home of many of the endemic plants, now cover only 10% of the country (Fuenzalida, 1984).

Checklists and Floras That part of Chile north of the Tropic of Capricorn (nearly 1500 km, from just north of Antofagasta) is covered by the family and generic monographs of *Flora Neotropica*, described in Appendix 1. Chilean Floras are:

Johow, F. (1948). Flora de Zapallar. *Rev. Chil. Hist. Nat.* 49: 1-566.

Muñoz P., C. (1959). *Sinopsis de la Flora Chilena: Claves para la identificación de familias y géneros.* Edit. Univ. Chile, Santiago. 500 pp. (Generic vascular flora, but including 248 illus. of species; includes botanical bibliography of Chile.)

Reiche, K.F. (1886-1911). *Flora de Chile.* Cervantes, Santiago. (6 vols, incomplete.)

See also:

Moore, D.M. (1983). *Flora of Tierra del Fuego.* Nelson, U.K., and Missouri Botanical Garden, U.S.A. 396 pp.

Navas Bustamente, L.E. (1973-1979). *Flora de la Cuenca de Santiago de Chile,* 3 vols. Edit. Univ. Chile, Santiago.

Rodríguez, R.R., Matthei, O. and Quezada, M. (1983). *Flora Arbórea de Chile.* Universidad de Concepción. 408 pp. (87 species described including their uses; vegetation types and their endemics.)

See also Boelcke, Moore and Roig (1985), under Argentina.

Field-guides

Donoso, C. (1974). *Dendrología: Arboles y Arbustos Chilenos.* Facultad de Ciencias Forestales, Universidad Austral de Chile. Manual 2. 142 pp.

Donoso, C. (1981). *Arboles Nativos de Chile: Guía de Reconocimiento.* Alborada, Valdivia, Chile. (51 species listed with distribution maps.)

Hoffmann, A. (1980, 1982). *Flora Silvestre de Chile: Zona Central* (1980); *Zona Austral* (1982). Fundación Claudio Gay, Santiago. 255 pp. (Colour illus., plants arranged by flower colour.)

Muñoz P., C. (1966). *Flores Silvestres de Chile.* Edit. Univ. Chile. 245 pp. (51 colour photos.)

Muñoz Schick, M. (1980). *Flora del Parque Nacional Puyehue.* Edit. Univ. Santiago. 557 pp.

Information on Threatened Plants See below, in particular Muñoz P. (1973):

Muñoz P., C. (1973). *Chile: Plantas en Extinción.* Edit. Univ. Chile, Santiago. 248 pp. (58 species described with illustrations, uses.)

Muñoz P., C. (1967). La extinción de especies vegetales en Chile. In *La Conservación de la Naturaleza y la Prensa en la America Latina.* Instituto Mexicano de Recursos Naturales Renovables, México. Pp. 75-82.

Muñoz P., C. (1975). Especies vegetales que se extinguen en nuestro pais. In Capurro, L. and Vergara, R. (Eds), *Presente y Futuro del Medio Humano.* Capítulo XI: 161-179. Edit. Cont. CECSA, México.

F.M. Schlegel of the Institute of Silviculture, Valdivia, has prepared a list of the areas high in endemism and diversity most urgently needing conservation. Threatened plants are mentioned in several papers in:

Prance, G.T. and Elias, T.S. (Eds) (1977), cited in Appendix 1. See in particular C. Muñoz Pizarro on endangered plants of Chile (pp. 252-256), H.E. Moore Jr. on endangerment in palms (pp. 267-282), P. Ravenna on threatened bulbous plants (pp. 257-266).

Unpublished lists also include:

Marticorena, C. (1980). Threatened plants and areas of Chile. Universidad de Concepción. (List of threatened plants of the continent and the Islas of Más á Tierra, Más Afuera, Santa Clara, San Felix and San Ambrosio.)

Muñoz P., C. (1975). I. Areas Naturales: localidas y regiones de Chile dignas de protección. II. La extinción de especies vegetales. In 2a *Jorn. Latinoam. de Parques Nacionales*. SAG, Minist. Agric., Viña del Mar. 23 pp.

Schlegel Sachs, F.M. (1982). Especies Chilenas Amenazadas. Univ. Austral de Chile. (List of threatened plants including Ex:9, E:53, V:15, R:42.)

Laws Protecting Plants Two plant species are protected as Natural Monuments: Araucaria (*Araucaria araucana*) under Law No. 29, 9 February 1976, published 26 April 1976, and Alerce (*Fitzroya cupressoides*) under Law No. 490, 1 October 1976, published 5 September 1977. Several laws on the exploitation of species are mentioned in Muñoz P. (1973), cited above. The U.S. Government has determined *Fitzroya cupressoides*, confined to Chile and Argentina, as 'Threatened' under the U.S. Endangered Species Act.

Voluntary Organizations

Comité Nacional pro Defensa de la Fauna y Flora, Casilla 3675, Huerfanos 972, Oficina 508, Santiago.

Instituto de Ecología y Evolución, Universidad Austral de Chile, Casilla 567, Valdivia.

Sociedad de Vida Silvestre, Valdivia.

Botanic Gardens

Arboretum (Institute of Silviculture), Universidad Austral de Chile, Casilla 853, Valdivia.

Jardín Botánico "Carl Skottsberg", Instituto de la Patagonia, Casilla 102-D, Punta Arenas, Magallanes.

Jardín Botánico Hualpen, Departamento de Botánica, Universidad de Concepción, Casilla 1367, Concepción.

Jardín Botánico (Instituto de Botánica), Universidad Austral de Chile, Casilla 567, Valdivia.

Jardín Botánico Nacional, Casilla 683, Viña del Mar.

Useful Addresses

Corporación Nacional Forestal (CONAF), Avenida Bulnes 285-5° Piso, Santiago. (Includes Departamento Areas Silvestre Protegidas.)

Facultad de Ciencias Biológicas y de Recursos Naturales, Universidad de Concepción, Casilla 2407, Apdo 10, Concepción.

Museo Nacional de Historia Natural, Casilla 787, Santiago.

Universidad Austral de Chile, Facultad de Ciencias Forestales, Casilla 853, Valdivia.

CITES Management Authority: Autoridad Administrativa de Chile para CITES, Servicio Agricola y Ganadero, Avda. Bulnes 285-5° Piso, Casilla 4088, Santiago.

CITES Scientific Authority: Comisión Nacional de Investigación Científica y Technológica (CONICYT), Canadá 308, Santiago.

Additional References

Börgel O., R. (1973). The coastal desert of Chile. In Amiran, D.H.K. and Wilson, A.W. (Eds), *Coastal Deserts: Their Natural and Human Environments*. Univ. Arizona Press, Tucson. Pp. 111-114.

Fuenzalida, M. (1984). Evaluation of native forest destruction in the Andes of South Central Chile: conservation alternatives. Project Proposal to IUCN from Comité Nacional Pro Defensa de la Fauna y Flora, Santiago, Chile.

Gajardo-Michell, R. (1983). *Sistema Básico de Clasificación de la Vegetación Nativa Chilena*. Corporación Nacional Forestal. Universidad de Chile. 4 partes. (Not seen.)

Pisano, E. and Fuenzalida, H. (1950). VIII. Biogeografía. *Geografía Económica de Chile CORFO* 1: 271-428. (Includes one vegetation map.)

Ramirez, C, (1984). *Bibliografía vegetacional de Chile*. Universidad Austral, Valdivia.

Veblen, T.T., Delmastro, R.J. and Schlatter, J.E. (1976). The conservation of *Fitzroya cupressoides* and its environment in southern Chile. *Envir. Conserv.* 3(4): 291-301.

Veblen, T., Schlegel, F. and Oltremari, J. (1983). Temperate broadleaved evergreen forest of South America. In Ovington, J.D. (Ed.), *Temperate Broadleaved Evergreen Forest*. Elsevier. Pp. 5-31.

Yudelevich, M., Brown, C.H., Elgueta, H. and Calderón, S. (1967). Clasificación preliminar del bosque nativo de Chile. *Inst. Forestal, Inf. Téc.* 27: 1-16. (2 maps.)

China

Area 9,597,000 sq. km

Population 1,051,551,000

Floristics About 30,000 vascular plant species (Yü, 1979), including about 7000 tree species (quoted in NCC, 1982). 15,000 species occur in tropical and subtropical regions, of which 7000 are in Yunnan (NCC, 1982). Of the 2980 flowering plant genera, 214 are endemic (including 9 gymnosperm genera). Centres of endemism include eastern Sichuan/western Hubei, south-east Yunnan/western Guangxi, and the western Sichuan/north-west Yunnan centre abutting Burma, Laos and Viet Nam (Ying Tsün-Shen and Zhang Zhi-Song, 1984). Western Yunnan and south-east Tibet are particularly rich in *Rhododendron* and *Primula* spp.

Vegetation Tropical evergreen rain forest in lowland parts of Yunnan and Guangdong Provinces and on eastern side of Hainan Island. Mangrove forests along the southern coasts (Chien, Wu and Chen, 1956). Temperate deciduous forests, subtropical evergreen and monsoon forests in the south; evergreen, semi-evergreen and mixed broadleaved deciduous forests on limestone in the tropical and subtropical zones of the south; various types of subarctic coniferous forest ('taiga') and cold temperate mixed forests in the north. The most extensive tracts of natural forest are in the north-east and the south-western provinces of Sichuan and Yunnan. Western China, the vast plains of north-east China and Inner Mongolia are largely semi-arid grassland whereas there are deserts and semi-deserts in the Gobi and Tibetan regions. North China, including southern Dongbei and parts of Inner Mongolia, are mainly arable. Northern Dongbei is steppe grassland though converted in part to farmland. Fuelwood cutting, overgrazing and deforestation have left only remnants of primary forest cover in more remote areas and on steep terrain. There are over 1.2 million sq. km of "fully-stocked forests" including afforested areas (FAO, 1982); probably more land has recently been afforested than in any other country in the world (see Smil, 1983, for a report on the afforestation programme).

For a more comprehensive account of vegetation, and map at scale 1:14,000,000, see Hou (1983). For more detailed vegetation map see:

China Vegetation Commission (1979). *Vegetation Map of China* (1:4,000,000) and accompanying booklet, *Legend to the "Vegetation Map of China"* (edited by H.Y. Hou). Chinese Academy of Sciences, Institute of Botany, Beijing. 12 pp.

Checklists and Floras

Chen, Y. (1957). *Illustrated Manual of Chinese Trees and Shrubs*. Science Technology
   Press, Shanghai. (Revised edition, in Chinese; accounts of 2100 native and
   introduced taxa.)

Chinese Academy of Sciences (1971-1976). *Iconographia Cormophytorum Sinicorum*,
   5 vols. Science Press, Beijing. (Keys, line drawings, descriptions of 8000 of the more
   common and economically important species; in Chinese. 2 Supplements – 1982,
   1983.)

Flora Plantarum Herbacearum Chinae Boreali-Orientalis (1958-). 8 vols so far. (Keys,
   line drawings, descriptions of herbaceous plants; in Chinese.)

Inner Mongolia Botanical Records Compiling Group (1977-1982). *Flora
   Intramongolica*, 6 vols. Typis Intramongolicae Popularis, Huhhot. (In Chinese.)

Institute of Botany, Academia Sinica (1959-). *Flora Reipublicae Popularis Sinicae*.
   Science Press, Beijing. (80 vols planned, 93 families treated so far in 33 volumes; in
   Chinese.)

Shun-Ching Lee (1935). *Forest Botany of China*. Commercial Press, Shanghai.
   Supplement (1973). Chinese Forestry Association Taipei, Taiwan. (In English.)

Xinjiang and Gansu are covered in Grubov (1963-), cited in Appendix 1. The north-west
steppe region is included in Norlindh (1949, cited in Appendix 1.

Information on Threatened Plants Sheng Cheng-kui, of the Hortus Botanicus
Nanjing, reports on a preliminary national threatened plant list compiled under the joint
auspices of the Environment Protection Agency, the Chinese Botanical Society and the
Editorial Commission of the Chinese Floras (in *Threatened Plants Committee – Newsletter*
No. 7: 5-6, 1981).

A *First National List of Chinese Threatened Plants* was published in April 1982 under the
auspices of the National Environmental Protection Agency and the Botanical Institute of
Academia Sinica. It lists 354 species of vascular plants, organized in three sections. The
first lists the species in conservation rating order (1, 2 or 3), the second in systematic order,
the third geographically, by Provinces.

A Red Data Book of Chinese rare and endangered species is due to be published in 1985
(English translation). The book will cover the 354 species (9 ferns, 68 gymnosperms and
277 angiosperms) with details on their distribution, ecology, present status in the wild and
conservation measures, with colour plates and a map for each species.

Voluntary Organizations

WWF-China Joint Committee, c/o The Environmental Protection Office of the State
   Council, Beijing.

Botanic Gardens

Beijing Botanical Garden, Academia Sinica, Beijing, Hebei.
Desert Botanical Garden, Minching, Gansu.
Gangnan Arboretum, Shongyu, Jiangxi.
Guilin Botanical Garden, Yanshan, Guilin, Guangxi.
Hangzhou Botanical Garden, Yuquan, Hangzhou, Zhejiang.
Heilongjiang Forestry Botanical Garden, Renjiaqiao, Harbin, Heilongjiang.
Kunming Botanical Garden, Academia Sinica, Heilongtan, Kunming, Yunnan.
Lushan Botanical Garden, Hanpoku, Lushan, Jiangxi.
Shanghai Botanical Garden, Longwu Road, Shanghai 201102.
Shenyang Botanical Garden, Shenyang, Liaoning.

South China Botanical Garden, Academia Sinica, Longyandong, Guangzhou, Guangdong.
Sun Yat-Sen Memorial Botanical Garden, Nanjing, Jiangsu.
Wuhan Botanical Garden, Wuhan, Hubei.
Xi'an Botanical Garden, Ciuhua Road, Xi'an, Shaanxi.
Xishuang Banna Tropical Botanical Garden, Mengla, Xishuang Banna, Yunnan.
Zhejiang Institute of Subtropical Crops, Wenzhou, Zhejiang.

For an illustrated account of Chinese botanical gardens and arboreta see:

Yu Dejun (Ed.) (1983). *The Botanical Gardens of China*. Science Press, Beijing. 319 pp. (Covers 21 gardens and arboreta.)

See also:

Sheng Cheng-kui (1981). Directory of Chinese botanical gardens. Hortus Botanicus Nanjingensis. (Lists 16 gardens with details of size, number of taxa, research interests, publications and associated herbaria. 15 other gardens, arboreta and plant introduction stations are also listed.)

Useful Addresses
CITES Management Authority: The People's Republic of China, Endangered Species of Wild Fauna and Flora Import and Export Administrative Office, Ministry of Forestry, Hepingli, Beijing.
CITES Scientific Authority: Endangered Species Scientific Commission of the People's Republic of China, 7 Zhongguancun Lu, Haidian, Beijing.

Additional References
Bohlin, B. (1949). *A Contribution to our Knowledge of the Distribution of Vegetation in Inner Mongolia, Kansu and Ching-Hai*. Report of the Scientific Expedition to NW Provinces, China, 33. Stockholm. 95 pp.
Chien, S.S., Wu, C.Y. and Chen, C.T. (1956). The vegetation types of China. *Acta Geogr. Sinica* 22: 87-92.
Duke, J.A. and Ayensu, E.S. (1985). *Medicinal Plants of China*, 2 vols. Reference Publications, Algonac, Michigan. 670 pp. (Covers 1240 species; includes English and Chinese names; notes on uses; line drawings.)
FAO (1982). *Forestry in China*, FAO Forestry Paper 35. FAO, Rome. 305 pp.
Hou, H.-Y. (1983). Vegetation of China with reference to its geographical distribution. *Annals Missouri Bot. Gard.* 70(3): 509-549.
NCC (1982). *Nature Conservation Delegation to China, 4-24 April 1982*. Nature Conservancy Council, London. 44 pp. (See in particular M.E.D. Poore on vegetation, pp. 29-33 and D.A. Ratcliffe on nature conservation, pp. 34-37.)
Smil, V. (1983). Deforestation in China. *Ambio* 12(5): 226-231. (Causes and extent of forest losses; analysis of afforestation projects.)
Walker, E.H. (1941). Plants collected by R.C. Ching in southern Mongolia and Kansu Province, China. *Contrib. U.S. Nat. Herb.* 38(4): 563-675.
Wang, Chi-Wu (1961). *The Forests of China with a Survey of Grassland and Desert Vegetation*. Harvard Univ., Cambridge, Mass. 313 pp.
Ward, F. Kingdon (1935). A sketch of the geography and botany of Tibet, being materials for a Flora of that country. *J. Linn. Soc. Bot.* 50: 239-265. (Useful introduction to botany and vegetation of Tibet.)
Wu Zheng-Yi *et al.* (Eds) (1980). *Vegetation of China*. Science Press, Beijing. (In Chinese.)

Ying Tsün-Shen and Zhang Zhi-Song (1984). Endemism in the flora of China – studies
on the endemic genera. *Acta Phytotaxonomica Sinica* 22(4): 259-268. (In Chinese,
with English summary; maps showing distribution of endemic genera.)
Yü, T.-T. (1979). Special report: status of the Flora of China. *Syst. Bot.* 4(3): 257-260.

Christmas Island

An External Territory of Australia, situated in the eastern Indian Ocean, at 10°30'S and
105°30'E. It consists of an elevated series of coral limestone and volcanic rocks. The
central plateau is mostly 150-250 m above sea-level but rises to 361 m in a series of cliffs
and terraces. 16 sq. km of the island were declared a National Park in 1979.

Area 135 sq. km

Population 3184 (1980 estimate, *Times Atlas*, 1983)

Floristics c. 380 flowering plant species, of which c. 280 native species; about 15
endemic flowering plant species (L. Forman, 1984, pers. comm.). The flora has affinities
with Java, S.E. Asia, Australia and the western Pacific Islands.

Vegetation Mixed closed forest above 180 m and occasionally extending down to
coastal terraces, with *Tristiropsis*, *Dysoxylon*, *Cryptocarya* and *Barringtonia* in the main
canopy, and *Eugenia*, *Planchonella* and *Hernandia* as emergents; *Celtis*, *Terminalia* and
Pisonia forest on coastal terraces. Mining of phosphate deposits has devastated much of
the island's rain forests (Mitchell, 1974).

Checklists and Floras Christmas Island will be covered in a forthcoming volume
of the *Flora of Australia* (1981-), cited under Australia. Forman *et al.* have prepared a
checklist of the flowering plants (1984, manuscript, Royal Botanic Gardens, Kew). See
also:

Baker, E.G., Rendle, A.B., Gepp, A., Blackman, V.H. and Lister, A. (1900). Botany.
   In Andrews, C.W. (Ed.), *A Monograph of Christmas Island*. British Museum
   (Natural History), London. Pp. 171-200.
Mitchell, B.A. (1974). The forest flora of Christmas Island. *Commonwealth Forestry
   Review* 53(1): 19-29.
Ridley, H.N. (1906). The botany of Christmas Island. *J. Straits Branch Royal Asiatic
   Soc.* 45: 156-271. (Lists 34 endemic species, many subsequently reduced to
   synonymy. For additional notes see *ibid.*, 48: 107-108.)

Information on Threatened Plants A preliminary list of endemic taxa with notes
on their conservation status is given in Leigh *et al.* (1981), cited under Australia. In 1983,
D. Powell prepared a preliminary list of endemics with notes on distribution and status.

Latest IUCN statistics: endemic taxa – E:1, V:1, R:11, I:1, K:2.

Additional References
Gray, H.S. (1981). *Christmas Island Naturally: The Natural History of an Isolated
   Oceanic Island*. Gray, Geraldton, W. Australia. (Covers fauna, vegetation, impact
   of man.)

Clipperton Island

The most easterly of the French Polynesian islands in the south-east Pacific. It is a remote uninhabited atoll of 5 sq. km at longitude 10°18'N and latitude 109°13'W. The nearest atoll is Pukapuka in the Tuamotu Archipelago, about 2500 km to the south-west. Clipperton is a low coral limestone ring enclosing a lagoon with a 29 m high volcanic plug. Parts of the island have been mined for phosphate (Douglas, 1969, cited in Appendix 1).

31 flowering plant taxa, all of which are widespread herbs, apart from introduced coconuts (Sachet, 1962 a & b).

References

Sachet, M.-H. (1962a). Flora and vegetation of Clipperton Island. *Proc. Calif. Acad. Sci. IV*, 31(10): 249-307. (Includes enumeration of native and introduced species; notes on localities and distributions.)

Sachet, M.-H. (1962b). Geography and land ecology of Clipperton Island. *Atoll Res. Bull*. 86. 115 pp. (Includes plant list.)

Coco, Isla del

Isla del Coco or Cocos Island is an uninhabited island of area 24 sq. km, with several offshore islets, 670 km south-west of Costa Rica and 630 km west of the Galápagos in the eastern Pacific Ocean, 5°32'N 87°04'W. It is an outcrop of the Cocos Ridge, comprising volcanic basalts and marine sediments which have been uplifted by tectonic activity. The highest point is 849 m (Cerro Iglesias). Isla del Coco is administered by Costa Rica.

Floristics 155 vascular plant taxa, including introductions; c. 10% species endemism (figures quoted by Fournier, 1966). The flora is related to that of Central America, the Caribbean and the Galápagos Islands.

Vegetation The main island has closed tropical rain forest with *Cecropia*, *Brosimum*, *Ochroma* (Balsa), epiphytes and lianas; cloud forest above 500 m with Lauraceae, bromeliads, orchids and ferns; littoral communities with *Erythrina* and introduced coconuts; extensive *Nephrolepis* scrub near the shore and small areas of brackish marshes. The offshore islets are sparsely vegetated.

Although Isla del Coco was declared a National Park in 1978, introduced plants and grazing animals continue to be a threat to the native flora.

Checklists and Floras

Fosberg, F.R. and Klawe, W.L. (1966). Preliminary list of plants from Cocos Island. In Bowman, R.I. (Ed.), *The Galápagos: Proceedings of the Galápagos International Project of 1964*. University of California Press, Berkeley. Pp 187-189. (Checklist of 148 vascular plant taxa; separate list of lower plants.)

Stewart, A. (1912). Expedition of the California Academy of Sciences to the Galapagos Islands, 1905-1906. V. Notes on the botany of Cocos Island. *Proc. California Acad. Sci.*, 4. 1: 375-404. (Enumeration of 22 taxa of ferns and 52 flowering plants collected during expedition; list of further 9 ferns and 6 flowering plants recorded from the island.)

Information on Threatened Plants None.

Useful Addresses

Servicio de Parques Nacionales, Ministerio de Agriculture y Ganaderia, CP 10094, San José, Costa Rica.

Additional References

Fournier, L.A. (1966). Botany of Cocos Island, Costa Rica. In Bowman, R.I. (Ed.), *The Galápagos: Proceedings of the Galápagos International Project of 1964*. University of California Press, Berkeley. Pp 183-186. (Notes on vegetation and origin of flora.)

Coco Islands

The Coco Islands comprise Table Island, Great Coco and Little Coco, c. 175 km north of the Andaman Islands in the Eastern Indian Ocean, between latitudes 13-15°N and at longitude 93°21'E. The highest point is 100 m, on Great Coco. Total land area c. 5 sq. km. The islands are dependencies of Burma.

Prain (1891) recorded dense tropical rain forest on all islands, with coconuts and *Pandanus* along the coasts, and mangrove swamps along creeks. No details of current status.

296 flowering plant taxa, one gymnosperm (*Cycas rumphii*) and 10 fern taxa; most are widespread throughout South East Asia (Prain, 1891).

References

Prain, D. (1891). The vegetation of the Coco group. *J. Asiatic Soc. Bengal* 60: 283-406. (Vegetation, floristic analyses, annotated list with notes on distributions.)

Cocos Islands

The Cocos Islands are an External Territory of Australia, situated in the east Indian Ocean. They are coral atolls 3700 km west of Darwin and 300 km south of Java. Area 14 sq. km; population 487 (1980 census, *Times Atlas*, 1983).

Cocos I. (12°S, 96°E) is mainly covered by coconut plantations; Keeling I. has a mixture of coconuts, *Pisonia grandis* and *Cordia subcordata* scrub. 57 vascular plant species of which c. 43 indigenous (Renvoize, 1979, quoting figures by Wood-Jones, 1912). The Cocos Islands will be included in a forthcoming volume of the *Flora of Australia* (1981–), cited under Australia.

References

Hemsley, W.B. (1885). List of plants from the Keeling or Cocos Islands. In *Report on the Scientific Results of the Voyage of H.M.S. Challenger During the Years 1873-76*. Botany, vol. 1, part 2. HMSO, London. (List of 19 flowering plant species, p. 113.)

Renvoize, S.A. (1979). The origins of Indian Ocean island floras. In Bramwell, D. (Ed.), *Plants and Islands*. Academic Press, London. Pp. 107-129.
Wood-Jones, F. (1912). *Coral and Atolls*. Reeve, London.

Colombia

Area 1,138,914 sq. km

Population 28,110,000

Floristics Prance (1977) estimates 45,000 species of flowering plants; 3000 species of Orchidaceae alone (Ospina, 1969). Areas rich in endemics are the Sierra Nevada de Santa Marta, the Guajira Peninsula, La Macarena, many parts of the Andes, and, above all, the Chocó Region in western Colombia; this is the wettest and possibly the richest rain forest in the neotropics, with both high species diversity and high endemism, concentrated in two or three distinct centres (Gentry, 1982). Some 1500 endemic species have been recorded and many new species are being discovered with additional exploration.

According to E. Forero (1984, pers. comm.), destruction or conversion of tropical forests is causing high rates of extinction in the Sierra de la Macarena, in the Pacific coastal region, in the Sierra Nevada d' Santa Marta from sea level to the páramo and the slopes of the Andes.

Vegetation Extending the length of the Pacific coast is very wet tropical rain forest (includes Chocó region). The Atlantic coast varies from humid forest near Panama to dry forest and desert at Guajira Peninsula. Extending inland to 250 km, vegetation ranges from mangroves (along Urabá coast), grassland/savanna to scattered thorn thickets and cactus scrub (Espinal and Montenegro, 1963). This region has been heavily disturbed by grazing and agriculture. Myers (1980, cited in Appendix 1) gives estimates for present forest extent; Myers also quotes Gentry to state that a considerable area has been disturbed at the southern end of the Chocó and where the northern extension borders Panama at the Darien Gap.

Central highlands: great variation in vegetation types; submontane forest to 2000 m; at 2400-3600 m cloud forests rich in epiphytes; above, very humid montane forest to páramo; above 4500 m, on the high peaks of the Central and Eastern Cordilleras, alpine tundra rich in endemics ('superpáramo').

Eastern plains: in the northern region to the Venezuelan border, tall grassland with broadleaved evergreen trees along river corridors.

In the south, Amazonian forest, little disturbed and little known botanically (Forero, pers. comm.). Myers (1980, cited in Appendix 1) quotes estimates of entent varying from 386,000 sq. km (1972) to 270,000 sq. km (1977). According to FAO/UNEP (1981), estimated rate of deforestation (for Colombia overall) of closed broadleaved forest 8200 sq. km/annum out of 464,000 sq. km.

Checklists and Floras Colombia is covered by the family and generic monographs of *Flora Neotropica* (cited in Appendix 1).

In 1982, the Instituto de Ciencias Naturales, Universidad Nacional, began a multi-volume *Flora of Colombia*. Published so far are Vol 1, Magnoliaceae, by G. Lozano-C. (1983) and Vol. 2, Connaraceae, by E. Forero (1983).

Individual family treatments include:

Cuatrecasas, J. (1958-). Prima Flora Colombiana. *Webbia* 12, 13, 15, 24.
(Burseraceae, Malphigiaceae, Compositae-Astereae.)
Idrobo, J.M. (1954). The Xyridaceae of Colombia. *Caldasia* 6: 183-260. (345 species.)
Leonard, E.C. (1951-1958). The Acanthaceae of Colombia. *Contr. U.S. Nat. Herb.* 31:
1-781.
Smith, L.B. (1957). The Bromeliaceae of Colombia. *Contr. U.S. Nat. Herb.* 33: 1-311.
Smith, L.B. and Fernandez-Perez, A. (1954). The Violaceae of Colombia. *Caldasia* 6:
83-181.
Smith, L.B. and Schubert, B.G. (1946). The Begoniaceae of Colombia. *Caldasia* 4:
3-38, 77-107, 179-209.
Standley, P.C. (1931). The Rubiaceae of Colombia. *Field Mus. Nat. Hist., Bot. Ser.* 7:
1-175.

See also:

Castaneda, R.R. (1965). *Flora del centro de Bolivar*. Universidad Nacional de
Colombia, Bogotá. 437 pp. (Line drawings.)
Espinal T., L.S. (1964). *Algunos Aspectos de la Vegetación del Oriente Antiogueño*.
I.G.A.C., Bogotá. 74 pp. (Describes 31 trees and shrubs of Antiogueño region,
photos, uses of woods.)
Garcia Barriga, H., Forero, E. *et al.* (1966-1979). *Catalogo Ilustrada de las Plantas de
Cundinamarca*, 7 vols. Universidad Nacional, Bogotá.
Machecha Vega, G. and Echeverri Restrepo, R. (1983). *Arboles del Valle del Cauca*.
Lithografia Arco, Bogotá. 208 pp.

Since 1973, the Missouri Botanical Garden and the Institute for Natural Sciences at the National University of Colombia have carried out collaborative field work in the western Chocó region. E. Forero and A.H. Gentry are now completing a plant checklist for the Chocó Department. The Institute, with the Rijksherbarium, Leiden, Netherlands, have been undertaking a survey of montane Colombia. Details of other floristic work in Colombia is given by Prance (1979, cited in Appendix 1). Floristic knowledge of the Chocó region is summarized by Gentry (1978), cited in Appendix 1.

Toledo (1985, cited in Appendix 1) refers to the following regional Floras as in progress: Páramo de Oroque by H. Garcia-Barriga, Santander del Sur by F. Llanos and A. Rentería, and Providencia by D.D. Soejarto.

In 1964 the governments of Colombia and Spain agreed to publish the *Flora de Mutis*, consisting of illustrations prepared by the Real Expedición Botánica del Nueva Reyno de Grenada between 1783 and 1816. 7 vols completed so far, an additional 93 vols expected by 1992.

Information on Threatened Plants The National University of Colombia is preparing an endangered species list for Colombia, described, with many examples, in:

Fernández-Pérez, A. (1977). The preparation of the endangered species list of
Colombia. In Prance, G.T. and Elias, T.S. (Eds) (1977), cited in Appendix 1.
Pp. 117-127.

Threatened plants are mentioned in several other papers in:

Prance, G.T. and Elias, T.S. (Eds) (1977), cited in Appendix 1. See in particular J.T. Mickel on rare and endangered ferns (pp. 323-328), H.E. Moore Jr. on endangerment in palms (pp. 267-282), P. Ravenna on rare and threatened bulbs (pp. 257-266).

Laws Protecting Plants Decreto Ley No. 2811 of 18 December 1974, the National Renewable Natural Resources and Environment Protection Code (D.O. 27 January 1975) authorizes INDERENA, the government conservation agency, to establish rules for the use, trade and conservation of wild plants (and animals). Acuerdo No. 38, promulgated by INDERENA on 10 September 1973, establishes rules governing conservation and utilization of wild plants, including licensing requirements for collection and commerce, and rules for propagation (Fuller and Swift, 1984, cited in Appendix 1).

Voluntary Organizations
Asociación Nacional para la Defensa de la Naturaleza, Apdo Aéreo 6227, Cali.
Sociedad Colombiana de Ecología, Calle 59 No. 13-83, Of. 302, Bogotá.

Botanic Gardens
Jardín Botánico "Guillermo Piñeres", Apto Aéreo 5456, Cartagena.
Jardín Botánico "Joaquin Antonio Uribe", Carrera 52 No. 73-298, Apto Aéreo 51-407, Medellin, Antioquia.
Jardín Botánico "José Celestino Mutis", Instituto de Ciencias Naturales, Carrera 66-A No. 56-84, Bogotá.
Jardín Botánico "Juan Maria Céspedes", Tuluá, c/o Instituto Vallecaucano de Investigaciones Cientificas (INCIVA), Apto Aéreo 5660, Cali.
Jardín Botánico Universidad de Tolima, Ibague, Colombia.
Jardín de la Facultad Agronomía, Universidad de Caldas, Apto Aéreo 275, Manizales, Caldas.

Useful Addresses
Asociación Colombiana de Herbarios, c/o Facultad de Ciencias Exactas y Naturales, Apto Aéreo 1226, Universidad de Antioquia, Medellin, Colombia. (19 Colombian herbaria are members.)
Instituto de Ciencias Naturales, Universidad Nacional, Apdo Aéreo 7495, Bogotá.
Instituto Nacional de los Recursos Naturales Renovables y del Ambiente (INDERENA), Gerente General, Diagonal 34, Numero 5-18, Apdo Aéreo 13458, Bogotá.
CITES Management Authority: INDERENA, see above.

Additional References
Cuatrecasas, J. (1958). Aspectos de la vegetación natural de Colombia. *Rev. Acad. Colombiana Ciencias Exactas, Físicas y Naturales* 10(40): 221-268. (Lists vegetation types, major representative species, photos.)
Espinal T., L.S. and Montenegro M., E. (1963). *Formaciones Vegetales de Colombia: Memoria Explicativa sobre el Mapa Ecológico*. Instituto Geográfico "Agustin Codazzi", Bogotá. 201 pp. (Includes descriptions of vegetation types, locations, photos, diagrams and 4 vegetation maps at 1:1,000,000, each covering a quarter of the country.) Updated 1977 as *Zonas de vida o Formaciones Vegetales de Colombia* (Map in 21 'planchas' at 1:500,000.)
Gentry, A.H. (1982). Phytogeographic patterns as evidence for a Chocó Refuge. In Prance, G.T. (Ed.) (1982), cited in Appendix 1. Pp. 112-136.

Ospina, M. (1969). Colombian orchids and their conservation. In Corrigan, M.J. (Ed.), *Proceedings of the 6th World Orchid Conference*. Sydney, New South Wales, Australia. Pp. 95-98.

Prance, G.T. (1977). Floristic inventory of the tropics: where do we stand? *Ann. Missouri Bot. Gard.* 64(4): 659-684.

Schultes, R.E. (1951). La riqueza de la flora Colombiana. *Rev. Acad. Colombiana Ciencias Exactas, Físicas y Naturales* 8(30): 230-242.

Comoro Islands

The Comoro Islands, an archipelago of four small islands (Njazidja, Mayotte, Anjouan, and Mohéli), together with numerous islets and coral reefs, lie between the east African coast and northern Madagascar, roughly 300 km from each, 11°20'-12°40'S, 43°-45°E. The islands are volcanic in structure; Mt Karthala on Njazidja (Grande-Comore) is still active and is the highest peak at 720 m. One of the islands, Mayotte, is a *collectivité territoriale* of France.

Area 2238 sq. km

Population 443,000

Floristics 935 species (416 indigenous), with 136 endemic (Voeltzkow, 1917). Floristic affinities with Madagascar.

Vegetation Native lowland plants almost all completely destroyed on all four islands. Very little intact upland forest remains on Anjouan and Mayotte islands. However, there is considerable forest on upper slopes of Njazidja and Mohéli, but much of this is badly degraded (Fosberg and Sachet, 1972, cited in Appendix 1).

Checklists and Floras The Comoros are included in the incomplete *Flore de Madagascar et des Comores*, cited under Madagascar.

Voeltzkow, A. (1917). Flora und Fauna der Comoren. In *Reise in Ostafrika in den Jahren 1903-1905*. Wiss. Ergeb. 3(5): 429-480. (Checklist with distributions; the only important inventory for the archipelago.)

Information on Threatened Plants None.

Additional References

Legris, P. (1969). La Grande Comore. Climats et végétation. *Trav. Sect. Sci. Techn. Inst. Franç. Pondichéry* 3(5): 1-28. (With coloured vegetation map 1:100,000.)

Congo

Area 342,000 sq. km

Population 1,695,000

Floristics Flora very poorly known; c. 4000 species (Bouquet, 1976); insufficient evidence for assessment of endemism, but likely to be comparable with Gabon (c. 22%).

The western portion of the northern forests are said to be especially diverse (Myers, 1980, cited in Appendix 1).

Floristic affinities Guinea-Congolian.

Vegetation Large areas of both lowland rain forest and swamp forest, and forest interspersed with secondary grassland and cultivation. Estimated rate of deforestation for closed broadleaved forest 220 sq. km/annum out of 213,400 sq. km (FAO/UNEP, 1981). However, Myers (1980, cited in Appendix 1), quoting Unesco (1978), gives a figure of 100,000 sq. km of forest, of which 30,000 sq. km is evergreen.

For vegetation map see White (1983), cited in Appendix 1.

Checklists and Floras

Descoings, B. (1961). Inventaire des plantes vasculaires de la République du Congo déposées dans l'herbier de l'Institut d'Etudes Centre-Africaines à Brazzaville. Institut d'Etudes Centre-Africaines (ORSTOM), Brazzaville. 63 pp. (Unpublished mimeograph; list of 1600 names.)

Bouquet (1976) mentions a new checklist in preparation, due to be finished in 1975 (his paper was presented in 1974), but it is not clear if this was ever published.

Information on Threatened Plants No published lists of rare or threatened plants; IUCN has records of 17 species and infraspecific taxa believed to be endemic; no categories assigned.

Useful Addresses

CITES Management Authority: Secrétariat Général aux Eaux et Forêts, B.P. 98, Brazzaville.
CITES Scientific Authority: Secrétariat Général à l'Economie Forestière, B.P. 98, Brazzaville.

Additional References

Bouquet, A. (1976). Etat d'avancement des travaux sur la Flore du Congo-Brazzaville. In Miège, J. and Stork, A.L. (1975, 1976), cited in Appendix 1. P. 581.
Farron, C. (1968). Congo-Brazzaville. In Hedberg, I. and O. (1968), cited in Appendix 1. Pp. 112-115.

Cook Islands

The Cook Islands, a self-governing territory associated with New Zealand, comprise 15 islands and atolls in the South Pacific Ocean, between latitudes 17-25°S and longitudes 155-165°W. The southern Cooks are over 1000 km south-east of the Samoan Archipelago. Area 241 sq. km; population 19,000.

The northern Cooks are low atolls, some of which still retain areas of native vegetation, e.g. Palmerston I., Penrhryn (Tongareva), Rakahanga and Suwarrow (Douglas, 1969, cited in Appendix 1). The southern Cooks are mainly volcanic, reaching 643 m at Te Manga on Rarotonga. Upland forests above 250 m are largely intact (Sykes, 1983). 'Makatea' (raised limestone) surrounds most of the low islands in the Ngaputoru Group and Mangaia, supporting open forest. Lowland forests almost totally destroyed. Coconuts abundant on all islands especially the atolls and lower areas of volcanic islands (W.R. Sykes, 1984, *in litt.*).

No overall figure for the Cook Islands but 560 vascular species recorded on Rarotonga (Wilder, 1931). No endemic genera (van Balgooy, 1970, cited in Appendix 1). Floristic affinities to the Society Islands. No information on threatened plants.

References

Brownlie, G. and Philipson, W.R. (1971). Pteridophyta of the southern Cook Group. *Pacific Science* 25: 502-511. (Annotated list of 80 taxa; notes on habitats, frequency.)

Cheeseman, T.F. (1903). The flora of Rarotonga, the chief island of the Cook Group. *Trans. Linn. Soc. Bot., Ser.* 2 6: 261-313.

Fosberg, F. R. (1972). List of vascular plants of Rarotonga. *Atoll Res. Bull.* 160: 9-14. (Checklist of 50 vascular plant taxa collected in 1969.)

Philipson, W.R. (1971). Floristics of Rarotonga. *Bull. Roy.* Soc. N.Z. 8: 49-54.

Stoddart, D.R. and Fosberg, F.R. (1972). Reef islands of Rarotonga. List of vascular plants. *Atoll Res. Bull.* 160. 14 pp.

Sykes, W.R. (1983). Conservation on South Pacific islands. In Given, D.R. (Ed.), *Conservation of Plant Species and Habitats*. Nature Conservation Council, Wellington, N.Z. Pp. 37-42.

Wilder, G.P. (1931). Flora of Rarotonga. *Bull. Bernice P. Bishop. Mus.* 86. 113 pp.

Coral Sea Islands

The Coral Sea Islands are a scattered group of 32 coral sand islands (cays) and coral reefs c. 300 km east of the Queensland coast and c. 200 km east of the Great Barrier Reef. The islands are situated in the Coral Sea between 147°-152°E and 12°-25°S. The highest point is 7 m, on Willis Island (150°E 16°S). None of the islands have a resident population, but there is a storm warning station on Willis Island (Douglas, 1969, cited in Appendix 1). The islands are an External Territory of Australia.

Most of the islands have no vegetation; a few are covered by scrub (Douglas, 1969, cited in Appendix 1). The only published checklist of the flora is that of 7 vascular plant taxa collected from Willis Island (Davis, 1923).

References

Davis, J.K. (1923). *Willis Island, a Storm Warning Station in the Coral Sea*, 5. 199 pp. Critchley Parker, Melbourne. (Geography, climatology; checklist of vascular plants with notes on distribution. Not seen; citation from Frodin.)

Costa Rica

Area 50,899 sq. km

Population 2,534,000

Floristics Costa Rica, for its size, may have the most diverse plant life anywhere in the world. It is a biogeographical land bridge where the floras of North and South America meet. Gentry (1978, cited in Appendix 1) estimates 8000 higher plant species; L.D. Gómez P. (1984, pers. comm.) estimates 10,000, of which 1500 are orchids and 1800 trees; 1393 taxa are believed endemic (IUCN figures). Even more staggering is that the 730 ha La Selva Reserve contains 1500 recorded species of vascular plants (Hammel and Grayum, 1982).

Vegetation Accounts of the vegetation are given by Janzen (1983), who distinguishes 14 major tropical plant formations, and Gómez (1983b), who identifies 40 vegetational units in Costa Rica. Natural vegetation is entirely forest and woodland except for the high paramo on the highest peaks of the Cordillera de Talamanca and for the savannas on unusual volcanic soils. Moist and wet tropical forests occupy 48% of the forested land, mainly along the cordilleras and on Peninsulas Nicoya and Osa. Other types of forest are the semi-deciduous tropical dry forest (in the northwestern Pacific lowlands) and the floristically diverse montane forests. Also present are mangrove and swamp forests.

By the 1960s logging and deforestation had destroyed over half the natural forest; estimated rate of deforestation for closed broadleaved forest 650 sq. km/annum out of 16,380 sq. km (FAO/UNEP, 1981); 5671 sq. km (11%) protected in parks and reserves, one of the highest percentages in the world.

Checklists and Floras Costa Rica is covered by the *Flora Mesoamericana* Project, described in Appendix 1, as well as by the family and generic monographs of *Flora Neotropica* (cited in Appendix 1). The country Floras are:

Burger, W. (Ed.) (1971-). Flora Costaricensis. *Fieldiana Bot.* 35, 40; and *Fieldiana Bot. new series* 4, 13. (34 families so far.)

Standley, P.C. (1937-1938). Flora of Costa Rica. *Field Mus. Nat. Hist., Bot. Ser.* 18(1-4). 1616 pp. (Complete systematic list of gymnosperms and flowering plants.)

See also:

Hartshorn, G.S. and Poveda, L.J. (1983). Checklist of Trees. In Janzen, D.H. (Ed.), *Costa Rican Natural History*. Univ. of Chicago Press, Illinois. Pp. 158-183. (Vegetation map and checklist of trees.)

Holdridge, L.R. and Poveda, L.J. (1975). *Arboles de Costa Rica*, Vol. 1. Centro Científico Tropical, San José. 546 pp.

Janzen, D.H. and Liesner, R. (1980). Annotated checklist of plants of lowland Guanacaste Province, Costa Rica, exclusive of grasses and non-vascular cryptograms. *Brenesia* 18: 15-90.

Field-guides

Allen, P.H. (1956). *The Rain Forests of Golfo Dulce*. Univ. of Florida Press, Gainesville. 417 pp. (Keys to 433 species, mainly trees, in southern Costa Rica, some illustrated.)

There are also a few illustrated field guides to trees in national parks.

Information on Threatened Plants There is no national Red Data Book. IUCN is preparing a threatened plant list for release in a forthcoming report *The List of rare, threatened and endemic plants of Middle America*. Latest IUCN statistics, based upon this work: endemic taxa – Ex:4, E:53, V:205, R:135, I:10, K:907, nt:79; non-endemics rare and threatened worldwide – E:7, V:33, R:18, I:2 (world categories).

43 threatened plants are listed in Organización de los Estados Americanos (1967), cited in Appendix 1. Threatened plants are also mentioned in several papers in:

Prance, G.T. and Elias, T.S. (Eds) (1977), cited in Appendix 1. See in particular W.G. D'Arcy on endangered landscapes in the region (pp. 89-104), J.T. Mickel on rare and endangered ferns (pp. 323-328), H.E. Moore Jr. on endangerment in palms (pp. 267-282).

Laws Protecting Plants Fuller and Swift (1984, cited in Appendix 1) outline legal controls on the export of ornamental plants and their parts. They also report that the Departamento de Vida Silvestre is currently reviewing draft legislation to regulate exports of wild orchids. The U.S. Government has determined *Jatropha costaricensis*, a plant confined to Guanacaste Province in Costa Rica, as 'Endangered' under the U.S. Endangered Species Act.

Voluntary Organizations
Asociación Costarricense para la Conservación de la Naturaleza (ASCONA), Apdo 8-3790, 1000 San José.
Centro Agronómico Tropical de Investigación y Ensenanza (CATIE), Turrialba.
Fundación de Parques Nacionales, Apdo 236, Cod. 1002, San José.
Organization for Tropical Studies (OTS), Universidad de Costa Rica, San Pedro, San José.
Programa Patrimonio Natural de Costa Rica, c/o Fundación de Parques Nacionales, Apdo 236, Cod. 1002, San José; and Apdo 103, Plaza González Víquez, San José.
Tropical Science Center, Apdo 8-3870, San José.

Botanic Gardens
Lankester Botanical Garden, Escuela de Biologia Universidad de Costa Rica, Ciudad Universitaria "Rodrigo Facio", San José.

Useful Addresses
Herbário Nacional de Costa Rica, Museo Nacional, P.O. Box 749, San José.
CITES Management Authority: Dirección General Forestal, Departamento de Vida Silvestre, Ministerio de Agricultura y Ganadería, Apdo 10094, San José.
CITES Scientific Authority: Colegio de Biologos de Costa Rica, Universidad de Costa Rica, Ciudad Universitaria "Rodrigo Facio", San José.

Additional References
Beebe, S. (1984). A model for conservation. *The Nature Conservancy News* 34(1): 4-7.
Burger, W.C. (1980). Why are there so many kinds of flowering plants in Costa Rica? *Brenesia* 17: 371-388. (Summary of recent papers on floristic diversity in Costa Rica, with references.)
Gómez P., L.D. (1983a). *Vegetation and Climate of Costa Rica*, 2 vols. Editorial Universidad Estatal a Distancia (EUNED), San José. (18 maps, 1:200,000.)
Gómez P., L.D. (1983b). *Vegetation map of Costa Rica, 1:200,000*. Fundación Parques Nacionales, San José.
Hammel, B.E. and Grayum, M.H. (1982). Preliminary report on the Flora Project of La Selva Field Station, Costa Rica. *Ann. Missouri Bot. Gard.* 69: 420-425.

Janzen, D. (1983). *Costa Rican Natural History*. Univ. of Chicago Press, Illinois. 816 pp.

Knight, P. (1964). The great oaks of Costa Rica. *IUCN Bull., new series* No. 13, Oct/Dec. Pp. 6-7.

Tosi, J.A., Jr. (1969). *República de Costa Rica Mapa Ecológico*. Instituto Geográfico Nacional, San José.

Cuba

Area 114,524 sq. km

Population 9,966,000

Floristics 7000 species of gymnosperms and flowering plants of which 4000 are endemic (Conde, 1952); 6000 species, almost 50% endemic (Alain, 1962, pers. comm., quoted in Prance, 1977).

Vegetation Semi-desert of thornbush and savannah vegetation up to the mountain edges; siliceous savannah in the west; montane evergreen and tropical cloud and rain forest with pine forest in the east and in the west. Species-rich pine forests on serpentine soils. Submontane rain forest along north-east, and semi-dry mountain forest along the south-east coast, with mangrove and tropical salt marsh along the west coast. 11% forested (FAO, 1974, cited in Appendix 1); estimated rate of deforestation for broadleaved closed forest 20 sq. km/annum, from a total of 12,550 sq. km (FAO/UNEP, 1981); according to Myers (1980, cited in Appendix 1), about 16,000 sq. km of tropical moist forest remain.

Checklists and Floras Covered by the family and generic monographs of *Flora Neotropica* (cited in Appendix 1). The Flora is:

Leon, H. and Alain, H. (1946-63). *Flora de Cuba*, 5 vols and suppl. (by A.H. Liogier). Published variously: Vols 1-3 – Cultural, SA La Habana; Vol. 4 – Museo de Historia Natural de la Salle, La Habana; Vol. 5 – Universidad de Puerto Rico, Río Piedras. (In Spanish; some black and white photographs.) Additional species are listed as a series 'Novedades de la Flora Cubana' in various publications including *Rev. Soc. Cubana Bot.* 5 (1948); *Phytology, Bulgarian Academy of Sciences* 11: 47-53 (by B.P. Kitanov, 1979); *Contr. Mus. Hist. Nat. Col. la Salle*, No. 9: 1-24 (1950); *Candollea* 17: 99-111, 113-121 (1960); there are also Polish, E. German, Hungarian and Russian references subsequent to Vol. 5 of the Flora.)

Muñiz, O. and Borhidi, A. (1982). Catálogo de las Palmas de Cuba. *Acta Botanica Academiae Scientiarum Hungaricae* 28(3-4): 309-345.

See also:

Borhidi, A. and Kerezty, Z. (1979). New names and new species in the Flora of Cuba resp. Antilles. Continued as: New names and new species in the Flora of Cuba II, by A. Borhidi (1980); New names and new species in the Flora of Cuba and Antilles III, by A. Borhidi (1983). *Acta Botanica Academiae Scientiarum Hungaricae* 25(1-2): 1-37; 26(3-4): 255-275; 29(1-4): 181-215.

Borhidi, A. and Muñiz, O. (1979). Notas sobre taxones criticos o nuevos de la flora de Cuba. *Acta Botanica Academiae Scientiarum Hungaricae* 25(1-2): 39-52.

Duek, J.J. (1971). Lista de las especies Cubanas de Lycopodiophyta, Psilotophyta, Equisetophyta y Polyodiophyta (Pteridophyta). *Adansonia*, ser. 2, 11: 559-578, 717-731.

For an account of botanical work in Cuba, with bibliography of botanical papers 1960-1976, addresses of herbaria and reports of a new Flora, see:

Howard, R.A. (1977). Current work on the flora of Cuba - A commentary. *Taxon* 26: 417-423.

Current work on the flora of Cuba is also published in the following journals:

Revista del Jardín Botánico Nacional, Habana. 1980, Vol. 1-.
Wiss. Zeitschr. Univ. Jena Mat. -Naturwiss. (28: 541-724 in particular).

The papers of the 3rd Symposium on the Flora of Cuba, and a working report, will be published in *Feddes Repertorium* 96(7-10), due out by the end of 1985.

Information on Threatened Plants

Borhidi, A. and Muñiz, O. (1983). *Catálogo de Plantas Cubanas Amenazadas o Extinguidas*. Edit. Academia. 85 pp. (Lists 959 species of gymnosperms and flowering plants threatened or extinct, including 832 endemics, with their distribution by provinces and assignment into categories 'rare', 'endangered' and 'extinct' – non-compatible with IUCN categories.)

The IUCN Plant Red Data Book has two data sheets for Cuba, on *Cereus robinii* and *Microcycas calocoma*. Threatened plant conservation is also discussed in:

Howard, R.A. (1977). Conservation and the endangered species of plants in the Caribbean islands. In Prance, G.T. and Elias, T.S. (Eds), cited in Appendix 1. Pp. 105-11

Botanic Gardens

Jardín Botánico de Cienfuegos, Apto 414, Cienfuegos.
Jardín Botánico de la Habana, Calle 26 e/c Puentes Grandes y Ave. Boyeros, Habana.
Jardín Botánico Nacional de Cuba, Universidad de la Habana, Carretera del Ricio Km 3.5, Calabazar, Habana.

Additional References

Borhidi, A. and Muñiz, O. (1980). Die Vegetationskarte von Kuba. *Acta Botanica Academiae Scientiarum Hungaricae* 26(1-2): 25-53. (In German.)
Borhidi, A., Muñiz, O. and Del Risco, E. (1979). Clasificación fitocenologica de la vegetación de Cuba. *Acta Botanica Scientiarum Hungaricae* 25(3-4): 263-301. (In Spanish.)
Borhidi, A., Muñiz, O. and Del Risco, E. (1983). Plant communities of Cuba, 1. *Acta Botanica Hungarica* 29(1-4): 337-376.
Conde, J.A. (1952). *La Flora de Cuba*. Memorias de la Sociedad Cubana de Historia Natural. Univ. Habana. Organo Oficial de Museo Poev. Facultad de Ciencias. Vol. 21 no. 1. (In Spanish.)
Friedrich Schiller Universität (1979). Zur Flora von Kuba. *Wiss. Zeitschr. Univ. Jena Mat.-Naturwiss* 28: 541-724. (Collection of 28 papers on aspects of the Cuban flora, in German and Spanish, and bibliography of papers published since 1975 by members of the Flora-Cuba Project.)
Muñiz, O. (1970). Endemismo en la Flora. In *Atlas Nacional de Cuba*. Havana. (The Atlas contains a vegetation map and a lengthy discursive description of the vegetation.)

Prance, G.T. (1977). Floristic inventory of the tropics: Where do we stand? *Ann. Missouri Bot. Gard.* 64(4): 659-684.

Samek, V. (1968). La protección de la naturaleza en Cuba. *Ser. Transform. Natur.* 7: 1-23. Acad. Cienc. Cuba. (Not seen.)

Smith, Earl, E. (1954). *The Forests of Cuba.* Maria Moors Cabot Foundation Publication No. 2. 98 pp., 3 maps. (Includes classification of forests based on floristic, edaphic and moisture criteria; descriptions of regional forests with species lists; illus.)

Cyprus

Area 9254 sq. km

Population 659,000

Floristics c. 2000 vascular plant species, including naturalized aliens, calculated from the *Flora of Cyprus* (Meikle, 1977, 1985). 116 endemic vascular taxa (IUCN figures).

Vegetation Dominant natural vegetation outside agricultural land is heavily grazed garigue, with occasional patches of taller maquis. Remaining forest (c. 17% of land area, Meikle, 1977, 1985) restricted to the mountains: the precipitous limestone Kyrenia (northern) range supports large stands of *Pinus brutia/Cupressus sempervirens* forest on the upper slopes and well-developed maquis on much of the northern slope; the predominantly igneous Troödos (southern) range is also well forested with *Pinus brutia* on lower slopes, replaced by *Pinus nigra* ssp. *pallasiana* at higher levels. The Troödos also supports many Cypriot endemics. Other important areas of floristic diversity: Akamas and Karpas Peninsulas, in extreme north-west and north-east of the island respectively. The few remaining wetlands, especially near Phasouri and Syrianokhori, support an interesting aquatic flora.

Checklists and Floras Cyprus has benefited from one of the most detailed and comprehensive Floras of recent years:

Meikle, R.D. (1977, 1985). *Flora of Cyprus*, 2 vols. Bentham-Moxon Trust, Royal Botanic Gardens, Kew, Richmond, Surrey, U.K. 832 pp, 1137 pp.

Cyprus will also be covered under the *Med-Checklist* (cited in Appendix 1). See also:

Chapman, E.F. (1949). *Cyprus Trees and Shrubs.* Cyprus Government Printing Office, Nicosia. 88 pp. (Keys and descriptions.)

Osorio-Tafall, B.F. and Seraphim, G.M. (1973). *List of the Vascular Plants of Cyprus.* Ministry of Agriculture and Natural Resources, Nicosia.

Field-guides

Matthews, A. (1968). *Lilies of the Field: A Book of Cyprus Wild Flowers.* Published by the author, P.O. Box 180, Limassol. 54 pp. (Colour photographs with text describing c. 50 species, mostly common, a few endemic.)

Megaw, E. and Meikle, D. (1973). *Wild Flowers of Cyprus.* Phillimore, London. (A handsome quarto book of 41 colour paintings of Cypriot plants by Elektra Megaw, with short descriptive text by D. Meikle.)

Information on Threatened Plants No national Red Data Book, but the Ministry of Agriculture and Natural Resources have prepared a list of 41 'rare plant species' (M.A Daniel, 1982, pers. comm.).

Cyprus is also included in the European threatened plant list (Threatened Plants Unit, 1983, cited in Appendix 1). The section on Cyprus is reprinted in Leon, 1983. Latest IUCN statistics, based on the former: endemic taxa – E:10, V:9, R:22, I:5, K:24, nt:46.

Laws Protecting Plants No legislation directly protects wild plants in Cyprus, except those in State Forests which are protected by Forest Law No. 14 of 1967. Section 13(2) of this Law prohibits cutting, uprooting, collecting, or removal from State Forests of any produce without authorization. 'Forest produce' includes timber and branches and all parts of wild plants, mosses, fungi and lichens. According to the Ministry of Agriculture and Natural Resources, "almost all" the 41 species on their rare plant list occur in State Forests and so receive protection.

Outside these Forests only 4 species of crop plants have been granted any form of protection: *Rhus coriaria* (Sumach), used in tanning Moroccan leather; 2 species of sage, *Salvia fruticosa* (*S. cypria*, *S. triloba*) and *S. willeana* (*S. grandiflora* auct.).

Voluntary Organizations
Association for the Protection of the Cyprus Environment, P.O. Box 2444, Chanteclair Building, Nicosia.

Useful Addresses
Cyprus Forest Association, c/o Forest Department, Ministry of Agriculture and Natural Resources, Nicosia.
Cyprus Herbarium, Department of Agriculture, Nicosia.
Ministry of Agriculture and Natural Resources, Forest Department, Nicosia.
CITES Management and Scientific Authority: Ministry of Agriculture and Natural Resources, Nicosia.

Additional References
Holmboe, J. (1914). Studies on the vegetation of Cyprus. *Bergens Museums Skrifter. NY Raekke* 1(2). 344 pp.
Ionnides, O. (1973). Nature conservation in Cyprus. *Nature in Focus* 14: 16-17.
Leon, C. (1983). [Cyprus:] Important Botanical Areas of High Conservation Value. 14 pp. Unpublished report, available from IUCN-CMC.

Czechoslovakia

Area 127,870 sq. km

Population 15,588,000

Floristics 2600-2750 native vascular species, estimated by D.A. Webb (1978, cited in Appendix 1) from *Flora Europaea*; 10 endemics (IUCN figures), 'at least 15' according to J. Holub (1984, *in litt.*). Centres of endemism include the Tatra Mts, Krkonoše Mts, Velká Fatra Mts and the karstic region of Slovenský Kras. Elements: Central European, Pannonian and alpine.

Vegetation Largely an agricultural and heavily industrialized landscape. The remaining area, supporting semi-natural vegetation, largely covered by forests, mostly of pine, oak and beech. Beechwoods well-developed, in Slovakia forming 32% of forest cover (Polunin and Walters, 1985, cited in Appendix 1). Altitudinal vegetation zones still apparent: at lowest levels relics of riverine forest give way to broadleaved deciduous woodland, but the latter extensively re-afforested by spruce and pine; at higher levels, mixed coniferous and deciduous woodland; in subalpine zone montane pine, giving way to alpine meadows. In warmer areas, steppe vegetation.

Checklists and Floras Covered by the completed *Flora Europaea* (Tutin *et al.*, 1964-1980, cited in Appendix 1). National Floras:

Dostál, J. (1948-1950). *Květena ČSR*. (Flora of Czechoslovakia.) Přírodovědecké Nakladatelství, Praha. 2269 pp. (An illustrated key for all vascular plants; revised edition in prep.)

Dostál, J. (1958). *Klíč k úplné květeně ČSR* (Key to the complete flora of Czechoslovakia), 2nd Ed. Československá Akademie Věd, Prague. 982 pp. (Essentially a revised and condensed version of Dostál, 1948-1950; illus.)

Floras are also being prepared for the 2 Socialist Republics that comprise Czechoslovakia, namely the Czech Socialist Republic (ČSR – Bohemia and Moravia) and the Slovak Socialist Republic (SSR – Slovakia):

Futák, J. (Ed.) (1966-). *Flóra Slovenska* (Flora of Slovakia), 4 vols. Slovenská Akadémia and Veda, Bratislava. (Incomplete; 1 – a morphological vocabulary; 2 – pteridophytes and gymnosperms; 3 – angiosperms, edited by J. Futák and L. Bertová; 4(1) – angiosperms, edited by L. Bertová.)

Hejný, S. and Slavík, B. (Eds) (in prep.). *Flóra ČSR* (Flora of the Czech Socialist Republic), 8 vols planned. Academia, Prague.

Checklists:

Dostál, J. (1982). *Seznam cévnatých rostlin květeny československé*. Pražská botanická zahrada, Praha. 408 pp.

Novacký, I.M. (1954). *Slovenská Botanická Nomenklatura*. Slovenská Akadémia Vied, Bratislava. 227 pp.

See also:

Holub, J. (1974). Taxonomic and floristic progress on the Czechoslovak flora and the contribution of Czechoslovak authors to knowledge of the European flora (1961-1972). *Mem. Soc. Brot.* 24(1): 173-352.

Slavík, B. (1972). Preparation of the phytogeographical atlas of the Czech Socialist Republic. *Acta Ecol. Natur. Region* 1: 24-28. (In Czech.)

Bibliographies:

Botanical Institute of the Czechoslovak Academy of Sciences (Ed.) (1978, 1980 and 1982). *Bibliographia botanica Cechoslovaca 1973-1974, 1975-1976 and 1977-1978*, 12 vols. Botanický Ústav ČSAV, Průhonice. 563, 272 and 590 pp. (2 consecutive authors: Z. Neuhčuslová-Novotná and D. Guthová-Jarkovská.)

Futák, J. and Domin, K. (1960). *Bibliografia k flóre ČSR*. Slovenská Akadémia Vied, Bratislava. 883 pp. (References to botanical literature published up to 1952.)

National botanical journals: *Preslia*, Journal of the Czechoslovak Botanical Society
(address below) and *Zprávy Československé botanické společnosti*, při ČSAV, Praha
(summaries in English).

Field-guides

Májovský, J. and Krejča, J. (1966-1977). *Obrázková květena Slovenska* (Illustrated
   Flora of Slovakia), 5 vols. Obzor, Bratislava. (Colour illus. for each species.)

Novák, F.A. and Svolinský, K. (1940-1946). *Rostliny* (Wild flowers), 2 vols. Vesmír,
   Praha.

   Information on Threatened Plants Published threatened plant lists for both ČSR
and SSR and a Red Data Book for ČSR are listed below. These will provide the basis for
the national Red Data Book.

Czech Socialist Republic:

Čeřovský, J., Holub, J. and Procházka, F. (1979). Červený seznam flóry ČSR (The
   Red List of the ČSR Flora). *Památky a Příroda* 4: 361-378. (First draft list of
   threatened vascular plants in the Czech Socialist Republic; includes 37 'extinct' and
   39 'missing' taxa, 267 'critically threatened' plants, 240 'strongly threatened', 239
   'threatened' and 330 rare taxa 'in need of further study'; English summary.)

Holub, J., Procházka, F. and Čeřovský, J. (1979). Seznam vyhynulých, endemických a
   ohrožených taxonů vyšších rostlin květeny ČSR (1. verze). (List of extinct, endemic
   and threatened taxa of vascular plants of the flora of the Czech Socialist Republic
   (first draft).) *Preslia* 51(3): 213-237. (English abstract and summary; same list as
   previous paper; reviewed in *Threatened Plants Committee – Newsletter*, No. 6: 13,
   1980.)

Procházka, F., Čeřovský, J. and Holub, J. (1983). *Chráněné a ohrožené druhy květeny
   ČSR* (Protected and endangered species in the flora of ČSR). UDPM, Praha.
   103 pp.

Slovak Socialist Republic:

Maglocký, S. (1983). Zoznam vyhynutých endemických a ohrozených taxónov vyšších
   rastlín flóry Slovenska (List of extinct, endemic and threatened taxa of vascular
   plants of the flora of Slovakia). *Biológia* 38: 825-852.

See also:

Čeřovský, J. and Podhájská, Z. (1981). *Registrace kriticky ohrožených druhů vyšších
   rostlin v ČSR* (Registration of critically endangered plant species in the Czech
   Socialist Republic). *Památky a Příroda* 6: 577-583.

Hendrych, R. (1977). Zaniklé nebo nezvěstné rostliny naší květeny (Extinct or missing
   plants of our flora). *Živa* 25(3): 84-85.

Holub, J. (Ed.) (1981). *Mizející flóra a ochrana fytogenofondu v ČSSR* (The vanishing
   flora and protection of the gene pool in Czechoslovakia). Proceedings from a
   conference. Studie ČSAV, 20. Academia, Prague. (See for example papers in Slovak
   by: J. Futák on endemic plants of the SSR (pp. 45-49); in Czech by E. Hadač on
   endemic plants of the ČSR (pp. 41-43); J. Holub on protection of the floristic
   diversity from the aspect of taxonomy and phytogeography (pp. 27-39); K. Kubát on
   threatened species in north-west Bohemia (pp. 133-137); F. Procházka on extinct
   species in the Czechoslovak flora (pp. 13-15); and L. Vaněčková on extinct and
   endangered species in the Moravian karst (pp. 139-141).)

Somšák, L. (1977). *Ohrozené a zriedkavé taxony horských a vysokohorských poloh Slovenska* (The threatened and rare taxa of the mountain range of Slovakia). Bratislava.

Since 1981, the Nature Conservation Section of the State Institute for Protection of Monuments and Conservation of Nature (address below) have co-ordinated a project entitled Conservation of Rare and Endangered Plants and Animal Species. Aims include issuing new species conservation decrees, ensuring all critically endangered species are safeguarded in protected areas, re-introduction, using rare and threatened plants in soil reclamation projects and *ex situ* conservation in botanic gardens (Čeřovský, 1982).

Included in the European threatened plant list (Threatened Plants Unit, 1983, cited in Appendix 1); latest IUCN statistics, based upon this work: endemic taxa – E:2, R:3, I:2, nt:3; non-endemics rare or threatened worldwide – E:1, V:19, R:11, I:6 (world categories).

Laws Protecting Plants Legislation prohibiting the uprooting of selected wild species was passed for ČSR by Decree No. 54 of 14 April 1958 and for SSR by Decree No. 211 of 23 December 1958. In the ČSR, 108 taxa of ferns and flowering plants are so protected; 100 of them receive complete protection and 8 partial protection. Some complete genera are also covered (e.g. *Aconitum, Orchis, Pulsatilla, Stipa*). In the SSR, 88 taxa receive complete protection and 8 partial protection. In addition 40 species and genera (*Oxytropis, Salix*) and one family (Orchidaceae) receive special legal protection within the High Tatra National Park.

Furthermore, the districts of Děčín, Litoměřice and Ústí nad Labem in the North Bohemian region have passed their own legislation, which includes the protection of several species not covered by the national legislation.

Relevant literature:

Bosacková, E. (n.d.). *Chránené rastliny na Slovensku a podmienky ich ochrany* (Protected plants in Slovakia and protective measures). Vydalo VPL pre Povereníctvo SNR pre kultúru a informácie. 6 pp.
Magic, D., Bosacková, E., Krejča, J. and Usák, O. (1979). *Atlas chránených rastlín* (Field-guide of protected plants). Obzor, Bratislava. 260 pp. (Slovakia only.)
Procházka, F. *et al.* (1983) cited in full above.
Randuska, D. and Krizo, M. (1983). *Chránené rastliny* (Protected plants). Príroda, Bratislava. 430 pp.
Somšák, L. and Slivka, D. (1981). *Chránené rastliny Slovenska* (Protected plants in Slovakia), 2nd Ed. Bratislava. (Illus.)
Veselý, J. (1961). *Chráněné rostliny* (Protected Plants), 2nd Ed. Orbis, Praha. 85 pp. (Lists protected species and includes conservation status; colour illus.)

Voluntary Organizations
Československá botanická společnost (Czechoslovak Botanical Society), Benátská 2, 128 01 Prague 2.
Český svaz ochránců přírody (Czech Union of Nature Conservationists), Staroměstské nám. 12, 110 00 Prague 1.
Slovenský zvaz ochrancov prírody a krajiny (Slovak Union of Nature and Landscape Conservationists), Leningradská 1, 811 01 Bratislava.

Botanic Gardens Many, as listed in Henderson (1983), cited in Appendix 1, although none subscribe to the IUCN Botanic Gardens Conservation Co-ordinating Body. Relevant references:

Setelová, V. *et al.* (1977). *Botanické zahrady* (Botanic Gardens). SPN, Prague. 280 pp.

Somšák, L. (1979). The role of botanic gardens in the conservation of rare and threatened plants in Slovakia. In Synge, H. and Townsend, H. (Eds), *Survival or Extinction. Proceedings of a Conference: The Practical Role of Botanic Gardens in the Conservation of Rare and Threatened Plants*, 11-17 September 1978. Bentham-Moxon Trust, Kew. Pp. 107-112.

Vyskočil, J. (1980). *Chráněné rostliny v botanických zahradách, jejich pěstování, zakládání sbírek a využívaní v kulturně výchovné činnosti* (Protected plants in botanic gardens, their cultivation and use in educational activities). Proceedings of a symposium 16-17 October 1980. Pražská botanická zahrada, Praha-Troja. 135 pp.

Useful Addresses

Botanický ústav ČSAV (Botanical Institute, Czechoslovak Academy of Sciences), 252 43 Průhonice.

Institute of Experimental Biology and Ecology, Slovak Academy of Sciences, 885 34 Bratislava, Sienkiewiczova 1.

Státní ústav památkové péče a ochrany přírody (State Institute for Protection of Monuments and Conservation of Nature), Valdštejnské nám. 4, 118 01 Prague 1.

Ustredie Štátnej ochrany prírody (Centre of State Nature Conservancy – Slovakia), 031 01 Liptovský Mikuláš.

Additional References

Čeřovský, J. (1982). Botanisch-okologische Probleme des Arten-schutzes in der CSSR unter Berucksichtigung der praktischen Naturschutzarbeit (Botanical and ecological problems of species preservation in the ČSSR with regard to practical conservation work). *Berichte der ANL* 6: 90-92.

Hendrych, R. (1981). Bemerkungen zum Endemismus in der Flora der Tschechoslowakei (Observations on endemism in the flora of Czechoslovakia). *Preslia* 53(2): 97-120. (In German; English abstract; maps.)

Holub, J., Hejný, S., Moravec, J. and Neuhčusl, R. (1967). Ubersicht der hoheren Vegetationseinheiten der Tschechoslowakei. *Rozprávy ČSAV, Rada Matematických a Přírodních Věd* 77(3): 1-75. (Phytosociological account.)

Plesník, P. (1976). *Die Vegetationsstufen in der Slowakei.* 18 pp. (Maps.)

Procházka, F. (1980). *Současné změny východočeské flóry a poznámky k rozšírení chráněných druhů rostlin* (Contemporary changes in the flora of eastern Bohemia and notes on the distribution of protected species). Krajské Muzeum Východních Čech, Hradec Králové. 134 pp.

Veselý, J. (1961). *Príroda Československa, její vývoj a ochrana* (Nature in Czechoslovakia, its development and conservation). Osveta, Bratislava. 146 pp.

Denmark

Area 43,075 sq. km

Population 5,141,000

Floristics c. 1000 native vascular species (Løjtnant, 1984, *in litt.*); c. 1350-1450 estimated by D.A. Webb (1978, cited in Appendix 1) from *Flora Europaea*; this discrepancy principally due to a recent assessment that many species in Denmark, hitherto believed native, are now recognized as longstanding introductions; 1 endemic species, 1 endemic subspecies (IUCN figures). Elements: Atlantic.

Vegetation 90% of land surface extensively modified, 70% by agriculture. Remaining pockets of semi-natural vegetation include forests of oak (*Quercus petraea* and *Q. robur*) and beech in the south and east, sand-dunes and salt-marshes mainly along the west coast, scattered inland heaths, peat bogs, swamps and lakes. Wetlands are considered to have suffered the greatest species loss and disturbance. Forests occupy about 10%, but most are spruce and pine plantations (Poore and Gryn-Ambroes, 1980, cited in Appendix 1).

Checklists and Floras Included in the completed *Flora Europaea* (Tutin *et al.*, 1964-1980, cited in Appendix 1). National Floras:

Christiansen, M.S. (1958-1961). *Danmarks Vilde Planter*, 2 vols. Branner og Korch, København. (Colour and line drawings.)

Hansen, K. (Ed.) (1984). *Dansk Feltflora*, 2nd Ed. Gyldendal, København. 757 pp. (Illus.)

Raunkiaer, C. (1950). *Dansk Ekskursions-Flora*, 7th Ed. by K. Wiinstedt. Gyldendal, København. 380 pp.

Rostrup, E. and Jørgensen, C.A. (1973). *Den Danske Flora, en Populaer Vejledning til at Laere de Danske Planter at Kende*, 20th Ed. revised by A. Hansen. Gyldendal, Copenhagen. 664 pp. (Line drawings.)

See also Lindman (1964), cited in Appendix 1, and:

Hagerup, O. and Petersson, V. (1956-1960). *Botanisk Atlas*, 2 vols. Munksgaard, København. (Line drawings only; 1 – angiosperms; 2 – bryophytes, pteridophytes, gymnosperms; English edition published in 1963, Copenhagen.)

For a regional plant atlas see Hultén (1971), cited in Appendix 1, and the results of the Danish Topographical-Botanical Investigation, published in *Botanisk Tidskrift*, 1935 onwards, containing dot maps for most Danish higher plants.

Relevant journals: *Botanisk Tidskrift*, København (now replaced by *Nordic Journal of Botany*); *Flora og Fauna*, Naturhistorisk Museum, Aarhus; *Urt* (popular journal published by the Danish Botanical Society; addresses below).

Information on Threatened Plants National plant Red Data Book:

Løjtnant, B. and Worsøe, E. (1977). *Foreløbig Status over den Danske Flora*. Reports from the Botanical Institute University of Aarhus, No. 2. 341 pp. (Detailed survey of status of over 200 native vascular plants in Denmark; line drawings; English summary.)

For a revision of the above see:

Løjtnant, B. (1985). *Rødliste over Danmarks Karplanter*. København. 23 pp. (A revised threatened plant list of Danish higher plants.)

See also:

Løjtnant, B. (1980). Status over den danske flora. In Møller, H.S. *et al.* (Eds), *Status over den Danske Plante – og Dyreverden*. Proceedings of a Symposium 18-20 April 1980. Miljøministeriet, Fredningsstyrelsen. Pp. 327-341. (Describes conservation status and threats to the flora.)

Denmark is included in the Nordic Council of Ministers' threatened plant list and supplements (Ovesen *et al.*, 1978 and 1982) and in the European list (Threatened Plants Unit, 1983), both cited in Appendix 1; latest IUCN statistics, based upon the latter: endemic taxa – R:2; non-endemics rare or threatened worldwide – V:6, R:1, I:2 (world categories).

In 1982 IUCN, under contract to the EEC through the U.K. Nature Conservancy Council, prepared a report (unpublished), *Threatened Plants, Amphibians and Reptiles, and Mammals (excluding Marine Species and Bats) of the European Economic Community*, which included a data sheet on *Thesium ebracteatum*, Extinct in Denmark, status unknown worldwide.

A computerized biological data-base is being developed by the National Agency for the Protection of Nature, Monuments and Sites, København (address below). This will include data about protected areas, threatened plants and habitats, as part of a larger biological conservation data centre.

Laws Protecting Plants The Conservation of Nature Act (No. 297 of 26 June 1975) provides total protection for 2 plant species and partial protection for 26 other plant taxa. The Ministry of the Environment may order the protection, throughout the country or in specified areas, of any plant species. Species in nature reserves may also be protected against picking, digging, etc., as part of nature reserve legislation. Some species are protected administratively where they grow on land owned by the State and managed by the National Agency for forests. For more details see Koester (1980), cited in Appendix 1.

Voluntary Organizations
Danmarks Botanisk Forening (Danish Botanical Society), Sølvgade 83, 1307 København.
Dansk Naturfredningsforening (Danish Nature Conservation Society), Frederiksberg Runddel 1, 2000 København.
Verdensnaturfonden (WWF-Denmark), H.C. Andersens Boulevard 31, 1553 København.

Botanic Gardens
Botanical Institute, University of Aarhus, 68 Nordlandsvej, 8240 Risskov.
Botanisk Have, Stadsgartnerens kontor, Viborgvej 144, 8210 Aarhus V.
Den Kgl. Veterinaer-og Landbohøjskoles Have, Bülowsvej 13, 1870 København V.
Forstbotanisk Have (Forest Botanic Garden), 2920 Charlottenlund.
Forest Botanical Garden, Aarhus.
Hørsholm Arboretum, 2970 Hørsholm.
Københavns Universitets Botaniske Have, Ø Farimagsgade 2B, 1353 København K.

Useful Addresses
National Agency for the Protection of Nature, Monuments and Sites, Ministry of the Environment, 13 Amaliegade, 1256 København.
Naturhistorisk Museum, Universitetsparken, 8000 Aarhus C.
CITES Management and Scientific Authorities: Fredningsstyrelsen, Miljøministeriet, 13 Amaliegade, 1256 København.

Additional References
Gravesen, P. (1976-1983). *Foreløbig Oversigt over Botaniske Lokaliteter*, 4 vols. Miljøministeriets Fredningsstyrelse i Samarbejde med Dansk Botanisk Forening, København. (1 – Sjaelland; 2 – Den Fynske Øgruppe; 3 – Lolland, Falster, Møn og Bornholm; 4 – Sønderjyllands Amt (S. Jutland); describes hundreds of botanical

localities and assesses their conservation value as part of a long-term monitoring programme; covers flowering plants, mosses, fungi, lichens and algae; protected species; photographs; maps.)

Hansen, A. (1981). Dansk Botanisk Litteratur i 1975, 1976, 1977, 1978 og 1979. *Bot. Tidssk*. 75(4): 221-275. (Review of recent botanical literature.)

D'Entrecasteaux Islands

Volcanic islands reaching 2400 m, situated c 30 km to the north of the eastern tip of Papua New Guinea. Area 3142 sq. km; population 34,400 (1971 estimate, *Encyclopedia Britannica*, 1974). The islands are part of Papua New Guinea.

The 3 principal islands of Normanby, Fergusson and Goodenough still have extensive primary tropical rain forests. The archipelago is included on the *Vegetation Map of Malaysia* (van Steenis, 1958) and on the vegetation map of Malesia (Whitmore, 1984), both covering the *Flora Malesiana* region at scale 1:5,000,000 and cited in Appendix 1.

No figure available for the size of the flora. No information on threatened plants.

Djibouti

Area 23,000 sq. km

Population 354,000

Floristics 534 species (Bavazzano, 1972). Degree of endemism unknown. In general flora is poorly known, but likely to be rich especially in the Goda Mountains (Verdcourt, 1968).

Floristic affinities with Somalia-Masai region.

Vegetation Mostly semi-desert grassland, shrubland and succulent scrub. Small areas of mangrove vegetation and coastal desert at the coast. Small patches of montane dry evergreen forest in Dai area of the Goda Mountains.

For vegetation map see White (1983), cited in Appendix 1.

Checklists and Floras Djibouti is included in *Enumeratio Plantarum Aethiopiae Spermatophyta* (Cufodontis, 1953-1972), and in *Adumbratio Florae Aethiopicae*, both cited in Appendix 1. See also:

Bavazzano, R. (1972). Contributo alla conoscenza della flora del Territorio Francese degli Afar e degli Issa. *Webbia* 26: 267-364. (Short diagnoses, specimen citations.)

Chevalier, A. (1939). La Somalie française. Sa flore et ses productions végétales. *Revue Bot. Appl. Agric. Trop.* 19: 663-687.

Information on Threatened Plants Two species which occur in Djibouti are included in *The IUCN Plant Red Data Book* (1978): *Dracaena ombet* and *Livistona carinensis* (Syn. *Wissmannia carinensis*).

Detailed information is lacking, but desertification is threatening the succulent scrub.

Additional References

Chedeville, E. (1972). La végétation du Territoire français des Afars et des Issas. *Webbia* 26: 243-266.

Pichi-Sermolli, R.E.G. (1957). Una carta geobotanica dell'Africa orientale (Eritrea, Ethiopia, Somalia). *Webbia* 13: 15-132. (Includes map 1:5,000,000.)

Verdcourt, B. (1968). French Somaliland. In Hedberg, I. and O. (1968), cited in Appendix 1. Pp. 140-141.

Dominica

The most mountainous island in the Lesser Antilles, in the Windward group, equidistant between Guadeloupe and Martinique, 153 km south of Antigua; 47 km long by 24 km wide; fertile and volcanic.

Area 751 sq. km

Population 77,000

Floristics About 1600 species of vascular plants (D.H. Nicolson, 1984, *in litt.*). Nicolson also reports 6 species and 2 varieties endemic to the island (1 fern, 1 monocotyledon and 6 dicotyledons). Other reported endemics have become synonyms of more widespread species or have been recently found on neighbouring islands.

Vegetation In the interior undisturbed, primary rain forest and lower montane rain forest, surrounded by a broad intermediate zone of cut-over secondary forest; on highest peaks elfin woodland; on steep slopes palm brakes; on the west (leeward) coast, a belt of dry scrub woodland and, north of St Joseph, grassland and open scrub with grass; at river mouths in north, swamp (*Lonchocarpus*) forest; mangrove rare, recently discovered in Cabrits swamp. (Mainly from Beard, 1949, cited in Appendix 1.) 46.7% forested, the highest percentage in the Caribbean, according to FAO (1974, cited in Appendix 1). The 6840-ha Morne Trois Pitons National Park, in the south of Dominica, includes elfin woodland, rain forest and secondary forest, and conserves the largest area of such forest in the Lesser Antilles (Protected Areas Data Unit).

Checklists and Floras Covered by the *Flora of the Lesser Antilles, Leeward and Windward Islands* (only monocotyledons and ferns published so far, Howard, 1974-, cited in Appendix 1) and by the family and generic monographs of *Flora Neotropica* (cited in Appendix 1).

A Flora of Dominica, covering dicotyledons, has been prepared by D.H. Nicolson and collaborators at the Smithsonian Institution, Washington, D.C. and has been submitted for publication; one family (Compositae) still to be written.

Ferns, gymnosperms and monocotyledons are covered in the checklist:

Hodge, W.H. (1954). Flora of Dominica, B.W.I., Part 1. *Lloydia* 17 (1-3): 1-238.

Also relevant:

Beard, J.S. (1944). Provisional list of trees and shrubs of the Lesser Antilles. *Caribbean Forester* 5(2): 48-67. (428 species assigned in a table to individual islands).

Hodge, W.H. (1953). The orchids of Dominica, BWI. *American Orchid Soc. Bull.* 22(12): 891-904.

Stehlé, H. and Stehlé, M. (1947). Liste complémentaire des arbres et arbustes des petites Antilles. *Caribbean Forester* 8: 91-123. (A further 328 species to Beard, 1944, in similar format.)

There are also various papers on the botany of Dominica in *Smithsonian Contributions to Botany*, particularly dealing with Algae, Lichens and Fungi.

Local botanical activity is centered at the facilities of the Dominica National Park Headquarters, who have produced articles on vegetation of the Park.

Field-guides

Honychurch, P.N. (1980). *Caribbean Wild Plants and Their Uses*. Published by the author, Roseau, Dominica. 163 pp. (Conspicuous plants only.)

Information on Threatened Plants Threatened plant conservation is discussed in:

Howard, R.A. (1977). Conservation and endangered species of plants in the Caribbean Islands. In Prance, G.T. and Elias, T.S. (Eds), cited in Appendix 1. Pp. 105-114.

Botanic Gardens

Botanic Gardens, Roseau. (Largely devastated by Hurricane David in 1978, now recuperating.)

Useful Addresses

Dominica National Park Headquarters, Botanic Gardens, Roseau.
Forestry Department, Botanic Gardens, Roseau.

Additional References

Anon. (1970). *Dominica: A chance for a choice*. The Conservation Foundation, Washington, D.C. 48 pp. Some considerations and recommendations on conservation of the island's natural resources.

Hodge, W.H. and Taylor, D. (1957). The ethnobotany of the Island Caribs of Dominica. *Webbia* 12(2): 513-644.

Shillingford, C.A. (1968). Climax Forest in Dominica. M.Sc. Thesis, University of the West Indies, Mona, Jamaica. (Comparative study of 2 examples of lowland rain forest at D'Leau Gommier and Terre Ferme.)

Thorsell, J.W. and Wood, G. (1976). Dominica's Morne Trois Pitons National Park. *Nature Canada* 5(4): 14-16, 33-34.

Weber, B.E. (1973). Dominica National Park. Dept. of Recreation Resources, College of Forestry and Natural Resources, Colorado State University. (Thesis.) (Lists some plants endemic to Dominica in Table 3, p. 57.)

Dominican Republic

A mountainous country consisting of the eastern two-thirds of the island of Hispaniola; west of Puerto Rico and east of Cuba.

Area 48,442 sq. km

Population 6,101,000

Floristics No figures for Dominican Republic; Hispaniola has an estimated 5000 species: 7 gymnosperms, 1087 monocotyledons and 3900 dicotyledons; with 1800 endemic species (Liogier, 1984).

Vegetation In the centre of the island, along the east-west mountain ranges moist forest, low moist forest and high mountain hardwood forest; *Pinus occidentalis* dominant along the central ridge; extensive dry forest along the northern and southern lowlands, arid in parts, with savannah type vegetation; stands of tree cacti and palms in places due to heavy logging of hardwoods. Mangrove swamps best developed along the north-east coast at Samana Bay where the low moist forest comes down to sea level. 22.7% forested (FAO, 1974, cited in Appendix 1); estimated rate of deforestation for closed broadleaved forest 25 sq. km/annum, out of a total of 4440 sq. km (FAO/UNEP, 1981); according to Myers (1980) (cited in Appendix 1), c. 11,000 sq. km of tropical moist forest, most disrupted or degraded.

Checklists and Floras Covered by the family and generic monographs of *Flora Neotropica* (cited in Appendix 1).

Liogier, A.H. (1981). Flora of Hispaniola. Part 1. *Phytologia Memoirs* 3: 1-218. (In Spanish, illus.)

Liogier, A.H. (1982, 1984). *La Flora de la Española*, 2 vols published, the third in press. San Pedro de Macorís. 317 pp., 420 pp., illus.

Moscoso, R.M. (1943). *Catalogus Florae Domingensis*. New York. 732 pp. (In Spanish; checklist of gymnosperms and flowering plants.)

The following provide additional information:

Alvarez, V. (1983). *Manglares de República Dominicana*. Contribuciones 53. CIBIMA/UASD – see Useful Addresses, below. (Describes mangroves.)

Dod, D.D. (1978-). Orquídeas Dominicanas Nuevas I-III. *Moscosoa* 1(1): 50-54; 1(2): 39-54; 1(3): 49-63.

Jiménez, J. de J. (1963-1967). Suplemento no. 1 al Catalogus Florae Domingensis del Prof. Rafael M. Moscoso. *Archiv. Bot. Biogeogr. Ital.* 39: 81-132; 40: 54-149; 41: 47-87; 42: 46-97 and 107-129; 43: 1-18.

Jiménez, J. de J. (1975). Apuntes para la flora de Santo Domingo (Hispaniola) Novedades III. *Anuario Acad. Ciencias República Dominicana* 1(1): 93-132a.

Liogier, A.H. (1971a). Novitates Antillanae. IV. *Mem. N.Y. Bot. Gard.* 21: 107-157.

Liogier, A.H. (1971b). Novitates Antillanae. V. Miscellaneous new species from the Dominican Republic. *Phytologia* 22(3): 163-174.

Liogier, A.H. (1973). Novitates Antillanae. VI. *Phytologia* 25(5): 265-280.

Liogier, A.H. (1976). Novitates Antillanae. VII. Plantas nuevas de la Española. *Moscosoa* 1(1): 16-49.

The botanical journal *Moscosoa* includes reports of new taxa, of new records and other papers on the flora and vegetation of the Dominican Republic and Haiti. It is published by the Jardín Botánico Nacional 'Dr Rafael M. Moscoso' – see Botanic Gardens, below.

Information on Threatened Plants

Jiménez, J. de J. (1978). Lista tentativa de plantas de la República Dominicana que deben protegerse para evitar su extinción. *Coloquio Internacional sobre la practica de la conservación*, Santo Domingo. CIBIMA/UASD – see Useful Addresses, below. (In Spanish; lists 133 species of threatened flowering plants, of which 49 are endemic.)

Dr A.H. Liogier has prepared a lengthy list of endangered plants; this is not published.

The IUCN Plant Red Data Book has one data sheet for the Dominican Republic, on *Pseudophoenix ekmanii*. Threatened plant conservation is also discussed in:

Howard, R.A. (1977). Conservation and the endangered species of plants in the Caribbean Islands. In Prance, G.T. and Elias, T.S. (Eds), cited in Appendix 1. Pp. 105-114.

Voluntary Organizations
Sociedad Dominicana de Orquidiología, c/o Jardín Botánico Nacional "Dr Rafael M. Moscoso", Apto 21-9, Santo Domingo.
Sociedad Ecológica del Cibao, Santiago.

Botanic Gardens
Jardín Botánico Nacional 'Dr Rafael M. Moscoso', Apto 21-9, Santo Domingo.

Useful Addresses
Centro de Investigaciones de Biologia Marina, Universidad Autonoma de Santo Domingo, República Dominicana (CIBIMA/UASD), Jonas E. Salk 56, Santo Domingo.
Herbario Dr José de Js. Jiménez Almonte, Universidad Católica Madre y Maestra, Santiago.

Additional References
Hartschorn, G. *et al.* (1981). *The Dominican Republic, country environmental profile, a field study*. AID Contract No. AID/SOD/PDC-C-0247. JRB Associates, 8400 Westpark Drive, Mclean, Virginia 22102, U.S.A. 109 pp.
Holdridge, L.R., (1945). A brief sketch of the Flora of Hispaniola. In Verdoorn F. (Ed.), cited in Appendix 1. Pp. 76-78.
Jiménez, J. de J. and Liogier, A.H. (1977). Adiciones a los nombros vulgares de las Plantas en la República Dominicana. *Moscosoa* 1(2): 9-21. (See Liogier, 1974.)
Liogier, A.H. (1974). *Diccionario botánico de nombres vulgares de la Española*. Jardín Botánico Dr R. Moscoso, Santo Domingo. 813 pp.
Liogier, A. (1984). La Flora de la Española: sus principales caraterísticas. *2da. Joranda Científica Academia de Ciencias de la República Dominicana*. Santo Domingo.
Zanoni, T.A., Long, C.R. and Mckiernan, G. (1984). Bibliografía de la flora y de la vegetación de la Isla Española. *Moscosoa* 3: 1-61. (Extensive annotated bibliography of the flora and the vegetation of Hispaniola.)

Easter Island

Easter Island (27°S, 109°30'W) is a triangular volcanic outcrop in the western Pacific Ocean 3700 km west of Chile, of which it is a dependency. It is also known as Rapa-Nui and Isla de Pascua. Area 117 sq. km; population 1400 (1971 estimate). The highest point is Mt Terevaka (601 m), part of the extinct Rano Aroi volcano in the north. Rana Kao (457 m) and Rano Raraku (427 m) form the south-west and south-east parts of the island.

Almost the entire population lives at Hanga-Roa on the west coast. Rapa-Nui National Park, established in 1935, covers 68 sq. km mainly around the coast.

30 native flowering plant species of which 3 grasses and *Sophora toromiro* endemic (Skottsberg, 1922). 12 species of ferns of which 2 endemic (Christensen and Skottsberg, 1920). Most genera and species have very wide distributions (van Balgooy, 1971, cited in Appendix 1).

The vegetation is mainly *Sporobolus* and *Stipa* grassland. *Sophora toromiro* is the only tree recorded on the island in historic times. Undoubtedly there were other trees before the natural vegetation was modified by fires, timber cutting and the introduction of sheep.

The IUCN Plant Red Data Book (1978) included *Sophora toromiro* as "probably Extinct", the last tree on the island having died before 1962. Subsequently it was discovered that small plants were in cultivation, principally at Göteborg Botanic Garden, Sweden (see *Threatened Plants Committee – Newsletter* 5: 1-2, 1980).

References
Skottsberg, C. (Ed.) (1920-1956). *The Natural History of Juan Fernandez and Easter Island*, 3 vols. Almqvist and Wiksells Boktryckeri AB, Uppsala. (1 – Physical features, geology; 2 – botany; 3 – zoology. See in particular C. Christensen and C. Skottsberg, 1920, on the ferns of Easter Island, *ibid*. 2: 47-53; C. Skottsberg, 1922, on phanerogams, *ibid*. 2: 61-84; C. Skottsberg, 1951, a supplement to the pteridophytes and phanerogams, *ibid*. 2: 763-792.)

Ecuador

Area 461,477 sq. km

Population 9,090,000

Floristics Dodson and Gentry (1978) quote estimates ranging from 10,000 to 20,000 species of vascular plants. Many scientists consider Ecuador to have more plants per unit area than any other country in South America; this is demonstrated by the over 1250 species from 136 families recorded in 100 of the 167 ha plot of Pacific lowland rain forest at Río Palenque Science Center; 43 are known only from the site (Dodson and Gentry, 1978) and, with subsequent work, about 100 are newly described (A. Gentry, 1984, pers. comm.). Río Palenque is within the Chocó phytogeographic region, "that part of the coastal lowlands of western Colombia and northwestern Ecuador covered by wet and moist forest vegetation" and believed to be exceptionally rich in both endemics and other species (Gentry, 1982).

Vegetation Between the Andes and the Pacific Ocean, desert and semi-desert, savanna, deciduous forest (dominated by thorny leguminous trees with cacti), semi-deciduous forest (mostly now destroyed) and in the north lowland rain forest. Gentry (1977) separates the lowland coastal forest into a narrow strip of wet forest along the base of the Andes (originally small in extent, now critically endangered, the minute Río Palenque site (see above) being a rare survivor), and the coastal moist forest, more extensive but with fewer endemics. In the Andes itself, lower montane rain forest (700-2500 m), cloud forest (2500-3400 m), grass páramos (3400-4000 m), shrub and cushion páramos (4000-4500 m) and desert páramos (4500 m to snow limit). In the

Interandean valley the length of Ecuador, little natural vegetation remains; most is now a mixture of steppe and scrub. In the lowlands east of the Andes, extensive lowland rain forest, covering 135,000 sq. km of the Amazonian forest (Unesco, 1981, cited in Appendix 1). Most of this section from Harling (1979), which has a useful bibliography.

Estimated rate of deforestation for closed broadleaved forest 3400 sq. km/annum out of 142,300 sq. km (FAO/UNEP, 1981).

Checklists and Floras Ecuador is covered by the family and generic monographs of *Flora Neotropica*, as described in Appendix 1. The country Flora is:

Harling, G. and Sparre, B. (1973-). *Flora of Ecuador*. 20 vols (28 families) published so far, 7 vols in prep. Department of Systematic Botany, University of Göteborg, and the Section of Botany, Riksmuseum, Stockholm, Sweden.

Floristic knowledge of the Pacific lowlands of Ecuador is summarized by Gentry (1978), cited in Appendix 1. Floras for part of Ecuador and countrywide family accounts include:

Dodson, C.H. and Gentry, A.H. (1978). Flora of the Río Palenque Science Center. *Selbyana* 4(1-6): 1-628. (Tropical Wet Forest site, all species illustrated with descriptions and keys, frequency, habitats and related species.)

Dodson, C., Gentry, A. and Valverde, F.M. (1984). *Flora of Jauneche*. Banco Central del Ecuador and *Selbyana* 8: 1-512. (Tropical Moist Forest site, all species illustrated with descriptions and keys, frequency, habitats and related species.)

Dodson, C. and Gentry, A. (in press for 1985). *Flora of Capeira and the Guayaquil region*. Banco Central del Ecuador and *Ann. Missouri Bot. Gard*. (Tropical Dry Forest site, all species illustrated with descriptions and keys, frequency, habitats and related species.)

Dodson, C. and Marmol, P. (1980-84). Orchids of Ecuador. *Icones Plantarum Tropicarum* 1-4: Plates 1-500; 5: Plates 501-600; 10: Plates 901-1000. (Illustrations of orchids of western Ecuador, upland Ecuador, and upland and eastern Ecuador, respectively, with descriptions and dot maps.)

Gilmartin, A.J. (1972). The Bromeliaceae of Ecuador. *Phanerogamarum Monographiae Tomus* 4: 1-255.

Hitchcock, A.S. (1927). The grasses of Ecuador, Peru and Bolivia. *Contr. U.S. Nat. Herb*. 24(8): 291-556.

Little, E.L. (1969). *Arboles comunes de la Provincia de Esmeraldas*. FAO/SF: 76/ECU 13, Rome. 536 pp.

Standley, P.C. (1931). The Rubiaceae of Ecuador. *Field Mus. Nat. Hist., Bot. Ser*. 7(2): 1-251.

Valverde, F.M. (1980). *Flora de la Península Santa Elena*. Univ. Guayaquil Press.

Information on Threatened Plants There is no national Red Data Book. An analysis of vegetation types with the most endangered species is:

Gentry, A.H. (1977). Endangered plant species and habitats of Ecuador and Amazonian Peru. In Prance, G.T. and Elias, T.S. (Eds) (1977), cited in Appendix 1. Pp. 136-148.

Threatened plants are mentioned in other papers in:

Prance, G.T. and Elias, T.S. (Eds) (1977), cited in Appendix 1. See in particular J.T. Mickel on rare and endangered ferns (pp. 323-328).

Other relevant literature:

Dodson, C. (1984). Orchids of Ecuador. (Unpublished list of 2200 orchids known to occur in Ecuador of which 25 are Vulnerable, 2 Endangered and 7 Rare.)

Laws Protecting Plants Ley Forestal y de Conservación de Areas Naturales y Vida Silvestre (Ley No. 74 of 14 August 1981, Registro Oficial 24 August 1981) governs conservation and includes plants. The Ministerio de Agricultura y Ganadería is responsible for implementation (Fuller and Swift, 1984, cited in Appendix 1).

Voluntary Organizations
Fundación Natura, Jorge Juan 481, Casilla 243, Quito.

Botanic Gardens The Ecuadorian Orchid Society is establishing a botanic garden outside Guayaquil. The Ministerio de Agricultura y Ganadería operate an orchid collection and sanctuary at Vilcabamba in Loja province; plans exist for a similar programme at Conocota near Quito.

Useful Addresses
Dept. de Biología, Universidad Católica, Apdo 2184, Quito.
Facultad de Ciencias Naturales, Universidad Central, Quito.
Facultad de Ciencias Naturales, Universidad de Guyaquil, Casilla 471, Guayaquil.
Museo Ecuatoriana de Ciencias Naturales, Casa de la Cultura, Quito.
CITES Management Authority: Director Ejecutivo del Programa Nacional Forestal, Ministerio de Agricultura y Ganadería, Casilla 2919, Quito.

Additional References
Acosta-Solis, M. (1968a). Naturalistas y Viajeros Científicos que han contribuido al conocimiento Florístico Fitogeográfico del Ecuador. *Inst. Ecuat. de Cienc. Nat. Contribución* 65: 1-138. (History of botanical collecting.)
Acosta-Solis, M. (1968b). *Divisiones fitográficas y formaciones geobotánicos del Ecuador*. Publ. Cient. de la Casa de la Cultura Ecuatoriana, Quito.
Gentry, A.H. (1982). Phytogeographic patterns as evidence for a Chocó Refuge. In Prance, G.T. (Ed.) (1982), cited in Appendix 1. Pp. 112-136.
Harling, G. (1979). The vegetation types of Ecuador – a brief survey. In Larsen, K. and Holm-Nielsen, L.B. (Eds), *Tropical Botany*. Academic Press, London. Pp. 165-174.
Putney, A.D. (1976). *Estrategia prelimina para la conservación de areas silvestres sobresalientes del Ecuador*. UNDP/FAO-ECU/71/527. 61 pp.
Svenson, H.K. (1945). Vegetation of the coast of Ecuador and Peru and its relation to the Galapagos Islands. *Am. J. Bot.* 33: 394-498.

Egypt

Area 1,000,250 sq. km

Population 45,657,000

Floristics 2085 species (Täckholm, 1974); 1095 species said to occur in the coastal strip (Boulos, 1975), but probably only 800-900 (M.N. el Hadidi, 1984, pers. comm.); 70 endemics (IUCN figures).

Predominantly Saharan flora, with Mediterranean elements along the north coast (mostly winter annuals); Irano-Turanian element in the Sinai. The Nile valley has a distinctive flora with Sudanian elements in the southern part. The Gebel Elba mountain block and the surrounding land has a Sahelian element, which also reaches south Sinai. Prominent centres of endemism are the mountains of Sinai, Gebel Elba and Gebel Uweinat, and some oases in the western desert. Oases often have a Mediterranean weed element.

Vegetation Mostly desert with little or no perennial vegetation except for scattered desert shrubs; oases consist mostly of the cultivated Date palm *Phoenix dactylifera*. Coastal strip of overgrazed and badly degraded land. Also of interest is the rich aquatic and riverine flora associated with the Nile.

For vegetation map see White (1983), cited in Appendix 1.

Checklists and Floras Egypt is included in the computerized *Atlas der Pflanzenwelt des Nordafrikanischen Trockenraumes* (Frankenberg and Klaus, 1980), *Flore du Sahara* (Ozenda, 1977), and is being covered in *Med-Checklist*, all of which are cited in Appendix 1.

Täckholm, V. (1974). *Students' Flora of Egypt*, 2nd Ed. Cairo Univ., Cairo. 888 pp. (Keys, diagnoses, distributions, line drawings.)

Täckholm, V. and Boulos, L. (1972). Supplementary notes to Students' Flora of Egypt, Second Edition. *Publ. Cairo Univ. Herb.* 5: 1-135. (16 plates of black and white photographs.)

Täckholm, V. and Boulos, L. (1977). Additions and corrections to the second edition of Students' Flora of Egypt. *Publ. Cairo Univ. Herb.* 7/8: 211-218.

Täckholm, V. and Drar, M. (1941-69). *Flora of Egypt*. Incomplete; 4 vols, principally monocotyledons. Bull. Fac. Sci. Cairo Univ. 17, 28, 30, 36.

A new multipart 'Flora of Egypt' is in preparation, under the direction of Professor M.N. el Hadidi of the Cairo Herbarium. Published so far are Amaranthaceae, by M.N. el Hadidi and A.M.H. el Hadidy, Globulariaceae, by A.A. Fayed, Santalaceae, by F.M. Sa'ad, and Vahliaceae, by D.M. Bridson; Plantaginaceae, by S. Snogerup, is in press (due out late 1984). Published in *Taeckholmia Additional Series* (1980-), and expected to take 10-15 years to complete.

Information on Threatened Plants No National Red Data Book published, but Professor M.N. el Hadidi has drafted one containing 112 species of Egyptian plants. Egypt is included in the draft list for North Africa and the Middle East produced by IUCN Threatened Plants Committee Secretariat (1980), cited in Appendix 1.

Abdallah, M.S. and Sa'ad, F.M. (1980). Proposals for conservation of endangered species of the flora of Egypt. *Notes Agric. Res. Centre Herb. Egypt* 5: 1-12. (Lists 54 rare or endemic species.)

Boulos, L. (1985). The arid eastern and south-eastern Mediterranean regions. In Gómez-Campo, C. (Ed.), *Plant Conservation in the Mediterranean Area*.

Latest IUCN figures: endemics: Ex:2, E:12, V:6, R:38, I:6, K:4, nt:2; non-endemics rare or threatened worldwide – E:2, V:9, R:13, I:2 (world categories).

Botanic Gardens

Botanic Garden, Botany Department, Faculty of Science, University of Alexandria, Moharram Bey, Alexandria.

El Saff Botanic Garden, El Saff, Upper Egypt.

Orman Botanic Garden, Giza-Orman, Cairo.

Qubba Botanic Garden, Qubba, Cairo.
Zohria Trial Gardens, Gezira, Cairo.

There is also a botanic garden in Asyût, but address not known.

Useful Addresses

Plant Protection Department, Agriculture College, Asyût.
The Herbarium, Cairo University, Giza.
CITES Authority: The Director, Flora and Phytotaxonomy Researches, Agricultural Research Centre, P.O. Box: Ministry of Agriculture, Dokki, Cairo.

Additional References

Batanouny, K.H. (1973). Habitat features and vegetation of deserts and semi-deserts in Egypt. *Vegetatio* 27(4-6): 181-199. (12 black and white photographs.)

Boulos, L. (1975). The Mediterranean element in the flora of Egypt and Libya. In CNRS (1975), cited in Appendix 1. Pp. 119-124.

Hassib, H. (1952). Distribution of plant communities in Egypt. *Bull. Fac. Sci. Cairo Univ.* 29: 59-261.

Kassas, M. *et al.* (1952-1970). Habitat and plant communities in the Egyptian desert. I. Introduction. *J. Ecol.* 40: 342-351 (with 6 black and white photographs); II. The features of a desert community. *Ibid.* 41: 248-256; III. The wadi bed ecosystem. *Ibid.* 42: 424-441 (with 6 black and white photographs); IV. The gravel desert. *Ibid.* 47: 289-310 (with 8 black and white photographs); V. The limestone plateau. *Ibid.* 52: 107-119 (with 8 black and white photographs); VI. The units of a desert ecosystem. *Ibid.* 53: 715-728 (with 8 black and white photographs); VII. Geographical facies of plant communities. *Ibid.* 58: 335-350 (with 8 black and white photographs).

Wickens, G.E. (1977). Some of the phytogeographical problems associated with Egypt. *Publ. Cairo Univ. Herb.* 7/8: 223-230.

El Salvador

Area 21,393 sq. km

Population 5,888,000

Floristics An estimated 2500 species of vascular plants (Gentry, 1978, cited in Appendix 1); 19 endemic taxa (IUCN figures).

Vegetation On the coastal plain and lower southern mountain slopes mostly savanna and broadleaved forest; in the mountains of the north and south temperate grassland, remnants of deciduous oak and pine forests; in the upland area around Cerro Montecristo, on the Guatemalan border (the wettest area), cloud forests, the last remaining primary forest in the country, now protected, but only 12 sq. km (Daugherty, 1973b). Less than 10% of the country has forest cover and very little wildlands are left. Estimated rate of deforestation for closed broadleaved forest 40 sq. km/annum out of 1010 sq. km (FAO/UNEP, 1981).

Checklists and Floras El Salvador is covered by the *Flora Mesoamericana* Project, described in Appendix 1, as well as by the family and generic monographs of

Plants in Danger: What do we know?

Flora Neotropica (cited in Appendix 1). Also "most plants" of El Salvador are included in the completed *Flora of Guatemala* and related articles in *Fieldiana*, outlined under Guatemala. Useful works specifically on El Salvador include:

Calderón, S. and Standley, P.C. (1944). *Lista Preliminar de Plantas de El Salvador*, 2nd Ed. San Salvador. 450 pp. (Annotated checklist.)

Carlson, M.C. (1948). Additional plants of El Salvador. *Bull. Torr. Bot. Club* 75(3): 272-281.

Guzman, D.J. (1950). *Flora Salvadorena*. Imprenta Nacional, San Salvador. 691 pp.

Hamer, F. (1974-1981). *Las Orquídeas de El Salvador*, 3 vols. Ministerio de Educación, Dirección de Publicaciones, San Salvador. 1140 pp. (Descriptions, drawings, colour plates of 362 species; in English, Spanish and German.)

Lötschert, W. (1953). Ferns of the Republic of El Salvador. *Ceiba* 4(1): 241-250. (List of 174 species.)

Seiler, R. (1980). *Una Guía Taxonómica Para Helechos de El Salvador*. Ministerio de Educación, San Salvador. 58 pp.

Information on Threatened Plants There is no national Red Data Book, but various lists have been prepared:

Reyna de Aguilar, M.L. (1981). Flora en vías de extinción. Servicio de Parques Nacional y Vida Silvestre. (Unpublished list of threatened trees, bromeliads, orchids and of endemic trees in protected areas.)

Witsberger, D. (1980). Tree species of El Salvador and their conservation status. (Unpublished list of trees of El Salvador with annotations for endemics, species of low population considered rare, and those in Montecristo National Park.)

IUCN is preparing a threatened plant list for release in a forthcoming report *The List of rare, threatened and endemic plants of Middle America*. Latest IUCN statistics, based upon this work: endemic taxa – V:4, R:6, I:2, K:7; non-endemic taxa rare or threatened worldwide – E:4, V:7, R:3 (world categories).

10 threatened plants are included in Organización de los Estados Americanos (1967), cited in Appendix 1, and 3 species included in *The IUCN Plant Red Data Book*. Threatened plants are also mentioned in several papers in:

Prance, G.T. and Elias, T.S. (Eds) (1977), cited in Appendix 1. See in particular W.G. D'Arcy on endangered landscapes in the region (pp. 89-104) and J.T. Mickel on rare and endangered ferns (pp. 323-328).

Laws Protecting Plants No wildlife legislation, but a draft law is under consideration (Fuller and Swift, 1984, cited in Appendix 1). The U.S. Government has determined *Abies guatemalensis* (El Salvador, Honduras, Guatemala, Mexico) as 'Threatened' under the U.S. Endangered Species Act.

Voluntary Organizations
Friends of the Earth, Edificio Comercial, 6° Piso, San Salvador.
National Committee for Ecology, Boulevard del Hipodromo 303, San Salvador.

Useful Addresses
Instituto Salvadoreno de Recursos Naturales, Ministerio de Agricultura y Ganadería, Canton El Matazano, Apdo Postal 2265, Soyapango, San Salvador.
Sección de Flora, Servicio de Parques Nacionales y Vida Silvestre, DIGERENARE, Apdo Postal 2265, San Salvador.

Additional References

Blutstein, H.I. *et al.* (1970). *El Salvador: A Country Study*. The American University, Washington, D.C. 260 pp.

Daugherty, H.E. (1973a). The Montecristo Cloud-forest of El Salvador – a chance for protection. *Biol. Conserv.* 5(3): 227-230.

Daugherty, H.E. (1973b). *Conservación Ambiental Ecológica de El Salvador con Recomendaciones para un Programa de Acción Nacional*. Artes Gráfica Publicitarias, San Salvador. 56 pp.

Holdridge, L.R. (1959). *Mapa Ecológico de El Salvador*. Instituto Interamericano de Ciencias Agrícolas de la Organización de los Estados Americanos (OEA), San José, Costa Rica.

Equatorial Guinea

Equatorial Guinea comprises mainland Mbini (Rio Muni) and five islands and islets in the Gulf of Guinea: Bioko (Fernando Po or Macias Nguema Biyogo) is the largest island in the Gulf, 32 km from Cameroon; Pagalu (Annobon) is the smallest of the offshore islands, 180 km SSW of S. Tomé and 340 km from the nearest mainland (Gabon); Corisco, Elobey Grande and Elobey Chico are small coastal islets. The other major islands in the Gulf of Guinea are São Tomé and Príncipe, *q.v.*

Area 28,051 sq. km, including Mbini (26,017 sq. km), Bioko (2017 sq. km) and Pagalu (17 sq. km).

Population 383,000

Floristics

Mbini No figures available, but flora likely to be rich (Brenan, 1978, cited in Appendix 1). Floristic affinities Guinea-Congolian.

Bioko 1105 species (Exell, 1973a); 49 endemic species (Brenan, 1978, cited in Appendix 1). (Exell, 1944 gives a figure of 99 endemic species, but this was before the revision of the *Flora of West Tropical Africa*.) Floristic affinities with mainland West Africa (particularly Mt Cameroun) and the other islands in the Gulf of Guinea.

Pagulu 208 species (Exell, 1973a), with 17 endemic species (out of a total of 115, Exell, 1944). Floristic affinities with the other islands in the Gulf of Guinea between it and the mainland.

Vegetation

Mbini Lowland rain forest, with small areas of mangrove forest at the coast.

Bioko Original low altitude vegetation: lowland rainforest, but very little left now, replaced by secondary and cultivation communities to meet the needs of the dense population. Afromontane communities of montane forest and grassland occur at higher altitudes.

Pagulu Difficult to assess original vegetation, since so little now remains, but predominantly lowland and submontane evergreen forest, with mist-forest on the upper slopes of the peaks. Most low and medium altitude vegetation now destroyed; replaced by savanna-like cultivated land with scattered bushes. Dry forest and mist forest are still quite well represented.

Estimated rate of deforestation for closed broadleaved forest 30 (27 in Mbini) sq. km/annum out of 12,950 (11,800 in Mbini) sq. km (FAO/UNEP, 1981).

For vegetation map see White (1983), cited in Appendix 1.

Checklists and Floras
Mbini
Guinea López, E. (1946). *Ensayo Geobotánico de la Guinea Continental Española.* Dirección de Agricultura de los Territorios Españoles del Golfo de Guinea, Madrid. 388 pp. (See especially pp. 218-368, where records of plants are given; illustrated throughout with maps, line drawings, paintings, and black and white photographs.)

Bioko Included in the *Flora of West Tropical Africa*, cited in Appendix 1.

Benl, G. (1978-1982). The Pteridophyta of Fernando Po. (Contributions to a Flora of the island.) *Acta Botanica Barcinonensia* 31: 1-31; 32: 1-34; 33: 1-46.
Escarré, A. (1968-1970). Aportaciones al conocimiento de la flora de Fernando Póo. *Acta Phytotax. Barcinonensia* 2 (1968), 15 pp.; 3 (1969), 23 pp.; 5 (1970), 32 pp. (by A. Escarré and T. Reinares). (Never completed; covers 5 families only.)

Pagulu
Exell, A.W. (1963). Angiosperms of the Cambridge Annobon Island expedition. *Bull. Brit. Mus. (Nat. Hist.) Bot.* 3(3): 93-118.

Pagalu is also included in the following Floras:

Exell, A.W. (1944). *Catalogue of the Vascular Plants of S. Tomé (with Principe and Annobon).* British Museum (Natural History), London. 428 pp. (Annotated checklist; line drawings.)
Exell, A.W. (1956). *Supplement to the Catalogue of the Vascular Plants of S. Tomé (with Principe and Annobon).* British Museum (Natural History), London. 58 pp.

Bioko and Pagalu are both included in:

Exell, A.W. (1973a). Angiosperms of the islands of the Gulf of Guinea (Fernando Po, Príncipe, S. Tomé and Annobon). *Bull. Brit. Mus. (Nat. Hist.) Bot.* 4(8): 325-411. London. (Checklist with distributions.)

Information on Threatened Plants No published lists of rare and threatened plants; IUCN has records of three species and infraspecific taxa believed to be endemic to Mbini, 58 endemic to Bioko and 17 endemic to Pagalu; no categories assigned.

Additional References
Exell, A.W. (1952/1955). The vegetation of the islands of the gulf of Guinea. *Lejeunia* 16: 57-66.
Exell, A.W. (1968). Príncipe, S. Tomé and Annobon. In Hedberg, I. and O. (1968), cited in Appendix 1. Pp. 132-134. (Includes lists of examples of endemic species for each of the three islands.)
Exell, A.W. (1973b). Relações florísticas entre as ilhas do golfo da Guiné e destas com o continente africano. *Garcia de Orta, Sér. Bot.* 1(1-2): 3-10.
Guinea, E. (1968). Fernando Po. In Hedberg, I. and O. (1968), cited in Appendix 1. Pp. 130-132.
Mildbraed, J. (1922). *Wissenschaftliche Ergebnisse der Zweiten Deutschen Zentral-Afrika-Expedition 1910-1911*, 2, Botanik. Klinkhardt and Biermann, Leipzig. 202 pp. (With 90 plates of black and white photographs.)

Ethiopia

Area 1,023,050 sq. km

Population 35,420,000

Floristics Cufodontis (1953-1972) includes 6283 species in his *Enumeratio*, cited in Appendix 1; also includes Somalia (c. 518 endemic species), but probably of the right order of magnitude for the flora of Ethiopia if Somalian endemics are excluded and new species and records included. Endemism fairly high in the mountains and in the sub-desert Ogaden in south-east Ethiopia; also the forests in the south-west. Brenan (1978, cited in Appendix 1) gives a value of almost 21% specific endemism from a sample of less than 3000 species; this is almost certainly too high, 10% being probably more accurate.

Large proportion of area in Afromontane region, with small pockets of the Afroalpine region at the highest altitudes; flora of these regions greatly impoverished, and related to the highland flora in other parts of Africa. Affinities also with the floras of South Africa, Europe and the Himalayas. Most of lowland southern Ethiopia belongs to the Somalia-Masai region with east African affinities, although forests of the south-west have links with the west African forests. Flora of western Ethiopia is Sudanian.

Vegetation The natural vegetation of the plateaux and highlands above 1800 m is largely coniferous forest; most has disappeared and is only found in the more inaccessible regions; there are also expanses of mountain grassland. Zonation in the mountains from forest through bamboo and heath thicket to tufted grass moorland is similar to that on the high Kenyan mountains, but less well marked. In the south-west higher rainfall and lower elevation has produced extensive broadleaved rain forests with a high diversity of species. In the lowlands, there is a range of dry-zone vegetation, from limited areas of desert through *Acacia-Commiphora* bushland to *Acacia* woodland. Estimated rate of deforestation for closed broadleaved forest 60 sq. km/annum out of 27,500 sq. km (FAO/UNEP, 1981).

For vegetation map see White (1983), cited in Appendix 1.

Checklists and Floras Ethiopia is included in *Enumeratio Plantarum Aethiopiae Spermatophyta* (Cufodontis, 1953-1972), and in *Adumbratio Florae Aethiopicae*, both cited in Appendix 1. See also:

Breitenbach, F. von (1963). *The Indigenous Trees of Ethiopia*, 2nd Rev. Ed. (1st Ed. 1960). Ethiopian Forestry Association, Addis Ababa. 306 pp. (Keys to families, genera; full descriptions; 129 line drawings.)

Burger, W.C. (1967). *Families of Flowering Plants in Ethiopia*. Experiment Station Bulletin No. 45, USAID, Oklahoma State Univ. Press, Oklahoma. 236 pp. (Keys to families; family descriptions; 74 line drawings.)

Fiori, A. (1909-1912). *Boschi e Piante Legnose dell'Eritrea*. Firenze. 428 pp. (Illustrated; rather old, but gives records for rare plants.)

Pirotta, R. (1903-1907). *Flora della Colonia Eritrea*, parts 1-3. Annuario del R. Istituto Botanico di Roma 8, Rome. 464 pp. (Never completed; final part lost by printers, according to Frodin.)

There is a new project to write a Flora of Ethiopia headed by Professor Tewolde-Berhan of the University of Asmara. It is expected to take 15-20 years to complete and will comprise 7 volumes. Volume 3 (including Leguminosae) and substantial parts of volume 2 are in manuscript.

111

Field-guides

Edwards, S. (1976). *Some Wild Flowering Plants of Ethiopia*. Addis Ababa.

Information on Threatened Plants

Hedberg, I. (1979), cited in Appendix 1. (List for Ethiopia, pp. 92-93, by M.G.
  Gilbert, contains 29 endemic succulent taxa – E:1, V:4, R:12, I:12.)

IUCN has records of c. 450 species and infraspecific taxa believed to be endemic; only a
few (mostly succulents) known to be rare or threatened.

Two species which occur in Ethiopia are included in *The IUCN Plant Red Data Book*
(1978).

Useful Addresses

Flora of Ethiopia project, P.O. Box 3434, Addis Ababa.

Additional References

Beals, E.W. (1968). Ethiopia. In Hedberg, I. and O. (1968), cited in Appendix 1.
  Pp. 137-140.

Friis, I. (1983). Phytogeography of the tropical north-east African mountains. In
  Killick, D.J.B. (1983), cited in Appendix 1. Pp. 525-532.

Friis, I., Rasmussen, F.N. and Vollesen, K. (1982). Studies in the flora and vegetation
  of southwest Ethiopia. *Opera Botanica* 63: 1-70.

Hedberg, I. (in prep.). Proceedings of a symposium on the Ethiopian flora held in
  Uppsala in May 1984. To be published in *Symb. Bot.*

Hedberg, O. (1983). Ethiopian Flora project. In Killick, D.J.B. (1983), cited in
  Appendix 1. Pp. 571-574.

Logan, W.E.M. (1946). An introduction to the forests of central and southern
  Ethiopia. *Inst. Pap. Imp. For. Inst.* 24. 64 pp. (Includes small-scale vegetation
  map.)

Pichi-Sermolli, R.E.G. (1957). Una carta geobotanica dell'Africa orientale (Eritrea,
  Ethiopia, Somalia). *Webbia* 13: 15-132. (Includes map 1:5,000,000.)

Faeroe Islands

Over 20 islands in the north Atlantic between Shetland and Iceland, forming a self-
governing community within the Kingdom of Denmark.

Area 1399 sq. km

Population 42,000

Floristics 262 native vascular species (Hansen, 1972); 1 endemic species. 3
floristic elements: Arctic (c. 25%), sub-Arctic (50%) and Atlantic (c. 25%).

Vegetation Mostly dwarf scrub with bog and grassy heath communities. Above
300 m alpine tundra covers the mountainous North Islands and the north-facing peaks of
the Central Islands group (Warming, 1901-1908).

Checklists and Floras Covered by *Flora Europaea* (Tutin *et al.*, 1964-1980, cited
in Appendix 1) and:

Ostenfeld, C.H. and Gröntved, J. (1934). *The Flora of Iceland and the Faeroes*. Levin and Munksgaard, Copenhagen. 195 pp.

Rasmüssen, R. (1952). *Føvoya Flora*, 2nd Ed. Jacobsens, Torshavn. 231 pp. (School and excursion manual with keys; line drawings.)

Field-guides

Bloch, D. (1980). *Farøflora*. Føroza Frodstaparfelag, Torshavn. 156 pp. (English edition also available.)

Information on Threatened Plants Only 1 non-endemic species is listed as threatened on the IUCN list, the orchid *Hammarbya paludosa*.

Useful Addresses

Museum of Natural History, 3800 Torshavn.

Additional References

Hansen, K. (1964). The botanical investigations of the Faroe Islands 1960-61 and some contributions to the Flora. *Bot. Tidssk*. 60(1-2): 99-107.

Hansen, K. (1966). Vascular plants in the Faeroes. Horizontal and vertical distribution. *Dansk Bot. Ark*. 24(3): 1-141. (Distribution maps for vascular plants.)

Hansen, K. (1972). Vertical vegetation zones and vertical distribution types in the Faeroes. *Saertryk Bot. Tidssk*. 67: 33-63. (Useful ecological description.)

Warming, E. *et al*. (1901-1908). *Botany of the Faeroes based upon Danish Investigations*, 3 parts. Gyldendalske, København. (Part 3 contains a detailed phytosociological description by C.H. Ostenfeld, pp. 867-1026.)

Falkland Islands (Islas Malvinas)

The Falkland Islands, an archipelago 520 km east of the straits of Magellan, comprise two main islands, East Falkland (5000 sq. km) and West Falkland (3500 sq. km), together with about 230 smaller islands. The highest point is Mt Usborne (705 m) on East Falkland. They are a Dependent Territory of the U.K.

Area 12,173 sq. km

Population 2000

Floristics 163 native species of flowering plants and pteridophytes, and 93 introduced species; 16 endemic species (Moore, 1968). *Phlebolobium* (Cruciferae) is an endemic genus. Floristic affinities with the southern Andes and Patagonia.

Vegetation Maritime tussock grassland, with *Poa flabellata*, now heavily overgrazed; *Hebe* and *Chiliotrichum* bush in places; dwarf shrub heath, dominated by *Empetrum*, on better drained ground; *Cortaderia* grassland in areas of poorer drainage; bog communities in very poorly drained areas; 'feldmark' formation above 600 m, in which there are large areas of open ground with cushion-forming vascular plants, mosses and lichens.

Checklists and Floras

Moore, D.M. (1968). *The Vascular Flora of the Falkland Islands*. British Antarctic Survey Scientific Report no. 60. NERC, London. 202 pp. (Includes description of vegetation.)

Moore, D.M. (1973). Additions and amendments to the vascular flora of the Falkland Islands. *Brit. Antarctic Survey Bull.* 32: 85-88.

Vallentin, E.F. and Cotton, E.M. (1921). *Illustrations of the Flowering Plants and Ferns of the Falkland Islands*. Reeve, London. (64 colour plates, with text.)

Information on Threatened Plants *Calandrinia feltonii* is included in *The IUCN Plant Red Data Book* (1978). Latest IUCN statistics: endemic taxa – Ex:1, E:1, R:3, nt:8.

Voluntary Organizations

Falkland Islands Foundation, Hon. Secretary, c/o WWF-United Kingdom, Panda House, 11-13 Ockford Road, Godalming, Surrey GU7 1QU, U.K.

Additional References

Correa Luna, H. *et al.* (1975). Campaña científica en las Islas Malvinas, 1974 (Noviembre 17 a Diciembre 2). *Anal. Soc. Científ. Argentina* 199: 51-180. (Articles on conservation, agronomy, physiognomy and fauna by visiting Argentine scientists.)

Erskine, P.J. (1985). Flowers of the Falklands. *Alpine Garden Society Bulletin* 53(1): 69-87. (Notes on vegetation and 19 flowering plant species.)

Skottsberg, C. (1913). A botanical survey of the Falkland Islands. *K. Svenska Vetensk Akad. Handl.* 50(3): 1-129.

Falkland Islands: South Georgia

South Georgia, a dependency of the Falkland Islands, is situated at latitude 54°S and longitude 36-38°W, 1287 km east of the Falkland Islands and 2000 km east of Tierra del Fuego. Area 3757 sq. km; the population comprises the staff of the British Antarctic Survey Station. Much of the land is permanently covered by ice.

There are 24 native vascular species (Smith and Walton, 1975). The vegetation consists of coastal tussock grassland (mainly of *Poa flabellata*); dry meadows of *Festuca contracta*; dwarf shrub (*Acacia magellanica*) and mire communities on higher ground, and sparsely vegetated fell-fields in the more exposed high areas.

References

Greene, S.W. (1964). *The Vascular Flora of South Georgia*. British Antarctic Survey Scientific Report no. 45. London. 58 pp. (Includes distribution maps.)

Greene, S.W. (1969). New records for South Georgian vascular plants. *Brit. Antarctic Survey Bull.* 22: 49-59.

Greene, S.W. and Walton, D.W.H. (1975). An annotated checklist of the sub-antarctic and antarctic vascular flora. *Polar Record* 17(110): 473-484.

Smith, R.I.L. and Walton, D.W.H. (1975). South Georgia, Subantarctic. *Ecol. Bull. (Stockolm)* 20: 399-423.

Walton, D.W.H. (1975). Nomenclatural notes on South Georgian vascular plants. *Brit. Antarctic Survey Bull*. 40: 77-79.

Falkland Islands: South Sandwich Islands

The South Sandwich Islands, dependencies of the Falkland Islands, are a chain of uninhabited islands, of area 310 sq. km, situated 756 km south-east of South Georgia. They have active volcanoes and support very scattered communities of crustaceous lichens, algae and mosses. 58 plant species recorded, but only one species of higher plant (*Deschampsia antarctica*).

References

Longton, R.E. and Holdgate, M.W. (1979). *The South Sandwich Islands: 4. Botany*. British Antarctic Survey Scientific Report no. 94. NERC, Cambridge. 53 pp. (Includes checklist and description of plant communities.)

Fiji

The Fiji group includes some 332 islands in the south-west Pacific Ocean, between latitudes 10° and 25°S, and longitudes 176°E and 173°W, about 2000 km north north-west of New Zealand. 3 types of islands: high volcanic islands, reaching 1323 m on Viti Levu; limestone islands; and low coral islands and atolls. About 97 islands permanently inhabited; most of the population live on the coast and along river valleys on Viti Levu and Vanua Levu.

Area 18,235 sq. km

Population 674,000

Floristics c. 1500 native vascular plant species, including 310 pteridophytes; in addition there are c. 1000 introduced flowering plant species (A.C. Smith, 1984, in *litt*.). About 40-50% of native species are endemic, including all 26 palm species (Smith, in *litt*.). One family and 11 genera endemic (van Balgooy, 1971, cited in Appendix 1). Floristic affinities with Malesia, New Hebrides, Samoa and Tonga.

Vegetation Rain forest (veikauloa) in south and east of larger islands and most parts of small volcanic islands, where not disturbed (Smith, 1951); montane rain forest up to 1735 m (Myers, 1980, cited in Appendix 1). Dry zone (talasinga) vegetation, including dry forests, savanna woodlands and grasslands on north and west slopes of large islands and inland to 450 m; dry forest mostly replaced by sugar cane plantations. Intermediate zone vegetation immediately leeward of wet forests. Mangrove forest still extensive along larger rivers and muddy coasts. Natural forest cover is estimated at 8650 sq. km (S. Siwatibau, 1984, in *litt*.). For an account of the vegetation see Schmid (1978).

Checklists and Floras The Flora is:

Smith, A.C. (1979-). *Flora Vitiensis Nova: A New Flora of Fiji*. Pacific Tropical Botanic Garden, Hawaii. (2 vols so far. 1 – Gymnosperms and monocotyledons except orchids, 495 pp.; 2 – dicotyledons, 810 pp.; 3,4 – dicotyledons and orchids, in prep.)

Also relevant:

Brownlie, C. (1977). *The Pteridophyte Flora of Fiji*. Cramer, FL-9490, Vaduz, Liechtenstein. 397 pp.

Parham, J.W. (1972). *Plants of the Fiji Islands*, 2nd Ed. Govt Printer, Suva. 462 pp. (Checklist with short descriptions and line drawings.)

Seemann, B. (1865-1873). *Flora Vitiensis: A Description of the Plants of the Viti or Fiji Islands, with an Account of their History, Uses and Properties*. London. 453 pp. (Reprinted 1977 by Cramer, FL-9490, Vaduz, Liechtenstein; many colour plates.)

Information on Threatened Plants No comprehensive list of threatened plants. An IUCN manuscript list of Fijian palms includes E:1, V:2, R:14, I:5. *Neoveitchia storckii* is included in *The IUCN Plant Red Data Book* (1978).

Voluntary Organizations

The National Trust of Fiji, P.O. Box 2089, Government Buildings, Suva. (Government statutory body with a voluntary membership.)

Botanic Gardens

Suva Botanical Gardens, Box 176, Suva, Fiji.

Additional References

Berry, M.J. and Howard, W.J. (1973). *Fiji Forest Inventory*, 3 vols. Land Resources Study no. 12. Overseas Development Administration, Tolworth, U.K.

Derrick, R.A. (1965). *The Fiji Islands: A Geographical Handbook*, 2nd Ed. Government Press, Suva. 336 pp.

Schmid, M. (1978) The Melanesian forest ecosystems (New Caledonia, New Hebrides, Fiji Islands and Solomon Islands). In Unesco/UNEP/FAO (1978), cited in Appendix 1. Pp. 654-683.

Smith, A.C. (1951). The vegetation and flora of Fiji. *Scientific Monthly* 73: 3-15.

Finland

Area 337,032 sq. km

Population 4,859,000

Floristics About 1100 native vascular species (Hämet-Ahti *et al.*, 1984); 1250-1450 species estimated by D.A. Webb (cited in Appendix 1) from *Flora Europaea*; no endemics. Entire country was glaciated so flora still young. Elements: mostly Boreal, with some Arctic/alpine influence in the mountains of the north.

Vegetation Extensive tracts of natural, coniferous forests cover about 70% of land surface, open mires about 10%, and treeless alpine areas 5%. In the north, a narrow lichen-tundra belt; in central northern Finland, extensive areas of peat bogs bordered by pine and spruce (Finland, Sweden and Norway contain 80% of Europe's peatlands); south of the Arctic Circle, pine is more widespread with heathlands; in the south, herb-rich meadows, once abundant, now disappearing, due to decline of traditional agriculture. c. 60,000 lakes throughout the country support extensive shore-line vegetation.

Checklists and Floras Covered by the completed *Flora Europaea* (Tutin *et al.*, 1964-1980, cited in Appendix 1) and the below:

Hiitonen, I. (1933). *Suomen Kasvio* (Flora of Finland). Kustannusosakeyhtiö Otava, Helsinki. 771 pp. (In Finnish; illus.)

Hjelt, H. (1888-1926). Conspectus florae Fennicae, 7 vols. *Acta Soc. Fauna Flora Fennica* 5: 1-562; 21: 1-261; 30: 1-140; 35: 1-411; 41: 1-502; 51: 1-450; 54: 1-397. (In Latin and Swedish.)

A regional plant atlas is Hultén (1971), cited in Appendix 1. For a bibliography of recent floristic work see:

Collander, R., Erkamo, V. and Lehtonen, P. (1973). Bibliographica Botanica Fenniae 1901-1950. *Acta Soc. Fauna Flora Fennica*. 646 pp.

Jalas, J. (1975). Progress in the study of vascular plants in Finland 1962-1971. *Mem. Soc. Brot.* 24(2): 395-462.

National botanical journals: *Annales Bot. Fennici*, Helsinki; *Memoranda Soc. Fauna Flora Fennica*, Helsinki; *Acta Bot. Fennica*, Helsinki.

The Botanical Museum of the University of Helsinki operates a computerized 'Flora Register' containing information about vascular plant species gathered from literature, herbarium specimens and other unpublished sources. At present it contains over 1 million records, including information on threatened plants.

Field-guides
Hämet-Ahti, L., Suominen, J., Ulvinen, T., Uotila, P. and Vuokko, S. (Eds), (1984). *Retkeilykasvio (Field Flora of Finland)*. Helsinki. 544 pp. (Keys; distribution maps at Province level; line drawings; in Finnish.)

Hiitonen, I. and Poijärvi, A. (1958). *Koulu-ja retkeilykasvio* (School and excursion Flora), 9th Ed. Helsinki. 472 pp. (In Finnish.)

Information on Threatened Plants A national threatened plant programme is being undertaken by the Committee for the Protection of Threatened Animals and Plants, in the Ministry of the Environment, Nature Conservation Division, Helsinki (address below). This programme includes the production of a national Red Data Book, for publication in 1985, and the development of a national protection and monitoring scheme. Preliminary lists of threatened plants have been compiled for both vascular and lower plants. For vascular plants both national and regional lists will be produced. They are presently available in the following reference book:

Vuokko, S. (1983). *Uhatut kasvimme* (Our threatened plants). Suomen Luonnonsuojelun Tuki Oy, Helsinki. 96 pp. (Popular book including lists of protected plants in Finland, Åland and the rest of Scandinavia; illus.)

The lists update the earlier national threatened plant list, which was produced in collaboration with WWF-Finland:

Borg, P. and Malmström, K. (1975). Suomen uhanalaiset eläin-ja kasvilajit (Threatened animals and plants in Finland). *Luonnon Tutkija* 79: 33-43. (Lists 62 vascular plant species threatened throughout the country.)

Also relevant:

Haeggström, C.-A., Haeggström, E. and Lindgren, L. (1982). *Rapport om fridlysta och sällsynta växter pa Åland* (Report on the protected and rare plants on the Åland Islands). Natö biologiska station. 137 pp.

Kaakinen, E., Salminen, P. and Ulvinen, T. (1979). Lapin kolmion lettojen tuho (Fenland loss in the Lapland Triangle). *Suomen Luonto* 38: 130-131. (Describes plant species on the decline.)

Murto, R. (1982). *Tutkimuksia Uudenmaan Läänin Uhanalaisista Kasveista. 1. Tammisaaren ja Inkoon saaristo.* (Studies on the threatened plants in the Province of Uusimaa. 1. Archipelago of Tammisaari and Inkoo.) Helsingin yliopiston kasvimuseo. 62 pp. (To be continued.)

Suominen, J. (1974). Tuloksia uhanalaisten kasvien tiedustelusta (Results from an enquiry about endangered plants in Finland). *Suomen Luonto* 33: 24, 29.

Tampereen seutukaavaliitto. (1982). Pirkanmaan uhanalaiset kasvit ja niiden esiintymisalueet (Threatened plants and their localities in the Province of Tampere). *Tampereen seutukaavaliiton julkaisu*, Ser. B, 116: 1-22.

Uotila, P. (1983). Project hotade växter i Nylands län (Projects about threatened plants in the Province of Uusimaa, S. Finland). *Memoranda Soc. Fauna Flora Fennica* 59: 106-112. (Maps.)

Finland is included in the Nordic Council of Ministers' threatened plant list and supplements (Ovesen *et. al*, 1978 and 1982) and in the European list (Threatened Plants Unit, 1983), all cited in Appendix 1; latest IUCN statistics, based upon the latter: non-endemics rare or threatened worldwide – V:7, R:6, I:1 (world categories).

In May 1984, WWF-Finland launched a national Plants Campaign in the Botanical Garden and Department of Botany in the University of Helsinki, as part of their contribution to the IUCN/WWF Plants Programme 1984-85. Further details available from WWF-Finland and the Garden (addresses below).

Laws Protecting Plants The recent 1983 law (Laki luonnonsuojelulain muuttamisesta) strengthens the earlier Nature Conservation Act of 1952. Under the new law, in the Statute on the Protection of Wild Plants, 94 vascular plant species receive complete protection, an additional 9 species receive complete protection in southern Finland only and 8 species in northern Finland only. It is prohibited to pick, damage or transport any of the species listed. It is also forbidden to use for trade purposes a further 7 species, listed in paragraph 4 of the Statute. Breaking the branches of *Hippophae rhamnoides* is also prohibited.

In the autonomous islands of Åland, the recent 1984 Statute (above), provides complete protection (stricter than that for the mainland species) for 52 vascular plant species. It is also forbidden to uproot *Dactylorhiza sambucina* and to cut down wild *Quercus robur* or *Juniperus communis* of a size specified in the Statute. For the list of protected species see Vuokko (1983).

Voluntary Organizations

Suomen Luonnonsuojeluliitto (Finnish Association for Nature Protection), P.O. Box 169, 00151 Helsinki.

WWF-Finland (Maailman Luonnon Säätiön Suomen Rahasto), Uudenmaankatu 40, 00120 Helsinki.

Botanic Gardens

Botanic Garden, University of Helsinki, Unioninkatu 44, 00170 Helsinki.

Botanic Garden, University of Joensuu, P.O. Box 111, 80101 Joensuu 10.

Botanic Garden, University of Jyväskylä, Yliopistonkatu 9, 40100 Jyväskylä.

Botanic Garden, University of Kuopio, P.O. Box 138, 70101 Kuopio.
Botanic Garden, University of Oulu, Box 191, 90101 Oulu 10.
Botanic Garden, University of Turku, 20500 Turku 50.

Useful Addresses

Maa-ja Metsalousministeriö, Ministry of Agriculture and Forestry, Bureau of Natural
Resources, Vuorikatu 16, Helsinki 10.
Ministry of the Environment, Nature Conservation Division, P.O. Box 306, 00531
Helsinki.
National Board of Forestry, P.O. Box 233, 00120 Helsinki. (11 provincial offices and
several local offices.)
CITES Management and Scientific Authority: Committee for the Protection of
threatened animals and plants, Ministry of the Environment, address above.

Additional References

Kalliola, R. (1970). Some features of nature and conservation in Finland. *Biol.
Conserv.* 2(2): 120-124.
Kalliola, R. (1973). *Suomen kasvimaantiede.* Porvoo. 308 pp. (The plant geography of
Finland, with good bibliography.)
Jalas, J. (1958, 1965, 1980). *Suuri Kasvikirja* (The Great Plant Book), 3 vols. Otava,
Keuruu-Helsinki. (A national account of floristics.)

France

Area 549,619 sq. km

Population 54,449,000

Floristics 4300-4450 native vascular species estimated by D.A. Webb (1978, cited
in Appendix 1) from *Flora Europaea*; 73 endemic taxa (IUCN figures). Diversity greatest
in montane areas: Pyrenees, Massif Central, Alps and Jura. Elements: Mediterranean,
Central European, Atlantic, Boreal and alpine.

Vegetation Largely an agricultural landscape, especially in north and west-central
regions. About 1/4 of total land area (c. 140,000 sq. km) under forest, comprising 2/3
deciduous broadleaved (2/3 of which is coppice) and 1/3 evergreen. The 4 main montane
areas (listed above) support a notable alpine flora. Dry grassland is still extensive, but is
shrinking fast due to agricultural change; valuable areas remain in the Jura, pre-Alps,
Quercy and the Causses (Wolkinger and Plank, 1981, cited in Appendix 1). On south
coast, Mediterranean influence present (*Quercus ilex, Q. pubescens, Q. mas*) with garigue
and diminishing areas of maquis. For a vegetation map see Rey and Dupias (1969).

Checklists and Floras France is covered by the completed *Flora Europaea* (Tutin
et al., 1964-1980, cited in Appendix 1). France will also be covered under the *Med-
Checklist* (cited in Appendix 1). For national Floras see:

Bonnier, G. and Douin, R. (1911-1935). *Flore Complète Illustrée en Couleurs de
France, Suisse et Belgique*, 13 vols. Neuchâtel. (Colour plates.)
Coste, H. (1901-1906). *Flore de la France*, 3 vols. Klincksieck, Paris. 5 supplements by
P. Jovet and R. de Vilmorin. Blanchard, Paris. (Reprinted 1937, 1950.)

Fournier, P. (1977). *Les Quatre Flores de la France*, 2nd Ed. 2 vols. Lechevalier, Paris. (1 – descriptions, 2 – line drawings.)

Guinochet, M. and Vilmorin, R. de (1973-1982). *Flore de France*, 4 vols. Centre National de la Recherche Scientifique, Paris. (Includes Corsica; line drawings; habitat and ecological details.)

Regional Floras:

Abbayes, H. des, Claustres, G., Corillion, R. and Dupont, P. (Eds) (1971). *Flore et Végétation du Massif Armoricain 1: Flore Vasculaire*. Presses Universitaires de Bretagne, Saint-Brieuc. 1226 pp. (Covers the Départements of Morbihan, Loire-Atlantique, Finistère, Côtes-du-Nord, Ille-et-Vilaine, most of Mayenne; line drawings.)

De Langhe, J.-E. *et al.* (1978). *Nouvelle Flore de la Belgique, du Grand-Duché de Luxembourg, du Nord de la France et des Régions Voisines*. Jardin Botanique National de Belgique, Meise. 899 pp.

Field-guides

Bournerias, M. (1979). *Guide des Groupements Végétaux de la Région Parisienne*, 2nd Ed. 510 pp.

Claustres, G. and Lemoine, C. (1980). *Connaître et Reconnaître la Flore et la Végétation des Côtes Manche-Atlantique*. Rennes. 331 pp. (Ecological information; illus.)

Guittonneau, A. and Huon, A. (1983). *Connaitre et Reconnaitre la Flore et la Végétation Mediterranénnes*. Rennes. 334 pp.

Jeanjean, A.F. (1961). *Catalogue des Plantes Vasculaires de la Gironde*. Bordeaux. 362 pp.

Rol, R. (1962-1965). *Flore des Arbres, Arbustes et Arbrisseaux*, 4 vols. La Maison Rustique, Paris. (1 – plaines et collines; 2 – montagnes, by R. Rol and P. Toulgouat; 3 – région mediterranéenne, by R. Rol and M. Jacamon; 4 – essences introduites, by R. Rol and P. Toulgouat; colour photographs.)

Romagnesi, H. and Weill, J. (1977). *Fleurs Sauvages de France et des Régions Limitrophes*, 2 vols. 288 pp.

Stefenelli, S. (1979). *Guide des Fleurs de Montagne: Pyrenees – Massif-Central – Alpes – Apennins* (French adaptation). Duculot, Paris-Gembloux. 160 pp. (Colour photographs and ecological data for each species.)

See also Grey-Wilson (1979) and Polunin and Smythies (1973), both cited in Appendix 1.

Information on Threatened Plants No national Red Data Book but a series of 3 unpublished papers compiled under the direction of the Ministère de la Qualité de la Vie:

Aymonin, G.G. (1974-1977). Etudes sur les régressions d'espèces végétales en France. Rapport No.1 – Espèces végétales considérées comme actuellement disparues du territoire; Rapport No.2 – Listes préliminaires des espèces endémiques et des espèces menacées en France; Rapport No.3 – Liste générale des espèces justifiant des mesures de protection. Museum National d'Histoire Naturelle, Paris. (Unpublished; reports 1 and 2 list over 1000 taxa; report 3 analyses the data in 1 and 2 and recommends levels of protection.)

See also:

Aymonin, G.G. (1973). Quelques raréfactions et disparitions d'espèces végétales en France. Causes possibles et consequences chorologiques. *C.R. Soc. Biogéogr.* 430: 49-64.

Aymonin, G.G. (1980a). Stratégies de sauvegarde pour les espèces végétales. Quelques aspects recents. *Bull. Soc. Et. Sc. Beziers, n.s.* 8(48): 24-37.

Aymonin, G.G. (1980b). Une estimation du degré de modification des milieux naturels: l'analyse des régressions dans la flore. *Bull. Soc. Bot. France* 127(2): 187-195.

Aymonin, G.G. (1981). Sur quelques espèces remarquables des complexes boisés de Bourgogne et leur situation de régression en Europe. *Bull. Soc. Bot. France* 128(3/4): 95-100.

Aymonin, G.G. (1982). Phénomènes de déséquilibres et appauvrissements floristiques dans les végétations hygrophiles en France. In Symoens, J.J., Hooper, S.S. and Compère, P. (Eds), *Studies on Aquatic Vascular Plants*, Proceedings of the International Colloquium on Aquatic Vascular Plants, 23-25 January 1981, Brussels. Société Royale de Botanique de Belgique, Brussels. Pp. 377-389.

Binet, P. and Provost, M. (1971). Les plantes rares en Normandie. *Sci. Nat.* 103: 2-6.

Bournerias, M. (1983). Espèces végétales protégées, espèces et biotopes à protéger dans le bassin de la Seine et le Nord de la France. *Nat. Par.* 39: 19-36.

Daunas, R. (1977). La protection des espèces végétales en France: plantes rares ou en voie de disparition en Poitou-Charentes et Régions limitrophes. *Bull. Soc. Bot. Centre-Ouest, n.s.* 8: 133-138. (Includes a list of plants in need of national protection.)

Deschatres, R. (1982). Plantes rares, plantes menacées, plantes protegées. *Rev. Scient. Bourb.* 3-24. (Not seen.)

Jovet, P. and Aymonin, G.G. (1980). Phénomènes d'appauvrissement dans une flore locale et leur signification générale: L'exemple du Pays Basque occidental français. *C.R. Soc. Biogéogr.* 489: 31-40.

Le Brun, P. (1959). Plantes rares et menacées de la France mediterranéene. In *Animaux et Végétaux de la Région Mediterranéene*, Proceedings of the IUCN 7th Technical Meeting, vol. 5. IUCN, Brussels. Pp. 103-111.

Mériaux, J.-L. (1982). Espèces rares ou menacées des biotopes lacustres et fluviatiles du nord de la France. In Symoens, J.J., Hooper, S.S. and Compère, P. (Eds), *Studies on Aquatic Vascular Plants*, Proceedings of the International Colloquium on Aquatic Vascular Plants, 23-25 January 1981, Brussels. Société Royale de Botanique de Belgique, Brussels. Pp. 398-402.

Royer, J.-M. (1971). Repartition et écologie de quelques plantes rares de la côte calcaire de Saône-et-Loire. *Bull. Mens. Soc. Linn. Lyon* 40(8): 243-249. (Maps.)

Included in the European threatened plant list (Threatened Plants Unit, 1983, cited in Appendix 1); latest IUCN statistics, based upon this work: endemic taxa – Ex:4, E:7, V:10, R:24, I:2, K:15, nt:11; doubtfully endemic taxa – V:1, R:1, K:1; non-endemics rare or threatened worldwide – E:3, V:39, R:22, I:8 (world categories).

In 1982 IUCN, under contract to the EEC through the U.K. Nature Conservancy Council, prepared a report (unpublished), *Threatened Plants, Amphibians and Reptiles, and Mammals (excluding Marine Species and Bats) of the European Economic Community*, which includes data sheets on 16 French Endangered plant species. *The IUCN Plant Red Data Book* (1978) includes 4 species for France.

Laws Protecting Plants Under the "Protection de la nature" Law, No. 76-629 of 1976, general protection is given to wild plants "where their conservation is considered justified". More recently (13 May 1982) a list of protected plant species was published (Anon, 1982 and 1983) granting 2 levels of protection under this law to c. 400 species of pteridophytes and angiosperms. Over 300 of these species receive complete protection throughout the country from picking, collection, uprooting and sale. For the remaining species, it is forbidden to destroy all or part of them; their collecting, harvesting or transport may be authorized by the Ministère de L'Environnement et du Cadre de Vie.

Anon (1982). Listes des espèces végétales protégées sur l'ensemble du territoire national. *J. Off. Rép. Française*, 13 May, 1982. Pp. 4559-4562.

Anon (1983). Listes départementales des espèces végétales protégées sur l'ensemble du territoire national. *Bull. Soc. Bot. Centre-Ouest. n.s.* 14: 13-16.

Voluntary Organizations

Fédération Française des Sociétés de Protection de la Nature (FFSPN), 57 rue Cuvier, 75005 Paris.

Société Botanique de France, rue J.-B. Clément, 92290 Châtenay-Malabry, C.C.P. Paris 1528.

Société Nationale de Protection de la Nature et d'Acclimation de France (SNPN). (Address as for the FFSPN.)

WWF-France (Association Française du World Wildlife Fund), 14 rue de la Cure, 75016 Paris.

Some members of FFSPN:

Fédération Rhone-Alpes pour la Protection de la Nature (FRAPNA), Univ. Claude Bernard, 43 bd, 69622 Villeurbane Cedex.

Société pour l'Etude et la Protection de la Nature en Bretagne (SEPNB), BP 32, 29276 Brest Cedex.

Botanic Gardens Many, as listed in Henderson (1983), cited in Appendix 1; only subscribers to the Botanic Gardens Conservation Co-ordinating Body are given below:

Conservatoire Botanique de Porquerolles, Parque National de Port-Cros, 50 Avenue Gambetta, 83400 Hyères. (Conservation activities described in *Threatened Plants Committee – Newsletter* No. 8: 11, 1981.)

Conservatoire Botanique du Stangelarc'h, 29200 Brest.

Jardins Botaniques de la Ville de Nice, 20 Traverse des Arboras, 06200 Nice.

Jardins Botaniques de Nancy, 100 rue du Jardin Botanique, 54600 Villers-Les-Nancy.

Useful Addresses

CITES Management Authority: Direction de la Protection de la Nature, Convention de Washington, Secretariat d'Etat auprés du Premier Ministre, charge de l'Environnement et de la Qualité de la Vie, 14 bd du General-Leclerc, 92524 Neuilly-sur-Seine.

CITES Scientific Authority: Secretariat Faune Flore, Museum National d'Histoire Naturelle, 57 rue Cuvier, 75231 Paris Cedex 05.

Additional References

Bordas (Ed.) (1979). *Guide de la Nature en France*. Paris. 504 pp. (Includes description of flora and vegetation.)

Olivier, L. (1979). Multiplication and re-introduction of threatened species of the littoral dunes in Mediterranean France. In Synge, H. and Townsend, H. (Eds), *Survival or Extinction*. Proceedings of a Conference 11-17 September 1978, Kew. Bentham-Moxon Trust, Kew. Pp. 91-93.

Rey, P. and Dupias, G. (Eds) (1969). *Carte de la Végétation de la France*, 1:200,000. Centre National de la Recherche Scientifique (CNRS), Toulouse, Paris.

France: Corsica

(A *département* of France)

Area 8723 sq. km

Population 230,100 (1981 estimate, *Times Atlas*, 1983)

Floristics 2159-2250 native vascular species estimated by D.A. Webb (1978, cited in Appendix 1) from *Flora Europaea*; 2500 species according to Gamisans (1982). 31 endemic taxa (IUCN figures); 170 according to Contandriopoulos (1964); as many as 250 estimated by M. Conrad (1984, *in litt.*), but this includes varieties, not normally included by IUCN (and *Flora Europaea*) for Europe, and transfrontier endemics. Mediterranean element dominant.

Vegetation Coastal and lowland vegetation much modified by agriculture and tourism but maquis still widespread up to 800 m, especially on siliceous soils, with scattered oaks (*Quercus ilex*, *Q. suber*) and pine (*Pinus halepensis*). In the supra-mediterranean zone (800-1000 m), mixed deciduous and evergreen woodland (*Q. pubescens*, *Pinus nigra* ssp. *laricio*, *Castanea sativa*); in the mountain zone (1000-1700 m), mainly forest of beech and pine (*P. nigra* ssp. *laricio*); in the north a subalpine zone (1600-2100 m) with white fir (*Abies alba*) or bushland with alder; in the south, between 1800-2200 m, shrub belt with juniper; alpine belt (above 2100 m) of species-rich grassland. For a vegetation map see Dupias *et al.* (1965).

Checklists and Floras See under France and:

Bouchard, J. (1977). Flore pratique de la Corse, 3rd Ed. Numéro spécial du *Bull. Soc. Sci. Hist. Nat. Corse*, No. 7. Société des Sciences Historiques et Naturelles de la Corse, Bastia. 407 pp. (Lists endemic taxa; phytogeography; maps; line drawings.)

Conrad, M. (1974-). *Flora Corsicana Iconographia: Flore de la Corse: Iconographie des Espèces et Variétés Endémiques Corses, Cyrno-sardes et Tyrrhéniennes*. L'Association pour l'Etude Ecologique du Maquis (APEEM), Laboratoire d'Ecologie de Pirio, Manso, Corse. 5 fascicles published, 2 in press. (Colour plates.)

Gamisans, J. (1982). Catalogue abrégé de la Flore de la Corse. *Trav. Sci. Parc Nation.* 8: 25-671.

Litardière, R. de and Briquet, J. (1936-1955). *Prodrome de la Flore Corse*, 3 vols.

Field-guides

Conrad, M. (1973). *Promenades en Corse parmi ses Fleurs et ses Forêts*. Archives départementales, Ajaccio.

Information on Threatened Plants See under France and:

Conrad, M. and Gamisans, J. (Eds) (n.d.). Les espèces végétales les plus menacées en Corse. Conservatoire Botanique de Porquerolles. (Unpublished.)

Gamisans, J., Conrad, M. and Olivier, L. (1981). Inventaire des espèces rares ou menacées de la Corse; la situation des espèces menacées de la Corse. Conservatoire Botanique de Porquerolles, Hyères. (2 unpublished reports; describe conservation status and habitats of over 300 rare or threatened taxa.)

A programme to monitor the status of rare and threatened plants in Corsica is being undertaken at the Conservatoire Botanique de Porquerolles, Hyères, in association with the Parc Naturel Régional de la Corse. This includes maintaining a list of rare and threatened taxa, protecting their localities in the wild, developing a seed bank and maintaining stocks in cultivation.

Included in the European threatened plant list (Threatened Plants Unit, 1983, cited in Appendix 1); latest IUCN statistics, based upon this work: endemic taxa – Ex:1, E:2, V:2, R:11, nt:15; doubtfully endemic taxa – E:1; non-endemics rare or threatened worldwide – E:2, V:5, R:4 (world categories).

Laws Protecting Plants See under France. A separate ordinance has recently been proposed for species threatened in Corsica only.

Voluntary Organizations

Association des Amis du Parc Naturel Régional de la Corse, Palais Lantivy, avenue du Général Fiorella, 20,000 Ajaccio.

Société des Sciences Historiques et Naturelles de la Corse, 36 rue César Campinchi, 20200 Bastia.

Botanic Gardens See under France. The following botanic gardens actively participate in the conservation of the Corsican flora:

Conservatoire Botanique de Porquerolles, 50 Avenue Gambetta, 83400 Hyères, France.
Conservatoire Botanique du Stangalarc'h, 29200 Brest, France.
Conservatoire et Jardin Botaniques, Case postale 21, 1211 Geneve 21, Switzerland.
Jardin Botanique de l'Université Liège, Sart Tilman, 4000 Liège, Belgium.

Useful Addresses

Association pour l'Etude Ecologique de Maquis (APEEM), Lycée Giocante de Casabianca, 20200 Bastia.

Comité pour l'inventaire des zones naturelles d'intérêt écologique, faunistique et floristique, Credec, 1 Avenue du Colonel Feracci, 20250 Corte.

Additional References

Aymonin, G.G. (1975). La nature Corse: menaces et espoirs (propos préliminaire). *Bull. Soc. Bot. France* 121: 5-8.

Brun, B., Conrad, M. and Gamisans, J. (1975). *La Nature en France: Corse.* Horizons de France, Paris.

Contandriopoulos, J. (1964). Recherches sur la flore endémique de la Corse et sur ses origines II. *Rev. Gén. Bot.* 71(845): 361-384.

Delvosalle, L. (1953). Aspects végétaux de la Corse. *Naturalistes Belges* 34(12): 234-248.

Dupias, J. (1976, 1978). La végétation des montagnes Corses. *Phytocoenologie* 3: 4; 4:1-4.

Dupias, G., Gaussen, H., Izard, M. and Rey, P. (1965). *Carte de la Végétation de la France.* No. 80-81 Corse. CNRS, Paris. (Text and map, 1:200,000.)

Gamisans, J. (1970-1983). Contribution à l'étude de la flore Corse. *Candollea* 25(1): 105-141 (1970); 26(2): 309-358 (1971); 27(1): 47-63 (1972); 27(2): 189-209 (1972); 28(1): 39-82 (1973); 29(1): 39-55 (1974); 32(1): 51-72 (1977); 36(1):1-17 (1981); 38(1): 217-235 (1983).

Gamisans, J. (1977). La végétation des montagnes Corses. *Phytocoenologie* 4(1): 35-131; 4(2): 133-179; 4(3): 317-376. (Several papers, giving detailed phytosociological accounts.)

Gamisans, J. (1980). Bibliographie Botanique Corse, 1955-1979. *Candollea* 35(1): 211-221.

Litardière, R. de (1928-1955). Nouvelles contributions à l'étude de la flore de la Corse. 9 fascicles. *Arch. Bot.* (1928-1930) and *Candollea* (1931-1955).

French Guiana

French Guiana is an overseas *département* of France on the Atlantic north-east coast of South America.

Area 91,000 sq. km

Population 72,000

Floristics de Granville (1982) estimates 6000-8000 species of vascular plants; J.C. Lindeman (1984, pers. comm.), however, estimates 8000 species of vascular plants for all 3 Guianas, implying the total for French Guiana is rather lower. Cremers (1984) estimates about 5000 plant species. Affinities with Amazonian forest flora; still imperfectly known.

Vegetation Over 90% of the country, undisturbed equatorial rain forest of Amazon type; above 500 m small areas of cloud forest, rich in endemics; along the coast a thin strip of mangrove; covering less than 1.7% of the land are coastal swamps and wet and dry savannas and rock savannas on granite outcrops (de Granville, 1982). Estimated rate of deforestation for closed broadleaved forest 10 sq. km/annum out of 89,000 sq. km (FAO/UNEP, 1981).

Checklists and Floras Covered by the family and generic monographs of *Flora Neotropica*, described in Appendix 1. The country Floras are:

Béna, P. (1966). *Essences forestières de Guyane.* Bureau Agricole et Forestier Guyanais, Imprimerie National, Paris. 488 pp. (Trees, illus.)

Benoist, R. (1933). *Les Bois de la Guyane Française.* Ed. des Archives de Botanique, Caen, France.

Granville, J.-J. de (1978). *Recherches Sur la Flore et La Vegetation Guyanaises.* Université des Sciences et Techniques du Languedoc, Montpellier. Thesis. 272 pp.

Lemée, A. (1952-56). *Flore de la Guyane Française,* 4 vols. Librairie Lechevalier, Paris. (Descriptions; keys only to genera, in selected families.)

A 30-year project to prepare the *Flora of the Guianas* is being coordinated by the Institute of Systematic Botany, University of the Utrecht, The Netherlands, and the Smithsonian Institution, Washington, D.C., in collaboration with Office de la Recherche Scientifique et Technique Outre-Mer, Cayenne, French Guiana, and other leading botanical institutions. Part 1 (Cannaceae, Musaceae and Zingiberaceae by P.J.M. Maas) is in press.

Field-guides

Cremers, G. (1982). *Végétation et Flore illustrée des savanes: l'example de la Savane Bordelaise.* Collection "La Nature de l'Homme en Guyane", ORSTOM, Cayenne.

Detienne, P., Jacquet, P. and Mariaux, A. (1982). *Manuel d'identification des Bois Tropicaux.* Tome 3: Guyane française. Centre Technique Forestier Tropical. Nogent sur Marne, France.

Granville, J.-J. de (1981). *Flore et Végétation.* Office Départemental du Tourisme de la Guyane. Cayenne.

Information on Threatened Plants J.-J. de Granville and G. Cremers have prepared a list of 90 very rare species endemic to French Guiana, 35 of them not yet described, and of 172 non-endemic species very rare in French Guiana. According to de Granville (1984, pers. comm.), the endemic list "will certainly increase in the future". These lists form the basis for a list of 14 botanical reserves proposed by de Granville (1975).

Voluntary Organizations

IBIS, Mouvement pour le Respect et la Conservation du Patrimoine Naturel Guyanais, 99 rue du Lieutenant Becker, 97300 Cayenne.

Société pour l'Etude, la Protection et l'Aménagement de la Nature en Guyane (SEPANGUY), c/o Services Vétérinaires, Avenue Pasteur, B.P. 411, 97300 Cayenne, and B.P. 120, 97310 Kourou.

Botanic Gardens

Institut de Botanique, ORSTOM, Ronte de Montabo, B.P. 165, 97301 Cayenne Cedex. (Very small.)

Jardin Botanique Municipal, 97300 Cayenne. (No plants from French Guiana.)

Useful Addresses

Délégation Régionale à l'Architecture et à l'Environnement (Guadeloupe-Guyane-Martinique), B.P. 1002, 97178 Pointe-à-Pitre Cedex, Guadeloupe.

Institute of Systematic Botany, University of Utrecht, Heidelberglaan 1, P.O. Box 80102, 3508 TC Utrecht, Netherlands.

Office de la Recherche Scientifique et Technique Outre-Mer (ORSTOM), Route de Motabo, B.P. 165, 97305 Cayenne Cedex.

CITES Management Authority: Direction de la Protection de la Nature Convention de Washington, Secrétariat d'Etat auprès du Premier Ministre, Chargé de l'Environnement et de la Qualité de la Vie, 14 bd du Général Leclerc, 92524 Neuilly-Sur-Seine, France.

CITES Scientific Authority: Secrétariat Faune et Flore, Muséum National d'Histoire Naturelle, 35 rue Cuvier, 75231 Paris Cedex 05, France.

Additional References

Atlas des Départements d'Outre-mer: 4 – La Guyane (1979). CNRS/ORSTOM, Paris, France. (Maps with chapters on topography, geology, geomorphology, pedology, hydrology, vegetation and climate.)

Benoist, R. (1924, 1925). La végétation de la Guyane Française. *Bull. Soc. Bot. France* 71: 1169-1177; 72: 1066-1078.

Cremers, G. (1984). L'Herbier du Centre ORSTOM de Cayenne à 25 ans. *Taxon* 33: 428-432.

Granville, J.-J. de (1975). Projets de réserves botaniques et forestières en Guyane. ORSTOM, Cayenne. 29 pp. (16 maps.)

Granville, J.-J. de (1978). *Recherches sur la flore et la végétation Guyanaises*. Doctor's Thesis, Univ. Languedoc, Montpellier. 277 pp.

Granville, J.-J. de (1982). Rain forest and xeric flora refuges in French Guiana. In Prance, G. (Ed.) (1982), cited in Appendix 1. Pp. 159-181. (Vegetation map.)

Hoock, J. (1971). *Les savanes guyanaises: Kourou. Essai de phytoécologie numérique.* Mémoire ORSTOM No. 44, Paris.

Gabon

Area 267,667 sq. km

Population 1,146,000

Floristics c. 8000 species in the forests (F.J. Breteler, 1984, *in litt.*); c. 6,000 species (Floret, 1976; Lebrun, 1976, cited in Appendix 1); no accurate figure for endemism available, but out of the 23 parts of the *Flore du Gabon* published by 1978, 243 species out of 1333 total (just over 22%) were endemic (Brenan, 1978, cited in Appendix 1). Floristic affinities Guinea-Congolian.

Vegetation Predominantly lowland rain forest, with mangrove and swamp forest at the coast and considerable areas of secondary grassland; forest as a whole covers 85% of the area. Estimated rate of deforestation for closed broadleaved forest 150 sq. km/annum out of 205,000 sq. km (FAO/UNEP, 1981). According to Myers (1980, cited in Appendix 1), who gives the same figure for the coverage of moist forest, evergreen rain forest covers 25,000 sq. km near the coast; most of the remainder is evergreen or semi-deciduous moist forest. As Gabon's forests are relatively intact and floristically rich, they are likely to become an important target area for plant conservation. The forests near the coast are the least well preserved, supporting the densest population.

For vegetation map see White (1983), cited in Appendix 1.

Checklists and Floras
Aubréville, A. *et al.* (Eds) (1961-). *Flore du Gabon*, 25 fasc. Muséum National d'Histoire Naturelle, Paris. (About a quarter completed; 62 families covered so far, including Caesalpiniaceae and Rubiaceae.)

Saint Aubin, G. de (1963). *La Forêt du Gabon*. Publication No. 21, Centre Technique Forestier Tropical, Nogent-sur-Marne. 208 pp. (Descriptions, distributions; black and white photographs throughout.)

Field-guides Letouzey (1969-1972), cited in Appendix 1, contains information about the forests of Gabon.

Information on Threatened Plants No published lists of rare or threatened plants; IUCN has records of 340 species and infraspecific taxa believed to be endemic: E:1, V:9, R:44, I:31, nt:6; the remainder are K.

Botanic Gardens
Arboretum de Sibange, near Libreville.

Useful Addresses
National Herbarium of Gabon, CENAREST, B.P. 842, Libreville.

Additional References
Catinot, R. (1978). The forest ecosystems of Gabon: an overview. In Unesco-UNEP/FAO (1978), cited in Appendix 1. Pp. 575-579.

Floret, J.J. (1976). Flore du Gabon. In Miège, J. and Stork, A.L. (1975, 1976), cited in Appendix 1, pp. 575-580.

Hallé, N. and Le Thomas, A. (1968). Gabon. In Hedberg, I. and O. (1968), cited in Appendix 1. Pp. 111-112.

Heitz, H. (1943). *La Forêt du Gabon*. Larose, Paris. 292 pp. (Descriptions, references; numerous line drawings and black and white photographs.)

Pellegrin, F. (1924-1938). La flore du Mayombe d'après les récoltes de M. Georges Le Testu. *Mém. Soc. Linn. Normandie* 26(2) (1924), 126 pp.; n.s. 1(3) (1928), 85 pp.; n.s. 1(4) (1938), 115 pp. (Covers only the south-central uplands.)

Galápagos Islands

45 volcanic islands and islets on the equator, in the Pacific Ocean c. 972 km west of Ecuador, of which they are a part. Most of the islands are relatively low; however, Isabela and Fernandina have volcanoes reaching 1500 m. The highest point is on Isabela (1707 m). About 92% of the land area is included in the National Park. In 1978, the islands were designated a World Heritage Site under the World Heritage Convention.

Area 7844 sq. km

Population 4037 (1974 census, *Times Atlas*, 1983)

Floristics 543 indigenous vascular taxa, of which 229 endemic (IUCN figures from Porter, 1978 and in prep.). Most of the endemics occur in the arid and *Scalesia* zones. The flora is mostly related to that of adjacent South America (Porter, 1984).

Vegetation Coastal mangroves; *Crytocarpus* and *Maytenus* forest up to 10 m altitude; arid zone up to 300 m, with cacti, *Acacia*, *Erythrina* and *Scalesia*; transition zone, mainly between 75-180 m, with *Pisonia*, *Tournefortia* and *Bursera*; humid zone above 180 m, with dense evergreen *Scalesia* forests from 180-550 m; closed *Miconia* scrub and evergreen *Xanthoxylum* forest between 400-700 m; pampa or fern-sedge zone from 550 m to the summits of most volcanoes. The summits of Cerro Wolf and Cerro Azul on Isabela are arid. For more detailed description of vegetation types see Hamann (1981), and Wiggins and Porter (1971).

All the larger islands, including Isabela, San Cristóbal, Santa Cruz and Santa María, have extensive areas of humid upland vegetation, threatened by overgrazing; the smaller islands are drier, and almost entirely covered by arid zone vegetation.

Checklists and Floras

Wiggins, I.L. and Porter, D.M. (1971). *Flora of the Galápagos Islands*. Stanford Univ. Press, California. 998 pp. (Treats 702 taxa; introduction covers geography, vegetation, fauna.)

Information on Threatened Plants The main list is:

Porter, D.M. (in prep.). Red Data Bulletin: Galápagos Islands. (232 endemic vascular plant taxa with notes on their distribution and conservation status.)

21 species are listed as threatened in Organización de los Estados Americanos (1967), cited in Appendix 1. Latest IUCN statistics: endemic taxa – E:9, V:15, R:111, I:15, K:2, nt:77.

An index of threatened plants in cultivation is:

Threatened Plants Unit, IUCN Conservation Monitoring Centre (1984). *The Botanic Gardens List of Rare and Threatened Species of the Galapagos and Juan Fernandez Islands*. Botanic Gardens Conservation Co-ordinating Body, Report No. 11. IUCN, Kew. 6 pp. (Lists 17 rare and threatened taxa, from the Galápagos, reported in cultivation, with gardens listed against each.)

Useful Addresses

Charles Darwin Foundation for the Galápagos, Casilla 3891, Quito, Ecuador.
(Publishes a journal, *Noticías de Galápagos*, on conservation issues and research on the islands.)

Charles Darwin Research Station, Bahia Académia, Isla Santa Cruz, Galápagos.

Superintendente Parque Nacional Galápagos, Puerto Ayora, Isla Santa Cruz, Galápagos.

Additional References

Bowman, R.I. (Ed.) (1966). *The Galápagos: Proceedings of the Galápagos International Scientific Project of 1964*. Univ. of California Press, Berkeley. 318 pp. (Covers physical environment, flora, fauna, evolution and adaptation of biota. See in particular I.L. Wiggins on the origins and relationships of the flora, pp. 175-182; E.Y. Dawson on cacti, pp. 209-214; and C.M. Rick on some plant-animal relationships, pp. 215-224.)

Carlquist, S. (1965), cited in Appendix 1. (Origin, evolution and adaptations of plants and animals.)

Carlquist, S. (1974), cited in Appendix 1. (Dispersal and evolution of plants and animals; separate chapter on flora.)

Hamann, O. (1979). The survival strategies of some threatened Galápagos plants. *Notícias de Galápagos* 30: 22-25.

Hamann, O. (1981). Plant communities of the Galápagos Islands. *Dansk Bot. Arkiv* 34(2). 163 pp. (Detailed analysis of plant communities; recent changes to vegetation.)

Kramer, P. (1983). The Galápagos: islands under siege. *Ambio* 12(3-4): 186-190.

Perry, R. (Ed.) (1984). *Key Environments: Galápagos*. Pergamon Press, Oxford. (Physical geography, fauna, flora, conservation problems.)

Porter, D.M. (1978). Galápagos Islands vascular plants. In Bramwell, D. (Ed.), *Plants and Islands*. Academic Press, London. Pp. 225-256.

Porter, D.M. (1984). Relationships of the Galápagos flora. *Biol. J. Linn. Soc.* 21: 243-251.

Schofield, E.K. (1973). Annotated bibliography of Galápagos botany, 1836-1971. *Ann. Missouri Bot. Gard.* 60: 461-477. (286 references.)

Schofield, E.K. (1973). A unique and threatened flora. *Gard. J. New York Bot. Gard.* 23: 68-73.

Schofield, E.K. (1973). Galápagos flora: the threat of introduced plants. *Biol. Conserv.* 5: 48-51.

Schofield, E.K. (1980). Annotated bibliography of Galápagos botany. Supplement 1. *Brittonia* 32(4): 537-547.

Werff, H.H. van der (1978). *The Vegetation of the Galápagos Islands: Proefschrift*. Lakenman and Ochtman, Zierikzee, Netherlands. 102 pp. (Includes checklist.)

Werff, H.H. van der (1979). Conservation and vegetation of the Galápagos Islands. In Bramwell, D. (Ed.), *Plants and Islands*. Academic Press, London. Pp. 391-404. (Describes vegetation types; conservation priorities.)

Gambia

Area 10,689 sq. km

Population 630,000

Floristics 530 species (Jarvis, 1980), with c. 3 endemics.

Flora in eastern half of country with Sudanian affinities; western half with Guinea-Congolian and Sudanian affinities.

Vegetation Most covered by Sudanian woodland without characteristic dominants. Coastal area with mangrove vegetation, and small area of evergreen forest interspersed with secondary grassland and cultivation. Estimated rate of deforestation for closed broadleaved forest 22 sq. km/annum out of 650 sq. km (FAO/UNEP, 1981).

For vegetation map see White (1983), cited in Appendix 1.

Checklists and Floras Gambia is included in the *Flora of West Tropical Africa*, cited in Appendix 1.

Jarvis, A.C.E. (1980). A checklist of Gambian plants. Cyclostyled. 30 pp. (530 species listed.)

Percival, D.A. (1968). The common trees and shrubs of the Gambia. Cyclostyled. 62 pp. (142 species described.)

Williams, F.N. (1907). Florula Gambica. *Bull. Herb. Boissier, Sér. 2*, 7: 81-96, 193-208, 369-386. (Annotated checklist of 285 species.)

Information on Threatened Plants No published lists of rare or threatened plants; IUCN has records of 3 species and infraspecific taxa believed to be endemic; no categories assigned.

Useful Addresses

CITES Management and Scientific Authorities: Wildlife Conservation Department, Ministry of Water Resources and the Environment, 5 Marina Parade, Banjul.

Additional References

Rosevear, D.R. (1937). Forest conditions of the Gambia. *Emp. For. J.* 16: 217-226.

Gambier Islands

The Gambier Islands (or Mangaréva) are a group of volcanic islands and atolls, 5600 km east of New Caledonia in the South Pacific Ocean, at 23°10'S and 135°W. The highest point is 441 m, on Mangaréva Island, the largest island in the Gambier group. The islands form part of the Tuamotu-Gambier administrative division of French Polynesia.

Area 25 sq. km

Population 585 (1983)

Floristics About 250 vascular plant species, including introductions. 41 native vascular plant species, of which 11 are endemic (Huguenin, 1974).

Vegetation Only fragments of the original forest remain, most having been decimated by burning and overgrazing by goats. Apart from small areas of forest on the precipitous southern slopes of Mt Mokoto, Mangaréva Island is mainly covered by *Miscanthus* grassland. Coconuts have been introduced on most islands in the group (Cooke, 1935; Douglas, 1969, cited in Appendix 1).

Checklists and Floras The *Flora of Southeastern Polynesia* (Brown and Brown, 1931-1935, cited in Appendix 1) includes only 29 indigenous flowering plants for the Gambier Islands and Pitcairn Island District. See also:

Copeland, E.B. (1932). Pteridophytes of the Society Islands. *Bull. Bernice P. Bishop Mus.* 93. 86 pp. (Descriptions and keys; notes on distribution.)

Information on Threatened Plants All endemics suspected to be threatened by overgrazing and fire (F.R. Fosberg, 1984, *in litt.*). *Achyranthes mangarevica* is included in *The IUCN Plant Red Data Book* (1978) as Extinct or possibly Endangered; *Gouania mangarevica* is only known from Mangaréva Island and probably also Extinct or Endangered.

Additional References
Cooke, C.M. (1935). Mangarevan expedition. *Bull. Bernice P. Bishop Mus.* 133: 33-71. (Includes description of vegetation.)
Huguenin, B. (1974). La végétation des Iles Gambier, relevé botanique des espèces introduites. *Cahiers du Pacifique* 18(2): 459-471.

German Democratic Republic

Area 108,177 sq. km

Population 16,658,000

Floristics 1842 native vascular species (Rauschert *et al.*, 1978). 3 endemic species, 1 of them extinct (IUCN figures). Areas of high floristic diversity: vicinity of Thüringer Becken in the south-west, the Harz mountains of the west (D. Benkert, 1984, *in litt.*). Elements: Atlantic, Central European, Boreal and subalpine.

Vegetation Mostly an agricultural landscape, especially in the glaciated north-central lowland depression. In the north, oaks, pine and beech constitute main woodland cover, but most now removed or replaced by conifer plantations. Scattered beechwoods still survive along Baltic coast. In the south, vertical zonation of oak and hornbeam forests, giving way to montane beech forests, and above 500 m, forests of beech, fir and spruce. Subalpine and alpine vegetation restricted to small area in Harz mountains. Habitats under greatest threat: grasslands, heathlands and wetlands (Benkert, *in litt.*).

Checklists and Floras The German Democratic Republic is included in the completed *Flora Europaea* (Tutin *et al.*, 1964-1980, cited in Appendix 1), although plants in G.D.R. are not distinguished from those in the Federal Republic. Also see Oberdorfer (1983), cited in Appendix 1, and:

Rothmaler, W. (1970-1984). *Exkursionsflora für die Gebiete der DDR und der BRD*, 4 vols. Volk und Wissen, Berlin. Covers both F.R.G. and G.D.R.; 1 - Niedere Pflanzen (Lower plants), 9th Ed. (1984) by R. Schubert, H.H. Handke and

H. Pankow; 2 – Gefässpflanzen (Vascular plants), 11th Ed. (1982) by W. Rothmaler, W. Meusel and R. Schubert; 3 – Atlas der Gefässpflanzen (Atlas of Vascular Plants), 5th Ed. (1970) by W. Rothmaler; 4 – Kritischer Band, 5th Ed. (1983) by R. Schubert, W. Vent and M. Bässler.

Information on Threatened Plants The national threatened plant list is:

Rauschert, S., Benkert, D., Hempel, W. and Jeschke, L. (1978). *Liste der in der Deutschen Demokratischen Republik Erloschenen und Gefährdeten Farn- und Blütenpflanzen.* Kulturbund der D.D.R., Berlin. 56 pp. (Lists over 500 threatened taxa and their status in individual Districts; colour photographs.)

District threatened plant lists include:

Benkert, D. (1978). Liste der in den brandenburgischen Bezirken erloschenen und gefährdeten Moose, Farn- und Blütenpflanzen (List of extinct and endangered mosses, ferns and flowering plants in the Brandenburg District). *Naturschutzarbeit in Berlin und Brandenburg* 14(2/3): 34-80. (Black and white photographs.)

Benkert, D. (1982). Vorläufige Liste der verschollenen und gefährdeten Grosspilzarten der DDR. *Boletus* 6(2): 21-32. (A preliminary list of missing and endangered fungi.)

Benkert, D. (1984). Die verschollenen und vom Aussterben bedrohten Blütenpflanzen und Farne der Bezirke Potsdam, Frankfurt, Cottbus and Berlin. (Extinct and threatened vascular plants and ferns in the Districts of Potsdam, Frankfurt, Cottbus and Berlin.) *Gleditschia* 11: 251-259.

Benkert, D., Succow, M. and Wisniewski, N. (1981). Zum Wandel der floristischen Artenmannigfaltigkeit in der DDR (On the changes in the floristic composition of the flora of the G.D.R.). *Gleditschia* 8: 11-30. (Results of a survey about problems of species protection with regard to the influence of man on the environment; briefly discusses degree of threat to individual species, especially orchids and threatened plant communities.)

Fukarek, F. (1980). Über die Gefährdung der Flora der Nordbezirke der DDR. *Phytocoenologia* 7: 174-182. (English abstract.)

Hempel, W. (1978). *Verzeichnis der in den Drei Sächsischen Bezirken (Dresden, Leipzig, Karl-Marx-Stadt) vorkommenden Wildwachsenden Farn- und Blütenpflanzen mit Angabe ihrer Gefährdungsgrade* (Index of native ferns and flowering plants in 3 districts and their conservation status). Bezirksnaturschutzorganen, Dresden. 65 pp.

Jeschke, L. *et al.* (1978). Liste der in Mecklenburg (Bezirke Rostock, Schwerin und Neubrandenburg) erloschenen und gefährdeten Farn- und Blütenpflanzen. *Botanischer Rundbrief für den Bezirk Neubrandenburg* 8: 1-29. (Lists over 600 extinct and endangered plant taxa.)

Rauschert, S. (1980). Liste der in den thüringischen Bezirken Erfurt, Gera und Suhl erloschenen und gefährdeten Farn- und Blütenpflanzen. *Landschaftspflege und Naturschutz in Thüringen* 17(1): 1-32.

Rauschert, S. *et al.* (1978). Liste der in den Bezirken Halle und Magdeburg erloschenen und gefährdeten Farn- und Blütenpflanzen. *Naturschutz und naturkundliche Heimatforschung in den Bezirken Halle and Magdeburg* 15(1): 1-31.

A list of endangered plant communities is in preparation (Benkert, *in litt.*)

Included in the European threatened plant list (Threatened Plants Unit, 1983, cited in Appendix 1); latest IUCN statistics, based upon this work: endemic taxa – Ex:1; E:1; nt:1; non-endemics rare or threatened worldwide – V:12, R:3, I:2 (world categories).

Laws Protecting Plants National legislation was passed on 6 July 1970: Anordnung zum Schutze von Wildwachsenden Pflanzen und Nichtjagdbaren Wildebenden Tieren (Order for the protection of wild growing plants and wild animals). This provides protection for 26 species of pteridophytes and angiosperms and the native species of 13 named genera. For more details see:

Weinitschke, H. (Ed.) (1971). *Gesetzliche Regelungen der Sozialistischen Landeskultur in der DDR*. Kulturbund der Deutschen Demokratischen Republik, Zentrale Kommission Natur und Heimat des Präsidialrates, Zentraler Fachausschluss Landeskultur und Naturschutz. 103 pp.

Botanic Gardens Many, as listed in Henderson (1983), cited in Appendix 1. Useful reference:

Ebel, F. and Rauschert, S. (1982). Die Bedeutung der Botanischen Gärten für die Erhaltung gefährdeter und vom Aussterben bedrohter heimischer Pflanzenarten (The importance of botanic gardens for the preservation of native plants which are endangered and threatened by extinction). *Arch. Naturschutz und Landschaftforsch.* 22(3): 187-199. (English summary.)

Useful Addresses

Centre for Protection and Improvement of the Environment, Schnellerstrasse 140, 1190 Berlin.

Institute of Landscape Research and Nature Conservation, 4020 Halle, Neuwerk 21.

Ministry of Environmental Protection and Water Conservation, Hans Beimler Street 70/52, 1020 Berlin.

CITES Management Authority: Ministerium für Land-Forst- und Nahrungsguterwirtschaft der D.D.R., Köpenicker Allee 39-57, 1157 Berlin.

CITES Scientific Authority: Zentrales Staatliches Amt für Pflanzenquarantäne beim Ministerium, Hermannswerder 20A, 15 Potsdam.

Additional References

Hueck, K. (1936). *Pflanzengeographie Deutschlands*. Berlin.

Rauschert, S. (1975). Floristic report on Germany (1961-1971). B. Deutsche Demokratische Republik. *Mem. Soc. Brot.* 24(2): 559-577.

Schlosser, S. (1982). Genressourcen für Forschung und Nutzung. *Naturschutz. Bezirken Halle und Magdeburg* 19: 1-96. (Contains a series of papers about the potential use of plant genetic resources in the G.D.R.; colour illus.; line drawings; maps.)

Germany, Federal Republic of

Area 248,744 sq. km

Population 61,214,000

Floristics 2476 native vascular species (Blab *et al.*, 1984); 3 endemics (IUCN figures). Elements: Central European, sub-Atlantic, sub-Mediterranean and alpine.

Vegetation Little natural vegetation due to industry, agriculture and plantation forestry. Most semi-natural vegetation survives on higher ground and in the south. Beech woods are the original natural vegetation of lowland and montane areas, together with semi-natural oak and hornbeam in the centre and south (Black Forest, Alps and Bavaria); today, however, largely replaced by pine and spruce plantations, especially in the north. Riverine woodlands replaced widely by poplar and maple plantations. Forest, including plantations, occupy about 20% of land area (Bundesamt Wiesbaden, 1983).

34% of forests threatened by acid rain (Agren, 1984). Other habitats under threat: grasslands, heathlands, peat-bogs, fens and other wetlands.

Checklists and Floras Included in the completed *Flora Europaea* (Tutin *et al.*, 1964-1980), although plants in F.R.G. are not distinguished from those in the German Democratic Republic. Covered by the *Illustrierte Flora von Mitteleuropa* (Hegi, 1935-). Both are cited in Appendix 1. Only recent national and regional Floras are listed below, Floras for individual Länder being too numerous.

Garcke, A. *et al.* (1972). *Illustrierte Flora, Deutschland und Angrenzende Gebiete*, 23rd Ed. by K. von Weihe. Parey, Berlin. 1607 pp. (Line drawings.)
Schmeil, O. and Fitschen, J. (1976). *Flora von Deutschland und seinen Angrenzenden Gebieten*, 86th Ed. by W. Rauh and K. Senghas. Quelle and Meyer, Heidelberg. 516 pp. (Illus.)

For a floristic bibliography, see Hamann and Wagenitz (1977), cited in Appendix 1, and:

Jäger, E.J. and Müller-Uri, C. (1981-1982.) Bibliographie: Gefässpflanzen Zentraleuropas. Wuchsform und Lebensgeschichte. *Terrestrische Ökologie* 1(1), 122 pp. and 1(2), 122 pp.
Merxmüller, H. and Lippert, W. (1975). Floristic Report on Germany (1961-1971). A. Bundesrepublik Deutschland. *Mem. Soc. Brot.* 24(2): 469-558. (In German.)

A floristic mapping scheme for the whole country is in progress (Zentralstelle für die Floristische Kartierung, address below). Details are given in the following papers:

Haeupler, H. *et al.* (1976). *Grundlagen und Arbeitsmethoden für die Kartierung der Flora Mitteleuropas. Anleitung für die Mitarbeiter in der Bundesrepublik Deutschland*, 2nd Ed. 75 pp. E. Goltae, Göttingen.
Niklfeld, H. (1971). Bericht über die Kartierung der Flora Mitteleuropas. *Taxon* 20(4): 545-571.
Schönfelder, P. (1983). Floristische Kartierung der Bundesrepublik Deutschland (Gefässpflanzen/Pteridophyta, Spermatophyta). *Natur und Landschaft* 58(6): 235-236.

For local atlases see:

Haeupler, H. (1976). Atlas zur Flora von Südniedersachsen. *Scripta Geobotanica* 10. 367 pp.
Haffner, P., Sauer, E. and Wolf, P. (1979). Atlas der Gefässpflanzen des Saarlandes. (Edited by the Minister für Umwelt, Raumordnung und Bauwesen.) *Wiss. Schr.-R. Obersten Naturschutzbehörde* 1.
Mergenthaler, O. (1982). Verbreitungsatlas zur Flora von Regensburg.-Hoppea, *Denkschr. Regensb. Bot. Ges.* 40(5-12): 1-297.
Seybold, S. (1977). Die aktuelle Verbreitung der höheren Pflanzen im Raum Württemberg. *Beih. z. d. Veröff. Natursch. und Landschpfl. in Baden-Württemberg* 9: 1-201.

Field-guides 1 national and 2 regional field-guides are listed in order below. See also Oberdorfer (1983), cited in Appendix 1:

Hegi, G., Merxmüller, H. and Reisigl, H. (1977). *Alpenflora: Die Wichtigeren Alpenpflanzen Bayerns, Österreichs und der Schweiz.* Parey, Berlin. 194 pp. (Covers Bavaria, Austria and Switzerland; introduction includes ecological descriptions of plant communities; lists protected plants; illus.; maps.)

Rothmaler, W. (1970-1984). *Exkursionsflora für die Gebiete der DDR und der BRD*, 4 vols. Volk und Wissen, Berlin. Covers both F.R.G. and G.D.R.; 1 - Niedere Pflanzen (Lower plants), 9th Ed. (1984) by R. Schubert, H.H. Handke and H. Pankow; 2 - Gefässpflanzen (Vascular plants), 11th Ed. (1982) by W. Rothmaler, W. Meusel and R. Schubert; 3 - Atlas der Gefässpflanzen (Atlas of Vascular Plants), 5th Ed. (1970) by W. Rothmaler; 4 - Kritischer Band, 5th Ed. (1983) by R. Schubert, W. Vent and M. Bässler.

Schauer, T. and Caspari, C. (1978). *Pflanzenführer.* BLV, München. 417 pp. (Covers F.R.G. only; over 1400 colour illus.)

See also Müller and Kast (1969) and Weber (1982).

Information on Threatened Plants Vast quantity of literature. Only the most recent national lists and those for individual Länder are given below. National lists:

Blab, J., Nowak, E., Trautmann, W. and Sukopp, H. (1984). *Rote Liste der Gefährdeten Tiere und Pflanzen in der Bundesrepublik Deutschland*, 4th Ed. Kilda-Verlag, Greven. 270 pp. (Lists threatened flowering plants, mosses, lichens, fungi and algae.)

Sukopp, H. (1974). 'Rote Liste' der in der Bundesrepublik Deutschland gefährdeten Arten von Farn- und Blütenpflanzen (1. Fassung). *Natur und Landschaft* 49(12): 315-322.

Below are lists for the Länder of Bayern, Niedersachsen, Baden-Württemberg, Hessen, Schleswig-Holstein, Nordrhein-Westfalen, Rheinland-Pfalz and Saarland, and for the Region of Senne and West Berlin:

Brinkmann, H. (1978). Schützenswerte Pflanzen und Pflanzengesellschaften der Senne. *Ber. d. Naturwiss. Ver. Bielefeld.* Pp. 36-38. (Includes a threatened fern and vascular plant list for the region of Senne.)

Foerster, E., Lohmeyer, W., Schumacher, W. and Wolff-Straub, R. (1982). Florenliste von Nordrhein-Westfalen. *Schr.-R. der LÖLF* 7. 89 pp.

Haeupler, H., Montag, A., Wöldecke, K. and Garve, E. (1983). Rote Liste Gefässpflanzen Niedersachsen und Bremen. *Fachbehörde für Naturschutz, Merkblatt* Nr. 18. 34 pp. (Edited by Niedersächsisches Landesverwaltungsamt.)

Haffner, P., Sauer, E. and Wolf, P. (1979). Atlas der Gefässpflanzen des Saarlandes. *Wiss. Schr.-R. der Obersten Naturschutzbehörde*, Vol. 1 with appendix: *Rote Liste der im Saarland ausgestorbenen und gefährdeten höheren Pflanzen.* 12 pp.

Harms, K.H., Philippi, G. and Seybold, S. (1983). Verschollene und gefährdete Pflanzen in Baden-Württemberg. Rote Liste der Farne und Blütenpflanzen (Pteridophyta et Spermatophyta), 2nd revision. *Beih. Veröff. Naturschutz Landschaftpflege Bad.-Württ.* 32. 160 pp.

Kalheber, H. *et al.* (1980). Rote Liste der in Hessen ausgestorbenen, verschollenen und gefährdeten Farn- und Blütenpflanzen. 2. *Hessische Landesanstalt für Umwelt.* 46 pp.

Korneck, D., Lang, W. and Reichert, H. (1984). *Rote Liste der in Rheinland-Pfalz ausgestorbenen, verschollenen und gefährdeten Farn- und Blütenpflanzen*, 2nd Ed. Ministerium für Soziales, Gesundheit und Umwelt.

135

Künne, H. (1974). Rote Liste bedrohter Farn- und Blütenpflanzen in Bayern. *Schr.-R. Natursch. Landschaftspfl.* 4: 1-44. (Lists 566 species; conservation categories not identical with those of IUCN; a revised list is in preparation.)

Landesamt für Naturschutz und Landschaftspflege Schleswig-Holstein (1982). Rote Liste der gefährdeten Pflanzen und Tiere Schleswig-Holsteins. *Schr.-R. Landesamt Natursch. Landschaftspfl.* 5. 149 pp.

Landesanstalt für Ökologie, Landschaftentwicklung und Forstplanung NRW (1979). Rote Liste der in Nordrhein-Westfalen gefährdeten Pflanzen und Tiere. *Schr.-R. der LÖLF Nordrhein-Westfalen* 4. 106 pp.

Raabe, E.-W. (1975). Rote Liste der in Schleswig-Holstein und Hamburg vom Aussterben bedrohten höheren Pflanzen. *Heimat* 82(7/8): 191-200. (Lists 148 threatened taxa with distribution maps.)

Schönfelder, P. *et al.* (in prep.). Entwurf zur Neufassung der Roten Liste der ausgestorbenen, verschollenen und gefährdeten Farn- und Blütenpflanzen in Bayern. 38 pp.

Sukopp, H. and Elvers, H. (Eds) (1982). Rote Liste der gefährdeten Pflanzen und Tiere in Berlin (West). *Landesentwicklung u. Umweltforschung* 11. 374 pp.

See also:

Raabe, W., Brockmann, C. und Dierssen, K. (1982). Verbreitungskarten ausgestorbener, verschollener und sehr seltener Gefässpflanzen in Schleswig-Holstein. *Mitt. Arb.-gem. Geobot. Schleswig-Holstein und Hamburg* 32. 317 pp.

Sukopp, H. (1972). Grundzüge eines Programms für den Schutz von Pflanzenarten in der Bundesrepublik Deutschland. *Schr.-R. Landschaftspfl. und Natursch.* 7: 67-79.

Sukopp, H. and Trautmann, W. (Eds) (1976). Veränderungen der Flora und Fauna in der Bundesrepublik Deutschland. Proceedings of a symposium 7-9 October 1975. *Schr.-R. Vegetationskde* 10. 409 pp. (Contains many relevant articles (in German with English summaries); see for example W. Trautmann on changes in the flora of woods and in woodland vegetation of the F.R.G. in recent decades (pp. 91-108) and J. Reichholf on ecological aspects of the changing flora and fauna in the F.R.G. (pp. 393-399).)

Sukopp, H. and Trautmann, W. (1981). Causes of the decline of threatened plants in the Federal Republic of Germany. In Synge, H. (Ed.) (1981), cited in Appendix 1. Pp. 113-116. (Identifies and discusses main threats to the flora.)

Sukopp, H., Trautmann, W. and Korneck, D. (1978). Auswertung der Roten Liste gefährdeter Farn- und Blütenpflanzen in der Bundesrepublik Deutschland für den Arten- und Biotopschutz. *Schr.-R. für Vegetationskunde* 12. 138 pp. (Detailed analysis of 2667 threatened plant taxa, threats, habitats, recommendations.)

Many plant conservation data-bases, some computerized, are underway in individual Länder. A data-base co-ordinating these activities is being developed at the Institut für Vegetationskunde, Bundesforschungsanstalt für Naturschutz und Landschaftsökologie (Federal Research Centre for Nature Conservation and Landscape Ecology) in Bonn (address below). For summary details of the individual projects in the Länder see:

Kohlhammer, W. (1983). Botanische und Zoologische Artenerbungen in der Bundesrepublik Deutschland. *Natur und Landschaft* 58(6). 255 pp.

F.R.G. is included in the European threatened plant list (Threatened Plants Unit, 1983, cited in Appendix 1); latest IUCN statistics, based upon this work: endemic taxa – E:2, nt:1; non-endemics rare or threatened worldwide – Ex:1, E:4, V:12, R:5, I:3 (world categories).

In 1982 IUCN, under contract to the EEC through the UK Nature Conservancy Council, prepared a report (unpublished), *Threatened Plants, Amphibians and Reptiles, and Mammals (excluding Marine Species and Bats) of the European Economic Community*, which includes data sheets on 10 plants Endangered in F.R.G., 3 of them Endangered on a world scale.

Two journals regularly containing plant conservation articles are *Natur und Landschaft* published by W. Kohlhammer, Köln, and *Schriftenreihe für Landschaftspflege und Naturschutz*, Bonn-Bad Godesburg. See also: *Dokumentation für Umweltschutz und Landschaftspflege*, Dokumentat. und Bibl. der Bundesforsch. Naturschutz und Landschaftsökologie.

Laws Protecting Plants The Federal Species Protection Order of 30 August, 1980 (Bundesartenschutzverordnung – BArtSchV), in accordance with Article 22 of the Federal Nature Protection Act of 1976 (Bundesnaturschaftzgesetz – BNatSchG), provides special protection for over 160 plant species. These laws are published as:

Der Bundesminister für Ernährung, Landwirtschaft und Forsten (1976). Gesetz über Naturschutz und Landschaftspflege (Bundesnaturschutzgesetz – BNatSchG) vom 20.12.76. *Bundesnaturschutzgesetz*. Der Bundesminister für Ernährung, Landwirtschaft und Forsten. 32 pp.
Der Bundesminister für Ernährung, Landwirtschaft und Forsten (1980). Verordnung über besonders geschützte Arten wildebender Tiere und wildwachsender Pflanzen (Bundesartenschutzverordnung – BArtSchV). *Bundesgesetzblatt* 1(54): 1565-1601.

See also:

Müller, T. and Kast, D. (1969). *Die Geschützten Pflanzen Deutschlands* (The Protected Plants of Germany). Schwäbischen Albvereins, Stuttgart. 348 pp. (Keys; brief morphological, biological and ecological descriptions; distribution data.)
Weber, H.C. (1982). *Geschützte Pflanzen, Merkmale, Blütezeit und Standort aller Geschützten Arten Mitteleuropas*. Belser, Stuttgart. 188 pp.

Voluntary Organizations

Bayerische Botanische Gesellschaft (Bavarian Botanical Society), Menzinger Strasse 67, 8000 München 19.
Deutsche Botanische Gesellschaft (German Botanical Society), Untere Karspüle 2, 3400 Göttingen.
Deutscher Natürschutzring e.V.(DNR), Bundesverband für Umweltschutz Kalkuhlstrasse 24, Postfach 32 02 10, 5300 Bonn 3.
Schützgemeinschaft Deutscher Wald e.V., Meckenheimer Allee 9, 5300 Bonn 1.
Stiftung zum Schutze gefährdeter Pflanzen (Institute for the Protection of Endangered Plants), Kalkuhlstrasse 24, 5300 Bonn 3.
WWF-Germany (Umweltstiftung WWF-Deutschland), Sophienstrasse 44, 6000 Frankfurt/Main 90.

Botanic Gardens Numerous, as listed in Henderson (1983), cited in Appendix 1; most gardens are engaged in species conservation activities either on a local or regional scale. Only those subscribing to the Botanic Gardens Conservation Co-ordinating Body are listed here:

Alter Botanischer Garten der Universität Göttingen, Untere Karspüle 1, 3400 Göttingen.
Botanischer Garten München, Menzingerstrasse 63-67, 8000 München.

Botanischer Garten und Museum Berlin-Dahlem, Königin-Luise-Strasse 6-8, 1000 Berlin 33.

Botanischer Garten Ruhr-Universität Bochum, Universitätstrasse 150, Postfach 102148, 4630 Bochum 1.

Botanischer Garten der Universität, Auf dem Lahnbergen, 3550 Marburg.

Botanischer Garten der Universität Düsseldorf, Universitätstrasse 1, 4000 Düsseldorf 1.

Botanischer Garten der Universität Heidelberg, Im Neuenheimer Feld 340, 6900 Heidelberg 1.

Botanischer Garten der Universität-Kiel, Olshausenstrasse 40-60, Biologiezentrum, 2300 Kiel.

Botanischer Garten der Universität Mainz, Saarstrasse 21, Postfach 3980, 6500 Mainz.

Botanischer Garten der Universität, Auf dem Lahnbergen, 3550 Marburg.

Botanischer Garten der Universität Oldenburg, Philosophenweg 41, 2900 Oldenburg.

Neuer Botanischer Garten der Universität Göttingen, Grisebachstrasse la, 3400 Göttingen.

Palmengarten der Stadt Frankfurt, Siesmeyerstrasse 61, 6000 Frankfurt/Main 1.

A seed bank for rare and threatened species has been established at the Institut für Pflanzenbau und Pflanzenzüchtung, Bundesforschungsanstalt für Landwirtschaft Braunschweig-Völkenrode, Bundesalle 50, 3300 Braunschweig.

Useful Addresses

Arbeitsgemeinschaft Beruflicher Pflanzen, Kalkuhlstrasse 24, 5300 Bonn 3.

Bundesministerium für Ernährung, Landwirtschaft und Forsten (Abt. 62 Umwelt, Naturschutz), Rochusstrasse 1, Postfach 140270, 5300 Bonn 1. (Each of the Länder have additional authorities for nature conservation.)

Institut für Vegetationskunde, Bundesforschungsanstalt für Naturschutz und Landschaftsökologie (Federal Research Centre for Nature Conservation and Landscape Ecology), Konstantinstrasse 110, 5330 Bonn 2.

TRAFFIC (Germany), WWF-Germany, address above.

Zentralstelle für die Floristische Kartierung. Bereich Nord: Ruhr-Universität, Spezielle Botanik, Postfach 102148, 4630 Bochum 1; Bereich Süd: Universität Regensburg, Botanisches Institut, Postfach 397, 8400 Regensburg.

CITES Management Authority: Bundesministerium für Ernährung, Landwirtschaft und Forsten, Referat 623, Postfach 140270, 5300 Bonn 1.

CITES Scientific Authority: Bundesamt für Ernährung und Forstwirtschaft, Postfach 180203, 6000 Frankfurt/Main 1.

Additional References

Agren, C. (1984). A report sounds the alarm: 34% of West German forest land damaged. *Acid News* 1: 6-7. (Newsletter from the Swedish and Norwegian NGO Secretariats on Acid Rain, address under Sweden.)

Bundesamt Wiesbaden (Ed.) (1983). *Statistisches Jahrbuch 1983 für die Bundesrepublik Deutschland*. W. Kohlhammer GmbH., Stuttgart. 780 pp.

Raabe, E.-W. (1978). Über den Wandel unserer Pflanzenwelt in neuerer Zeit. *Kieler Notizen zur Pflanzenkunde* 10(1/2): 1-24.

Ghana

Area 238,305 sq. km

Population 13,044,000

Floristics 3600 species (quoted in Lebrun, 1976, cited in Appendix 1); 43 endemic species (Brenan, 1978, cited in Appendix 1); diversity greatest in evergreen forests of south-west.

Flora with Sudanian (northern c. 1/2, mainly woodland and grassland flora) and Guinea-Congolian (south-west corner, mainly forest flora) affinities.

Vegetation Small patches of rain forest are all that remain of the vast area of forest which used to cover rather less than a third of Ghana in the south-west, but which have been replaced by cultivation and secondary grassland; various types of grassland and wooded grassland/woodland with abundant *Isoberlinia* cover most of the remainder; a coastal band of strand and mangrove.

Estimated rate of deforestation for closed broadleaved forest 220 sq. km/annum out of 17,180 sq. km (FAO/UNEP, 1981). According to Myers (1980, cited in Appendix 1), the remaining forest totals 19,864 sq. km in reserves, plus about 500 sq. km outside reserves; the rate of depletion of forest by shifting cultivation has been estimated to be as high as 5000 sq. km/annum.

For vegetation map see White (1983), cited in Appendix 1.

Checklists and Floras Ghana is included in the *Flora of West Tropical Africa*, cited in Appendix 1. See also:

Irvine, F.R. (1961). *Woody Plants of Ghana: with Special Reference to their Uses.* Oxford Univ. Press, London. 868 pp. (Plates, some in colour; line drawings.)

Information on Threatened Plants
Hedburg, I. (Ed.) (1979), cited in Appendix 1. (List for Ghana, pp. 88-91, by J.B. Hall, contains 210 species and infraspecific taxa divided between five categories of endangerment, not IUCN-compatible.)

IUCN has records of 73 species and infraspecific taxa believed to be endemic, of which 16 are known to be rare or threatened – E:1, V:5, R:9, I:1.

Botanic Gardens
Aburi Botanic Garden, Aburi.
University Botanic Garden, c/o Department of Botany, University of Ghana, P.O. Box 55, Legon.
University Botanic Garden, University of Science and Technology, Kumasi.

Useful Addresses
Forestry Department, P.O. Box 527, Accra.
National Herbarium, University of Ghana (address above).
CITES Management and Scientific Authorities: Department of Game and Wildlife, P.O. Box M 239, Ministries Post Office, Accra.

Additional References
Ahn, P.M. (1959). The principal areas of remaining original forest in western Ghana, and their potential value for agricultural purposes. *J. West Afr. Sci. Assoc.* 5(2): 91-100. (With small-scale vegetation map.)

Asibey, E.O.A. and Owusu, J.G.K. (1982). The case for high-forest national parks in Ghana. *Envir. Conserv.* 9(4): 293-304. (With five black and white photographs.)

Hall, J.B. and Swaine, M.D. (1981). *Distribution and Ecology of Vascular Plants in a Tropical Rain Forest: Forest Vegetation in Ghana.* Junk, The Hague. 383 pp. (Geobot. 1.)

Lawson, G.W. (1968). Ghana. In Hedberg, I. and O. (1968), cited in Appendix 1. Pp 81-86.

Taylor, C.J. (1952). The vegetation zones of the Gold Coast. *For. Dep. Bull.* 4: 1-12. Govt Printer, Accra. (With coloured vegetation map 1:1,500,000.)

Taylor, C.J. (1960). Vegetation. In *Synecology and Silviculture in Ghana.* Thomas Nelson, London. Pp. 31-73.

Gibraltar

Gibraltar is a Dependent Territory of the United Kingdom. 'The Rock', a limestone ridge 3 km long and 426 m high, is the dominant feature.

Area 6.5 sq. km

Population 31,000

Floristics 587 native vascular species (Wolley-Dod, 1914). 1 endemic taxon (IUCN figure). A Mediterranean flora.

Vegetation A large proportion of original vegetation on 'The Rock' cleared for water catchment constructions or replaced by conifer plantations; remaining semi-natural vegetation includes: high maquis with *Phillyrea*, *Rhamnus* and the palm *Chamaerops humilis* (on the middle slopes); and a rich chasmophytic flora containing species of *Dianthus*, *Iberis*, *Scilla* and *Saxifraga* on the inaccessible ledges and east slope.

Checklists and Floras Plants from Gibraltar are included in *Flora Europaea* (Tutin *et al.*, 1964-1980, cited in Appendix 1), but are not distinguished from plants in Spain. There are no other recent publications about the flora, so see:

Wolley-Dod, A.H. (1914). A Flora of Gibraltar and the neighbourhood. *J. Bot.* 52. Supplement. 131 pp. (Native and naturalized vascular plants of The Rock of Gibraltar and neighbouring parts of Andalucia; habitat notes.)

Also relevant:

Hamilton, A.P. (1970). *The Flowers of Gibraltar.* Gibraltar Tourist Office.

Stocken, C.M. (1969). *Andalusian Flowers and Countryside.* Privately published by the author, Devon, U.K. 184 pp. (General desciption of many aspects of Gibraltar, including vegetation.)

Field-guides

Anon (1968). *The Wild Flowers of Gibraltar and Neighbourhood.* Committee of the Gibraltar Garrison Library, Gibraltar. 79 pp. (Illus. only.)

See also Polunin and Smythies (1973), in Appendix 1.

Information on Threatened Plants None locally. The section for Gibraltar in the European threatened plants list (Threatened Plants Unit, 1983, cited in Appendix 1) contains one endemic (Rare) and 3 non-endemics rare or threatened on a regional scale, all insufficiently known on a world scale.

Useful Addresses
The Gibraltar Society, John Mackintosh Hall, Gibraltar.

Glorieuses, Iles

Two coralline islands 183 km WNW of Madagascar in the Indian Ocean, 11°34'S, 47°19'E. Grande Glorieuse is by far the larger; area c. 4 sq. km. 43 species of plant (including introductions) are listed in Battistini and Cremers (1972). The smaller island, Ile du Lys, has only 8 species.

The vegetation of Grande Glorieuse consists of a littoral dune belt with grasses, a tufted sedge turf behind it, and, inland, dense woodland 2-4 m high with no understorey. Much of the island is given over to coconut plantations. Ile de Lys has very little vegetation, and it is not zoned.

References
Battistini, R. and Cremers, G. (1972). Geomorphology and vegetation of Iles Glorieuses. *Atoll Res. Bull*. 159. 10 pp. (With vegetation map, 19 black and white photographs.)

Hemsley, W.B. (1919). Flora of Aldabra: with notes on the flora of neighbouring islands. *Bull. Misc. Inf. Kew* 1919: 108-153. (Checklist, with descriptions of new species.)

Renvoize, S.A. (1979). The origins of Indian Ocean island floras. In Bramwell, D. (Ed.), *Plants and Islands*. Academic Press, London. Pp. 107-129.

Stoddart, D.R. (Ed.) (1967). Ecology of Aldabra Atoll, Indian Ocean. *Atoll Res. Bull*. 118. 141 pp. (Includes 41 black and white photographs, list of endemic plant species, bibliography of Aldabra; see especially paper by Stoddart, pp. 53-61, on the ecology of coral islands north of Madagascar, but excluding Aldabra.)

Great Barrier Reef Islands

The Great Barrier Reef, the largest coral reef in the world, extends for more than 2000 km along the north-east coast of Australia, between 24°30'-10°41'S and 145°-154°E. At its northern end it is c. 160 km from the mainland. The reef consists of c. 2500 individual coral reefs, some with coral sand islands (or cays). The total area is c. 207,200 sq. km. The Great Barrier Reef is part of Australia.

The highest point is 40 m, but most of the reef is less than 5 m above sea-level and has no vegetation. Heron Island (23°25'S 151°55'E) has *Casuarina*, *Cordia* and *Pisonia* forest; mangrove swamps are found in the Low Islands (16°18'S 145°35'E); several sand cays support scrub vegetation (Douglas, 1969, cited in Appendix 1).

Parts of the Great Barrier Reef have been gazetted as a national park (e.g. parts of Green Island (16°43'S 146°E) and Heron Island in 1937 and 1943, respectively). The first part of the Great Barrier Reef Marine Park was proclaimed in 1979, and the whole of the reef was accepted as a World Heritage Site in 1981.

References

Bennett, I. (1973). *The Great Barrier Reef*. Frederick Warne, London.

Frankel, E. (1978). *Bibliography of the Great Barrier Reef Province*. Great Barrier Reef Marine Park Authority, Canberra. (Lists 4444 publications.)

Greece

Area 131,986 sq. km

Population 9,884,000

Floristics One of the richest floras in Europe; c. 5500 species and subspecies, of which 20% are endemic (Rechinger, 1965); excluding Crete and the East Aegean, 3950-4100 native vascular species, estimated by D.A. Webb (1978, cited in Appendix 1) from *Flora Europaea*. 763 endemic taxa (IUCN figures), including Crete which, because of its size and separate treatment in *Flora Europaea*, is treated separately below. Elements: Mediterranean, alpine.

Most species occur in the lowlands, many as part of the widespread pan- and especially east-Mediterranean element, but endemism concentrated on the mountains and on many islands of the Aegean and Ionian Seas. Mt Olimbos (Olympus) alone supports c. 1700 vascular plant species from sea-level to its summit, almost 20 endemic to the mountain (Strid and Papanicolaou, in press). Other centres of endemism include Crete (see below), the Peloponnese and central Greece (which includes Mt Parnassós).

Vegetation Although formerly well wooded, forest clearance, fire and centuries of overgrazing by sheep and goats have created large areas of maquis, phrygana and secondary steppe. 3 main vegetation zones: coastal plains and hills, mostly now with evergreen scrub and in the south degraded phrygana or garigue but formerly with dry evergreen forest; the middle slopes of mountains, now cultivated but still supporting large areas of forest dominated by conifers (*Pinus, Abies, Juniperus*), chestnut, oak and beech; above the tree-line a variety of alpine habitats, mainly of rock and scree, with grassland in the north. Today, forests occupy c. 250,000 sq. km (c. 19% of Greece), of which c. 6% is broadleaved and the remainder coniferous.

Checklists and Floras Greece, excluding the East Aegean Islands, is included in the completed *Flora Europaea* (Tutin *et al.*, 1964-1980, cited in Appendix 1). The East Aegean Islands are included in Davis (1965-). All of Greece will be covered in the *Med-Checklist* (cited in Appendix 1). No recent national Flora is available but the most important reference work, although rather old and relating to Greece's pre-1913 frontiers, is:

Halácsy, E. von (1901-1908). *Conspectus Florae Graecae*, 3 vols, 2 supplements in *Magyar Bot. Lapok* (1912) 11: 114-202. Engelmann, Lipsiae. (Parts of northern and eastern Greece not covered; reprinted 1968 by Cramer, Lehre, F.R.G.)

Two new Floras:

Runemark, H. and Greuter, W. (in prep.). *Flora of the South Aegean*.

Strid, A. (1986-). *A Mountain Flora of Greece*. Vol. 1. Cambridge Univ. Press.
   (Includes all higher plants above 1800 m, and in open habitats above 1500 m.)

See also:

Cavadas, D.S. (1957-1964). *Illustrated Botanical Phytological Dictionary*, 9 vols.
   Athens. (Species descriptions arranged alphabetically by genus; illus; in Greek.)
Davis, P.H. (1965-1985). *Flora of Turkey and the East Aegean Islands*, 9 vols.
   Edinburgh Univ. Press, Edinburgh. Includes ferns, gymnosperms, dicotyledons; a
   supplement is in prep. — East Aegean islands covered include Rhodes, Samos,
   Ikaria, Chios and Lesbos.
Greuter, W. and Rechinger, K.H. (1967). Flora der Insel Kythera. *Boisierra* 13. 206 pp.
Phitos, D. (1967). Florula Sporadum. *Phyton* 12(1-4): 102-149. (Covers the Sporades.)
Rechinger, K.H. (1943). *Flora Aegaea*. Vienna. *Akad. Wiss. Wien Math.-Naturwiss.*
   Denkschr. 105(1). 924 pp. Supplement (1949) in *Phyton* 1: 194-228. (Includes
   vascular plants of the foreshores of the Aegean Sea and of the Aegean islands,
   including Crete; German text; Latin keys; illus.; maps; reprinted 1973 by Koeltz,
   Koenigstein-Taunus.)
Rechinger, K.H. (1961). Die Flora von Euboea. *Bot. Jahrb. Syst.* 80(3): 294-465.
   (Maps; illus.)
Strid, A. (1980). *Wild Flowers of Mount Olympus*. Goulandris Natural History
   Museum, Kifissia. 362 pp. (Keys; descriptions; illus.)

Two Greek journals, recently founded, include many papers on the taxonomy and
floristics of Greek plants: *Annales Musei Goulandris*, published by the Goulandris Natural
History Museum, and *Botanika Chronika*, published by the Botanical Institute, University
of Patras. For details on the state of floristic research, see:

Greuter, W. (1975). Floristic studies in Greece. In Walters, S.M. (Ed.), *European
   Floristic and Taxonomic Studies, a Conference at Cambridge, 29 June to 2 July
   1974*. BSBI Conference Report No. 15. E.W. Classey, U.K. Pp. 18-37.
Greuter, W., Phitos, D. and Runemark, H. (1975). Greece and the Greek Islands. A
   report on the available floristic information and floristic and phyto-taxonomic
   research. In CNRS, 1975, cited in Appendix 1. Pp. 67-89. (Describes floristic
   exploration and main floristic works; lists botanical institutes, herbaria and current
   research projects.)
Phitos, D. (1975). Taxonomic and floristic research in Greece during the last decade,
   1961-1971. *Mem. Soc. Brot.* 24(2): 579-597.
Rechinger, K.H. (1968). Bericht über die botanische Erforschung von Griechenland.
   Webbia 18: 234-259.
Stearn, W.T. (1982). The "Flora Europaea" and the Greek flora. *Ann. Mus.*
   Goulandris 5: 123-129.

Field-guides
Goulimis, C.N. (1968). *Wild Flowers of Greece*. Goulandris Botanical Museum,
   Kifissia. 206 pp. (Colour paintings by N.A. Goulandris of about 120 Greek plants
   with botanical notes by C.N. Goulimis, edited by W.T. Stearn.)
Huxley, A. and Taylor, W. (1977). *Flowers of Greece and the Aegean*. Chatto and
   Windus, London. 185 pp. (Colour photographs.)
Phitos, D. (1965). *Wild Flowers of Greece*. Athens Society of the Friends of the Trees,
   Athens. 64 pp. (Translated from the Greek by P. Haritonidou; describes 46 species
   with colour illus.)

See also Polunin (1980), cited in Appendix 1.

Information on Threatened Plants The Botanical Institute and Botanical Museum of the University of Patras, with support from the Hellenic Society for the Protection of Nature, intend to prepare a national Red Data Book. IUCN recently prepared a national threatened plant list:

IUCN Threatened Plants Committee Secretariat (1982). The rare, threatened and endemic plants of Greece. *Ann. Mus. Goulandris* 5: 69-105. (Lists over 900 taxa of international conservation concern; introduction discusses history of data and threats to the species listed.)

See also:

Broussalis, P. (1977). The protection of the flora in Greece and its problems. *Ann. Mus. Goulandris* 3: 23-30.

Diapoulis, C. (1959). Conservation measures for the plants of the Greek flora. In *Animaux et Végétaux Rares de la Région Méditerranéenne*. Proceedings of the IUCN 7th Technical Meeting, 11-19 September 1958, Athens, vol. 5. IUCN, Brussels. Pp. 189-191. (Lists 72 species "in danger of disappearing".)

Goulimis, C. (1959). Report on species of plants requiring protection in Greece and measures for securing their protection. *loc. cit.* Pp. 168-188. (Includes threatened plant list and suggested remedies, including proposed reserve sites.)

Sfikas, G. (1979). Threatened plants of our mountains, I. *Fusis (Nature – Bull. Hellenic Soc. Protection Nature)* 18: 42-44.

Snogerup, S. (1979). The Aegean endemics, distribution and present situation. I. Preliminary list of some of the most suitable sites for conservation. Unpublished manuscript presented to IUCN. 22 pp. (Maps showing floristically diverse areas in the Aegean.)

Some individual case studies on endangered plants have been published in *Fusis* (Nature – the Bulletin of the Hellenic Society for the Conservation of Nature), e.g. *Linaria hellenica* in 12: 13-16, 34-35 (1977) and *Tulipa goulimyi* in 14: 5-9, 36-37 (1978), both by A. Yannitsaros, and *Cephalanthera cucullata* (from Crete, see below) by J. Kalopissis in 18: 26-30, 46 (1979).

Greece is included in the European threatened plant list (Threatened Plants Unit, 1983, cited in Appendix 1); latest IUCN statistics, based upon this work, for all of Greece: endemic taxa – Ex:5, E:26, V:36, R:359, I:40, K:72, nt:225; non-endemics rare or threatened worldwide – V:10, R:54, I:5 (world categories).

In 1982 IUCN, under contract to the EEC through the U.K. Nature Conservancy Council, prepared a report (unpublished), *Threatened Plants, Amphibians and Reptiles, and Mammals (excluding Marine species and Bats) of the European Economic Community*, which included data sheets on 33 plants Endangered in Greece.

Laws Protecting Plants Presidential Decree No. 67, No. 23A (1980), on the "Protection of natural vegetation and wildlife and the establishment of the procedure, co-ordination and control of the research on them", gives protection to over 700 endemic and non-endemic taxa, from "collecting, transplanting, uprooting, cutting, transporting, selling or exporting". These restrictions refer to all parts of the plant.

Voluntary Organizations

Ellenikos Oreibatikos Syndesmos (Hellenic Alpine Club), Pheidiou 18 (Athens branch), Athens.

Friends of the Trees, 22 Anagnostopoulou Str., 106 73 Athens.

Hellenic Society for the Protection of Nature, 24 Nikis Street, 105 58 Athens.

Botanic Gardens

Botanical Garden of Julia and Alexander Diomides, 405 Iera Street, Dafni, Athens.
University of Athens Botanical Garden, Panepistimiopolis, Athens 621.

Useful Addresses

Goulandris Botanical Museum, Levidou 13, Kifissia, Athens.
CITES Management and Scientific Authorities: Wildlife Management Department, Ministry of Agriculture, Hippokratous 3/5, Athens.

Additional References

Antipas, B. and Müller, G. (1974). Conservation in Greece. Problems and achievements. *Nature in Focus* 19: 15, 18-21.

Dafis, S. and Landolt, E. (Eds) (1971, 1976). *Zur Vegetation und Flora von Griechenland, Ergebnisse der 15. Internationalen Pflanzengeographischen Exkursion (IPE) durch Griechenland*, 2 vols. Veröff. Geobot. Inst. ETH Stiftung Rübel, Zürich. (Includes a bibliography on Greek floristic work.)

Hellenic Society for the Protection of Nature (1979). *Proceedings of a Conference on the Protection of the Flora, Fauna and Biotopes in Greece, 11-13 October*. Hellenic Society for the Protection of Nature, Athens. 262 pp.

Rechinger, K.H. (1965). Der Endemismus in der griechischen Flora. *Rev. Roum. Biol. (Sér. Bot.)* 10(1-2): 135-138.

Strid, A. and Papanicolaou, K. (1985). The Greek Mountains. In Gómez-Campo, C. (Ed.), cited in Appendix 1.

Greece: Crete

Area 8,331 sq. km

Population 456,642 (1971 census, *Times Atlas*, 1983)

Floristics 1600-1800 native vascular plant species estimated by D.A. Webb (1978) from *Flora Europaea*; 150 endemic taxa (IUCN figures), including 1 endemic genus (*Petromarula*); c. 137 endemics according to C. Barclay (1984, *in litt.*). Represents highest level of endemism in the Aegean and probably for any comparable area in Europe (Critopoulos, 1975). Areas of high endemism are the Levka Ori (White Mts), including the Samaria Gorge, the Lassithi plateau and the coastal area of Akrotiri north of Khania. Floristic elements: Mediterranean, alpine.

Vegetation In the lowlands, mostly agricultural land, with extensive maquis and garigue; on steep slopes and cliffs in the lowlands and more widely in the mountains, a chasmophyte flora rich in endemics. In places, small scattered stands of near natural forest survive, dominated by oaks (*Quercus pubescens, Q. macrolepis*) and conifers (*Pinus brutia, Cupressus sempervirens*). Crete also supports one of best examples of Kermes Oak (*Q. coccifera*) woodland, between 350-1000 m. Elsewhere in the Mediterranean this habitat almost completely grazed out of existence or converted to garigue.

Checklists and Floras See under Greece. C. Barclay (1984, *in litt.*) is compiling a checklist of the Cretan flora in collaboration with W. Greuter and D. Meikle. For other floristic accounts see:

Greuter, W. (1973). Additions to the flora of Crete, 1938-1972. *Ann. Mus. Goulandris* 1: 15-83. (Annotated list of 250 taxa new to Crete or rediscovered between 1938 and 1972.)

Greuter, W. (1974). Floristic report on the Cretan area. *Mem. Soc. Brot.* 24(1): 131-171. (Describes the taxonomic, biosystematic, phytosociological, floristic and phytogeographical literature and provides corrections to *Flora Europaea*, vols. 1 and 2; extensive bibliography.)

Greuter, W., Matthäs, U. and Risse, H. (1984). Additions to the flora of Crete, 1973-1983 – I. *Willdenowia* 14(1): 27-36. (Pteridophytes, dicotyledons.)

Field-guides See under Greece.

Information on Threatened Plants See under Greece and:

Greuter, W. (1979). The endemic flora of Crete and the significance of its protection. In Hellenic Society for the Protection of Nature, *Proceedings of a Conference on the Protection of the Flora, Fauna and Biotopes in Greece, 11-13 October*. Hellenic Society for the Protection of Nature, Athens. Pp. 91-97.

Laws Protecting Plants See under Greece.

Additional References

Critopoulos, P. (1975). The endemic taxa of Crete. In Jordanov, D. *et al.* (Eds), *Problems of Balkan Flora and Vegetation*. Proceedings of the 1st International Symposium on Balkan Flora and Vegetation, Varna, June 7-14 1973. Bulgarian Academy of Sciences, Sofia. Pp. 169-177.

Gradstein, S.R. and Smittenberg, J.H. (1977). The hydrophilous vegetation of western Crete. *Vegetatio* 34(2): 65-86.

Greuter, W. (1971a). L'apport de l'homme à la flore spontanée de la Crète. *Boissiera* 19: 329-337.

Greuter, W. (1971b). Betrachtungen zur Pflanzengeographie der Südägais (Considerations on the plant geography of the south Aegean). *Opera Bot.* 30: 49-64. (English abstract.)

Greuter, W. (1975). Die Insel Kreta – eine geobotanische Skizze. In Dafis, S. and Landolt, E. (Eds), *Zur Vegetation und Flora von Griechenland, Ergebnisse der 15 Internationalen Pflanzengeographischen Exkursion (IPE) durch Griechenland*, 2 vols. (Describes plant geography and phytosociology.)

Rechinger, K.H. (1943). Neue Beiträge zur Flora von Kreta. *Akad. Wiss. Wien Math.-Naturwiss., Denkschr.* 105, No.2 (1). 184 pp. (Botanical report of a field excursion by the author.)

Zaffran, J. (1976). *Contributions à la Flore et à la Végétation de la Crete. 1. Floristique*. Marseilles.

Zohary, M and Orshan, G. (1965). An outline of the geobotany of Crete. *Israel J. Bot.* 14 (supplement). 49 pp. (Summary of vegetation; map; illus.)

Greenland

(Part of Denmark)

Area 2,175,600 sq. km

Population 54,000

Floristics 497 species of vascular plants; 15 endemic species (Böcher *et al.*, 1978). Elements: Arctic/alpine, Boreal.

Vegetation Much of Greenland is covered in permanent ice. In the southern coastal areas, sub-Arctic dwarf-shrub heaths dominated by *Empetrum hermaphroditum*; in the interior of the ice-free coastal strip, similar heaths are dominated by *Betula nana*. In the north, high Arctic *Cassiope* heaths in the coastal part, with very open *Dryas* communities further inland.

More than 700,000 sq. km protected by the North East Greenland National Park, the world's largest protected area.

Checklists and Floras
Böcher, T.W., Fredskild, B., Holmen, K. and Jakobsen, K. (1978). *Grønlands Flora*, 3rd Ed. Haase, København. 326 pp. (Translated by T.T. Elkington and M.C. Lewis from Danish 2nd Ed.; illus.)

Danish Arctic Station (1968). *Check-list of the Vascular Plants of Greenland*. Godhavn, Disko. 39 pp. (Compiled from 'The Flora of Greenland'.)

Jorgensen, C.A., Sorensen, T. and Westergaard, M. (1958). *The Flowering Plants of Greenland: a Taxonomical and Cytological Survey*. Munksgaard. 172 pp.

For a plant atlas see Hultén (1971), cited in Appendix 1.

Field-guides
Feilberg, J., Fredskild, B. and Holt, S. (1984). *Grønlands Blomster* (Flowers of Greenland). Regnbuen, Denmark. 96 pp. (Illus.)

Foersom, T., Kapel, F.O. and Svarre, O. (1982). *Nunatta Naasui, Grønlands Flora*, 3rd Ed. Haasa, København. 326 pp. (Illus.)

Information on Threatened Plants No publications known. IUCN statistics: endemic taxa – R:3, I:1, nt:3, no data for remainder.

Laws Protecting Plants Although there is no legislation for the protection of plant species in Greenland, a list of species for protection has been proposed for inclusion in the 1980 Act on the Protection of Nature in Greenland.

Additional References
Böcher, T.W., Holmen, K. and Jakobsen, K. (1959). A synoptical study of the Greenland flora. *Med. Grønland* 163(1). 32 pp.

Grenada

Grenada, in the Windward chain of the Lesser Antilles, 131 km north of Trinidad, is about 33.8 km long and 19.3 km broad. Mountains reach 839 m. There is no coastal plain. The state of Grenada also includes some of the 600 small islands of the Grenadines to the north, in particular Carriacou (the largest), St Andrew, St David, St John, St Mark and St Patrick.

Area 345 sq. km (including the Grenadan Grenadines)

Population 112,000

Floristics None.

Vegetation Palm break on steep slopes of Mt St Catherine in the central massif; elfin woodland on summits; remnants of high forest on SW ridge of Mt Sinai; secondary cut-over rain forest on the lower hills; dry scrub woodland along the extreme south coast; rough grazing land with thorn bush over most of the Point Saline peninsula. The forests have been profoundly modified by timber felling in the 19th Century (Beard, 1949, cited in Appendix 1.) 11.8% forested according to FAO (1974, cited in Appendix 1).

On the Grenadines predominantly deciduous and semi-deciduous forests; dry evergreen littoral stunted vegetation on windward slopes (Howard, 1952).

Checklists and Floras Covered by the *Flora of the Lesser Antilles, Leeward and Windward Islands* (only monocotyledons and ferns published so far, Howard, 1974- , cited in Appendix 1) and by the family and generic monographs of *Flora Neotropica* (cited in Appendix 1). See also:

Beard, J.S. (1944). Provisional list of trees and shrubs of the Lesser Antilles. *Caribbean Forester* 5(2): 48-67. (428 species assigned in a table to individual islands separating Grenada from the Grenadines.)
Stehlé, H. and Stehlé, M. (1947). Liste complémentaire des arbres et arbustes des petites Antilles. *Caribbean Forester* 8: 91-123. (A further 328 species to Beard, 1944, in similar format.)

Information on Threatened Plants None.

Botanic Gardens
Botanic Gardens, Department of Agriculture, St George's.

Additional References
Groome, J.R. (1970). *A Natural History of the Island of Grenada, West Indies.* Caribbean Printers Limited, O'Meara Rd., Arima, Trinidad. 115 pp. (About 40 pages deal with plants to which there is an index of common names annotated with cross-references and uses. The catalogue of plants is alphabetical by families.)
Howard, R.A. (1952). The vegetation of the Grenadines, Windward Islands, British West Indies. *Contr. Gray Herb. Harv. Univ.* 174: 1-129, 29 plates.

Guadeloupe and Martinique

Guadeloupe in the Leeward islands of the West Indies consists of two islands joined by a mangrove swamp: Grande Terre, limestone, flat and intensively cultivated; and Basse Terre, volcanic and mountainous – the Soufrière volcano at 1464 m is the highest peak in the Lesser Antilles. Just south and east of Guadeloupe are the limestone islets of La Désirade and Marie Galante (the largest at c. 160 sq. km) and the volcanic Iles des Saintes.

Martinique in the Windward islands is 200 km south of Guadeloupe; Dominica is between them. It is much cultivated with three regions: low hills in the south, a central massif, and the active volcano of Mt Pelée in the north.

Guadeloupe and Martinique are French overseas *départements*. The small island of St Barthélémy, and part of neighbouring St Martin, at the north end of the Leeward Islands, are dependencies of Guadeloupe. (For the other part of St Martin see Netherlands Antilles.) Their flora is small; see Questel (1941) and the account for Antigua and Barbuda.

Area Guadeloupe: 1779 sq. km; Martinique: 1079 sq. km

Population Guadeloupe: 319,000; Martinique: 312,000

Floristics c. 2800 species of gymnosperms and flowering plants (c. 1700 indigenous and 1100 introduced) (Fournet, 1978). Early figures for endemism (e.g. 5% for Guadeloupe and 4% for Martinique in Stehlé and Quentin, 1937) are now known to be too high, as many of the species have been found on neighbouring islands. Up-to-date figures not available.

Vegetation
Guadeloupe On Grand Terre little forest remains, the only natural growth being man-induced scrub woodland. Basse Terre has untouched rain forest and lower montane rain forest. At the junction of the islands are large expanses of mangrove and *Pterocarpus* swamp. 34.8% forest cover according to FAO (1974, cited in Appendix 1).

Martinique No natural rain forest remains. In the centre and at low elevations there is secondary forest, at higher elevations montane thicket, palm brake and elfin woodland. 25.5% forest cover according to FAO (1974, cited in Appendix 1).

Checklists and Floras Covered by the *Flora of the Lesser Antilles, Leeward and Windward Islands* (only monocotyledons and ferns published so far, Howard, 1974- , cited in Appendix 1) and by the family and generic monographs of *Flora Neotropica* (cited in Appendix 1). Island floras are:

Fournet, J. (1978). *Flore Illustrée des Phanérogames de Guadeloupe et de Martinique.* Institut National de la Recherche Agronomique, Paris. 1654 pp.
Questel, A. (1951). *1 – La Flore de la Guadeloupe et Dépendances (Antilles Françaises). Géographie Générale de la Guadeloupe et Dépendances (Antilles Françaises).* L. le Charles, Paris. 327 pp. (With description of the vegetation, illus. and maps.)
Stehlé, H. and M. and Quentin, L. (1935-1949). *Flore de la Guadeloupe et Dépendances et de la Martinique.* Several vols. Catholic Press, Basse-Terre.

See also:

Beard, J.S. (1944). Provisional list of trees and shrubs of the Lesser Antilles. *Caribbean Forester* 5(2): 48-67. (428 species assigned in a table to individual islands.)
Stehlé, H. and Stehlé, M. (1947). Liste complémentaire des arbres et arbustes des petites Antilles. *Caribbean Forester* 8: 91-123. (A further 328 species to Beard, 1944, in similar format.)

For St Barthélémy, see:

Monachino, J. (1940-41). A check-list of the spermatophytes of St. Bartholomew: part i-ii. *Caribbean Forester* 2: 24-66.

Questel, A. (1941). *La Flore de Saint-Barthélemy (Antilles Françaises) et son Origine*. Imprimerie Catholique, Basse-Terre. 224 pp. (In French, also an English version.)

Field-guides

Chauvin, G. (1977, 1978). Etude illustrée des familles de plantes à fleurs de la Martinique. *Les cahiers documentaires éducation et enseignement*, no. 16: les Gamopétales and no. 18: les Dialypétales. C.D.D.P. Fort-de-France.

Fournet, J. (1976). *Fleurs et plantes des Antilles*. Cited in Appendix 1.

Information on Threatened Plants

Sastre, C. (1978). Plantes menacées de Guadeloupe et de Martinique. 1. Espèces altitudinales. *Bull. Mus. natn. Hist. nat., Paris, 3e sér. no. 519, Ecologie générale* 42: 65-93. (Description of vegetation, sheets on 13 rare and threatened species with illustrations and habitat photographs.)

Sastre, C. and Mestoret, L. (1978). Plantes rares ou menacées de Martinique. *Le courrier du parc naturel régional de la Martinique* no. 2: 20-22.

C. Sastre has also written popular papers on threatened plants of Guadeloupe and Martinique, e.g. in *L'Orchidophile* 13(52): 83-90 (1982) and in an unnumbered issue of *Panda*, the magazine of WWF-France (pp. 6-7).

The IUCN Plant Red Data Book has three data sheets for Guadeloupe and Martinique.

Voluntary Organizations

Association des Amis du Parc Naturel de la Guadeloupe et de l'environnement, Préfecture de la Guadeloupe, Basse-Terre.

Useful Addresses

Délégation Régionale à l'Architecture et à l'environnement (Guadeloupe, Guyane, Martinique), B.P. 1002, 97178 Pointe-à-Pitre Cedex.

Office National des Forêts, Jardin des plantes, 97100 Basse-Terre.

Additional References

Fiard, J.P., Association des Amis du Parc Naturel Régional (1979). *La forêt martiniquaise: présentation et propositions de mesures de protection*. Fort de France, Parc Naturel Régional Ex-Caserne Bouille. 65 pp. (Illus., maps.)

Portecop, J. (1979). Phytogéographie, cartographie écologique et aménagement dans une île tropicale: le cas de la Martinique. *Doc. Cart. Ecol. Univ. Grenoble* 21: 1-78. (With map, 1:75,000.)

Sastre, C. (1979). Considérations phytogéographiques sur les sommets volcaniques Antillais. *C.R. Soc. Biogéogr.* 484: 127-135.

Stehlé, H. (1980). Modifications écologiques récentes dans la végétation des Antilles françaises et leurs causes essentielles (42e contribution). *Bull. Soc. Bot. Fr., 127, Lettres Bot.* 3: 275-287.

Guam

Guam (13°20'N, 144°45'E) is the largest and southernmost of the Mariana Islands, and is located in the west Pacific Ocean, c. 2030 km east of the Philippines. It is an unincorporated territory of the United States. Most of the northern part is a raised limestone plateau 152 m high, separated from the volcanic south, which reaches 407 m, by a narrow neck of land 8 km wide.

Area 450 sq. km

Population 119,000

Floristics 931 vascular plant species of which c. 330 are native and 20 doubtfully native (Stone, 1970). 69 vascular species occurring on Guam are endemic to the Marianas Group. Indomalaysian-Pacific elements account for over a third of the total vascular flora.

Vegetation Rain forest with *Artocarpus*, *Elaeocarpus*, *Pandanus*, *Ficus* and *Guamia* originally covered most of the island; much has been logged and cleared for coconut plantations; mixed forests on old volcanic soils completely destroyed (Fosberg, 1973, cited in Appendix 1). Ravine forests occur along river valleys and on some volcanic and limestone hill slopes; small areas of poorly developed mangroves.

Checklists and Floras

Stone, B.C. (1970). The Flora of Guam. *Micronesica* 6. 659 pp. (Keys, descriptions; notes on distributions; introductory chapters on floristics, vegetation, forests and other plant resources.)

Wagner, W.H. and Grether, D.F. (1948). Pteridophytes of Guam. *Occ. Papers Bernice P. Bishop Mus.* 19(2): 25-99.

Guam is included in *Flora Micronesica* (Kanehira, 1933), the regional checklists of Fosberg, Sachet and Oliver (1979, 1982), and will also be covered by the *Flora of Micronesia* (1975-), all of which are cited in Appendix 1.

Information on Threatened Plants There are about 20 vascular plant taxa, endemic to Guam and the Marianas which are 'endangered'; a further 30 taxa, not confined to Guam or the Marianas are 'endangered' on Guam (Moore, 1980). *Heritiera longipetiolata* and *Serianthes nelsonii* are included in *The IUCN Plant Red Data Book* (1978). See also:

Moore, P.H. (1980). Notes on the endangered species of Guam. *Notes from Waimea Arboretum* 7(1): 14-17. (Notes on 3 'endangered' endemic taxa; checklist by C. Daguio of 58 mostly non-endemic taxa, in cultivation at Waimea.)

Moore, P., Raulerson, L., Chernin, M. and McMakin, P. (1977). Inventory and mapping of wetland vegetation in Guam, Tinian and Saipan, Mariana Islands. Mimeo. Univ. of Guam. (Lists 5 non-endemics threatened on Guam.)

Additional References

Lee, M.A.B. (1974). Distribution of native and invader plant species of the island of Guam. *Biotropica* 6(3): 158-164.

Guatemala

Area 108,888 sq. km

Population 8,165,000

Floristics An estimated 8000 species of vascular plants (Gentry, 1978, cited in Appendix 1); 1171 endemic species (IUCN figures); over 550 orchids (Lizama, 1981); according to D'Arcy (1977), 70% of the high mountain vascular flora is endemic.

Vegetation Predominantly tropical broadleaved moist forests (83% of forest cover), mostly in the Department of Petén in the northeast; mangroves on tidal flow areas on the Pacific coast; coniferous montane forests in the west, restricted to the highlands, extending c. 10,000 sq. km (Myers, 1980, cited in Appendix 1). Estimated rate of deforestation 900 sq. km/year out of a total area of 49,020 sq. km; figures for broadleaved closed forest are 720 and 37,850 sq. km respectively (Nations and Komer, 1984, from FAO/UNEP, 1981).

Checklists and Floras Guatemala is covered by the *Flora Mesoamericana* Project, described in Appendix 1, as well as by the family and generic monographs of *Flora Neotropica*, also in Appendix 1. The country Flora is:

Standley, P.C., Steyermark, J.A. and Williams, L.O. (1946-1977). Flora of Guatemala. *Fieldiana, Bot.* 24 (1-13). (Complete except for orchids and ferns, covered separately, see below.)

Also relevant:

Ames, O. and Correll, D.S. (1952-53). Orchids of Guatemala. *Fieldiana*, Bot. 26(1-2).
Correll, D.S. (1965). Supplement to the orchids of Guatemala and British Honduras. *Fieldiana, Bot.* 31(7): 177-221.
Record, S.J. and Kuylen, H. (1926). Trees of the Lower Río Motagua Valley, Guatemala. *Trop. Woods* 7: 10-29.
Stolze, R.S. (1976, 1981, 1983). Ferns and fern allies of Guatemala. *Fieldiana, Bot.* 39: 1-130; *Fieldiana, Bot., New Series* 6: 1-522 (Part II: Polypodiaceae); 12: 1-91 (Part III: Marsileaceae, Salviniaceae and the fern allies).

Information on Threatened Plants 24 species are listed as threatened in Organización de los Estados Americanos (1967), cited in Appendix 1. 24 species, mostly different ones, are listed in the Annex to the Convention on Nature Protection and Wildlife Preservation in the Western Hemisphere (1940). 5 species are listed as threatened in Nations and Komer (1984). 93 threatened species, most not threatened on a world scale, are covered in a list by the Instituto Nacional Forestal (INAFOR) - this is:

Rodas Zamora, J. and Aguilar Cumes, J. (1980). Lista de algunas especies vegetales en vía en extinción. INAFOR, Guatemala City. (Unpublished.)

IUCN is preparing a threatened plant list for release in a forthcoming report *The list of rare, threatened and endemic plants of Middle America*. Latest IUCN statistics, based upon this work: endemic taxa - E:14, V:37, R:90, I:35, K:974, nt:21; non-endemics rare or threatened worldwide - E:4, V:25, R:38, I:11 (world categories).

Threatened plants are mentioned in several papers in:

Prance, G.T. and Elias, T.S. (Eds) (1977), cited in Appendix 1. See in particular W.G. D'Arcy on endangered landscapes in the region (pp. 89-104), J.T. Mickel on rare and endangered ferns (pp. 323-328), H.E. Moore on endangerment in palms (pp. 267-282), P. Ravenna on rare and threatened bulbs (pp. 257-266).

Laws Protecting Plants Government Decree 13-79, Emergency Law, National Rainforest Campaign, includes provisions for re-afforestation and for the prevention of felling of trees to collect seeds, especially of *Pimenta diaica*. Governmental law of 9 August 1946 prohibits the collection and export of the orchid and national flower, *Lycaste virginalis* var. *alba*; collection may only be authorized by the Ministerio de Agricultura. Governmental Resolution of 29 August 1950 prohibits the use of the bark of *Pinus ayacahuite* for tanning, and Resolution of 18 August 1958 prohibits the export of fresh roots and seeds of 6 *Dioscorea* species and one *Agave* species (J.M. Aguilar Cumes, *in litt.*, 1984). The U.S. Government has determined *Abies guatemalensis* (El Salvador, Honduras, Guatemala, Mexico) as 'Threatened' under the U.S. Endangered Species Act.

Voluntary Organizations
Asociación Guatemalteca de Historia Natural (AGHN), c/o Jardín Botánico, Avenida de la Reforma 0-43, Zona 10, Ciudad de Guatemala.

Botanic Gardens
Jardín Botánico, Avenida de la Reforma 0-43, Zona 10, Ciudad de Guatemala.

Useful Addresses
Centro de Estudios Conservacionistas (CECON), Avenida de la Reforma 0-43, Zona 10, Ciudad de Guatemala.

Empresa Nacional de Fomento y Desarrollo Económico del Petén (FYDEP), Santa Elena, Petén.

Escuela de Biología, Universidad de San Carlos de Guatemala, Calle Mariscal Cruz 1-56, Zona 10, Ciudad de Guatemala.

Instituto de Antropología e Historia y Historia Natural (IDAEH), 6a Calle 7-30, Zona 13, Ciudad de Guatemala.

Instituto Nacional Forestal (INAFOR), 5a Avenida 12-31, Zona 9, Edificio "El Cortez", Ciudad de Guatemala.

Museo Nacional de Historia Natural, Apdo Postal 987, Ciudad de Guatemala.

CITES Management and Scientific Authorities: Instituto Nacional Forestal (INAFOR), Edificio Galerías España, 6° Nivel, 7a Avenida 11-63, Zona 9, Ciudad de Guatemala.

Additional References
D'Arcy, W.G. (1977). Endangered landscapes in Panama and Central America: the threat to plant species. In Prance, G.T. and Elias, T.S. (Eds) (1977), cited in Appendix 1. Pp. 89-104.

Holdridge, L.R., Lamb, F.B. and Mason, B. (1950). *Los Bosques de Guatemala*. Turrialba, Costa Rica. Instituto Interamericana de Ciencias Agrícolas. 174 pp.

INAFOR (1981). *Estudio sobre Exportaciónes de Fauna y Flora Silvestre de Guatemala de Enero/78 a Diciembre/80*. Departamento de Parques Nacionales y Vida Silvestre, Instituto Nacional Forestal. 24 pp.

Lizama, C. (1981). Orchids of Guatemala. In Stewart, J. and van der Merwe, C.N. (Eds), *Proceedings of the 10th World Orchid Conference*. Durban, South Africa. Pp. 109-110.

Lundell, C.L. (1937). *The vegetation of Petén*. Carnegie Institute, Washington, D.C. 244 pp. (Publication No. 478.)

Nations, J.D. and Komer, D.I. (1984). *Conservation in Guatemala: Final report, presented to WWF-US*. Center for Human Ecology, Box 5210, Austin, Texas 78763,

U.S.A. Mimeo. 170 pp. (From WWF Project US-269, Development of a
conservation program for Guatemala; extensive report listing conservation
organizations, individuals and other useful contacts in Guatemala.)
Veblen, T.T. (1976). The urgent need for forest conservation in highland Guatemala.
Biol. Conserv. 9: 141-154.

Guinea

Area 245,855 sq. km

Population 5,301,000

Floristics Size of flora unknown; 88 endemics (Brenan, 1978, cited in Appendix
1), but see below. Mt Nimba, shared with Liberia and Ivory Coast, has over 2000 species.

Floristic affinities range from Sudanian in the extreme north-east to Guinea-Congolian in
the south and south-west. Afromontane elements occur on the Fouta Djallon and Mt
Nimba which are important centres of endemism. The forests also have numerous
endemics.

Vegetation Over most of the country a mosaic of patches of lowland rain forest
interspersed with secondary grassland and cultivated land; extensive areas of forest still
survive near the borders with Liberia and Ivory Coast. Considerable areas of mangrove
along coast. Sudanian woodland occurs in north-eastern sector. Also, transitional rain
forest (between lowland and montane) on Mt Nimba and the Fouta Djallon. Estimated
rate of deforestation for closed broadleaved forest 360 sq. km/annum out of 20,500 sq.
km (FAO/UNEP, 1981).

For vegetation map see White (1983), cited in Appendix 1.

Checklists and Floras Guinea is included in the *Flora of West Tropical Africa*.
The Guinean portion of Mt Nimba is included in *Flore Descriptive des Monts Nimba*. Both
are cited in Appendix 1.

Information on Threatened Plants No published lists of rare or threatened
plants; IUCN has records of 99 species and infraspecific taxa believed to be endemic,
including V:10, R:27, nt:10.

Useful Addresses
CITES Management Authority: Direction Générale des Eaux, Forêts et Chasses,
Secrétariat d'Etat aux Eaux et Forêts, B.P. 624, Conakry.

Additional References
Adam, J.G. (1958). *Eléments pour l'Etude de la Végétation des Hauts Plateaux du
Fouta Djalon (Secteur des Timbis), Guinée Française. 1. La Flore et ses
Groupements*. Gouvernement Général de l'AOF, Bureau des Sols, Dakar. 80 pp.
(With coloured vegetation map 1:50,000.)
Adam, J.-G. (1970). Etat actuel de la végétation des monts Nimba au Libéria et en
Guinée. *Adansonia*, Sér. 2, 10: 193-211. (With 10 black and white photographs.)
Lamotte, M. (1983). The undermining of Mount Nimba. *Ambio* 12(3-4): 174-179.
(Photographs, maps.)

Pobéguin, H. (1906). *Essai sur la Flore de la Guinée Française*. Challamel, Paris.
  392 pp. (Numerous black and white photographs.)
Schnell, R. (1968). Guinée. In Hedberg, I. and O. (1968), cited in Appendix 1.
  Pp. 69-72.

Guinea-Bissau

Area 36,125 sq. km

Population 875,000

Floristics c. 1000 species (quoted in Lebrun, 1976, cited in Appendix 1). No
endemics given in Brenan (1978, cited in Appendix 1), but IUCN has records of 12 species
and infraspecific taxa believed to be endemic; five of these are undescribed species.

Flora with Guinea-Congolian and Sudanian affinities.

Vegetation Large areas of mangrove around coast and offshore islands. Inland,
original vegetation lowland rain forest, but much now destroyed and replaced by
cultivation and secondary grassland. Estimated rate of deforestation for closed
broadleaved forest 170 sq. km/annum out of 6600 sq. km (FAO/UNEP, 1981).

For vegetation map see White (1983), cited in Appendix 1.

Checklists and Floras Guinea-Bissau is included in the *Flora of West Tropical
Africa*, cited in Appendix 1.

D'Orey, J. and Liberato, M.C. (1972-). *Flora da Guiné Portuguesa*. Ministério do
  Ultramar, Lisboa. (5 fascicles so far, covering most of Leguminosae plus two other
  smaller families. Descriptive keys, distributions, etc.)
Pereira de Sousa, E. (1946-1963). *Contribuições para o Conhecimento da Flora da
  Guiné Portuguesa*. Vols 1-8 published by Ministério das Colonias, Lisboa in *Anais
  Junta Invest. Colon.*, and *Anais Junta Invest. Ultram.*; vols 9-10 by Junta de
  Investigações do Ultramar, Lisboa. (Annotated checklist. Frodin gives more
  publication details.)

Information on Threatened Plants No published lists of rare or threatened
plants. No categories assigned to the 12 taxa believed to be endemic.

Additional References
Espirito Santo, J. do (1949). Contribição para o conhecimento fitogeográfico da Guiné
  portuguesa. *Bol. Cult. Guiné Portug.* 4(13): 95-129.
Malato-Beliz, J. (1963). Aspectos da investigação geobotânica na Guiné Portuguesa.
  Estud. Agron. 4(1): 1-20.
Malato-Beliz, J. and Alves Pereira, J. (1965). Constituição e ecologia das pastagens
  naturais da Guiné Portuguesa. *Garcia de Orta* 13: 1-7. (With 6 black and white
  photographs.)

Guyana

Area 214,970 sq. km

Population 936,000

Floristics No figures available for number of species; likely to be higher than French Guiana (estimated at 6000-8000 species), because of wider range of vegetation. J.C. Lindeman (1984, pers. comm.), however, estimates 8000 species of vascular plants for all 3 Guianas, implying the total for Guyana is rather lower. Floristic affinities with neighbouring countries, in particular the dry savanna with that of Brazil and the rain forests with Amazonia and Venezuela through the Guayana Highland sandstone mountains.

Vegetation On the coast mangrove and swamp forests, with pockets of seasonal evergreen forests, now largely destroyed. Most of the population and cultivated land are on the coast. In the interior, equatorial rain forests, lowland and submontane, covering 85% of the country and forming 2.9% of the Amazon forest. From the Demerara River along the coast to the Surinam border wet savanna; in the south on and around the Kanuku Mts dry (Rupununi) savanna. In the west are the spectacular Pakaraima Mts, reaching 2810 m on Mt Roraima and forming part of the Guayana Highland which covers much of southern Venezuela; sandstone capped by granite, with elfin forest, bog and swamp on the top, mainly forest but some grassland lower down; very rich in endemics.

According to FAO/UNEP (1981), estimated rate of deforestation for closed broadleaved forest 25 sq. km/annum out of 184,750 sq. km; according to Myers (1980, cited in Appendix 1), montane rain forest covers 47,500 sq. km, lowland evergreen rain forest 134,000 sq. km, swamp and marsh forest 5300 sq. km: "there seems little prospect that Guyana's primary forests will be much modified within the foreseeable future".

Checklists and Floras Guyana is covered by the family and generic monographs of *Flora Neotropica* (cited in Appendix 1). Country accounts are:

Fanshawe, D.B. (1949). Check-list of the indigenous woody plants of British Guiana. Forestry Bulletin No. 3 (New Series), Forest Dept, British Guiana. 244 pp. Unpublished typescript, copy at Kew.

Graham, E.H. (1934). Flora of the Kartabo Region, British Guiana. *Ann. Carnegie Mus.* 22: 17-292.

Maguire, B. *et al.* (1953-). The botany of the Guayana Highland. *Mem. New York Bot. Gard.* 12 parts, between vols. 8 and 38. Various family treatments resulting from field activities begun in 1944. Parts 13 and 14 (in prep.) will conclude the systematic treatment of the flora of the Roraima Formation in Guyana; other reports will be issued as separate papers.

A 30-year project to prepare the *Flora of the Guianas* is being coordinated by the Institute of Systematic Botany, University of Utrecht, The Netherlands, and the Smithsonian Institution, Washington, D.C., in collaboration with Office de la Recherche Scientifique et Technique Outre-Mer, Cayenne, French Guiana, and other leading botanical institutions. Part 1 (Cannaceae, Musaceae and Zingiberaceae by P.J.M. Maas) is in press.

Information on Threatened Plants There is no national Red Data Book. Threatened plants are mentioned in:

Mickel, J.T. (1977). Rare and endangered pteridophytes in the New World and their prospects for the future. In Prance, G.T. and Elias, T.S. (Eds) (1977), cited in Appendix 1. Pp. 323-330.

Botanic Gardens

Botanical Gardens, Guyana Forestry Commission, Water Street, Georgetown.

Botanical Gardens, Ministry of Agriculture, Turkeyen, Greater Georgetown.

Useful Addresses

Institute of Sytematic Botany, University of Utrecht, Heidelberglaan 1, P.O. Box 80102, 3508 TC Utrecht, Netherlands.

National Science Research Council, University Campus, Turkeyen, Greater Georgetown.

CITES Management Authority: The Permanent Secretary, Ministry of Agriculture, P.O. Box 1001, Georgetown.

CITES Scientific Authority: The National Science Research Council, 44 Pere Street, Kitty.

Additional References

Dalfelt, A. (1978). *Nature Conservation Survey of the Republic of Guyana*. IUCN, Switzerland. 55 pp.

Fanshawe, D.B. (1952). *The Vegetation of British Guiana: A Preliminary Review*. Institute Paper No. 29, Imperial Forestry Institute, Oxford. 95 pp.

Maguire, B. (1970). On the flora of the Guayana Highland. *Biotropica* 2(2): 85-100.

Haiti

The western third of the island of Hispaniola, bordered by the Dominican Republic; three quarters mountainous.

Area 27,749 sq. km

Population 6,419,000

Floristics No figures for Haiti; Hispaniola has an estimated 5000 species: 7 gymnosperms, 1087 monocotyledons and 3900 dicotyledons; with 1800 endemic species (Liogier, 1984).

Vegetation Vegetation greatly modified; what remains is similar to that of the neighbouring Dominican Republic; only a few pine forests survive at the higher altitudes and also small areas of mahogany, rosewood and cedar; alpine vegetation above 1463 m; coastal mangrove; estimated rate of deforestation for closed broadleaved forest 12 sq. km/annum, out of a total of 360 sq. km (FAO/UNEP, 1981); earlier FAO figures estimate only 1.8% forested (FAO, 1974, cited in Appendix 1).

Checklists and Floras Covered by the family and generic monographs of *Flora Neotropica* (cited in Appendix 1). For Haiti see:

Barker, H.D. and Dardeau, W.S. (1930). *Flore d'Haiti*. Service technique de la Département de L'Agriculture et L'Enseignment professionel. Port-au-Prince. 456 pp. (Angiosperms only; keys to genera; species mostly listed.)

The following works refer to Hispaniola:

Liogier, A.H. (1982, 1983). *La Flora de la Española*. 2 vols published, the third in press. San Pedro de Macorís. 317 pp., 420 pp., illus.

Moscoso, R.M. (1943). *Catalogus Florae Domingensis*. New York. 732 pp. (In Spanish; checklist of gymnosperms and flowering plants. Includes reports from Haiti as well as Dominican Republic.)

Also relevant:

Jiménez, J. de J. (1963-1967). Suplemento no. 1 al Catalogus Florae Domingensis del Prof. Rafael M. Moscoso. *Archiv. Bot. Biogeogr. Ital.* 39: 81-132; 40: 54-149; 41: 47-87; 42: 46-97 and 107-129; 43: 1-18.

Liogier, A.H. (1976). Novitates Antillanae. VII. Plantas nuevas de la Española. *Moscosoa* 1(1): 16-49.

Urban, I. (1922-1932). Plantae Haitienses novae vel rariores a cl Er. L. Ekman 1917 lectae. *Arkiv för Botanik* 17(7)-24A(4), series of ten papers. Uppsala. (In German.)

Urban, I. (1920, 1921). Flora domingensis. *Symbolae Antillanae* 8(1): 1-480; 8(2): 481-860.

The botanical journal *Moscosoa* includes reports of new taxa, of new records and other papers on the flora and vegetation of the Dominican Republic and Haiti. It is published by the Jardín Botánico Nacional 'Dr Rafael M. Moscoso', Apdo 21-9, Santo Domingo.

Information on Threatened Plants No national list or report. Threatened plant conservation is discussed in:

Howard, R.A. (1977). Conservation and the endangered species of plants in the Caribbean islands. In Prance, G.T. and Elias, T.S. (Eds), cited in Appendix 1. Pp. 105-114.

Additional References

Ekman, E.L. (1926). Botanizing in Haiti. *U.S. Naval Med. Bull.* 24: 483-497. Ekman also wrote accounts (in English) of the Hispaniola islands, Tortue, Navassa and Gonave. See *Arkiv för Botanik* 22A(9): 1-61; 22A(16): 1-12 (both in 1929) and *Ark. Bot.* 23A(6): 1-73 (1930).

Holdridge, D.R. (1945). A brief sketch of the Flora of Hispaniola. In Verdoorn F. (Ed.), cited in Appendix 1. Pp. 76-78.

Liogier, A.H. (1974). *Diccionario botánico de nombres vulgares de la Española*. Jardín Botánico Dr R. Moscoso, Santo Domingo. 813 pp.

Liogier, A. (1984). La Flora de la Española: sus principales caraterísticas. *2da Joranda Científica Academia de Ciencias de la República Dominicana*. Santo Domingo.

Zanoni, T.A., Long, C.R. and Mckiernan, G. (1984). Bibliografía de la flora y de la vegetación de la Isla Española. *Moscosoa* 3: 1-61. (An extensive annotated bibliography of the flora and the vegetation of Hispaniola.)

Hawaii

A group of volcanic islands in the central Pacific Ocean. Hawaii became the 50th State of the United States in 1959.

Area 16,641 sq. km

Population 965,000 (1980 census, *Times Atlas*, 1983)

Floristics About 950 vascular plant species (P.H. Raven, 1986, pers. comm., quoting Wagner, Herbst and Sohmer, in prep.), most endemic.

Vegetation Coastal forest of *Scaevola* and *Pandanus*, with *Santalum* between 600-800 m; lowland dry forest with *Myoporum*, almost entirely cleared for cultivation, grazing and settlements; upper dry forest, mainly open Koa (*Acacia koa*) woodland, on lower mountain slopes and occasionally on mountain ridges; Ohia (*Metrosideros*) rain forest – the richest community – in highland areas with more than 1750 mm annual rainfall (Carlquist, 1980). All the larger volcanic islands, except Kahoolawe and Niihau, retain some natural forests in uplands; Hawaii and Maui islands, in particular, have large areas of intact rain forest. Small patches of forest ('kipukas') have been isolated by lava flows and contain many endemics. 2 National Parks and several other protected areas have been established, mostly in uplands.

Checklists and Floras A new Flora, entitled *Manual of the Flowering Plants of Hawai'i*, is being prepared by W.L. Wagner, D.R. Herbst and S.H. Sohmer at the Bishop Museum, Honolulu. It will include all known native and naturalized alien species, with keys and descriptions of families, genera and species; introductory chapters to cover vegetation. Expected publication date – 1988. Published works are:

Degener, O. and I. (1932-). *Flora Hawaiiensis or The New Illustrated Flora of the Hawaiian Islands*. J. Pan-Pacific Research Institute, Honolulu. (7 loose-leaf fascicles, each dealing with c. 100 taxa.)
Hillebrand, W.F. (1888). *Flora of the Hawaiian Islands*. Heidelberg. 673 pp. (According to Frodin, treats 999 species. Reprinted 1965, by Hafner, New York.)
Rock, J.F. (1913). *The Indigenous Trees of the Hawaiian Islands*. Honolulu. 518 pp. (Revised by D.R. Herbst, 1974; Tuttle, Rutland.)
St John, H. (1973). *List and Summary of the Flowering Plants in the Hawaiian Islands*. Pacific Tropical Botanical Garden, Hawaii. 519 pp. (Comprehensive checklist with distributions.)

Field-guides The following guides contain short descriptive accounts and colour photographs of c. 70 taxa, including introductions:

Lamoureux, C.H. (1976). *Trailside Plants of Hawaii's National Parks*. Hawaii Natural History Assoc. and U.S. National Parks Service, Hawaii. 78 pp.
Merlin, M.D. (1976). *Hawaiian Forest Plants: A Hiker's Guide*. Oriental Publ. Co., Honolulu. 68 pp.
Merlin, M.D. (1977). *Hawaiian Coastal Plants and Scenic Shorelines*. Oriental Publ. Co., Honolulu. 68 pp.

Information on Threatened Plants Hawaii is covered in the Federal U.S. lists (U.S. Fish and Wildlife Service, 1980, 1983, cited under United States); Ayensu and DeFilipps (1978) list 270 'Extinct', 646 'Endangered' and 197 'Threatened' taxa, most of which are endemic. According to Wagner, Herbst and Sohmer (in prep.), about 10% of the native flora is presumed extinct and about 40% threatened (P.H. Raven, 1986, *in litt.*). Publications specifically on Hawaiian threatened plants are:

Fosberg, F.R. and Herbst, D. (1975). Rare and endangered species of Hawaiian vascular plants. *Allertonia* 1(1). 72 pp. (Estimates 70% of flora is threatened; lists 1186 taxa, of which 273 'extinct', 800 'endangered'.)

Kimura, B.Y. and Nagata, K.M. (1980). *Hawaii's Vanishing Flora*. Oriental Publ. Co., Honolulu. 88 pp.
St John, H. and Corn, C.A. (1981). *Rare Endemic Plants of the Hawaiian Islands*, Book 1. Dept of Land and Natural Resources, Div. of Forestry and Wildlife, Honolulu. (68 threatened taxa giving status and threats.)

5 species are included in *The IUCN Plant Red Data Book* (1978). Latest IUCN statistics: endemic taxa – Ex:62, E:830, V:45, R:66; I:784, K:28. This is the highest recorded number of Extinct and Endangered taxa for any country in the world, let alone an island group the size of Hawaii, but has not yet been brought into line with the new Flora.

Laws Protecting Plants See under United States.

Voluntary Organizations See under United States. The Nature Conservancy (TNC) has a particularly active programme in Hawaii. Local address:

The Nature Conservancy of Hawaii, 1026 Nuuanu Avenue, Suite 201, Honolulu, Hawaii 96817.

Botanic Gardens The principal gardens are:

Foster Botanic Garden, 50 N. Vineyard Boulevard, Honolulu, Hawaii 96817.
Harold L. Lyon Arboretum, University of Hawaii, 3860 Manoa Road, Honolulu, Hawaii 96822.
Pacific Tropical Botanic Garden, P.O. Box 340, Lawai, Kauai, Hawaii 96765.
Waimea Arboretum and Botanical Garden, Park Office, 59-864 Kamehameha Highway, Haleiwa, Oahu, Hawaii 96712.

A *Checklist of Hawaiian Endemic, Indigenous, Food Plants and Polynesian Introductions in Cultivation in Hawaii* was compiled in 1983 at the Waimea Arboretum and Botanical Garden for the Council of Botanical Gardens and Arboreta, and published by the Waimea Arboretum Foundation. It lists Hawaiian plants in cultivation in Hawaiian collections. It adds the following gardens to those listed above:

Amy Greenwell Ethnobotanical Garden, Hawaii.
Kapalua Botanic Garden, Maui.
Keanae Arboretum, Maui, Hawaii.
Koko Crater Botanic Garden, Oahu.
Lo'i Botanic Garden, Oahu.
Maui Zoo and Botanical Garden, Maui, Hawaii.
Wahiawa Botanic Garden, Oahu.
Waikamoi Arboretum, Maui, Hawaii.

Index of threatened plants in cultivation:

Threatened Plants Unit, IUCN Conservation Monitoring Centre (1985). *The Botanic Gardens List of Rare and Threatened Plants of the Hawaiian Islands*. Botanic Gardens Conservation Co-ordinating Body, Report No. 14. IUCN, Kew. 21 pp. (Lists 274 rare and threatened endemic taxa, reported in cultivation, with gardens listed for each.)

Useful Addresses
Endangered Species Office, U.S. Fish and Wildlife Service, 300 Ala Moana Boulevard, P.O. Box 50167, Honolulu, Hawaii 96850.
Hawaii State Department of Forestry and Wildlife, 1179 Punchbowl Street, Honolulu 96813.

Additional References

Carlquist, S. (1965), cited in Appendix 1. (Origin, evolution and adaptations of plants and animals.)

Carlquist, S. (1974), cited in Appendix 1. (Dispersal and evolution of plants and animals; separate chapter on flora.)

Carlquist, S. (1980). *Hawaii: A Natural History*, 2nd Ed. Pacific Tropical Botanical Garden, Honolulu. 468 pp. (Geology, fauna, vegetation types.)

Fosberg, F.R. (1975). The deflowering of Hawaii. *National Parks and Conservation Mag.* 49(10): 4-10.

Kay, E.A. (1972). *A Natural History of the Hawaiian Islands: Selected Readings*. Univ. of Hawaii, Honolulu. 653 pp. (Covers physical geography, flora, fauna. See in particular F.R. Fosberg on the derivation of the flora, pp. 396-408; H. St John on endemism, pp. 517-519.)

Honduras

Area 112,087 sq. km

Population 4,232,000

Floristics An estimated 5000 species of vascular plants (Gentry, 1978, cited in Appendix 1); 148 endemic species (IUCN figures).

Vegetation Tropical moist forest, covering slightly less than half the country's forested area; the remainder mainly coniferous forest; other vegetation types include montane wet forests, moist subtropical forests, wet tropical forests and cloud forests. According to FAO/UNEP (1981), estimated rate of deforestation for closed broadleaved forest 480 sq. km/annum out of 18,550 sq. km; Myers (1980), presumably including the coniferous forests, records 70,500 sq. km as forested according to "recent government documentation", of which "rather more than 40,000 sq. km" are moist forests, mostly in the eastern part of the country and including the relict Mosquitia forest.

Checklists and Floras Honduras is covered by the *Flora Mesoamericana* Project, described in Appendix 1, as well as by the family and generic monographs of *Flora Neotropica* (cited in Appendix 1). Also "most plants" of Honduras are included in the completed *Flora of Guatemala* and related articles in *Fieldiana*, outlined under Guatemala. Floras and papers specifically on Honduras include:

Gilmartin, A.J. (1965). Las Bromeliácias de Honduras. *Ceiba* 11(2): 1-81. (97 species listed.)

Molina, A. (1975). Enumeración de las plantas de Honduras. *Ceiba* 19(1): 1-118. (List of species names; no information on each.)

Nelson, C. (1976-1979). Plantas nuevas para la flora de Honduras, I-III. *Ceiba* 20: 58-68; 21: 51-55; 23: 85-92.

Nelson, C. (1978). Contribuciónes a la Flora de la Mosquitia, Honduras. *Ceiba* 22(1): 41-64. (338 species listed.)

Record, S.J. (1927). Trees of Honduras. *Trop. Woods* 10: 10-47. (Description of trees and their uses.)

Standley, P.C. (1930). A second list of the Trees of Honduras. *Trop. Woods* 21: 9-41. (c. 480 species listed.)

Standley, P.C. (1931). Flora of the Lancetilla Valley, Honduras. *Field Mus. Nat. Hist., Bot. Ser.* 10: 1-418. (Description of habitats and annotated list of species for the Tela area.)

Standley, P.C. (1934). Additions to the Trees of Honduras. *Trop Woods* 37: 27-39. (55 species listed.)

Yuncker, T.G. (1938). A contribution to the Flora of Honduras. *Field Mus. Nat. Hist., Bot. Ser.* 17(4): 287-407. (List of species for the Tela area and also Siguatepeque.)

Information on Threatened Plants There is no national Red Data Book. IUCN is preparing a threatened plant list for release in a forthcoming report *The list of rare, threatened and endemic plants of Middle America*. Latest IUCN statistics, based upon this work: endemic taxa – Ex:1, E:2, V:5, R:5, I:8, K:124, nt:3; non-endemics rare or threatened worldwide – V:7, R:5, I:2 (world categories).

Threatened plants are mentioned in:

D'Arcy, W.G. (1977). Endangered landscapes in Panama and Central America: the threat to plant species. In Prance, G.T. and Elias, T.S. (Eds), cited in Appendix 1. Pp. 89-104.

Stolze, R.G. (1979). Ferns new and rare in Honduras. *Brenesia* 16: 139-141. (5 new records to the flora; 3 species found to be rare.)

Laws Protecting Plants No information. The U.S. Government has determined *Abies guatemalensis* (El Salvador, Guatemala, Honduras and Mexico) as 'Threatened' under the U.S. Endangered Species Act.

Botanic Gardens
Escuela Agrícola Panamericana, El Zamorano, Francisco Morazán.
Jardín Botánico, Lancetilla, Tela.

Useful Addresses
Asociación Hondurena de Ecológia para la Conservación de la Naturaleza, Apto T-250, Tegucigalpa D.C.
Departamento de Biología, Universidad Nacional Autonoma de Honduras, Ciudad Universitaria, Tegucigalpa.
Departamento de Vida Silvestre, Dirección General de Recursos Naturales Renovables (DIGERENARE), Secretaría de Recursos Naturales, Tegucigalpa, D.C.
Herbario Paul C. Standley, Escuela Agrícola Panamericana, Apto 93, Tegucigalpa.

Additional References
Campanella, P. *et al.* (1982). *Honduras. Perfil Ambiental del País. Un estudio de Campo. Resumen Ejecutivo.* AID Contract No. AID/SOD/PDC-C-0247. JRB Associates. McLean, U.S.A. 201 pp.

Holdridge, L.R. (1962). *Mapa Ecológico de Honduras.* Organización de los Estados Americanos. Lith. A. Hoen & Co., Baltimore, Md, U.S.A.

Molina, A. (1974). Vegetación del Valle de Comayagua. *Ceiba* 18: 47-80.

Yuncker, T.G. (1945). The vegetation of Honduras. In Verdoorn, F. (Ed.), cited in Appendix 1. Pp. 55-56. (Short descriptive account.)

Hong Kong

Hong Kong consists of the New Territories to the south of the Chinese province of Guangdong, and more than 200 islands, one of which is Hong Kong Island. Rugged hills comprise much of the territory; the highest peaks include Tai Mo Shan (957 m), Lantau Peak (934 m) and Kowloon Peak (602 m). More than 80% of the population live in urban areas covering only 20% of the land area.

Area 1062 sq. km

Population 5,498,000

Floristics About 2500 vascular plant species of which about 1800 taxa are native (Hong Kong Herbarium, 1978); species endemism perhaps as low as 1% (C.C. Lay, 1984, *in litt.*). Many species are also found in south China, India, Japan, Taiwan and Vietnam.

Vegetation Semi-deciduous broadleaved forest throughout Hong Kong has been greatly modified by man; remnants on steep ravines, hillsides and around some villages and temples, particularly in the New Territories; scrubland and exotic plantations also found on hill slopes; grassland on hilltops especially on many offshore islands (Hong Kong Herbarium, 1978).

Checklists and Floras

Bentham, G. (1861). *Flora Hongkongensis: A Description of the Flowering Plants and Ferns of the Island of Hongkong*. Reeve, London. 482 pp. (The only comprehensive Flora, but rather dated; for additions see the Supplement by H.F. Hance, 1872, London, 59 pp.)

Edie, H.H. (1978). *Ferns of Hong Kong*. Hong Kong Univ. Press. 285 pp.

Hong Kong Herbarium (1978). *Check List of Hong Kong Plants*. Dept of Agriculture and Fisheries Bulletin no. 1 (revised). Govt Printer, Hong Kong. (Checklist of 2502 vascular species, including introductions.)

Field-guides

Thrower, S.L. (1971). *Plants of Hong Kong*. Longman, Hong Kong. 192 pp.

Urban Services Department (1975, 1977). *Hong Kong Trees*, 2 vols. Govt Printer, Hong Kong.

Urban Services Department (1976). *Hong Kong Herbs and Vines*. (Revised Edition.) Govt Printer, Hong Kong. 114 pp.

Urban Services Department (1976). *Hong Kong Shrubs*, 2nd Ed. Govt Printer, Hong Kong. 112 pp.

Urban Services Department (1978). *Hong Kong Freshwater Plants*. Govt Printer, Hong Kong. 89 pp.

Urban Services Department (1980). *Hong Kong Orchids*. Govt Printer, Hong Kong. 108 pp.

Walden, B.M. and Hu, S.Y. (1977). *Wild Flowers of Hong Kong Around the Year*. Sino-American Publ. Co., Hong Kong. 83 pp.

Information on Threatened Plants No national list of threatened plants. *Ailanthus fordii*, *Camellia crapnelliana* and *C. granthamiana* are included in *The IUCN Plant Red Data Book* (1978).

Laws Protecting Plants All wild plants are protected by law. Written permission from the Director of Agriculture and Fisheries is needed for the collection of any wild plants from unleased Crown land. Special protection is given to "threatened" plants which include *Camellia* spp., *Rhododendron* spp., *Magnolia* spp., and all orchids (C.C. Lay, 1984, *in litt.*).

Voluntary Organizations

WWF-Hong Kong, 10th Floor, Wing on Life Building, 22 Des Voeux Road, Central, Hong Kong.

Botanic Gardens

Hong Kong Zoological and Botanic Gardens, Urban Services Department, Hong Kong. (Offices at 12th Floor, Central Government Offices, West Wing, 11 Ice House Street, Hong Kong.)

Kadoorie Experimental and Extension Farms and Botanic Gardens (Kadoorie Agricultural Aid Association), Lam Kam Road, Tai Po, New Territories, Hong Kong.

Ocean Park Botanic Garden, Aberdeen, Hong Kong.

Useful Addresses

Agriculture and Fisheries Department, 12th Floor, Government Offices, 393 Canton Road, Kowloon, Hong Kong.

Department of Biology, CUHK, Shatin, New Territories, Hong Kong.

Department of Botany, Hong Kong University, Pokfulam, Hong Kong.

Hungary

Area 93,032 sq. km

Population 10,786,000

Floristics c. 2300-2500 vascular species, of which 40-45 are endemic (F. Németh, 1984, pers. comm.); 11 endemics according to IUCN figures. D.A. Webb (1978, cited in Appendix 1) estimates 2250-2450 native vascular species from *Flora Europaea*. Rich in Tertiary and Pleistocene relicts. Areas of high endemism: the Central Hungarian Mts and the Carpathian range. Elements: Mediterranean c. 35%, Eurasian c. 23%, Central European c. 16% (including North Carpathian, Pannonian and Balkan), Atlantic, sub-Mediterranean and alpine (Németh, 1979; Németh and Seregélyes, n.d.).

Vegetation Much of natural vegetation replaced by agriculture, especially on the central Great Hungarian Plain; semi-natural vegetation restricted to c. 10%. 4 main vegetation types still apparent: (a) mountain bog on peat with sedges and rushes (*Carex*, *Eriophorum*); (b) mountain meadows rich in grass species (especially *Festuca*, *Poa*, and *Bromus*); (c) steppe or 'puszta', an alkaline and very saline grassland rich in annuals; (d) broadleaved and coniferous woodland. Scots Pine (*Pinus sylvestris*) forms extensive stands in western Hungary, together with beech and hornbeam/oak forests on dry grasslands and rocky steppes in the lowlands (e.g. Szatmar-Bereg Plain in the Bodrog and Kiskun areas) (Vajda, 1956; Németh and Seregélyes, n.d.).

Checklists and Floras Hungary is covered by the completed *Flora Europaea* (Tutin *et al.*, 1964-1980, cited in Appendix 1). National Floras are:

Jávorka, S. and Csapody, V. (1929-1934). *A Magyar Flóra Képekben: Iconographia Florae Hungaricae*, 19 vols. Studium, Budapest.

Soó, R. and Kárpáti, Z. (1968). Magyar Flóra: Harasztok (Pteridophytes) Virágos Növények (Anthophytes). In *Növényhatározó*, 4th Ed. Tankönyvkiadó, Budapest. 846 pp. (Illustrated key to native, naturalized and commonly cultivated vascular plants; phytosociology and habitat details.)

See also:

Jávorka, S. and Csapody, V. (1979). *Iconographia Florae Partis Austro-orientalis Europae Centralis*, revised edition. Fischer, Stuttgart. 704 pp. (Atlas of vascular plants of Hungary and neighbouring areas; illus.)

Soó, R. de (1975). Hauptergebnisse der Floristischen-Geobotanischen und Systematischen Forschungen in Hungarn, 1961-1972. *Mem. Soc. Brot.* 24(2): 599-613.

Field-guides

Jávorka, S. and Csapody, V. (1972). *Erdö Mezö Virágai, A Magyar Flora Szines Kisatlasza*. Mezögazdasagi Kiadó, Budapest. 246 pp.

Information on Threatened Plants Recently published national plant Red Data Book:

Németh, F. and Seregélyes, T. (n.d.). *Hüte die Blumen*. Hungarian State Office for Environment and Nature Conservation, with MTI Publishing, Budapest. 127 pp. (Includes distribution and conservation data for 52 rare and threatened taxa; lists over 300 protected taxa; maps; English edition (*Save the Wild Flowers: Some Rarities Growing in Hungary*); also in German; colour photographs; line drawings.)

Included in the European threatened plant list (Threatened Plants Unit, 1983, cited in Appendix 1); latest IUCN statistics, based upon this work: endemic taxa – E:1, V:8, R:1, I:1; non-endemics rare or threatened worldwide – E:1, V:13, R:5, I:4 (world categories).

Laws Protecting Plants Decree on Nature Conservation (1982) and Ordinance No.1 (1982) provides protection for 172 plant taxa, 24 genera and 2 families. The 1982 Act is published in:

Anon (1983). *Nature Conservation Legislation in Hungary*. National Authority for Environment Protection and Nature Conservation, Budapest. 55 pp., 5 annexes. (Annexes 1 and 3 list protected and specially protected plant species.)

See also:

Borhidi, A. and Jánossy, D. (1984). Protected Plants and Animals in Hungary. *Ambio* 13(2): 106.

Csapody, I. (1982). *Védett Növényeink* (Our protected plants). Gondolat, Budapest. 346 pp. (In Hungarian; black and white photographs and colour drawings.)

Botanic Gardens

Agrobotanic Garden, University of Agricultural Sciences, 2103 Gödöllö.

Botanic Garden of the Hungarian Academy of Sciences, 2163 Vacratot.

Budapest Fövaros Allat-es Növenykertje, Varosliget, 1371 Budapest XIV.

Erdeszeti és Faipari Egyetem Botanikus Kertjke, 9401 Sopron.

Hortus Botanicus, Instituti Plantarum Medicinalium, 2011 Budakalasz, Pf 11.

Institutum Botanicum et Hortus Botanicus, 1502 Budapest pf53, 1118 Budapest XI, Menesi UT44.

Kamoni Arboretum, Institutum Scientiarum Silviculturae Hungariae, Vöröszaszlo u 102, 9707 Szombathely.

Research Centre for Agrobotany, NIAVT, 2766 Tapioszele.

Soroksár Botanical Garden, Budapest.

Szarvas Arboretum, 5540 Szarvas.

University of Budapest Botanical Garden, Illés Utca 25, 1083 Budapest.

Useful Addresses

Department of Nature Conservation, Ministry of Agriculture and Food, Kossathajostev 11, 1860 Budapest 5.

Hungarian State Office for Environmental and Nature Conservation, Országos Természetvédelmi Hivatal, Költö utca 21, 1121 Budapest.

National Office for Nature Conservation, Tulipan Koz 10, 9400 Sopron.

Additional References

Németh, F. (1979). The vascular flora and vegetation on the Szabadszállás-Fülöpszállás territory of the Kiskunság National Park (KNP), I. *Stud. Bot. Hungarica* 13: 79-105. (In English; checklist of vascular plants in the National Park; includes valuable table showing phytogeograhical composition of entire Hungarian flora.)

Soó, R. (1964-1980). *A Magyar Flóra és Vegetáció Rendszertani-Növényföldrajzi Kézikönyve*, 6 vols. Akadémiai Kiadó, Budapest. (A systematic geobotanical work; detailed phytosociological classification; includes bryophytes.)

Vajda, E. (1956). *A Magyar Növényvilág Képeskönyve*. English translation by E. Rácz (*Wild Flowers in Hungary: The Origin and Development of Plant Communities*). Corvina, Budapest. 49 pp. (Illus.)

Iceland

Area 102,819 sq. km

Population 239,000

Floristics c. 470 species of indigenous and naturalized vascular plants, of which nearly 20% believed introduced by man during the past 1100 years (Einarsson, 1984, *in litt.*). 1 endemic species (IUCN figure). Elements: circumpolar; amphi-atlantic (plants distributed almost equally on both sides of the Atlantic); eastern element; and western or American element (Einarsson, *in litt.*).

Vegetation Original spruce and birch forests once occupied coastal areas up to 400 m; now completely cleared due to extensive sheep grazing. Today forest occupies only c. 1250 sq. km in the more sheltered lowland valleys where willow, birch and rowan (*Sorbus aucuparia*) survive. Arctic/alpine tundra in centre and north of country, with dwarf shrubs (*Juniperus*, *Betula nana* and *Arctostaphylos*). Elsewhere, large areas of almost bare rock, gravel and sand, sparsely colonized by mosses, lichens and vascular plants. Extensive wetlands, but many in the lowlands now drained.

Checklists and Floras Iceland is covered by the completed *Flora Europaea* (Tutin *et al.* 1964-1980, cited in Appendix 1). There is no up-to-date national Flora. The most recent account is:

Gröntved, J. (1942). The Pteridophyta and Spermatophyta of Iceland. In Rosenvinge, L.K. *et al.* (Eds), *The Botany of Iceland* (cited under 'Additional References').

(Detailed introduction about vegetation, phytogeography, botanical exploration and research; in English.)

Also relevant:

Kristinsson, H. (1973-1978). Recent literature on the botany of Iceland. *Acta Bot. Islandica* 2: 67-76; 3: 102-104; 4: 67-74; 5: 63-70.
Löve, A. (1963). Taxonomic botany in Iceland since 1945. *Webbia* 18: 277-301.
Löve, A. (1970). Emendations in the Icelandic flora. *Taxon* 19(2): 298-302.

Field-guides

Löve, A. (1981). *Islenzk Ferdaflóra*, 2nd Ed. Almenna Bókafélagid, Reykavík. 429 pp. (In Icelandic; lists protected species; colour plates; English edition, 1983.)
Ostenfeld, C.H. and Gröntved, J. (1934). *The Flora of Iceland and the Faeroes*. Levin and Munksgaard, Copenhagen. 195 pp. (Standard English Flora of Iceland.)
Stefánsson, S. (1948). *Flóra Islands*, 3rd Ed. by S. Steindórsson. Islenzka Náttúrufraedífélag, Akureyri. 407 pp. (In Icelandic.)
Wolseley, P. (1979). *A Field Key to the Flowering Plants of Iceland*. Thule Press, Sandwick, Shetland. 64 pp.

Information on Threatened Plants No plant Red Data Book. An unpublished threatened plant list has been prepared for the Council of Europe by the Nature Conservation Council and the Department of Botany in the Icelandic Museum of Natural History (addresses below). 44 taxa are listed.

Included in the European threatened plant list (Threatened Plants Unit, 1983, cited in Appendix 1); latest IUCN statistics, based upon this work: endemic taxa – R:1; non-endemics rare or threatened worldwide – none.

Laws Protecting Plants The Nature Conservation Act 1956, amended 1971, provides protection for plant species in Article 23, which states, "The Nature Conservation Council can declare the protection of scientifically or culturally important plants or animals in order to prevent their disturbance, decrease or extinction. Protection can be applied locally or to the whole country." At present 31 taxa of vascular plants are protected in the whole country. It is absolutely forbidden to pick the leaves or flowers, uproot or damage any of these plants. For the list of protected plants see Löve (1981).

Voluntary Organizations None relate specifically to plants but the main nature conservation organizations are:

Icelandic Association of Nature Conservation Societies, Sundstraeti 24, 400 Isafjördur.
Icelandic Environment Union, Skólavördustig 25, 101 Reykjavík.

Botanic Gardens

Grasagardur Reyjavíkur (Botanic Gardens of Reykjavík), Skúlatún 2, 105 Reykjavík.
Lystigardur Akureyrar (Botanic Section, Public Gardens of Akureyri), Hafnarstraeti 81, P.O. Box 95, 600 Akureyri.

Useful Addresses

Icelandic Museum of Natural History, P.O. Box 5320, 125 Reykjavík.
Institute of Biology, University of Iceland, Grensásvegur 12.
Landvernd (Icelandic Environment Union), Skolavördustigur 25, 101 Reykjavík.
Museum of Natural History, P.O. Box 580, 602 Akureyri.
Nature Conservation Council, Hverfisgata 26, 101 Reykjavík.

Additional References

Löve, A. and D. (1956). Cytotaxonomical conspectus of the Icelandic Flora. *Acta Horti Gotoburgensis* 20(4): 65-291.

Rosenvinge, L.K. *et al.* (Eds) (1912-1949). *The Botany of Iceland*, 5 vols, 9 parts. J. Frimodt and E. Munksgaard, Copenhagen. (1 and 2 – physical geography, diatoms, bryophytes; 3 – vegetation studies, fungi, genus *Taraxacum*, by J. Gröntved; 4 – pteridophytes, spermatophytes, habitat accounts; 5 – flora of Reykjanes Peninsula, south-west Iceland, by J. Gröntved and E. Hadac.)

India

Area 3,166,828 sq. km

Population 746,742,000

Floristics An estimated 15,000 vascular plant species (Botanical Survey of India, 1983b) including c. 600 pteridophytes. About 5000 endemic vascular plant species; c. 140 endemic genera, but no endemic families. Areas rich in endemism are north-east India, the southern parts of peninsular India, the Western Ghats and the north-western and eastern Himalayas. Tropical S.E. Asian and Malayan elements comprise c. 35% of the flora; also temperate Asian elements (8%), Mediterranean-Iranian elements (5%) (Nayar, 1977).

Vegetation Tropical moist deciduous or monsoon forests are the natural vegetation cover over much of India between the Himalayas, Thar and Western Ghats. Tropical evergreen rain forest up to 1200 m, in north-east, and along seaward side of the Western Ghats in the States of Maharashtra, Karnataka, Tamil Nadu and Kerala, mostly cleared below 500 m; mangrove forests most extensive along the south coast of West Bengal, particularly the Sunderban region; tropical semi-evergreen forests and subtropical broadleaved hill forest below 1500 m on the Himalayan foothills of Assam, and in the Western Ghats. Tropical dry deciduous forest with Teak (*Tectona grandis*) and tropical moist deciduous forest with Sal (*Shorea robusta*) in central and northern India at 450-600 m, but depleted; extensive areas of bamboo forests, especially in south. Montane and temperate forests grade into coniferous forests and alpine scrub in Himalayas over 3000 m. Desert or near-desert conditions in western Rajasthan and Gujarat; extensive thorn scrub in Maharashtra, Andra Pradesh, Karnataka and Tamil Nadu.

Much of India's natural vegetation has been greatly modified by various forms of agriculture, forestry and urbanization. Over 50% of the land area is cultivated, with rice the most important crop. Estimated rate of deforestation of closed broadleaved tropical forests 1320 sq. km/annum out of a total of 460,440 sq. km (FAO/UNEP, 1981). However, according to sources quoted in Myers (1980, cited in Appendix 1), only as little as c. 260,000 sq. km can be considered to be "adequately stocked forestlands", comprising 21,040 sq. km of tropical evergreen rain forest, 8340 sq. km of semi-evergreen rain forest, 102,000 sq. km of tropical moist deciduous forest and 138,750 sq. km of tropical dry deciduous forest. All forests, particularly moist forest types, are rapidly being degraded as a result of population pressure and shifting cultivation.

See Champion and Seth (1968) for a comprehensive account of vegetation, and the summary accounts for each State in *Bull. Bot. Survey India* (1977), 19(1-4). 336 pp.

A series of vegetation maps has been prepared for Peninsular India at 1:1,000,000, showing degradation status, available from the Scientific Section, French Institute, Pondicherry, India. See also:

Anon (1976). *Atlas of Forest Resources of India*. National Atlas Organization,
 Calcutta. (Major forest types based on classification of Champion and Seth, 1968.)

Checklists and Floras India is covered by the *Flora of British India* (Hooker, 1872-1897), and is included in the *Flora of Eastern Himalaya* (1966, 1971, 1975), both cited in Appendix 1. The Sikkim Himalaya is included in Grierson and Long (1980) and (1983-), cited under Bhutan. For ferns see Beddome (1892) and the companion volume by Nayar and Kaur (1972), cited in Appendix 1.

A national Flora is being published:

Botanical Survey of India (1978-). *Flora of India*. Botanical Survey of India, Howrah.
 (18 fascicles so far, most covering a small family or single genus.)

The *Flora of India* project was re-organized in 1984 with a target of 15-20 volumes to be published over a period of 15 years, with collaboration between the Botanical Survey of India and the Royal Botanic Gardens, Kew. Each volume will treat c. 1000 species. The Himalayas will be covered as a single geographical unit, with records of plants found in Bhutan, Nepal and the Sikkim Himalayas. A checklist of c. 18,000 flowering plant taxa will be prepared in 1986.

There are many Floras at State and regional level. Only a selection are cited here. For a comprehensive bibliography see the proceedings of the Symposium on Status of Floristic Studies in India in *Bull. Bot. Survey India* (1977), vol. 19. 336 pp. The *Flora Malesiana Bulletin*, cited in Appendix 1, also includes a bibliographic section covering India. Among the more recent Floras are the following:

Bhandari, M.M. (1978). *Flora of the Indian Desert*. Scientific Publishers, Jodhpur.
 471 pp. (Introduction covers physical geography, floristics and vegetation of the
 desert areas of north-west India; 592 species treated.)
Chowdhery, H.J. and Wadhwa, B.M. (1984). *Flora of Himachal Pradesh: Analysis, 1.*
 Flora of India, Ser. 2. Botanical Survey of India, Howrah. 340 pp. (Enumeration of
 1202 flowering plant species, including Ranunculaceae to Caprifoliaceae (85
 families). Covers north-western and western Himalayas; notes on distributions.)
Cooke, T. (1901-1903). *The Flora of the Presidency of Bombay*, 3 vols. London. (1 –
 Ranunculaceae to Rubiaceae; 2 – Elaeagnaceae to Gramineae; 3 – Compositae to
 Thymelaeaceae. Reprinted in 1958 by the Botanical Survey of India, Calcutta.)
Dhar, U. and Kachroo, P. (1983). *Alpine Flora of Kashmir Himalaya*. Scientific Publ.,
 Jodhpur. 280 pp. (Includes annotated checklists, distribution maps, floristic
 analyses.)
Haines, H.H. (1921-1925). *The Botany of Bihar and Orissa*, 6 parts. Govt of Bihar and
 Orissa. (Reprinted 1961 by the Botanical Survey of India.)
Kanjilal, U.N. *et al.* (1934-1940). *Flora of Assam*, 5 vols. Shillong. (Covers mainly
 woody species.)
Maheshwari, J.K. (1963). *The Flora of Delhi*. Council of Scientific and Industrial
 Research, New Dehli. 447 pp. (Covers 478 out of a total of 531 indigenous and
 naturalized species of angiosperms.)
Matthew, K.M. (1981-1983). *The Flora of Tamilnadu Carnatic*, 3 vols. Rapinat
 Herbarium, Tiruchirapalli. (1 – Materials for the Flora, documentation of 32,000

vascular plant specimens; notes on forest types, ethnobotany; 2 – detailed accounts of 2260 species; 3 – illustrations.)

Nair, N.C. (1977). *Flora of Bashahr Himalayas*. International Bioscience Publications, Hissar. 360 pp. (Enumeration of 1629 species of angiosperms and gymnosperms found between 650-6930 m in Kinaur and Mahasu districts of Himachel Pradesh.)

Nair, N.C. and Henry, A.N. (1983-). *Flora of Tamil Nadu, India. Series 1: Analysis, 1*. Botanical Survey of India, Coimbatore. (3 vols planned in Series 1, the first includes enumeration of c. 2000 angiosperms covering Ranunculaceae to Sambucaceae; economic plants, endemics, rare and endangered plants indicated.)

Puri, G.S., Jain, S.K., Mukherjee, S.K., Sarup, S., and Kotwal, N.N. (1964). Flora of Rajasthan – West of the Aravallis. *Rec. Bot. Survey India* 19(1). 159 pp. (Covers 750 species in 90 families.)

Raizada, M.B. (1976). *Supplement to Duthie's 'Flora of the Upper Gangetic Plain and the Adjacent Siwalik and Sub-Himalayan Tracts'*. Bishen Singh Mahendra Pal Singh, Dehra Dun. 355 pp.

Rao, R.R. and Razi, B.A. (1981). *A Synoptic Flora of Mysore District*. International Bioscience Series 7. Today and Tomorrow's Printers, New Delhi. 674 pp.

Santapau, H. (1953). The flora of Khandala on the Western Ghats of India. *Rec. Bot. Survey India* 16(1). 396 pp.

Sharma, B.M. and Kachroo, P. (1981). *Flora of Jammu and Plants of Neighbourhood*, 2 vols. Bishen Singh Mahendra Pal Singh, Dehra Dun.

Sharma, S. and Tiagi, B. (1979). *Flora of North-East Rajasthan*. Kalyani Publishers, New Dehli. 540 pp. (Treats 612 species of flowering plants in 95 families.)

Varma, S.K. (1981). *Flora of Bhagalpur: Dicotyledons*. Today and Tomorrow's Printers. 414 pp.

Information on Threatened Plants In 1980, a 5-year Project on Study, Survey and Conservation of Endangered Flora (POSSCEF) with financial support from the U.S. Fish and Wildlife Service, Washington, D.C., was initiated in the Botanical Survey of India (address below). Illustrated accounts and lists of rare, threatened and endemic species are in preparation. The most comprehensive list so far is:

Botanical Survey of India (1983a). *Materials For a Catalogue of Threatened Plants of India*. Dept of Environment, Government of India, Calcutta. 69 pp. (Lists c. 900 rare and threatened taxa together with their distributions. Prepared by the POSSCEF team under S.K. Jain for the IUCN Plants Programme. Reviewed in *Threatened Plants Newsletter* 12: 18 (1983), where H. Synge predicts as many as 3000-4000 Indian plants might be threatened (see also *ibid*. 9: 1-3 (1982)).

The first volume of a Plant Red Data Book has recently been published:

Jain, S.K. and Sastry, A.R.K. (Eds) (1984). *Indian Plant Red Data Book, 1*. Calcutta. (Data sheets on 125 species, with illustrations.)

POSSCEF also issues a *Plant Conservation Bulletin*, edited by S.K. Jain and A.R.K. Sastry, containing numerous papers on threatened plants; in particular see:

Hajra, P.K. (1983). Rare, threatened and endemic plants of the western Himalayas – monocotyledons. *Ibid*. 4: 1-13. (Annotated list of c. 100 species.)

Raghavan, R.S. and Singh, N.P. (1983). Endemic and threatened plants of western India. *Ibid*. 3: 1-16. (Annotated list of 207 species.)

Vajravelu, E. (1983). Rare, threatened and endemic flowering plants of South India (Part 1). *Ibid*. 4: 14-30. (Annotated list of 212 species.)

A seminar on threatened plants of India was organized at Dehra Dun in September 1981. The proceedings have been published in:

Jain, S.K. and Rao, R.R. (Eds) (1983). *An Assessment of Threatened Plants of India.* Botanical Survey of India, Howrah. 334 pp. (Includes 60 papers presented at the seminar; many include lists of threatened plants with IUCN categories for various regions. See for example N.C. Shah on threatened medicinal plants of Uttar Pradesh Himalaya, pp. 40-49; R.P. Pandley *et al.* on threatened plants of Rajasthan, pp. 55-62; S.D. Sabnis and K.S.S. Rao on threatened plants in south-east Kutch, pp. 71-77; R.R. Rao and K. Haridasan on threatened plants of Meghalaya, pp. 94-103; Sandhyajyoti Das and N.C. Deori on endemic orchids of north-east India, pp. 104-109; S.K. Kataki on rare plants in the Khasi and Jaintia Hills, pp. 146-150; A.R.K. Sastry and P.K. Hajra on rare and endemic rhododendrons, pp. 222-231; K.N. Bahadur and S.S. Jain on rare bamboos, pp. 263-271; R.K. Arora and E. Roshini Nayar on the distibution of wild relatives and related species of economic plants in India, pp. 285-291.)

Other papers and publications including lists are:

Abraham, Z. and Mehrotra, B.N. (1982). Some observations on endemic species and rare plants of the montane flora of the Nilgiris, South India. *J. Econ. Taxonomic Botany* 3(3): 863-867. (Lists 26 rare endemics and 2 rare non-endemics.)

Bahadur, K.N. and Jain, S.S. (1981). Rare bamboos of India. *Indian J. Forestry* 4(4): 280-286. (Preliminary review of 26 rare bamboos.)

Chandra, P. (1983). Observations on the rare and endangered ferns of India. *New Botanist* 10: 41-47. (Lists 49 taxa; notes on distribution and conservation status.)

Cook, C.D.K. (1980). The status of some Indian endemic plants. *Threatened Plants Committee - Newsletter* 6: 17-18. (Mentions 5 threatened wetland species.)

Henry, A.N., Vivekananthan, K. and Nair, N.C. (1978). Rare and threatened flowering plants of south India. *J. Bombay Nat. Hist. Soc.* 75(3): 684-697. (Lists 224 angiosperms.)

Jain, S.K. and Sastry, A.R.K. (1980). *Threatened Plants of India: A State-of-the-Art Report.* Botanical Survey of India, Howrah. 48 pp. (Short accounts of 134 species, many with colour photographs; reviewed at some length in *Threatened Plants Committee - Newsletter* 6: 15-16 (1980).)

Kataki, S.K. (1976). Indian orchids – a note on conservation. *American Orchid Soc. Bull.* 46(2): 117-121. (Lists threatened orchids.)

Kataki, S.K., Jain, S.K. and Sastry, A.R.K. (1984). *Threatened and Endemic Orchids of Sikkim and North-eastern India.* Botanical Survey of India, Howrah. 95 pp. (Descriptions, distributions, illustrations of over 100 species.)

Sahni, K.C. (1979). Endemic, relict, primitive and spectacular taxa in eastern Himalayan flora and strategies for their conservation. *Indian J. Forestry* 2(2): 181-190. (Mentions 30 taxa rare or threatened in the Himalayan region; notes on vegetation.)

Santapau, H. (1970). Endangered plant species and their habitats. In IUCN, *11th Technical Meeting Papers and Proceedings, 2. Problems of Threatened Species.* IUCN New Series 18, Switzerland. Pp. 83-88. (Includes list of threatened medicinal plants and orchids in need of protection.)

A number of papers on plant conservation in India are included in:

Jain, S.K. and Mehra, K.L. (Eds) (1983). *Conservation of Tropical Plant Resources.* Proceedings of the Regional Workshop on Conservation of Tropical Plant

Resources in South East Asia, New Delhi, March 8-12, 1982. Botanical Survey of India, Howrah. (Workshop reviewed in *Threatened Plants Newsletter* 9: 1-3 (1982) and book in *ibid.* 13: 19-20 (1984).)

In particular see:

Gupta, R. and Sethi, K.L. Conservation of medicinal plants resources in the Himalayan region. *Ibid.*, pp. 101-109. (Lists 8 Endangered, 12 Vulnerable and 8 Rare medicinal plants.)

Husain, A. Conservation of genetic resources of medicinal plants in India. *Ibid.*, pp. 110-117. (Notes on 15 taxa threatened by overcollecting.)

5 species are included in *The IUCN Plant Red Data Book* (1978). Latest IUCN statistics, principally derived from Botanical Survey of India (1983a): endemic taxa – Ex:4, E:18, V:2, R:3, I:541.

Laws Protecting Plants

The Wildlife (Protection) Act, 1972. Govt of India, Ministry of Law, Justice and Company Affairs. (Appendices have lists of 'endangered' species to which plants are being added; S.K. Jain, 1984, *in litt.*)

Voluntary Organizations

Bombay Natural History Society, Hornbill House, Shahid Bhagat Singh Road, Bombay 400023.

Friends of Trees, Tata Building, Choringhee Road, Calcutta 17.

Indian Society of Naturalists (INSONA), c/o Maharaja Fatehsingh Zoo Trust, Indumati Mahul, Jawaharlal Nehru Marg, Baroda 390001.

WWF-India, c/o Godrej & Boyce Mfg. Co. Private Ltd., Lalbaug, Parel, Bombay 400012.

Botanic Gardens

The Botanical Survey of India have prepared 2 reports on Indian botanic gardens (1983):

A Directory of Botanic Gardens in India (A Preliminary Account of History, Organisation and Holdings of Some Government University and Public Gardens of India). 131 pp. (Entries for 55 Indian botanic gardens and botanical institutions. The largest garden is the Indian Botanic Garden, Sibpur, Howrah 71103, West Bengal.)

Materials for a Green Book of Botanic Gardens in India. 88 pp. (Lists 100 rare, endangered and endemic plants known to be cultivated in the 8 botanic gardens run by the Botanical Survey.)

Useful Addresses

Botanical Survey of India, P.O. Botanic Garden, Howrah 71103. (Includes POSSCEF programme.)

Department of the Environment, Bikaner House, Shahjahan Road, New Delhi 110011.

National Bureau of Plant Genetic Resources, New Dehli 110012.

CITES Management Authority: The Director of Wildlife Preservation, Government of India, Ministry of Environment, Room 240, Krishi Bharan, New Delhi 110001.

CITES (for Orchidaceae): The Deputy Director of Wildlife Preservation, Government of India, 97/18 Hazra Road, Calcutta, West Bengal.

CITES Scientific Authority: Botanical Survey of India, P.O. Botanic Garden, Howrah 71103.

Additional References

Botanical Survey of India (1983b). *Flora and Vegetation of India - An Outline*.
Botanical Survey of India, Howrah. 24 pp. (Introduction to the flora and vegetation
of India and its phytogeographical affinities; review of the District Flora
Programme and threats to plant life. Prepared for the IUCN Plants Programme.)

Champion, H.G. and Seth, S.K. (1968). *A Revised Survey of the Forest Types of
India*. Govt of India Press, Delhi. 404 pp.

Chatterjee, D. (1939). Studies on the endemic flora of India and Burma. *J. Royal
Asiatic Soc. Bengal Sci*. 5: 19-67.

Mani, M.S. (Ed.) (1974). *Ecology and Biogeography in India*. Junk, The Hague.
773 pp. (Chapters on vegetation, flora, biogeography.)

Nayar, M.P. (1977). Changing patterns of the Indian flora. *Bull. Bot. Survey India* 19:
145-155. (Origin and distribution of the flora; floristic relationships.)

Singh, J.S., Singh, S.P, Saxena, A.K. and Rawat, Y.S. (1984). India's Silent Valley
and its threatened rain-forest ecosystems. *Envir. Conserv.* 11(3): 223-233.

For useful background information on the Himalayan region see Lall and Moddie (1981),
cited in Appendix 1. For an account of the alpine flora of the Sikkim Himalaya see
Bulletin of the Alpine Garden Society 52(3), September 1984 (No. 217).

Indonesia

An archipelago of 13,667 islands of which about 600 are inhabited. A chain of high
mountains stretch in an arc from western Sumatra, through southern Java and parts of the
Lesser Sunda Islands.

Area 1,919,443 sq. km

Irian Jaya: 412,981 sq. km; Java: 134,044 sq. km; Kalimantan: 550,203 sq. km; Maluku:
74,504 sq. km; Nusa Tenggara: c. 80,000 sq. km; Sulawesi: 227,654 sq. km; Sumatra:
524,097 sq. km.

Population 147,673,800

Irian Jaya: 1,173,800 (1980); Java: 94,000,000 (1981); Kalimantan: 6,700,000 (1980);
Maluku: 1,400,000 (1980); Nusa Tenggara: 6,000,000 (1980); Sulawesi: 10,400,000 (1980);
Sumatra: 28,000,000 (1980).

Floristics One of the richest floras in the world, with about 10,000 trees alone
(FAO, 1982). The archipelago forms the greater part of the botanical region of Malesia.
Floristic affinities are with Asia, and to a lesser extent Australia; about 40% of genera are
either endemic or have their centre of development in Malesia. There are floristic
subdivisions between Sumatra and Java and between Sulawesi and the island of Borneo (of
which Kalimantan forms the greater part). The richest areas are the primary lowland rain
forests of Borneo and Irian Jaya (Jacobs, 1974).

Irian Jaya Good (1960) estimates that the island of New Guinea, of which Irian
Jaya is the western portion, has c. 9000 angiosperm species, of which 90% endemic, and
Parris (1985), cited in Appendix 1, estimates that it has c. 2000 fern species. There are 1465
genera in New Guinea, of which 124 are endemic (van Balgooy, in Paijmans, 1976). The

Tamrau-Arfak mountains of the Volgelkop are important centres of endemism. The flora is related to both Asia and Australia.

Java 5011 vascular plant species of which 4598 indigenous; includes 497 ferns (Backer and Bakhuizen van den Brink, 1963-1968). Only 10 genera endemic. Dipterocarps less abundant in the seasonally dry monsoon forests with only 10 species on the island (Jacobs, 1981; P. Ashton in *Flora Malesiana* 9(2), 1982, cited in Appendix 1).

Kalimantan No figure for Kalimantan but Borneo, floristically the richest of the Sunda islands, has c. 10,000-11,000 vascular species (based on Merrill, 1921); Borneo (whole island) has c. 1000 fern species (Parris, 1985, cited in Appendix 1). Endemism is high with c. 34% of vascular species and 59 genera restricted to the island. Especially diverse are the primary lowland rain forests below 300 m, particularly on sandy yellow soils (FAO, 1981). Borneo, with 267 species, is the centre of diversity of Dipterocarpaceae, the most important family of commercial trees in the region; 158 dipterocarps are endemic to the island (Jacobs, 1981; P. Ashton in *Flora Malesiana* 9(2), 1982, cited in Appendix 1).

Maluku (The Moluccas) A relatively impoverished flora with low endemism, with western (Sundaland) and eastern (Sahul) elements.

Nusa Tenggara (The Lesser Sunda Islands) Less rich than other parts of Indonesia; 12% species endemism. Most endemics found on Lombok and Timor (Kalkman, 1955). Floristic affinities mainly Asian, although in the drier monsoon forests of the east there are Australian elements (van Steenis, 1979).

Sulawesi Floristically poor compared with neighbouring Borneo. Australasian elements in high mountains; otherwise Malesian.

Sumatra Comparable in richness to Kalimantan and Irian Jaya; richer than Java, Sulawesi and smaller islands (FAO, 1982). Species endemism about 12%; 17 endemic genera. Dipterocarps dominate lowland rain forests; 96 species in all, of which 11 endemic (Jacobs, 1981; P. Ashton in *Flora Malesiana* 9(2), 1982, cited in Appendix 1). The Bukit Barisan Range contains Himalayan elements (van Steenis, 1934).

Vegetation Tropical moist forests are the dominant climax vegetation. Tropical evergreen rain forest is the most extensive formation, of which Indonesia has an estimated 1,018,000 sq. km, nearly 10% of world total. Deciduous monsoon forests and fire-maintained savanna grasslands in seasonally dry areas, particularly in southern and eastern islands. Clearance for agriculture, shifting cultivation, logging and transmigration programmes are the main causes of deforestation. Mangroves occupy c. 25,000 sq. km (FAO/UNEP, 1981).

Estimated rate of deforestation of closed broadleaved forests in Indonesia 6000 sq. km/annum out of a total of 1,135,750 sq. km (FAO/UNEP, 1981); however, Myers (1980, cited in Appendix 1), estimates the amount of primary forest remaining is probably well below 1,000,000 sq. km and possibly as low as 800,000 sq. km.

Indonesia is included on the *Vegetation Map of Malaysia* (van Steenis, 1958) and on the vegetation map of Malesia (Whitmore, 1984), both covering the *Flora Malesiana* region at scale 1:5,000,000 and cited in Appendix 1. For a general description of the forests of Indonesia see Whitmore (1975b), cited in Appendix 1. See also:

Direktorat Bina Program (1980). *Peta Tegakan Hutan Indonesia, 1:2,750,000*. Bogor. (Map of forest stands of Indonesia.)

Laumonier, Y., Gadrinab, A. and Purnajaya (1983). *Southern Sumatra: International Map of the Vegetation and of Environmental Conditions*. Institute de la Carte

International du Tapis Végétal and SEAMEO/BIOTROP, Toulouse. (Scale 1:1,000,000; maps of north and central Sumatra in preparation.)

Irian Jaya Large tracts of primary tropical evergreen rain forest, rich in tree ferns, palms, bamboos, lianas; dry evergreen forests, with *Tristania, Syzygium* and *Acacia*, in the monsoonal south-east; lower montane forests between 1000-3000 m, with *Araucaria, Podocarpus, Agathis* and *Nothofagus*; upper montane forests up to 4000 m, with tree ferns, conifers, and rhododendrons; above 4000 m, alpine heathland with low shrubs, bryophytes and lichens. The Fakfak Mountains have limestone forest and large areas of anthropogenic grassland. Swamp forests, with sago palm (*Metroxylon sagu*), and extensive mangrove forests mainly along the southern coast, and in the north between the Mamberamo delta westwards to Teluk Cenderawasih; beach forests share most of the species of similar habitats in Malesia, but are better developed than anywhere else (FAO, 1981). Closed broadleaved forests of all kinds were estimated to cover 380,050 sq. km at the end of 1980 (FAO/UNEP, 1981). This represents 92% of the total land area.

Java All lowland forests have been cleared, with the exception of patches near the south coast of East Java; in West and East Java, evergreen rain forests are restricted to isolated patches on south-facing mountain sides; monsoon forests (tropical moist deciduous forests) with Teak (*Tectona grandis*), *Bombax* and *Tetrameles* in centre and east; plantations of teak have been established in cleared areas where soils are unsuitable for cultivation; Tjemera (*Casuarina junghuhniana*) forests mainly on the northern slopes of mountains in East Java above 1400 m. Where fire is excluded a succession to mixed oak-laurel forest begins. Subalpine vegetation above 2400 m, dominated by Ericaceae with temperate herbaceous species (Backer and Bakhuizen van den Brink, 1963-1968); extensive montane grasslands following forest destruction by fire (van Steenis, 1972). Limestone karst with a distinctive flora occurs along Java's southern and north-eastern coasts, most of which is now planted with teak. Freshwater swamp forests and mangroves occur in a few isolated patches. Closed broadleaved forests were estimated to cover 11,800 sq. km at the end of 1980 (FAO/UNEP, 1981). This represents only 9% of the land area. Most of Java is intensively cultivated (FAO, 1982), and on the island of Madura there is no extant forest at all.

Kalimantan Tropical lowland evergreen rain forest up to 1300 m; extensive hill dipterocarp forests and various montane forest formations with Fagaceae, Lauraceae and Myrtaceae up to 2300 m. Large areas of mangroves, peat swamps and freshwater non-peaty swamps, and the most extensive heath forests (kerangas) in S.E. Asia. Extensive secondary forests (blukar) and Alangalang (*Imperata cylindrica*) grassland as a result of past forest clearance. Closed broadleaved forests were estimated to cover 353,950 sq. km at the end of 1980 (FAO/UNEP, 1981). This represents c. 65% of the total land area. A huge area (c. 30,000 sq. km) of Kalimantan, including 8000 sq. km of primary forest, was destroyed by fire in 1983.

Maluku Transition from evergreen rain forest in the north-west of Halmahera and Seram to seasonal monsoon forests in south Halmahera, in Obi and the north-east of Buru and Banda Sea islands. Small areas of mangroves; freshwater swamps with important stands of Sago (*Metroxylon sagu*); lowland forest formations with *Melaleuca* on drier soils. Rich montane forests occur on Seram and Halmahera. Closed broadleaved forests were estimated to cover 47,150 sq. km at the end of 1980 (FAO/UNEP, 1981). The northern islands are being logged and most forest is already parcelled out in timber concessions (FAO, 1981).

Nusa Tenggara Savanna woodland with *Eucalyptus* and *Casuarina* now covers most of the island (K. Kartawinata, 1984, *in litt.*); evergreen rain forest only surviving in isolated patches in steep valleys on south-facing sides of mountain ranges; elsewhere, there are monsoon forests and extensive grasslands. Timor has some of the finest natural Sandalwood (*Santalum album*) forests in the world (FAO, 1981). Closed broadleaved forests of all kinds were estimated to cover 25,150 sq. km at the end of 1980 (FAO/UNEP, 1981). This represents c. 30% of the total land area.

Sulawesi Extensive tracts of primary hill and montane variants of tropical evergreen rain forest, with few dipterocarps; *Syzygium* (Myrtaceae) sometimes dominates forests at all altitudes (FAO, 1982). Forests on limestone and ultrabasic rocks also present. Small areas of inland heath forest occur in central Sulawesi; mangroves occur in isolated patches in the south. Large areas in the south and some parts of the north have been cleared for shifting cultivation (FAO, 1982). Closed broadleaved forests were estimated to cover 95,250 sq. km at the end of 1980 (FAO/UNEP, 1981). This represents c. 40% of the total land area.

Sumatra Tropical evergreen rain forest dominated by dipterocarps, and with Ironwood (*Eusideroxylon zwageri*) abundant in some forests in the south; heath forests in east; lowland peat swamp forest and mangroves along eastern coasts. Drier mountain areas in north support the only natural pine (*Pinus merkusii*) forests in Indonesia (FAO, 1982). According to the 1978 Bina Programme, forests cover 57% of the land area (figures quoted in FAO, 1982); however, estimates from satellite imagery indicate only 42% still covered by primary forest (FAO, 1982). The total area of closed broadleaved forests was estimated to be 222,400 sq. km at the end of 1980 (FAO/UNEP, 1980).

Checklists and Floras Indonesia is included in the incomplete but very detailed *Flora Malesiana* (1948-), cited in Appendix 1. See, in particular, the extensive bibliography and history of plant collecting in Series 1, vol. 4, pp. 71-161, and the annotated selected bibliography in Series 1, vol. 5, pp. i-cxliv. Other floristic accounts include:

Backer, C.A. and Bakhuizen van den Brink, R.C. (1963-1968). *Flora of Java (Spermatophytes Only)*, 3 vols. Noordhoff (Vols 1, 2) and Wolters-Noordhoff, Groningen. (Keys and descriptions for all taxa; vegetation types described in vol. 2.)
Handbooks of the Flora of Papua New Guinea (1978-), 2 vols so far. Melbourne Univ. Press. (Includes Irian Jaya. 1 – vegetation, keys, treatments of Combretaceae, Magnoliaceae, Meliaceae and many smaller families; edited by J.S. Womersley; 2 – Elaeocarpaceae, Juglandaceae, Loranthaceae and others; edited by E.E. Henty.)
Kalkman, C. (1955). A plant geographical analysis of the Lesser Sunda Islands. *Acta Bot. Neerl.* 4: 200-225. (Lists 480 species in 51 families with occurrence by island.)
Merrill, E.D. (1921). *A Bibliographic Enumeration of Bornean Plants*. Fraser and Neave, Singapore. 637 pp. (Systematic enumeration with notes on distribution; introduction covers vegetation, history of botanical investigation.)
Steenis, C.G.G.J. van (1972). *The Mountain Flora of Java*. Brill, Leiden. 90 pp. (Contains 57 plates with pictures of 456 native plants; lists 68 species, including 29 endemics, known only from one mountain in Java; chapters on plant geography, vegetation types, dispersal and distribution.)

There is extensive information on Indonesian botany in the *Flora Malesiana Bulletin*, cited in Appendix 1, which includes a bibliography section.

Contributions to the flora and vegetation of New Guinea (including Irian Jaya) have been published in the journal *Nova Guinea* (Contributions to the anthropology, botany, geology and zoology of the Papuan region).

Field-guides

Kartawinata, K. (1983). *Jenis-jenis Keruing*. LBN-LIPI, Bogor. (Illustrated popular account of Dipterocarpaceae.)

Meijer, W. (1974). *Field Guide to Trees of West Malesia*. Univ. of Kentucky. 328 pp.

Steenis, C.G.G.J. van, Den Hoed, G. and Eyma, P.J. (1951). *Flora voor de Scholen in Indonesië*. Noordhoff-Kolff NV, Djakarta. 407 pp. (Indonesian translation, 1978, by M. Soerjowinoto *et al.*)

For Irian Jaya, see also the publications listed under Papua New Guinea.

Information on Threatened Plants Little data. 6 species are included in *The IUCN Plant Red Data Book* (1978). IUCN has an unpublished list of 22 orchids endemic to Java, most of which are Rare, as well as a full list of palms, some of which have conservation categories. Also relevant:

Anon (1978). Endangered species of trees. *Conservation Indonesia* 2: 4. (Newsletter of WWF Indonesia Programme; lists 9 Indonesian trees.)

Voluntary Organizations

Institute for Nature Conservation, Lembaga Pengawetan Alam, Djl. Pledang 30, Bogor, Java.

Yayasan Indonesia Hijau (Green Indonesia Foundation), P.O. Box 208, Bogor, Java.

Botanic Gardens

Arboreta and Experimental Gardens of Silviculture Division, Forest Research Institute, Bogor, Java.

Botanical Gardens of Indonesia, Kebun Raya Bogor, Jalan Ir. H. Juanda 11, Bogor, Java.

Branches of Kebun Raya Bogor are:

Botanic Garden, Cibodas, Sindanglaya, West Java.
Botanic Garden, Purwodadi, Lawang, East Java.
'Eka Karya' Botanic Garden, Bedugul, Bali.

Useful Addresses

Directorate General of Forest Protection and Nature Conservation (PHPA), Jalan Ir. H. Juanda 9, P.O. Box 133, Bogor, Java.

Lembaga Biologi Nasional (LBN), LIPI, Jalan Juanda 18, Bogor, Java.

WWF/IUCN Conservation for Development Programme, Jalan Ir. H. Juanda 9, P.O. Box 133, Bogor, Java.

CITES Management Authority: Director General of Forest Protection and Nature Conservation (Perlindungan Hutan dan Pelestarian Alam), Departemen Kehutanan, Jalan Ir. H. Juanda No. 9, Bogor, Java.

CITES Scientific Authority: Indonesian Institute of Science (LIPI), Jalan Tenku Chik Ditiro 43, P.O. Box 250 JKT, Jakarta, Java.

Additional References

FAO (1981, 1982). *National Conservation Plan for Indonesia*. Field Report of UNDP/FAO National Parks Development Project Ins/78/061, 8 vols. Bogor, Indonesia. (1 - Introduction; 2 - Sumatra; 3 - Java and Bali; 4 - Lesser Sundas; 5 - Kalimantan; 6 - Sulawesi, 7 - Maluku and Irian; 8 - General topics.)

Gibbs, L.S. (1917). *A Contribution to the Phytogeography and Flora of the Arfak Mountains etc.* Taylor and Francis, London. 226 pp. (Covers vegetation types and systematic account of 330 plants collected in Arfak Mts.)

Good, R. (1960). On the geographical relationships of the angiosperm flora of New Guinea. *Bull. British Museum Nat. Hist. Bot.* 2: 205-226.

Gressitt, J.L. (Ed.) (1982). *Biogeography and Ecology of New Guinea*, 2 vols. Junk, Hague. (1 – Physical background, man's impact, vegetation and flora; 2 – fauna, conservation.)

Jacobs, M. (1958). Contribution to the botany of Mount Kerintji and adjacent area in west central Sumatra, 1. *Ann. Bogor.* 3: 45-104. (Plant collections now total 3977; many species collected and named by author.)

Jacobs, M. (1974). Botanical panorama of the Malesian archipelago (vascular plants). In Unesco, *Natural Resources of Humid Tropical Asia*. Natural Resources Research 12. Unesco, Paris. Pp. 263-294.

Jacobs, M. (1981). Dipterocarpaceae: the taxonomic and distributional framework. *Malaysian Forester* 44: 168-189.

Jacobs, M. (1982). Assessment of the deforestation problem in Malesia. Rijksherbarium, Leiden. 7 pp. (Typescript.)

Jacobs, M. and de Boo, T.J.J. (1982). *Conservation Literature on Indonesia: Selected Annotated Bibliography*. Rijksherbarium, Leiden. 274 pp. (850 entries covering Dutch, English, French, German and Indonesian literature from c. 1900 to 1979.)

Meijer, W. (1981). Sumatra as seen by a botanist. *Indonesian Circle* 25: 17-27.

Ochse, J.J. and Bakhuizen van den Brink, R.C. (1931). *Vegetables of the Dutch East Indies (edible tubers, bulbs, rhizomes and species included)*. Buitenzorg. 1006 pp. (Reprinted 1977. 389 species in 241 genera; notes on uses, habitat requirements, distribution, propagation.)

Paijmans, K. (Ed.) (1976). *New Guinea Vegetation*. Elsevier, Amsterdam. 213 pp. (Includes lists of medicinal and other useful species.)

Petocz, R.G. (1984). Conservation and development in Irian Jaya: a strategy for rational resource utilization. WWF/IUCN Conservation for Development Programme in Indonesia (address above). 279 pp. Mimeo.

Steenis, C.G.G.J. van (1934). On the origin of the Malaysian mountain flora, 1. *Bull. Jard. Bot. Buitenzorg, Ser. 3*, 13: 135-262.

Steenis, C.G.G.J. van (1979). Plant-geography of east Malesia. *Bot. J. Linn. Soc.* 79: 97-178. (Floristic analysis of the Lesser Sunda Islands.)

Whitten, A.J., Damanik, S.J., Anwar, J. and Hisyam, N. (1984). *The Ecology of Sumatra*. Gadjah Mada Univ. Press. 583 pp. (Vegetation types; flora and fauna; effects of disturbance on plant and animal communities.)

WWF/IUCN are supporting field surveys in existing and potential reserve sites identified in the FAO/UNDP report *A National Conservation Plan for Indonesia* (FO:INS/78/061, Field Report 17) with the aim of developing management plans.

Iran

Area 1,648,000 sq. km

Population 43,799,000

Floristics c. 7000 species (Parsa, 1943-1952) of which c. 20% endemic (Zohary, 1963). Most of the endemics are found in the mountains; centres of endemism include the peaks of the Elburz and Zagros Mountains, solitary peaks in the Central Plain, mountain ridges south of Kashan and Yazd, and to the north and south of Kerman (Zohary, 1973, cited in Appendix 1). The central plateau is species-poor. The Irano-Turanian element comprises about 69% of the flora. Euro-Siberian and Sudanian elements each make up 5% of the flora. There are also Mediterranean and Saharo-Arabian elements (Zohary, 1963).

Vegetation Deserts cover about 60% of Iran. Hot desert in south-east with sparse open scrub, including *Ziziphus*, *Acacia* and *Prosopis* on rocky slopes; herbaceous communities with *Atriplex* and *Heliotropium* in sandy depressions; steppes and deserts with *Artemisia* and *Astragalus* over most of centre and east; dry deciduous forest in west and *Pistacia – Amygdalus* steppe forest in south and west; *Juniperus* steppe forests in north; broadleaved temperate forest (with *Alnus*, *Quercus*, *Fagus* and *Carpinus*) in north up to 2500 m (Zohary, 1963). Small areas of mangroves on northern Qeshm Island (Kunkel, 1977).

Checklists and Floras Iran is included in the incomplete *Flora Iranica* (1963-), cited in Appendix 1. Floras covering Iran and offshore islands include:

Léonard, J. (1981). *Contribution a l'Etude de la Flore de la Vegetation des Deserts d'Iran*. Jardin Botanique National Belgique, Meise. (4 fascicles so far. 1 – Introduction, ferns, gymnosperms, monocotyledons; 2-4 – Compositae, Cruciferae, Labiatae and many smaller families.)

Parsa, A. (1943-1952). *Flore de L'Iran*, 12 vols. Tehran. (1 – Physical geography, ecology, ferns, gymnosperms. Ministry of Science and Higher Education, Tehran; 2-4 – dicotyledons; 5 – monocotyledons, ferns; 6 – Supplement; 7-12 – dicotyledons. See also the revised, English translation, *Flora of Iran* (1978-) by the same author and publishers.)

Sabeti, H. (1976). *Forests, Trees and Shrubs of Iran*. Min. Agriculture and Natural Resources, Tehran. 810 pp. in Persian; 64 pp. in English. (Includes nearly 1000 species, distribution maps.)

Termeh, F. and Moussavi, M. (1980). *Plants of Kish Island*. Dept of Botany Publ. no. 15. Tehran. (104 species collected on Kish; includes checklist, short descriptions and line drawings.)

Wendelbo, P. (1976). Annotated checklist of the ferns of Iran. *Iran J. Bot.* 1: 11-17.

Information on Threatened Plants None, except for 7 threatened plants mentioned in:

Wendelbo, P. (1978). Endangered flora and vegetation, with notes on some results of protection. In IUCN, *Ecological Guidelines for the Use of Natural Resources in the Middle East and South-West Asia*. IUCN, Switzerland. Pp. 189-195.

Botanic Gardens

Botanical Garden of the Botanical Institute of Iran, Karaj Road, P.O. Box 8-6096, Tehran.

Karadj College Botanical Gardens, Faculty of Agriculture, University of Tehran, Karadj, Tehran.

Useful Addresses

CITES Management Authority: Department of Environment, P.O. Box 1430, Tehran.

Additional References

Kunkel, G. (1977). *The Vegetation of Hormoz, Qeshm and Neighbouring Islands (Southern Persian Gulf Area)*. Cramer, FL-9490, Vaduz, Liechtenstein. 186 pp. (Includes annotated checklist giving local distributions; notes on 339 plants collected on islands.)

Wendelbo, P. (1972). Some distributional patterns within the Flora Iranica area. In Davis, P.H., Harper, P.C. and Hedge, I.C. (Eds), *Plant Life of South-West Asia*. Botanical Society of Edinburgh. Pp. 29-41.

Zohary, M. (1963). *On the Geobotanical Structure of Iran*. Bull. Research Council Israel Vol. 11D, Suppl. 113 pp. (Includes a 'Geobotanical Outline Map of Iran', scale 1:4,000,000.)

Iraq

Area 438,446 sq. km

Population 15,158,000

Floristics 2937 vascular plant species (A.H. Al-Khayat, 1984, *in litt.*). 190 endemic species (according to Zohary, 1950). Of the endemics, 95% belong to the Irano-Turanian floral element and 5% to the Saharo-Sindian element. There are also small numbers of Mediterranean and Eurosiberian-Boreoamerican species (Zohary, 1950). Centres of endemism include the montane and subalpine zones of the Kurdish Mountains, particularly the western slopes (Zohary, 1973, cited in Appendix 1).

Vegetation About 400,000 sq. km is desert or semi-desert, mainly in south, with dry *Poa, Carex* and *Artemisia* steppe; moist steppe zone to north with open savanna mainly with *Pistacia*; extensive marshlands with alluvial vegetation in the Mesopotamian Plain, NW of Basra, between the Tigris and Euphrates; temporarily inundated 'ahrash' forest, with *Tamarix* and *Populus*, on more stable soils and islands; *Quercus aegilops* and *Pinus brutia* forests on northern mountains between 500-2750 m, much disturbed or completely destroyed; thorn cushion open shrub formation between 1750-3000 m; alpine vegetation above 1750 m (Townsend and Guest *et al.*, 1966). Natural forest covers only 4% of the country, almost entirely restricted to north (Kurdistan), mostly overexploited and overgrazed (Nasser, 1984).

Checklists and Floras The main Floras are:

Rechinger, K.H. (1964). *Flora of Lowland Irak*. Cramer, Weinheim. 746 pp. (Selected bibliography.)

Townsend, C.C. and Guest, E. *et al.* (Eds) (1966-). *Flora of Iraq*, 9 vols planned, 5 published so far. Min. of Agriculture, Baghdad. (1 – Geology, vegetation, ecology, selected bibliography; 2 – ferns, gymnosperms, Rosaceae; 3-9 – angiosperms continued.)

See also:

Al-Rawi, A. (1964). *Wild Plants of Iraq with their Distribution*. Technical Bulletin no. 14. Min. of Agriculture, Baghdad. 248 pp. (Introductory notes on vegetation; checklist of ferns, gymnosperms and angiosperms with distributions.)

Gillett, J.B. (1948). Provisional list of trees and shrubs found in Iraq. (Unpublished report.)

Zohary, M. (1950). *The Flora of Iraq and its Phytogeographical Subdivision*. Bulletin no. 31. Ministry of Economics, Iraq. 201 pp. (Annotated checklist, distribution and phytogeographical relationships indicated.)

The highlands of northern Iraq are included in *Flora Iranica* (1963-), cited in Appendix 1.

Field-guides

Agnew, A.D.Q. (Ed.) (1962). *Flora of the Baghdad District. Part 1, Monocotyledons*. College Science Bulletin Suppl. 6, Baghdad. 170 pp. (Line drawings; introductory notes on vegetation.)

Al-Saad, H.A. and Al-Mayah, A.-R.A. (1983). *Aquatic Plants of Iraq*. Univ. of Basra.

Karim, F.M. (1978). *Flowering Parasitic Plants of Iraq*. Min. of Agriculture and Agrarian Reform, Abu-Ghraib. 90 pp. (Describes about 30 parasitic plants; keys and line drawings.

Information on Threatened Plants None.

Botanic Gardens

Za'faraniyah Botanical Garden, Horticultural Experiment Station, Abu-Ghraib, Baghdad.

Additional References

Guest, E.R. and Blakelock, R.A. (1954). Bibliography of Iraq. *Kew Bull.* 9(2): 243-249.

Nasser, M.H. (1984). Forests and forestry in Iraq: prospects and limitations. *Commonwealth Forestry Review* 63(4): 299-304.

Wendelbo, P. (1971). Some distributional patterns within the Flora Iranica area. In Davis, P.H., Harper, P.C. and Hedge, I.C. (Eds), *Plant Life of South-West Asia*. Botanical Society of Edinburgh. Pp. 29-41.

Ireland

(For Northern Ireland see United Kingdom)

Area 68,895 sq. km

Population 3,555,000

Floristics Size of flora for entire island: 1000-1150 native vascular species, estimated by D.A. Webb (1978, cited in Appendix 1) from *Flora Europaea*; one endemic species (IUCN figure). In Republic of Ireland only, c. 21 species less than figure above (E. Ni Lamha, 1984, *in litt.*). Elements: North American, Atlantic, Mediterranean, Holarctic, Eurasian and Arctic/alpine.

Vegetation Over much of the country agricultural land, moorland and bog. Most of the original broadleaved deciduous woodland destroyed; what remains consists mostly of semi-natural oakwoods with birch and holly. Plantations of pine, spruce and larch now cover c. 5% of the country (D.A. Webb, 1984, *in litt.*). Extensive areas of heath and heathy grassland on mountains near the coast. The rocky, limestone grasslands of the Burren region of Co. Clare are of special interest, as are the raised bogs; the latter now under threat.

Checklists and Floras Most publications make no distinction between species occurrence in the Republic of Ireland and in Northern Ireland, as in the case with the completed *Flora Europaea* (Tutin *et al.*, 1964-1980, cited in Appendix 1) and with Clapham, Tutin and Warburg's *Flora of the British Isles* (1962, 1968, cited under U.K.). The standard Irish Checklist and Flora are, respectively:

Scannell, M.J.P. and Synnott, D.M. (1972). *Census Catalogue of the Flora of Ireland.* Stationery Office, Dublin. 127 pp. (Checklist for both the Republic and Northern Ireland; natives and aliens; new edition in prep.)

Webb, D.A. (1977). *An Irish Flora*, 6th Ed. Dundalgan Press, Dundalk. 277 pp.

'County Floras', in effect, detailed checklists with localities, include:

Booth, E.M. (1979). *The Flora of County Carlow.* Royal Dublin Society, Dublin. 172 pp.

Brunker, J.P. (1950). *Flora of the County Wicklow.* Dundalgan Press, Dundalk. 310 pp. (Introduction includes history of the flora, geography, climate, botanical sub-divisions; pteridophytes, gymnosperms.)

Colgan, N. (1904). *Flora of the County Dublin. Flowering plants, higher Cryptogams and Characeae.* Hodges and Figgis, Dublin. 324 pp. (Supplement, 1961, published by the National Museum of Ireland, 95 pp.)

Hart, H.C. (1898). *Flora of the County Donegal.* Dublin. 391 pp.

Scully, R.W. (1916). *Flora of County Kerry.* Hodges and Figgis, Dublin. 406 pp. (Introduction describes geology and geography; pteridophytes and angiosperms.)

Webb, D.A. and Scannell, M.J.P. (1983). *Flora of Connemara and the Burren.* Cambridge Univ. Press, Cambridge, and Royal Dublin Society, Dublin. 322 pp. (History, climate, geology, vegetation description; gymnosperms, angiosperms and cryptogams; illus.)

The Irish Biological Records Centre (address below) is preparing a national atlas to illustrate the distribution of the 52 taxa protected by the 1976 Wildlife Act. Ireland is also covered by Perring and Walters' *Atlas of the British Flora* (1982, cited under U.K.).

Relevant journals: *Bulletin of the Irish Biogeographical Society*; *Irish Naturalists' Journal*; *Journal of Life Sciences, Royal Dublin Society*; *Proceedings of the Royal Irish Academy.*

Field-guides Most of the field-guides covered under U.K. could be used in the Republic of Ireland, since there are only a handful of plants that occur in the Republic and not in the U.K. Those specifically covering Ireland include Fitter, Fitter and Blamey (1974) and Page (1982), both cited under U.K.

Information on Threatened Plants No national plant Red Data Book or published threatened plant list except for the schedule of protected plants (see 'Laws Protecting Plants') and the section on Ireland in the list of rare species not to be collected, in:

Richards, A.J. (1972). The code of conduct: a list of rare plants. *Watsonia* 9(1): 67-72. (Lists 70 species for protection in Ireland; whole island.)

There is a protected species cultivation programme in Trinity College Botanic Gardens (address below), to bring into cultivation the 52 nationally protected species. A seed bank is also being set up. For details see:

Wyse Jackson, P. (1984). Irish rare plant conservation in the Trinity College Botanic Gardens, Dublin. In Jeffrey, D.W. (Ed.), *Nature Conservation in Ireland; Progress and Problems*. Proceedings of a Seminar, 24-25 February 1983. Royal Irish Academy, Dublin. 175 pp.

Ireland is included in the European threatened plant list (Threatened Plants Unit, 1983, cited in Appendix 1); latest IUCN statistics, based upon this work: endemic taxa – R:1; non-endemics rare or threatened worldwide – E:1, V:2, R:1 (world categories).

In 1982 IUCN, under contract to the EEC through the U.K. Nature Conservancy Council, prepared a report (in press), *Threatened Plants, Amphibians and Reptiles, and Mammals (excluding Marine Species and Bats) of the European Economic Community*, which included a data sheet on one Irish Endangered plant.

Laws Protecting Plants The Flora (Protection) Order of 1980, in accordance with the Wildlife Act 1976, provides protection for 52 plant species throughout the State. Under this Order, it is an offence to cut, pick, uproot or otherwise take, purchase, sell or be in possession of any of these plants whether whole or part, or wilfully to alter, damage, destroy or interfere with the habitat of these species. The list of 52 protected taxa is currently under review by a sub-committee of the BSBI (address below). For a summary see:

White, J. (1981). Irish plants – protection at last. *BSBI News* 27: 6-8. (Includes extract from the 1976 Wildlife Act and lists taxa protected.)

The Forest and Wildlife Service (address below) is responsible for implementing and enforcing the Wildlife Act, the main legislation relating to conservation.

Voluntary Organizations
An Taisce (The National Trust for Ireland), The Tailor's Hall, Back Lane, Dublin 8.
Botanical Society of the British Isles – BSBI (Irish Branch), c/o Irish Biological Records Centre (An Foras Forbartha), address below.
Dublin Naturalists' Field Club, c/o Trinity College Botanic Gardens, Palmerston Park, Dublin 6.
Irish Alpine Garden Society, c/o Ivanhoe, 28 Spencer Villas, Glasthule, Co. Dublin. (One of its aims is the cultivation and conservation of endangered wild plants.)
Irish Wildlife Federation, 22 Grafton Street, Dublin 2.

Botanic Gardens
Botanic Gardens, University College, Cork.
National Arboretum, John F. Kennedy Park, New Ross, Co. Wexford.
National Botanic Gardens, Glasnevin, Dublin 9.
Trinity College Botanic Gardens, Palmerston Park, Dublin 6.

Useful Addresses
Forest and Wildlife Service, Department of Fisheries and Forestry, 2 Sidmonton Place, Bray, Co. Wicklow.
Irish Biological Records Centre (An Foras Forbartha), St Martin's House, Waterloo Road, Dublin 4.
Wildlife Advisory Council, c/o Department of Fisheries and Forestry, Leeson Lane, Dublin 2. (Representatives from many voluntary conservation bodies and government agencies; appointed by the Government to advise the Minister for Fisheries and Forestry about the workings of the Wildlife Act.)
CITES Management and Scientific Authorities: Wildlife Advisory Council, see above.

Additional References

Doyle, J. (1958). Irish floristics since the I.P.E. of 1949. *Veröff. Geobot. Inst. Rübel.,* *Zürich* 33: 33-46. (I.P.E.: International Phytogeographical Excursion.)

Praeger, R.L. (1901). *Irish Topographical Botany.* Royal Irish Academy, Dublin. 410 pp.

Praeger, R.L. (1934). *The Botanist in Ireland.* Hodges and Figgis, Dublin. 587 pp. (Physical and botanical descriptions; maps; black and white photographs; line drawings; reprinted 1974 by E.P. Publishing, Wakefield.)

Webb, D.A. (1975). Floristic report for Ireland. *Mem. Soc. Brot.* 24(2): 615-622.

Webb, D.A. (1983). The flora of Ireland in its European context. The Boyle Medal Discourse, 1982. *J. Life Sc. R. Dublin Soc.* 4: 143-160.

White, J. (Ed.) (1982). *Studies on Irish Vegetation. Contributions from Participants in the Vegetation Excursion to Ireland,* July 1980. Organized by the International Society for Vegetation Science. *J. Life Sciences,* Royal Dublin Society. 408 pp. Papers include G.F. Mitchell on the influence of man on vegetation in Ireland, pp. 7-14; J. White on a history of Irish vegetation studies, pp. 15-42; J. White on a key for the identification of Irish plant communities, pp. 65-110; J. White and G. Doyle on the vegetation of Ireland – a catalogue raisonné, pp. 289-368.

Israel

Area 20,705 sq. km

Population 4,216,000

Floristics 2317 native species; 155 are endemic (Shmida, 1984, pers. comm.). Most of the endemics are found on the coastal plains in the transitional zone between the Mediterranean and desert regions, and in the high mountains of the desert region. 800 species belong to the Mediterranean element, over 300 species to both the Irano-Turanian and the Saharo-Arabian elements. In addition, there is a small Euro-Siberian element, and a Sudano-Zambezian element occupying favourable sites in the south (Zohary, 1982).

Vegetation Most of the south covered by deserts. Sandy desert with *Retama,* *Artemisia* and *Stipagrostis* in the western Negev and with *Anabasis, Hammada* and *Haloxylon* in the Arava Valley. Stony desert with *Artemisia, Gymnocarpos* and *Zygophyllum* scrub; open dwarf shrub steppes occupy large areas of the Judean Desert, northern Negev and parts of the Mediterranean territory in the north. Evergreen forests and maquis, dominated by *Quercus calliprinos,* throughout the Mediterranean territory, with *Pistacia, Crataegus,* and *Ziziphus* steppe forests along its eastern and south-western borders; deciduous *Quercus/Pistacia* forest in north and north-west (Zohary, 1982).

Checklists and Floras An up to date Flora of Israel is provided by *Flora Palaestina* (1966-). Also relevant may be the *Flora of Syria, Palestine and Sinai* (Post, 1932), and Eig, Zohary and Feinbrun-Dothan (1931); Israel will also be covered by the *Med-Checklist*; all of these works are cited in Appendix 1. See also:

Zohary, M. (1976). *A New Analytical Flora of Israel.* Am Oved, Tel Aviv. 540 pp. (Text in Hebrew.)

Field-guides

Duvdevani, S. and Osherov, S. (1969). *Analytical Key for Identification of Wild and Cultivated Plants of Israel by their Vegetative Characters*. Massada, Tel Aviv. 254 pp. (In Hebrew.)

Feinbrun-Dothan, N. (1960). *Wild Plants in the Land of Israel*. Hakibbutz Hameuchad and Massada, Israel. 185 pp. (94 species illustrated; text in English.)

Plitmann, U., Heyn, C., Danin, A., and Shmida, A. (1982). *Pictorial Flora of Israel*. Massada, Givatayim. 338 pp. (Covers 750 species; text in Hebrew with English preface; distribution maps.)

Shmida, A. and Daron, D. (in press). *Field Guide to the Common Plants of Israel*. Keter Publ., Jerusalem.

Information on Threatened Plants A Botanical Information Centre – ROTEM (the Hebrew word for the broom *Retama raetam*) – has a database on rare and endangered plants of Israel. The Centre is a joint project of the Society for the Protection of Nature in Israel and the Hebrew University Department of Botany, at Har-Gillo Field Study Centre, south of Jerusalem. Apart from computer listings giving distributions and status of plants, there is an Ecological Mapping Program which uses the Rotem database to produce computer-generated maps of species distributions. In addition, the Nature Reserves Authority are planning a Red Data Book of Israel, to cover flora and fauna.

Israel is included in the draft list for North Africa and the Middle East produced by IUCN Threatened Plants Committee Secretariat (1980, cited in Appendix 1), but the coverage for Israel is known to be very incomplete. The *IUCN Plant Red Data Book* (1978) has sheets for *Iris lortetii* and *Rumex rothschildianus*.

Dafni, A. and Agami, M. (1976). Extinct Plants of Israel. *Biol. Conserv.* 10: 49-52.

Voluntary Organizations

Society for the Protection of Nature in Israel, 4 Hashfela Street, Tel Aviv 66183.

Botanic Gardens

Ben Gurion University of the Negev, Research and Development Authority, P.O. Box 1025, Beer Sheva.

Botanic Garden of Tel Aviv University, Ramat Aviv, Tel Aviv.

Botanic Gardens of the Hebrew University, Dept of Botany, Jerusalem 91000.

Botanical Garden "Mikveh-Israel", Holon.

Havath-Noy Garden, Ministry of Agriculture Research Post, Ruppin.

Useful Addresses

Nature Reserves Authority, 78 Yirmeyahu Street, Jerusalem 94467.

ROTEM, Har Gillo F.S.C. Sak Na'ul, Jerusalem 91999.

CITES Management Authority: Nature Reserves Authority, 78 Yirmeyahu Street, Jerusalem 94467.

Additional References

Danin, A. (1983). *Desert Vegetation of Israel and Sinai*. Cana Publ. House, Jerusalem. 148 pp.

Gómez-Campo, C. (Ed.) (1985). *Plant Conservation in the Mediterranean Area*. (See in particular L. Boulos on the arid eastern and south-eastern Mediterranean regions.)

Rabinovitz, D. (1981). *Nature Conservation and Environmental Protection in the Negev Desert. A Challenge for Israel in the 1980's*. Anglo-Israel Assoc. Pamphlet no. 62. 16 pp.

Shmida, A. (in press). Endemism in the flora of Israel. *Bot. Jahrb.*

Waisel, Y. and Alan, A. (1980). *Trees of the Land of Israel*. Division of Ecology, Tel Aviv. 126 pp.

Zohary, M. (1959). Wild life protection in Israel (flora and vegetation). In *Animaux et Végétaux Rares de la Région Méditérranéenne*. Proceedings of the IUCN 7th Technical Meeting, 11-19 September 1958, Athens, vol. 5. IUCN, Brussels. Pp. 199-202.

Zohary, M. (1962). *The Plant Life of Palestine: Israel and Jordan*. Ronald Press, New York. 262 pp. (Includes useful vegetation map of Palestine.)

Zohary, M. (1982). *Vegetation of Israel and Adjacent Areas*. Reichert, Wiesbaden. 166 pp. (Includes vegetation maps, bibliography.)

Zohary, M. and Wood, H. (1975). *Bouquet of Protected Wild Flowers*. Nature Conservation Authority, Tel Aviv. 79 pp. (Coloured plates of 37 species; text in Hebrew.)

Italy

(Mainland)

Area 251,447 sq. km

Population 56,724,000

Floristics 4750-4900 native vascular species, for peninsula Italy only, according to D.A. Webb (1978, cited in Appendix 1) estimated from *Flora Europaea*; endemic taxa: 142 (IUCN figure) principally based upon *Flora Europaea*; 712 endemics, including subspecies and other infraspecific taxa, and including Sardinia and Sicily (Pignatti, 1982). Central European element well-developed in northern Italy and south to the Apennines, with the typical Mediterranean flora becoming dominant southwards. Areas of high endemism concentrated in parts of the northern, central and southern Apennines and in Calabria (S. Pignatti, 1984, *in litt.*). Elements: Mediterranean, Central European, alpine.

Vegetation Much of country modified by agriculture. Central European vegetation of broadleaved and coniferous forests, with pines (*Pinus sylvestris*, *P. cembra*), oaks and beech, along the foothills of the Italian Alps and in the Apennines. These once extensive forests now largely modified by grazing and forest plantations or, in the northwest, replaced by subalpine heaths. Alpine meadows abundant at higher altitudes; up to 4000 m in the Alps, and 2200 m in the Apennines. In the lowlands and coastal areas, especially in the south, original cover of sclerophyllous forests (dominated by *Pinus halepensis*) largely replaced by maquis and farmland. Almost all of the formerly extensive wetlands have disappeared, although relict aquatic communities survive in the Po valley.

Checklists and Floras Italy is included in the completed *Flora Europaea* (Tutin *et al.*, 1964-1980) and will also be covered under the *Med-Checklist* (both cited in Appendix 1). For a floristic bibliography see Hamann and Wagenitz (1977), cited in Appendix 1. The most comprehensive and modern national checklist and Flora are:

Pignatti, S. *et al.* (1980). *Check-list of the Flora of Italy, with Codified Plant Names for Computer use*. Consiglio Nazionale delle Ricerche, Rome. 256 pp.

Pignatti, S. (1982). *Flora d'Italia*, 3 vols. Edagricole, Bologna. (1 – history of Floras, ecology, gymnosperms, pteridophytes, dicotyledons; 2 and 3 – remainder of angiosperms; line drawings and distribution maps for each species.)

Other works:

Baroni, E. (1969). *Guida Botanica d'Italia*, 4th Ed. Cappelli, Bologna. 545 pp.
(Revised by S. Baroni Zanetti; covers mainland Italy, Corsica, Sardinia, Sicily, Istria
and the French Riviera; illus.)

Fiori, A. (1923-1933). *Nuova Flora Analitica d'Italia*, 3 vols. Edagricole, Bologna.
(Covers mainland Italy, Corsica, Sardinia, Sicily, Pantellaria and nearby smaller
islets; 1 – pteridophytes, gymnosperms and angiosperms (Gramineae to
Leguminosae); 2 – Myrtaceae to Compositae; 3 – line drawings only, by A. Fiori
and G. Paoletti; reprinted 1969 and 1974.)

Zángheri, P. (1976). *Flora Italica*, 2 vols. Cedam, Padova. (1 – gymnosperms,
pteridophytes, angiosperms; 2 – line drawings.)

See also:

Moggi, G. (1975). Données disponibles et lacunes de la connaissance floristique de
l'Italie. In CNRS (1975, cited in Appendix 1). Pp. 53-63. (Describes present
situation of floristic and systematic research in Italy; lists main herbaria and centres
of floristic study.)

Pichi Sermolli, R.E.G. and Moggi, G. (1975). Report on the progress of floristic
research in Italy since 1961. *Mem. Soc. Brot.* 24(2): 623-746.

A computerized floristic mapping scheme, under the direction of S. Pignatti
(Dipartimento di Biologia Vegetale, Città Universitaria I, 00100 Rome), is in progress.
Based essentially on Pignatti's *Flora d'Italia* (1982), it will include species and distribution
data for the whole country, threatened plant data, biotopes containing threatened species
and areas of high endemism (Anon, 1985, cited in Appendix 1).

Field-guides

Dalla Fior, G. (1963). *La Nostra Flora (Guida alla Conoscenza della Flora della
Regione Trentino)*. Casa Editrice G.B. Monauni, Trento. (Not seen.)

Fenaroli, L. (1971). *Flora delle Alpi Vegetazione e Flora delle Alpi e degli altri Monti
d'Italia*, 2nd Ed. Aldo Martello, Milano. 428 pp. (Keys, colour and black and white
drawings.)

Fenaroli, L. and Gambi, G. (1976). *Alberi: Dendroflora Italica*. Museo Tridentino di
Scienze Naturali, Trento. 717 pp. (Trees – colour and black and white drawings;
photographs; maps.)

Rasetti, F. (1980). *I Fiori delle Alpi*. Accademia Nazionale dei Lincei, Roma. 316 pp.
(Illus.)

Information on Threatened Plants There is no national plant Red Data Book. A
very preliminary threatened plant list was published in 1972:

Anon (1972). Specie della Flora italiana meritevoli di protezione (Gruppo di Lavoro per
la Floristica, Società Botanica Italiana). *Inform. Bot. Ital.* 4(1): 12-13. List also in
Webbia 29(1): 361-363 (1974). (Lists 41 species in need of protection with
explanatory text in Italian, French, English and German.)

In 1971 and 1979, the Società Botanica Italiana published 2 large volumes documenting
563 sites considered to be of high botanical interest and in need of conservation:

Pedrotti, F. *et al.* (Eds) (1971, 1979). *Censimento dei Biotopi di Rilevante Interesse
Vegetazionale Meritevole di Conservazione in Italia*, 2 vols. Società Botanica
Italiana, Camerino. (Site details – description, threats, proposed protection, maps.)

See also:

Corti, R. (1959). Specie rare o minacciate della flora Mediterranea in Italia. In
  Animaux et Végétaux Rares de la Région Méditerranéenne, cited in Appendix 1.
  Pp. 112-129. (Brief distribution and status details on 65 threatened plant taxa.)
Filipello, S. (Ed.) (1981). *Problemi Scientifici e Tecnici della Conservazione del
  Patrimonio Vegetale*. Proceedings of a conference, 18-19 December 1979, Firenze.
  Consiglio Nazionale delle Ricerche, Pavia. 146 pp. (*OPTIMA Leaflet* No. 114.)
  (Contains many relevant articles in Italian with English abstracts, e.g. S. Filipello on
  plant species to protect (pp. 13-18); G.G. Lorenzoni on a census of vegetation types
  under threat (pp. 39-46); A. Robecchi-Majnardi on plant and vegetation
  conservation (pp. 33-37); F. Pedrotti on the conservation of wetland vegetation
  (pp. 63-80); P.L. Nimis on a data bank for Italian flora and vegetation (pp. 83-86)
  and F.M. Raimondo on Italian species in threatened biotopes (pp. 103-125).)
Filipello, S. and Gardini-Peccenini, S. (1985). The Italian Peninsular and Alpine
  Regions. In Gómez-Campo, C. (Ed.) (1985), cited in Appendix 1. Pp. 71-88.
  (Includes lists of threatened plants, species case-histories and details of laws and
  protected areas.)

Included in the European threatened plant list (Threatened Plants Unit, 1983, cited in
Appendix 1); latest IUCN statistics for mainland Italy, based upon this work: endemic
taxa – E:6, V:17, R:48, I:5, K:16, nt:50; doubtfully endemic taxa – V:1, K:1, nt:2; non-
endemics rare or threatened worldwide – E:4, V:34, R:38, I:4 (world categories).

In 1982 IUCN, under contract to the EEC through the U.K. Nature Conservancy Council,
prepared a report (unpublished), *Threatened Plants, Amphibians and Reptiles, and
Mammals (excluding Marine Species and Bats) of the European Economic Community*,
which included data sheets on 31 Italian Endangered plant species (including 8 in Sicily and
6 in Sardinia). *The IUCN Plant Red Data Book* includes 4 Italian threatened species.

For details of computerized threatened plant data see under Checklists and Floras.

Laws Protecting Plants There is no national legislation giving protection to wild
plant species except those regulating the collection of truffles and plants registered under
the official flora – plants of medicinal or traditional economic value. 13 out of 21 Regions
and Autonomous Provinces have passed local legislation to protect their flora, in
particular their rare or characteristic species. Moreover, Law No. 984 of 27 December 1977
obliged those regions who had not already done so to legislate for the protection of their
flora by 24 June 1978. Existing Regional and Provincial laws are:

Regional:

Abruzzo	No.66 of 1980.
Basilicata	No.42 of 22 May 1980
Emilia-Romagna	No.2 of 24 January 1977.
Friuli-Venezia Giulia	No.44 of 18 August 1972.
Lazio	No.61 of 19 September 1974.
Liguria	No.9 of 30 January 1984.
Lombardia	No.58 of 17 December 1973.
Marche	No.6 of 22 February 1973.
Piedmonte	No.24 of 13 August 1974.
Umbria	No.40 of 11 August 1978.
Valle-d'Aosta	No.6 of 8 November 1956 and special decree no. 43 of 31 January 1957.
Veneto	No.53 of 15 November 1974.

Provincial:

Bolzano	No.13 of 28 July 1972.
Trento	No.17 of 25 July 1973.

Bortolotti, L. (1975). Sulle leggi per la protezione della flora emanate dalle Regioni a statuto speciale e ordinario dalle Province autonome. *Boll. Soc. Bot. Ital.* 7(2): 132-139.

Filipello, S. *et al.* (Eds) (1979). *Repertorio delle Specie della Flora Italiana Sottoposte a Vincolo di Protezione nella Legislazione Nazionale e Regionale.* Consiglio Nazionale delle Ricerche, Pavia. (Includes taxa protected at Regional and Provincial levels.)

Peyronel, B. (1973). Considerazione su una legge regionale per la conservazione della flora: Italia. *Inf. Bot. Ital.* 5(2): 151-154.

Region Marche (Ed.) (1979). *Flora Protetta delle Marche.* Region Marche. 96 pp. (Maps; illus.)

Region Veneto (Ed.) (1975). *Fauna Inferiore Flora e Funghi Natura da Salvare.* 71 pp. (Describes 48 protected species; illus.)

Sonnino, P.F. (1975). Protezione delle flora alpina e legislazione. *Natura e Montagna (Italy)* 22(2): 41-47.

Voluntary Organizations

Associazione Italiana per il World Wildlife Fund (WWF-Italy), Via P.A. Micheli 50, 00197 Rome.

Italia Nostra, Via N. Porpora 22, 00100 Rome.

Società Botanica Italiana, Via La Pira, 4-50121 Firenze.

Botanic Gardens Numerous; outlined in Henderson (1983), cited in Appendix 1; only those that subscribe to the Botanic Gardens Conservation Co-ordinating Body listed here:

Ente Giardini Botanici Villa Taranto, 28048 Verbania Pallanza, Lago Maggiore.

Istituto e Orto Botanico dell' Università di Pavia, Via San Epifanio 14, 27100 Pavia.

Useful Addresses

Federazione Nazionale Pro Natura, Via Marchesana 12, 40124 Bologna.

Food and Agriculture Oganization of the U.N. (FAO), Via delle Terme di Caracalla, 00100 Roma.

CITES Management and Scientific Authorities: Ministero dell'Agricoltura e delle Foreste, Direzione generale per l'Economia montana e per le Foreste, Divisione II, Via G.Carducci 5, 00187 Roma.

Additional References

Filipello, S. (1979). Projets, problèmes et aboutissements de la conservation de la flore et de la végétation en Italie. In Proceedings of the 2nd OPTIMA meeting, 23-29 May 1977. *Webbia* 34(1): 63-69.

Società Botanica Italiana (1975). Aufruf zum Schutze der Italienischen Flora. *Willdenowia* 7(3): 537-538. (Lists 43 protected species.)

Toschi, A. (1959). Etablissement des réserves pour la protection de la faune et de la flore en Italie. In *Animaux et Végétaux Rares de la Région Méditerranéenne.* Proceedings of the IUCN 7th Technical Meeting, 11-19 September 1958, Athens, vol. 5. IUCN, Brussels. Pp. 58-63.

Italy: Sardinia

Second largest island in the Mediterranean after Sicily, c. 255 km long, 90 km wide, with over 1200 km of coastline.

Area 24,090 sq. km

Population 1,594,175 (1981 census)

Floristics 1900-2000 native vascular species, estimated by D.A. Webb (1978, cited in Appendix 1) from *Flora Europaea*. 27 endemic taxa (IUCN figures). Affinities with flora of Corsica rather than Sicily. Flora entirely Mediterranean.

Vegetation Little natural vegetation, especially around the coast. Inland, a zone of Holm Oak (*Quercus ilex*) is dominant, although much has been replaced by dry pastures and on the lower ground it has largely been degraded to garigue. Natural formations of thorny shrubs are widespread in mountainous areas (S. Pignatti, 1984, in *litt*.).

Checklists and Floras See under Italy, and also a series of papers by different authors (B. Corrias, P.V. Arrigoni, I. Camarda, M. Rafaelli and F. Valsecchi) in *Boll. Soc. Sarda Sci. Nat.* entitled 'Le piante endemiche della Sardegna'. Vols 16 (1977): 259-280, 287-313; 17 (1978): 177-225, 227-241, 243-328. (Reprinted in *OPTIMA Leaflets* 49-54 (1977) and 73-79 (1978); case-studies on individual taxa, with details of distribution and ecology; maps; line drawings.)

Cossu, A. (1968). *Flora Pratica Sarda*. Gallizi, Sassari. 365 pp. (Includes distribution, habitat and cultivation details; illus.)

Information on Threatened Plants See under Italy, and:

Arrigoni, P.V. (1971). Nuovi reperti di alcune species rare o notevoli della flora sarda (New records for some rare or interesting species in Sardinia). *Giorn. Bot. Ital.* 105(4): 177-178.

IUCN statistics: endemic taxa – E:5, V:3, R:10, K:1, nt:8; non-endemics rare or threatened worldwide – E:2, V:5, R:6, I:1 (world categories).

Laws Protecting Plants See under Italy.

Additional References

Arrigoni, P.V. (1968). Fitoclimatologia della Sardegna. *Webbia* 23(1): 1-100. (English summary.)

Camarda, I. and Valsecchi, F. (1984). *Alberi e arbusti spontanei della Sardegna*. Gallizzi. 480 pp. (Illus.)

Italy: Sicily

Sicily, the largest island in the Mediterranean, is separated from mainland Italy to the north-east by the 3-km straits of Messina.

Area 25,708 sq. km

Population 4,906,878 (1981 census)

Floristics 2250-2450 native vascular species estimated by D.A. Webb (1978, cited in Appendix 1) from *Flora Europaea*. 41 endemic taxa (IUCN figures). Floristic diversity and endemism highest in the north-west, especially the mountains of the Madonie and Nebrodi area and the slopes of Mt Etna. A Mediterranean flora.

Vegetation Little natural vegetation. Most of the land cultivated. The forest cover of Sicilian Fir (*Abies nebrodensis*), once almost continuous in the northern mountain range, now confined to tiny fragments in the Madonie area; some broadleaved, deciduous forest of oak, chestnut and beech in the Nebrodi and Madonie Mountains and the northern slopes of the Rocca Busambra; maquis confined to the drier areas, especially the lower slopes of the mountains. The volcanic Mt Etna (3323 m), in north-east Sicily, supports oak, birch and chestnut forests, with fragments of beech (1000-1450 m), but forest degradation widespread; at higher altitudes, Laricio Pine (*Pinus nigra* ssp. *laricio*), giving way to low scrub communities rich in endemics e.g. *Genista aetnensis*; lower slopes are heavily cultivated (Poli Marchese, 1984). For a vegetation map see Gentile *et al.* (1968).

Checklists and Floras See under Italy and:

Di Martino, A. and Raimondo, F.M. (1979). Biological and chorological survey of the Sicilian flora. In Proceedings of the 2nd OPTIMA meeting, 23-29 May 1977. *Webbia* 34(1): 309-335. (English summary.)

Information on Threatened Plants See under Italy.

Case studies have been written about individual threatened and endemic species, e.g. by F. Garbari and A. Di Martino on *Leopoldia gussonei* in *Webbia* 27(1): 289-297 (1972). (English summaries.)

IUCN statistics: endemic taxa – E:6, V:5, R:13, I:4, K:3, nt:10; doubtfully-endemic taxa – R:1, nt:1; non-endemics rare or threatened worldwide – E:3, V:5, R:9, I:1 (world categories).

Laws Protecting Plants See under Italy.

Voluntary Organizations See under Italy.

Botanic Gardens
Istituto Botanico e Giardino Coloniale, Via A. Lincoln 2, 90133 Palermo.

Useful Addresses See under Italy and:

Istituto Sperimentale per la Selvicoltura, Viole S. Margherita, 80/82, 52100 Arezzo, Italy. (Involved with a conservation programme for *Abies nebrodensis*.)

Additional References
Gentile, S., Tomaselli, R., Pirola, A. and Balduzzi, A. (1968). *Carta della Vegetazione Naturale Potenziale della Sicilia, 1/500,000.* No. 40. Quaderni, Pavia. 114 pp.
Poli Marchese, E. (1984). Excursion au M. Etna (10 Juin 1983): une vue synthétique du paysage végétal de l'Etna. In Proceedings of the 4th OPTIMA meeting, 6-14 June 1983, Palermo, Sicily. *Webbia* 38: 69-78.
Raimondo, F.M. (1983). On the natural history of the Madonie Mountains. In Proceedings of the 4th OPTIMA meeting, 6-14 June 1983, Palermo, Sicily. *Webbia* 38: 29-61. (A floristic and ecological account with comments on conservation.)
Raimondo, F.M., Rossitto, M. and Villari, R. (1982). *Bibliografia Geobotanica Siciliana.* Consiglio Nazionale delle Ricerche, Palermo. 159 pp. (Includes algae, lichens, bryophytes and angiosperms.)

Riggio, S. and Massa, B. (1974). Problemi di conservazione della natura in Sicilia. 1. Contributo. *Atti IV Simp. Naz. Conservazione Nat. Bar.* 2: 299-425. (Not seen.)

Ivory Coast

Area 322,463 sq. km

Population 9,474,000

Floristics 3660 species of vascular plants (Aké Assi, 1984); Aké Assi (1971) gives 4892 species; 4700 species (Lebrun, 1976, cited in Appendix 1); 2770 species in the forest zone (Aubréville, 1959). 62 endemic angiosperms (Aké Assi, 1984); 41 endemic species (Brenan, 1978, cited in Appendix 1); 89 endemic taxa (IUCN figures, see below).

Floristic affinities predominantly Guinea-Congolian, but flora in north with Sudanian affinities. Taï Forest (868 species, Aké Assi and Pfeffer, 1975) and Mt Nimba (shared with Guinea and Liberia, 2000 species) are especially important floristically.

Vegetation Northern quarter covered by Sudanian woodland with *Isoberlinia*. Remainder of country lowland rain forest interspersed with secondary grassland and cultivation; transitional rain forest (between lowland and montane) on Mt Nimba. Small area of mangrove and swamp forest at coast.

Estimated rate of deforestation for closed broadleaved forest 2900 sq. km/annum out of 44,580 sq. km (FAO/UNEP, 1981). However, Myers (1980, cited in Appendix 1) quotes coverage of primary moist forest to be 30,000 sq. km or less (World Bank), which is being opened up at a rate of 4000-5000 sq. km/annum.

For vegetation map see White (1983), cited in Appendix 1.

Checklists and Floras Ivory Coast is included in the *Flora of West Tropical Africa*. The Ivorian portion of Mt Nimba is included in *Flore Descriptive des Monts Nimba*. Both works are cited in Appendix 1.

Aké Assi, L. (1964). *Contribution à l'Etude Floristique de la Côte d'Ivoire et des Territoires Limitrophes*. Lechevalier, Paris. 321 pp. (Annotated checklist with extensive specimen citations; line drawings.)

Aké Assi, L. (1984). *Flore de la Côte d'Ivoire: Etude Descriptive et Biogéographique, avec Quelques Notes Ethnobotaniques*, 3 parts in 6 vols. Thesis presented to University of Abidjan. 1206 pp. (Part 1 – notes on families, genera, species; numerous line drawings; part 2 – checklist of species; part 3 – analysis of the flora; list of ailments and plants used in their cure; bibliography.)

Aké Assi, L. and Pfeffer, P. (1975). *Inventaire Flore et Faune du Parc National de Taï*. BDPA/SEPN, Abidjan.

Aubréville, A. (1959). *La Flore Forestière de la Côte d'Ivoire*, 3 vols. 2nd Ed. (1st Ed. 1936). Publication No. 15 of the Centre Technique Forestier Tropical, Nogent-sur-Marne. (Keys, descriptions, broad distributions, line drawings.)

Guillaumet, J.-L. (1967). *Recherches sur la Végétation et la Flore de la Région du Bas-Cavally (Côte d'Ivoire)*. ORSTOM, Paris. 247 pp. (Includes vegetation map 1:1,000,000; 39 black and white photographs.)

Information on Threatened Plants No published lists of rare or threatened plants; IUCN has records of 89 species and infraspecific taxa believed to be endemic – E:6, V:36, R:17, nt:2, K:28.

Botanic Gardens

Laboratoire de Botanique, ORSTOM, B.P. 20, Abidjan.

Additional References

Adjanohoun, E., Aké Assi, L. and Guillaumet, J.L. (1968). La Côte d'Ivoire. In Hedberg, I. and O. (1968), cited in Appendix 1. Pp. 76-81.

Aké Assi, L. (1971). Progrès dans la préparation de la flore de la Côte d'Ivoire. In Merxmüller, H. (1971), cited in Appendix 1. Pp. 27-29.

Lamotte, M. (1983). The undermining of Mount Nimba. *Ambio* 12(3-4): 174-179. (Photographs, maps.)

Lanly, J.P. (1969). Régression de la forêt dense en Côte d'Ivoire. *Bois Forêts Trop.* 127: 45-59.

Mangenot, G. (1971). Une nouvelle carte de la végétation de la Côte d'Ivoire. In Merxmüller, H. (1971), cited in Appendix 1. Pp. 116-121. (With vegetation map 1:4,000,000.)

Jamaica

Jamaica lies south of the eastern extremity of Cuba, in the Caribbean Sea. 235 km long and 82 km wide, it consists of coastal plains, divided by the Blue Mountain Range in the east which reaches 2256 m, and hills and limestone plateaux in the centre and west.

Area 11,425 sq. km

Population 2,290,000

Floristics 3003 species of flowering plants, with 27.6% endemism (C.D. Adams pers. comm., from Proctor, 1982); 579 species of ferns, 82 (13.5%) endemic (Proctor, in press). In Bromeliaceae and Orchidaceae, both richly represented in Jamaica, endemism is 30.7% (Adams, 1972).

Vegetation Much of lowlands cleared for agriculture; natural vegetation in littoral mangrove swamps and salt pans; xeric woodlands, varying from cactus-thorn scrub to high forest, on limestone; secondary woodland common on dry alluvial soils of southern plains. Native forest, on the limestone hills and plateaux of the interior, modified and receding steadily; the largest extent of natural forest is in the Cockpit Country in the NW where 101 endemic species have been described. Some well-developed lower montane rain forest on limestone in the John Crow Mountains, at the wet NE corner of the island; extensive montane rain forest in the upper reaches of the Blue Mountains, steadily receding; elfin woodland on the summits and ridges of the Blue and John Crow Mountains. 44.9% forested (FAO, 1974, cited in Appendix 1); estimated rate of deforestation for closed broadleaved forest 20 sq. km/annum, out of a total of 670 sq. km (FAO/UNEP, 1981).

Checklists and Floras Covered by the family and generic monographs of *Flora Neotropica* (cited in Appendix 1). The Flora is:

Adams, C.D. (1972). *Flowering Plants of Jamaica*. University of the West Indies, Mona. 848 pp.

See also:

Proctor, G.R. (1982). More additions to the Flora of Jamaica. *J. Arnold Arbor*. 63(3): 199-315. (115 native species further to Adams, 1972.)
Proctor, G.R. (in press). *Ferns of Jamaica*. British Museum (Natural History), London.

Field-guides

Hawkes, A.D. and Sutton, B.C. (1974). *Wild Flowers of Jamaica*. Collins. 96 pp. (An introduction and guide to 174 taxa, each illustrated.)

Information on Threatened Plants D.L. Kelly (1985, pers. comm.) estimates 363 endemic species, 48.8% of the total are rare, very rare or extinct; 90 of them are known in recent times only from single sites and 40 only from old collections of which majority are probably now extinct.

Proctor, G.R. Conservation of Jamaican plants: Partial list of endangered species. Undated manuscript.

Threatened plant conservation is discussed in:

Howard, R.A. (1977). Conservation and the endangered species of plants in the Caribbean islands. In Prance, G.T. and Elias, T.S. (Eds) (1977), cited in Appendix 1. Pp. 105-114.

Laws Protecting Plants Existing legislation:

Bark of Trees Act – regulation of commercial bark removal for specific species.
Forest Act – declaration of forest reserves.
Town Planning Act – declaration of Tree Preservation Orders.

Proposed legislation:

Wild Life Protection Act: redefinition of 'Wild Life' to include plants.
Trade Law: Inclusion of certain plants under various schedules to regulate export.

Voluntary Organizations

Jamaica Orchid Society, c/o Mr. A. Gloudon, 4A Wai Rua Road, Gordon Town, St Andrew.
Natural History Society of Jamaica, c/o Institute of Jamaica, Duke St., Kingston.

Botanic Gardens

Bath Garden, Bath, St Thomas.
Castleton Gardens, St Mary.
Royal Botanic Gardens (Hope), Hope Road, Kingston 6.
The Hill Gardens, Cinchona, Hall's Delight, St Andrew.

See:

Eyre, A. (1966). *The Botanic Gardens of Jamaica*. André Deutsch, London. 96 pp., 16 plates. (A guide to the gardens, remarks on the areas in which they occur, and their history.)

Useful Addresses

Department of Botany and Herbarium, University of the West Indies, Mona, Kingston 7.

Forestry Department, 173 Constant Spring Road, Kingston 8.

Institute of Jamaica, 12 East Street, Kingston.

Natural Resource Conservation Division, Ministry of Science, Technology and
Environment, P.O. Box 305, Kingston 10.

The Herbarium, Institute of Jamaica, Duke St., Kingston.

Additional References

Adams, C.D. (1971). *The Blue Mahoe & Other Bush: an Introduction to Plant Life in Jamaica*. Sangster's Bookstores Ltd., 97 Harbour Street, Kingston, Jamaica and McGraw-Hill Far Eastern Publishers Ltd., Singapore. 157 pp.

Asprey, G.F. and Loveless, A.R. (1957). The dry evergreen formations of Jamaica. *J. Ecol.* 45: 799-822.

Asprey, G.F. and Robbins, R.G. (1953). The vegetation of Jamaica. *Ecol. Monog.* 23: 359-412.

Grubb, P.J. and Tanner, E.V.J. (1976). The montane forests and soils of Jamaica: a reassessment. *J. Arnold Arbor.* 57: 313-368.

Thompson, D.A., Bretting, P. and Humphries, M. (Eds) (in press). *Forests of Jamaica*. Institute of Jamaica Publications.

Woodley, J.D. (Ed.) (1971). *Hellshire Hills Scientific Survey 1970*. University of the West Indies and Institute of Jamaica. 168 pp.

Japan

Area 369,698 sq. km

Population 119,492,000

Floristics 4022 vascular plant species in 1098 genera (excluding Ogasawara-Gunto and Ryukyu Retto); about 500 fern species (Ohwi, 1965). 1371 endemic species (based on Ohwi, 1965, quoted in Nishida, 1972); many occur in the high altitude zones. Floral elements from Siberia, Manchuria, Korea, southern China, Taiwan and Malesia.

Vegetation Subtropical broadleaved evergreen forest and warm temperate broadleaved evergreen forest near south and east coasts, and in the lowlands of south-west Honshu, Shikoku and Kyushu; cool temperate broadleaved forest in low mountains and highlands of the coastal hinterlands; subarctic coniferous forests on mountains higher than 1400-1500 m, in the north, on Shikoku and in the lowlands of Hokkaido. Alpine zone with scrub, grassland and rocky desert, above 2500 m in Central Honshu, above 1900-2000 m in the Tohoku district and above 1400-1500 m in Hokkaido. Many areas of lowland vegetation, especially near coasts, cleared for agriculture and urbanization.

Checklists and Floras The principal Floras are:

Nakaike, T. (1982). *New Flora of Japan. Pteridophyta*. Shibundo, Tokyo. 808 pp.
(About 850 taxa described in Japanese; many photographs.)

Ohwi, J. (1965). *Flora of Japan*. Smithsonian Institution, Washington, D.C. 1067 pp.
(Revised and extended English translation of *Nihon Shokubutsu-shi*, 1953 and *Flora of Japan – Pteridophyta*, 1957, by the same author. Japanese revision, 1983, by M. Kitagawa *et al.*, published by Shibundo, Tokyo.)

See also:

Hara, H. and Kanai, H. (1958, 1959). *Distribution Maps of Flowering Plants in Japan*, 2 vols. Inoue, Tokyo. (Dot maps of 200 taxa; endemics indicated.)

Horikawa, Y. (1972, 1976). *Atlas of the Japanese Flora: An Introduction to Plant Sociology of East Asia*, 2 vols so far; 5 planned (according to Frodin). Gakken, Tokyo. (Dot maps showing distribution and altitudinal range of 800 taxa; short descriptions; vegetation map at scale 1:5,000,000.)

Kurata, S. and Nakaike, T. (Eds) (1979-). *Illustrations of Pteridophytes of Japan*, 4 vols so far. Univ. Press, Tokyo. (Each volume describes about 100 taxa in Japanese; distribution maps; photographs.)

Field-guides The following illustrated guides cover most of the flora; Japanese text includes notes on distribution and habitats for each species:

Coloured Illustrations of Herbaceous Plants of Japan. Hoikusha, Osaka. Vol. 1 (1958) by S. Kitamura, M. Hori and G. Murata (Sympetalae); vol. 2 (1961) by S. Kitamura and G. Murata (Choripetalae); vol. 3 (1964) by S. Kitamura, G. Murata and T. Koyama (monocotyledons).

Coloured Illustrations of the Pteridophyta of Japan (1962), by M. Tagawa. Hoikusha, Osaka. 207 pp.

Coloured Illustrations of Wild Plants of Japan (1957-1959), 4 vols by S. Okuyama. Seibundo-Shinkosha, Tokyo. (Line drawings, colour photographs, distribution maps.)

Coloured Illustrations of Woody Plants of Japan (1973, 1979), 2 vols by S. Kitamura and G. Murata. Hoikusha, Osaka. (Over 1200 taxa described, many illustrated.)

Satake, Y., Ohwi, J., Kitamura, S., Watari, S. and Tominari, T. (Eds) (1981). *Wild Flowers of Japan: Herbaceous Plants (Including Dwarf Subshrubs)*, 3 vols. Heibonsha, Tokyo. (In Japanese.)

Shimizu, T. (1982, 1983). *The New Alpine Flora of Japan in Color*, 2 vols. Hoikusha, Osaka. (About 800 taxa described in Japanese; keys in English; many colour plates.)

Takeda, H. and Tanebe, K. (1951). *Illustrated Manual of Alpine Plants of Japan*. Hokuryu-Kan, Tokyo. 347 pp. (Short descriptions, line drawings of 432 species.)

Information on Threatened Plants Japan has no national Red Data Book. IUCN has a preliminary list of endemic Japanese trees, including E:4, V:4, R:5. See also:

Shimizu, T. and Satomi, N. (1976). A preliminary list of the rare and critical vascular plants of Japan, 2 parts. *J. Fac. Liberal Arts, Shinshu Univ. Nat. Sci.* 10: 3-16; 11: 43-54. (Annotated list of ferns, gymnosperms, monocotyledons and a number of dicotyledons; distribution details for Hokkaido, Honshu, Kyushu and Shikoku.)

Laws Protecting Plants The conservation of plant life in Japan was first covered by law under an act of 1919, which designated various plants as "national monuments". This category also includes a number of natural forests and special plant communities. The National Park Law and the Nature Conservation Law protect a number of plants and vegetation types.

Voluntary Organizations
Nature Conservation Society of Japan, 2-8-1 Toranomon, Minato-ku, Tokyo 105.
WWF-Japan, 6F 39, Mori Building, 2-4-5 Azabudai, Minato-ku, Tokyo 106.

Botanic Gardens Japan has 106 botanic gardens, but none subscribe to the IUCN Botanic Gardens Conservation Co-ordinating Body. For a full list of them see Henderson (1983), cited in Appendix 1.

Useful Addresses

Biological Institute and Herbarium, Faculty of Liberal Arts, Shinshu University, Matsumoto 390.

Japan Society of Plant Taxonomists, c/o Department of Botany, National Science Museum, Hyakunin-cho, Shinjuku-ku, Tokyo.

TRAFFIC Japan, 6F 39 Mori Building, 2-4-5 Azabudai, Minato-ku, Tokyo 106.

CITES Management Authority: Ministry of International Trade and Industry, International Economic Affairs Division, International Economic Affairs Department, International Trade Policy Bureau, 3-1, Kasumigaseki 1-chome, Chiyoda-ku, Tokyo.

Additional References

Nishida, M. (1972). An outline of the distribution of Japanese ferns. In Graham, A. (Ed.), *Floristics and Palaeofloristics of Asia and Eastern North America*. Elsevier, Amsterdam. Pp. 101-105. (Discussion of distribution patterns, checklists of ferns of various floral zones.)

Numata, M. (Ed.) (1974). *The Flora and Vegetation of Japan*. Kodansha, Tokyo and Elsevier, Amsterdam. 294 pp. (Includes simplified vegetation map.)

Numata, M., Yoshioka, K. and Kato, M. (Eds) (1975). *Studies in Conservation of Natural Terrestrial Ecosystems in Japan. Part 1: Vegetation and its Conservation*. Japanese Committee for IBP. 157 pp. (Not seen.)

Johnston Island

Johnston Island (area 129.5 sq. km; population 327, 1980 census) is an unincorporated territory of the United States, c. 1150 km WSW of Honolulu in the Pacific Ocean, at latitude 16°45'N, longitude 169°31'W. There are 2 highly modified sand and coral islands (Johnston and Sand Islands), and 2 completely man-made islands (Akau and Hikina).

No original vegetation remained on the atoll by 1946 due to military operations (Fosberg, 1949). A few species have arrived by natural means, but the majority have been intentionally or accidentally introduced by man (Christophersen, 1931). 127 vascular plant species have so far been recorded; no endemics (Amerson and Shelton, 1976).

References

Amerson, A.B. (1973). *Ecological Baseline Survey of Johnston Atoll, Central Pacific Ocean*. Technical Report, Environment Programme, Smithsonian Institution, Washington, D.C. 365 pp. (Plants on pp. 48-61.)

Amerson, A.B. and Shelton, P.C. (1976). The natural history of Johnston Atoll, central Pacific Ocean. *Atoll Res. Bull.* 192. 479 pp. (Lists 127 vascular species; origin and distribution within Johnston Atoll indicated.)

Christophersen, E. (1931). Vascular plants of Johnston and Wake Islands. *Occ. Papers Bernice P. Bishop Mus.* 9(13). 20 pp. (3 vascular species recorded.)

Fosberg, F.R. (1949). Flora of Johnston Island, central Pacific. *Pacific Science* 3: 338-339. (Includes annotated checklist of 27 vascular plants.)

Jordan

Area 97,668 sq. km

Population 3,375,000

Floristics c. 2200 vascular plant species so far recorded from eastern Jordan, and an additional 100-200 species likely to be found to the west of the Dead Sea (D.M. Al-Eisawi, 1985, pers. comm.). No figure for endemics to Jordan; 150 species are endemic to Palestine (Shmida, in press). The flora of Jordan has Mediterranean, Irano-Turanian, Saharo-Arabian and Sudanian elements. The high plateaux of Edom in Trans-Jordan include limestone and sandstone areas rich in endemics.

Vegetation About 88% is desert, less than 1% forested (Kasapligil, 1956). The Jordan River Valley, a branch of the African Rift Valley system, divides Jordan into two regions. The hilly West Bank area is mainly hammada (stony) desert supporting sparse thorn scrub, particularly in the Upper Jordan Valley (Zohary, 1973, cited in Appendix 1). The East Bank, and land to the east of the Dead Sea, is the edge of a high plateau which supports dwarf shrub steppes with *Artemisia*, and deciduous steppe forests with *Amygdalus*, *Crataegus* and *Pistacia*; *Pinus halepensis* and evergreen oak forests, with *Quercus calliprinos*, to the north-east of the Dead Sea, between Irbid and Amman, above 700 m; deciduous oak forests, with *Quercus aegilops* at lower altitudes; juniper forests on the southern mountains above 1000 m, greatly modified by overgrazing. Most of the area further east is an extension of the Syrian and North Arabian Desert. There are extensive areas of saline marshes to the north and south of the Dead Sea, with *Tamarix*, *Salsola* and *Atriplex*.

Checklists and Floras The first volume of the *Flora of Jordan* by D. Al-Eisawi is in preparation. 3-4 volumes are projected over a period of 10-15 years. A recent checklist of the flora is:

Al-Eisawi, D. (1983). List of Jordan vascular plants. *Mitt. Bot. München* 18: 79-182.
(Covers mainly the area to the east of the Dead Sea; no distribution details.)

Part of Jordan is covered by the *Flora of Syria, Palestine and Sinai* (Post, 1932); *Flora Palaestina* (1966-); and Eig, Zohary and Feinbrun-Dothan (1931); the whole country will be included in the *Med-Checklist*. All of these are cited in Appendix 1.

A number of papers in the series 'Studies on the flora of Jordan' have been published in the journal *Candollea* since 1975, each describing new species or listing plants in a given region. See in particular:

Boulos, L. (1977). Studies on the flora of Jordan, 5. On the flora of El Jafr-Batir Desert. *Ibid*. 32(1): 99-110.

Boulos, L. and Al-Eisawi, D. (1977). Studies on the flora of Jordan, 6. On the flora of Ras en Naqb. *Ibid*. 32(1): 111-120.

Boulos, L. and Lahham, J. (1977a). Studies on the flora of Jordan, 3. On the flora of the vicinity of the Aqaba gulf. *Candollea* 32(1): 73-80. (Includes annotated checklist of 91 angiosperms.)

Boulos, L. and Lahham, J. (1977b). Studies on the flora of Jordan, 4. On the desert flora north-east of Aqaba. *Ibid*. 32(1): 81-98. (Includes annotated checklist of 250 vascular plants, mainly collected in 1974 and 1975, in the area between Wadi Yutum and Wadi Rum.)

The series is to continue in *Kew Bulletin*; papers in press include D. Al-Eisawi on orchids of Jordan.

Information on Threatened Plants Jordan is included in the draft list for North Africa and the Middle East produced by IUCN Threatened Plants Committee Secretariat (1980), cited in Appendix 1. Coverage for Jordan is very incomplete.

Voluntary Organizations
Royal Society for the Conservation of Nature, P.O. Box 6354, Amman.

Useful Addresses
University of Jordan, Biology Department, Irbid.
CITES Management Authority: Royal Society for the Conservation of Nature (address above).

Additional References
Al-Eisawi, D.M. (1983). Vegetation in Jordan. Paper presented at the Second International Conference on the History and Archaeology of Jordan. 20 pp. Mimeo.
Gómez-Campo, C. (Ed.) (1985). *Plant Conservation in the Mediterranean Area*. (See in particular L. Boulos on the arid eastern and south-eastern Mediterranean regions.)
Kasapligil, B. (1956). *Report to the Government of the Hashemite Kingdom of the Jordan on an Ecological Survey of the Vegetation in Relation to Forestry and Grazing*. FAO, Rome. 39 pp.
Mountfort, G. (1966). *Portrait of a Desert: the Story of an Expedition to Jordan*. Collins, London. 192 pp. (Mainly covers fauna.)
Nelson, B. (1973). *Azraq: Desert Oasis*. Allen Lane, London. 436 pp. (Physical geography, vegetation, fauna.)
Shmida, A. (in press). Endemism in the flora of Israel. *Bot. Jahrb*. (Analysis of endemism includes references to Jordanian flora.)
Zohary, M. (1962). *The Plant Life of Palestine: Israel and Jordan*. Ronald Press, New York. 262 pp. (Includes useful vegetation map of Palestine.)
Zohary, M. (1983). *Vegetation of Israel and Adjacent Areas*. Reichert, Wiesbaden. 166 pp.

Juan Fernández

The Juan Fernández, or Robinson Crusoe Islands, consist of 3 precipitous volcanic islands – Más á Tierra (Isla Robinson Crusoe), Más Afuera (Isla Alejandro Selkirk) and Isla Santa Clara – situated in the South Pacific Ocean, 665 km west of Chile, between 33-34°S and 78-81°W. The highest point is El Yunque (916 m), on Más á Tierra. The islands are administered by Valparaiso province, Chile.

Area 93 sq. km

Population 650-700

Floristics 147 native species including 54 ferns (Skottsberg, 1920-1956); 118 endemic taxa (IUCN figures). 10 endemic genera (of which 5 in Compositae) and one endemic family, the monotypic Lactoridaceae. Of the endemics, 50% are confined to Más á Tierra, 33% to Más Afuera. *Chenopodium santa-clarae* is restricted to Isla Santa Clara (Perry, 1984).

Vegetation The Juan Fernández were originally covered by forests dominated by *Drimys*, *Fagara* and *Nothomyrica*; however, the slopes of eastern Más á Tierra below 100 m receive less rainfall and may have always been treeless. Throughout the islands, native vegetation is now restricted to ridges and cliffs due to overgrazing and competition from introduced plants (Sanders *et al.*, 1982). Remnants of temperate evergreen forest, with tree ferns abundant on slopes above 500 m; cloud forest and alpine meadows above 700 m; secondary scrub with invasive *Acaena*, *Rubus* and maqui scrub (*Aristotelia*) up to montane zone; the summit of El Yunque is covered by *Ugni*, *Blechnum* and *Dendroseris* scrub. Santa Clara is mainly grassland. For sketch maps showing principal plant communities see Skottsberg (1920-1956), vol. 2.

It is predicted that little of the flora will remain if nothing is done to reduce the abundant introduced cattle, sheep, goats and horses. IUCN/WWF plan a rescue programme with the Chilean authorities as part of the IUCN/WWF Plant Conservation Programme. Although the islands were declared a National Park in 1935 and accepted as a Biosphere Reserve in 1977, little has been done so far to save the flora.

Checklists and Floras
Nishida, H. (1979). Plants of the Robinson Crusoe Islands. *Plant and Nature* 13(2): 27-32; 13(4): 29-33, 35. (In Japanese.)

Skottsberg, C.J.F. (Ed.) (1920-1956). *The Natural History of the Juan Fernández and Easter Island*, 3 vols. Almqvist and Wiksell, Uppsala. (See in particular, 1: 193-438, derivation of the flora and fauna; 2: 1-46, pteridophytes; 2: 95-240, phanerogams; 2: 763-792, supplement to the pteridophytes and phanerogams; 2: 793-960, vegetation.)

Information on Threatened Plants 6 species are included in *The IUCN Plant Red Data Book* (1978). See also Marticorena (1980), cited under Chile.

Perry, R. (1984). Juan Fernández Islands: a unique botanical heritage. *Envir. Conserv.* 11(1): 72-76. (Lists 60 threatened endemic species giving distribution by islands.)

Latest IUCN statistics: endemic taxa – Ex:1 (*Santalum fernandezianum*), E:52, V:32, R:9, I:1, K:17, nt:6.

An index of threatened plants in cultivation is:

Threatened Plants Unit, IUCN Conservation Monitoring Centre (1984). *The Botanic Gardens List of Rare and Threatened Species of the Galapagos and Juan Fernandez Islands*. Botanic Gardens Conservation Co-ordinating Body, Report No. 11. IUCN, Kew. 6 pp. (Lists 14 rare and threatened taxa, from the Juan Fernández Islands, which are in cultivation, with gardens listed against each.)

Useful Addresses
Corporación Nacional Forestal de Chile (CONAF), Av. Bulnes, 285 Santiago, Chile; (park management), V Region, 3 Norte 541, Vina del Mar, Chile.

Additional References
Gutierrez, A., Mann, G., Merino, R., Thelen, K.D. and Dalfelt, A. (1976). Plan de manejo Parque Nacional Juan Fernández. Documento Técnico de Trabajo 22. Proyecto FAO/RLAT tf-199. Santiago.

Hemsley, W.B. (1885). Report on the botany of Juan Fernandez and Masafuera. In *Report on the Scientific Results of the Voyage of H.M.S. Challenger During the Years 1873-76*. Botany vol. 1, part 2. HMSO, London. Pp. 1-96. (Includes annotated checklist of ferns and flowering plants; botanical history.)

Kunkel, G. (1956). Über den Waldtypus der Robinson-Insel. *Forschungen und Fortschritte* 30(5): 129-137. (Forest types of Robinson Crusoe Island; notes on distribution of indigenous plants.)

Kunkel, G. (1968). Robinson Crusoe's Islands. *Pacific Discovery* 21: 1-8.

Muñoz P., C. (1969). El Archipiélago de Juan Fernández y la conservación de sus recursos naturales renovables. *Bol. Acad. Cien.* Instituto de Chile, Ser. 1(2): 83-103. (Reprinted, 1974, in *Serie Educativa*. Museo Nac. Hist. Nat., Santiago 9: 17-47.)

Nishida, H. and M. (1979). The vegetation of the Más a Tierra (Robinson Crusoe) Island, Juan Fernández. In Nishida, M. (Ed.), *A Report of the Palaeobotanical Survey to Southern Chile by a Grant-in-Aid for Overseas Scientific Survey, 1979*. Faculty of Science, Chiba Univ., Japan. Pp. 41-48. (Lists 55 taxa collected during botanical survey 1976-1979, includes vegetation map.)

Sanders, R.W., Stuessy, T.F. and Marticorena, C. (1982). Recent changes in the flora of the Juan Fernández Islands, Chile. *Taxon* 31(2): 284-289.

Kampuchea

Area 181,940 sq. km

Population 7,149,000

Floristics No figure for size of flora or number of endemics.

Vegetation Closed broadleaved forests cover 71,500 sq. km (FAO/UNEP, 1981). About 40% of the forest cover is probably deciduous monsoon forest, including dry dipterocarp and semi-evergreen dipterocarp forests, mostly in the north, and extensively modified by burning. About 30% of the forest cover is hill evergreen rain forest, mostly in southern uplands and along Annamite Chain (Myers, 1980, cited in Appendix 1). Pine forests on Kirikom Plateau; seasonally inundated "flood forest" around Great Lake (Legris, 1974). Much of Mekong Basin converted to rice cultivation.

Kampuchea's forests have been greatly modified over many centuries; little can be described as primary forest (Myers, 1980, cited in Appendix 1). Estimated rate of deforestation of closed broadleaved forest 250 sq. km/annum out of a total of 71,500 sq. km (FAO/UNEP, 1981).

Checklists and Floras No national Flora. Kampuchea is included in *Flore du Cambodge, du Laos, et du Vietnam* (1960-) and *Flore Générale de L'Indo-Chine* (1907-1951), both cited in Appendix 1.

Information on Threatened Plants None.

Additional References

Legris, P. (1974). Vegetation and floristic composition of humid tropical continental Asia. In Unesco, *Natural Resources of Humid Tropical Asia*. Natural Resources Research 12. Paris. Pp. 217-238.

Vidal, J.E. (1979). Outline of ecology and vegetation of the Indochinese Peninsula. In Larsen, K. and Holm-Nielsen, L.B. (Eds), *Tropical Botany*. Academic Press, London. Pp. 109-123.

Kazan Retto

Kazan Retto, or the Volcano Islands, comprise 3 volcanic islands – Iwo Jima (18 sq. km), Kita Iwo Jima (5 sq. km) and Minami-Iwojima (4 sq. km). The islands are c. 1250 km south of Japan, of which they are a dependency. The highest point is 916 m, on Minami-Iwojima. The population consists of personnel of the military base on Iwo Jima. Douglas (1969, cited in Appendix 1) describes Minami-Iwojima as "practically inaccessible" and "one of the least disturbed islands in the world". It was designated a Wilderness Area in 1975.

The natural vegetation is broadleaved evergreen forest, but much of that on Iwo Jima and Kita Iwo Jima has been destroyed by military activities, or else cleared for settlements and crops in the past. Minami-Iwojima, on the other hand, still has intact forest dominated by *Machilus kobu* (H. Ohba, 1985, *in litt.*).

257 flowering plant species (including introduced species) of which 9 are endemic to Kazan Retto and 33 are restricted to Kazan Retto and Ogasawara-Gunto (Ohba, *in litt.*). Minami-Iwojima has 118 vascular plant taxa of which 4 are endemic to the island and a further 5 are endemic to Kazan Retto (Ohba in Okutomi, 1982a). The flora is related to that of eastern Asia and Ogasawara-Gunto.

No information on threatened plants.

References
Okutomi, K. (Ed.) (1982a). *Conservation Reports of the Minami-Iwojima Wilderness Area*. Nature Conservation Bureau, Environment Agency of Japan, Tokyo. 403 pp. (In Japanese with English summary. See in particular H. Ohba on vascular plants, with floristic analyses and distribution maps of selected species, pp. 61-143; and H. Okutomi, H. Ohba, N. Ishii, Y. Tsukamoto and M. Sato on the endemic flora and fauna, pp. 393-403.)
Okutomi, K. (Ed.) (1982b). *Science Report on Nature and Natural Resources in Minami-Iwojima*. Min. of Environment, Tokyo. 174 pp. (In Japanese.)

Kenya

Area 582,644 sq. km

Population 19,761,000

Floristics Just under 6000 species, plus about 500 ferns and fern-allies (J.B. Gillett, 1984, pers. comm.); 8000-9000 species of flowering plant (Blundell, 1982), but this estimate too high. Brenan (1978, cited in Appendix 1), from a sample of the *Flora of Tropical East Africa*, estimates 265 endemic species, but that is probably an under-estimate.

Largely within the Somalia-Masai region; the area from Lake Turkana and the Tana River to the Ethiopian and Somalian border is especially rich in regional endemics. Coastal band occupied by Zanzibar-Inhambane regional mosaic; forest fragments, including some on limestone, are remarkably rich, diverse, and of exceptional biological interest; recognized as a major target for conservation effort. Afromontane region mostly on volcanic mountains; not notably rich in local species. South-west of Kenya within Lake Victoria regional mosaic; Kakamega Forest is the easternmost part of the Guinea-Congolian rain forest, and has distinct West African affinities.

Vegetation Most of the low and medium altitude parts of Kenya are covered with bushland, with species of *Acacia* and *Commiphora* dominant, including some semi-desert with many ephemerals and succulents. Vegetation nearer the coast lusher, with coastal bushland, grassland, wooded grassland and small patches of evergreen and dry semi-deciduous forest still remaining. Large expanses of wooded grassland, grassland and cultivation surrounding the highland areas. High altitudes covered with forest and forest-grassland mosaic, with clear altitudinal zonation from forest through bamboo thicket and heath thicket to tufted grass moorland above about 3500 m.

Estimated rate of deforestation for closed broadleaved forest 110 sq. km/annum out of 6900 sq. km (FAO/UNEP, 1981). However, Myers (1980, cited in Appendix 1) gives a figure of 16,702 sq. km total forest, of which 10,521 sq. km is primary moist deciduous forest.

For vegetation map see White (1983), cited in Appendix 1.

Checklists and Floras Kenya is included in the incomplete *Flora of Tropical East Africa*. Kenya's plants of high altitudes are listed in *Afroalpine Vascular Plants* (Hedberg, 1957). Both works are cited in Appendix 1.

Agnew, A.D.Q. (1974). *Upland Kenya Wild Flowers: a Flora of the Ferns and Herbaceous Flowering Plants of Upland Kenya*. Oxford Univ. Press, London. 827 pp. (Excludes grasses and sedges; keys, short descriptions, representative specimens, line drawings.)

Dale, I.R. and Greenway, P.J. (1961). *Kenya Trees and Shrubs*. Buchanan's Kenya Estates, Nairobi. 654 pp. (Keys, short descriptions, representative specimens; 110 line drawings, 80 black and white photographs, 31 colour plates.)

Gillett, J.B. and McDonald, P.G. (1970). *A Numbered Check-List of Trees Shrubs and Noteworthy Lianes Indigenous to Kenya*. Govt Printer, Nairobi. 67 pp.

Field-guides A very useful key to families is included in Lind and Tallantire, (in press), cited under Uganda.

Blundell, M. (1982). *The Wild Flowers of Kenya*. Collins, London. 160 pp. (Short descriptions; 310 species illustrated by colour photographs.)

Information on Threatened Plants

Hedberg, I. (1979), cited in Appendix 1. (List by J.B. Gillett for Kenya, pp. 93-94, includes examples of taxa threatened in each of several major vegetation types, and includes E:11, V:20, R:4, I:1.)

Mungai, G.M., Gillett, J.B., and Eagle, C.F. (1980). *Plant Species in Kenya: Survival or Extinction*. Bulletin of Wildlife Clubs of Kenya, Nairobi. 6 pp. (Lists over 20 species as threatened.)

IUCN holds records of 44 species and infraspecific taxa believed to be endemic; most are succulents. (E:15, V:16, R:3, I:3.)

Data sheets are published in *The IUCN Plant Red Data Book* (1978) of two species occurring in Kenya and Tanzania, and of three species endemic to Kenya.

Botanic Gardens

Mazeras Nurseries, c/o Municipal Council of Mombasa, P.O. Box 90440, Mombasa.
Mutomo Hill Plant Sanctuary, Kitui.

Nairobi Arboretum, The Chief Conservator of Forests, Forest Dept, P.O. Box 30513, Nairobi.

National Museums of Kenya, P.O. Box 40658, Nairobi. (Surrounding grounds planted with many named indigenous trees and shrubs.)

Voluntary Organizations

African Wildlife Foundation, P.O. Box 48177, Nairobi.

East Africa Natural History Society, P.O. Box 44486, Nairobi.

Kenya Orchid Society, P.O. Box 241, Nairobi.

Wildlife Clubs of Kenya Association, P.O. Box 40658, Nairobi.

Useful Addresses

East African Herbarium, P.O. Box 45166, Nairobi.

Environment Liaison Centre, P.O. Box 72461, Nairobi.

IUCN/WWF Programme Representative for Eastern Africa, c/o African Wildlife Foundation, P.O. Box 48177, Nairobi.

Kenya Agricultural Research Institute (KARI), P.O. Box 30148, Nairobi.

Kenya Rangeland Ecological Monitoring Unit (KREMU), P.O. Box 47146, Nairobi.

United Nations Environment Programme (UNEP), P.O. Box 30552, Nairobi.

CITES Management Authority: Wildlife Conservation and Management Dept, Ministry of Tourism and Wildlife, P.O. Box 40241, Nairobi.

Additional References

Edwards, D.C. (1940). A vegetation map of Kenya with particular reference to grassland types. *J. Ecol.* 28: 377-385. (With small-scale vegetation map.)

Kuchar, P. (1981). *The Plants of Kenya: a Handbook of Uses and Ecological Status.* Technical Report Series, Kenya Rangeland Ecological Monitoring Unit, Ministry of Environment and Natural Resources, Nairobi.

Lucas, G.Ll. (1968). Kenya. In Hedberg, I. and O. (1968), cited in Appendix 1. Pp. 152-166.

Trapnell, C.G. *et al.* (1966-1969). *Kenya Vegetation*, sheets 1-3 (maps 1:250,000). Directorate of Overseas Surveys, Tolworth, U.K.

Kermadec Islands

The Kermadec Islands (30°S, 178°30'W) are an outlying volcanic island group, in the South Pacific Ocean. They are 976 km north-east of New Zealand, of which they are a dependency. Raoul, or Sunday Island (34 sq. km), is the only inhabited island in the group. It attains 520 m at the rim of the central crater. Curtis (0.5 sq. km) lies to the south of Macauley (3 sq. km). The remaining islets are stacks and rocks scattered around the main islands. The island group is now a Nature Reserve.

Area 33.5 sq. km

Population 10 (Douglas, 1969, cited in Appendix 1)

Floristics 195 vascular plant species of which 113 native (Sykes, 1977). Raoul has c. 120 vascular plant species (*Flora of New Zealand*, 1961, cited under New Zealand); 23 endemic vascular plant taxa (figures quoted in Given, 1981a, cited under New Zealand). About 100 flowering plants and ferns on the Kermadecs are shared with mainland New

Zealand; affinities also with Norfolk and Lord Howe Islands. 45 taxa are found in Polynesia (Given, 1981a, cited under New Zealand).

Vegetation Coastal scrub on talus at the foot of cliffs; dry forest dominated by *Metrosideros*, below 240 m; wet forest also dominated by *Metrosideros* with tree ferns, on higher slopes. The islands are still volcanically active; crater floors almost unvegetated.

Checklists and Floras The Kermadecs are included in the *Flora of New Zealand* (1961, 1970, 1980), cited under New Zealand. See also:

Sykes, W.R. (1977). *Kermadec Islands Flora. An Annotated Checklist*. DSIR Bulletin no. 219. Wellington. 216 pp. (Enumeration of native and naturalized plants; chapters on physical geography.)

Information on Threatened Plants Given (1976, 1977, 1978, cited under New Zealand) includes 5 Kermadec endemic taxa, of which 4 are now Endangered and *Hebe breviracemosa* is probably Extinct. Latest IUCN statistics: endemic taxa – Ex:1, E:4, R:2, nt:2.

Kiribati

Kiribati (area 684 sq. km; population 62,000) comprises the Gilbert Group (17 islands), the Phoenix Islands (8) and the Equatorial (Line) Islands (8); mostly small coral islands and atolls, many only a few metres wide and less than 6 m above sea level; spread over 5 million sq. km in the south-west central Pacific Ocean. Banaba (Ocean Island), to the west of the main Gilbert group, is an elevated limestone island reaching 81 m. Most of the islands are uninhabitated. The Equatorial Islands and Banaba have been worked for guano. Christmas Island has been greatly modified by testing nuclear weapons.

Floristics c. 100 vascular plant species recorded from the Gilbert Islands, of which c. 60 are indigenous (Allerton and Herbst, 1973); most are widespread throughout the Pacific. Fanning Island, in the southern Line Island group, has 102 taxa of which only 22 indigenous, including 2 endemic (St John, 1974). Vostok Island (0.25 sq. km), in the northern Line Island group, has only 2 vascular plant species (Clapp and Sibley, 1971).

Vegetation Most of the natural vegetation of the larger islands (*Cordia*, *Tournefortia* and *Scaevola* scrub) has been replaced by plantations of coconuts, breadfruit and *Pandanus*. Some areas of *Pemphis* scrub and mangroves (Catala, 1957; Fosberg, 1973, cited in Appendix 1).

Checklists and Floras No complete Flora; the following checklists have been published for individual islands:

Chock, A.K. and Hamilton, D.C. (1962). Plants of Christmas Island. *Atoll Res. Bull.* 90. 7 pp. (Lists 41 species.)
Christophersen, E. (1927). Vegetation of Pacific Equatorial Islands. *Bull. Bernice P. Bishop Mus.* 44. 79 pp. (Includes annotated checklist for Palmyra, Line Islands.)
Clapp, R.B. and Sibley, F.C. (1971). Notes on the vascular flora and terrestrial vertebrates of Caroline Atoll southern Line Islands. *Atoll Res. Bull.* 145. 18 pp. (Includes annotated checklist of 35 taxa, many widespread throughout the Pacific.)
St John, H. (1974). The vascular flora of Fanning Island, Line Islands, Pacific Ocean. *Pacific Science* 28(3): 339-355.

The Gilbert Islands and Banaba are included in the regional checklists of Fosberg, Sachet and Oliver (1979, 1982), cited in Appendix 1, and will be covered by the *Flora of Micronesia* (1975-), also cited in Appendix 1.

Information on Threatened Plants None.

Additional References

Allerton, J.G. and Herbst, D. (1972, 1973). Report from the Gilbert and Ellice Islands. *Bull. Pacific Tropical Botanic Garden* 2(4): 63-68; 3(1): 2-6.

Catala, R.L.A. (1957). Report on the Gilbert Islands: some aspects of human ecology. *Atoll Res. Bull*. 59. 187 pp. (Includes list of plants collected, including introductions with notes on localities and uses.)

Christophersen, E. (1927). Vegetation of the Pacific Equatorial Islands. *Bull. Bernice P. Bishop Mus*. 44. 79 pp. (Includes annotated checklist of vascular plants.)

Clapp, R.B. and Sibley, F.C. (1971). The vascular flora and terrestrial vertebrates of Vostok Island, south-central Pacific. *Atoll Res. Bull*. 144. 10 pp.

Luomala, K. (1975). Ethnobotany of the Gilbert Islands. *Bernice P. Bishop Mus*. 213. 129. (List of plants with uses; arranged by vernacular names.)

Korea, Democratic People's Republic of

(NORTH KOREA)

Area 122,312 sq. km

Population 19,630,000

Floristics No figure for North Korea, but in the Korean Peninsula 2898 vascular plant species (T.B. Lee, 1976). The Korean Peninsula has 407 endemic vascular taxa of which 107 restricted to North Korea (Lee, 1983).

Vegetation Extensive mixed deciduous-coniferous forests between 700-1700 m (Sun, 1974b). 'Taiga' forest in uplands with larch, pine, fir forests and scrub; lowlands mainly cleared for cultivation. Alpine vegetation above 2000 m (Sun, 1974a).

Checklists and Floras

Lee, T.B. (1976). Vascular plants and their uses in Korea. *Bull. Kwanak Arboretum* 1. 137 pp. (Checklists and statistics of useful plants.)

Lee, T.B. (1983). Endemic plants and their distribution in Korea. *Bull. Kwanak Arboretum* 4: 71-113. (Lists Korean endemic ferns, gymnosperms and angiosperms; notes on distribution.)

Lee, Y.N. (1966). *Manual of the Korean Grasses (Excluding Bambuseae)*. Ewha Womens Univ. Press, Seoul. 300 pp. (120 taxa described; notes on distribution.)

Lee, Y.N. (1976). *Illustrated Flora and Fauna of Korea. 18: Flowering Plants*. 893 pp. Samhwa, Seoul. (In Korean; appendix includes short notes in English on 889 taxa.)

Mori, T. (1922). *An Enumeration of Plants Hitherto Known from Corea*. Govt of Chosen, Seoul. 546 pp. (Checklist of 2904 species, 506 varieties; endemics to Korean Peninsula indicated; separate indices of Japanese and Chinese names.)

Nakai, T. (1915-1939). *Flora Sylvatica Koreana*, 22 parts. Govt of Chosen, Seoul. (In Latin and Japanese; all known woody species recorded for Korea listed before each family treatment.)

Nakai, T. (1952). A synoptical sketch of Korean flora, or the vascular plants indigenous to Korea, arranged in a new natural order. *Bull. Tokyo Nat. Sci. Mus.* 31. 152 pp. (Systematic list of 3176 vascular plant taxa, with summary.)

Park, M.K. (1975). *Illustrated Encyclopedia of Flora and Fauna of Korea, 16: Pteridophyta.* 549 pp. Samhwa, Seoul. (Descriptions with notes on distribution, habitats; floristic summary and statistical table, includes 272 species.)

Uyeki, H. (1926). *Corean Timber Trees, 1. Ginkgoales and Coniferae.* Forestry Expt Station, Govt of Chosen, Japan. (In Japanese, maps showing distribution in Korean Peninsula.)

Information on Threatened Plants None.

Botanic Gardens

The Central Botanical Garden of DPRK, Pyongyang.

Additional References

Sun, C.I. (1974a). Taiga, a major flora community in our country. *Korean Nature* 2(33): 30-32.

Sun, C.I. (1974b). Coniferous-deciduous mixed forest zone, a major plant community in our country. *Korean Nature* 3(34): 25-27.

Korea, Republic of

(SOUTH KOREA)

Area 98,447 sq. km

Population 40,309,000

Floristics No figure for South Korea, but Korean Peninsula has 2898 vascular plant species (T.B. Lee, 1976). 407 taxa endemic to the Peninsula, of which 224 restricted to South Korea (Lee, 1983).

Vegetation Warm temperate, broadleaved evergreen forests, with *Quercus*, *Camellia* and bamboos, along southern coasts and on offshore islands; temperate forests containing *Quercus*, *Carpinus* and *Pinus densiflora* in south; *Quercus/Abies* forest and cold temperate *Abies/Betula* forest in north and at high elevations in Taebaek Mts. Rhododendrons commonly found in understorey of all forest types (Hagman *et al.*, 1978). Forests cover about two-thirds of South Korea (Hagman *et al.*, 1978); about 25% is under cultivation.

Checklists and Floras

Lee, T.B. (1973). *Illustrated Woody Plants of Korea.* Forest Expt Station, Seoul. 262 pp. (Short descriptions of 755 taxa, with line drawings and keys; in Korean.)

Lee, T.B. (1976). Vascular plants and their uses in Korea. *Bull. Kwanak Arboretum* 1. 137 pp. (Checklists and statistics of useful plants.)

Lee, T.B. (1979, 1982). *Illustrated Flora of Korea*, 2 vols. Hyangmunsa, Seoul. (Atlas flora covering 3160 taxa with descriptions in Korean; no details of distribution or ecology; not seen, citation based on Frodin.)

Lee, T.B. (1983). Endemic plants and their distribution in Korea. *Bull. Kwanak Arboretum* 4: 71-113. (Lists Korean endemic ferns, gymnosperms and angiosperms; notes on distribution.)

Lee, Y.N. (1966). *Manual of the Korean Grasses (Excluding Bambuseae)*. Ewha Womens Univ. Press, Seoul. 300 pp. (240 taxa described; notes on distribution.)

Lee, Y.N. (1976). *Illustrated Flora and Fauna of Korea. 18: Flowering Plants*. 893 pp. Samhwa, Seoul. (In Korean; appendix includes short notes in English on 889 taxa.)

Mori, T. (1922). *An Enumeration of Plants Hitherto Known from Corea*. Govt of Chosen, Seoul. 546 pp. (Checklist of 2904 species, 506 varieties; endemics to Korean Peninsula indicated; separate indices of Japanese and Chinese names.)

Nakai, T. (1915-1939). *Flora Sylvatica Koreana*, 22 parts. Govt of Chosen, Seoul. (In Latin and Japanese; all known woody species recorded for Korea listed before each family treatment.)

Nakai, T. (1952). A synoptical sketch of Korean flora, or the vascular plants indigenous to Korea, arranged in a new natural order. *Bull. Tokyo Nat. Sci. Mus.* 31. 152 pp. (Systematic list of 3176 vascular plant taxa, with summary.)

Park, M.K. (1975). *Illustrated Encyclopedia of Flora and Fauna of Korea, 16: Pteridophyta*. 549 pp. Samhwa, Seoul. (Descriptions with notes on distribution, habitats; floristic summary and statistical table, includes 272 species.)

Uyeki, H. (1926). *Corean Timber Trees, 1. Ginkgoales and Coniferae*. Forestry Expt Station, Chosen. (In Japanese, maps showing distribution in Korean Peninsula.)

Information on Threatened Plants

Choi, K.-C., Kim, C.-H., Lee, Y.-N., Won, P.-O. and Yoon, I.B. (1981). *Rare and Endangered Species of Animals and Plants of Republic of Korea*. Korean Assoc. for Conservation of Nature. 293 pp. (Lists 118 plant taxa, including widespread non-endemic species.)

Lee, T.B. (1980). Rare and endangered species in the area of Mt Sorak. *Bull. Kwanak Arboretum* 3: 197-201. (Mentions 12 taxa with notes on distribution.)

Lee, T.B. (1984). Endemic and rare plants of Mt. Sorak. *Bull. Kwanak Arboretum* 5: 1-6. (Enumeration of 114 vascular plant taxa of which 65 are endemic; 5 taxa are 'endangered', 12 taxa are 'rare'.

Preliminary IUCN statistics, mainly based on Choi *et al.* (1981), cited above: endemic taxa – Ex:1, E:8, V:2, R:20.

Laws Protecting Plants The Cultural Properties Protection Law (1973) provides protection for a number of plant species and their habitats by designating them as natural monuments. The law covers 13 taxa at the northern limit of their distribution, and 6 endemic and threatened taxa (T.B. Lee, 1984, *in litt.*).

Voluntary Organizations A committee has been set up to protect the natural habitat of *Abeliophyllum* (Lee, *in litt.*).

Botanic Gardens

Chollipo Arboretum, Uihangni 1-gu, Sosan Gun, Chungchong Namdo.

Hongnung Arboretum, Forest Research Institute, Chongnyangni, Tongdaemun-gu, Seoul.

Kumkang Botanic Garden, San 43-1, Changjon 2-Dong, Tongnaegu, Pusan.

Kwanak Arboretum, College of Agriculture, Seoul National University, Suwon.

Useful Addresses

Forest Research Institute, Chung-Ryang-Ri, Tong dae mun-Ku, Seoul.

Additional References

Hagman, M., Feilberg, L., Lagerström, T. and Sanda, J.E. (1978). *The Nordic Arboretum Expedition to South Korea 1976*. Forest Research Institute, Helsinki. 102 pp. (Expedition report, useful background notes on vegetation, forestry research in South Korea.)

Lee, T.B. (1980). Conservation of threatened plants in Korea. *Bull. Kwanak Arboretum* 3: 190-196. (Includes notes on plant re-introductions; summary in English.)

Kuwait

Area 24,281 sq. km.

Population 1,703,000

Floristics About 300 species of vascular plants estimated (quoted in Dickson, 1955); Halwagy and Macksad (1972) record a further 56 species not previously known from Kuwait. Affinities with the flora of Iraq.

Vegetation Mostly sparse scrub with perennial herbs and ephemerals; in the south-east and north-west, principally of the Chenopod *Haloxylon salicornicum*, in the west of the dwarf shrub *Rhantherium epapposum*, and immediately south and south-west of Kuwait City a zone dominated by the sedge *Cyperus conglomeratus* (Halwagy, 1974).

Checklists and Floras The late Professor Daoud prepared a *Flora of Kuwait*, now partly in press, edited by Ali al-Rawi (T.A. Cope, 1984, pers. comm.). Works relating to the Arabian peninsula as a whole are outlined under Saudi Arabia. See also:

Burtt, B.L. and Lewis, P. (1949-1954). On the Flora of Kuweit. *Kew Bull.* 4: 273-308 (1949); 7: 333-352 (1952); 9: 377-410 (1954).

Deeb, M. and Salim, K. (1974). *Wild and Ornamental Plants of Kuwait*. Kuwait. (In Arabic.)

Dickson, V. (1955). *The Wild Flowers of Kuwait and Bahrain*. Allen and Unwin, London. 144 pp. (Notes on species; some illustrated.)

Dickson, V. and Macksad, A. (1973). *Plants of Kuwait*. Ahmadi Natural History and Field Studies Group, Kuwait. 13 pp. (Computer checklist of 395 plant names.)

Halwagy, R. and Macksad, A. (1972). A contribution towards a Flora of the State of Kuwait and the Neutral Zone. *Bot. J. Linn. Soc.* 65: 61-79. (Lists 100 species of flowering plants.)

Field-guides

Husain, S.M. and Mirza, J.H. (1979). *A Field Key for the Identification of Common Trees, Shrubs and Climbers of Kuwait*. Newsletter Supplement No. 1, Botany and Microbiology Dept, Univ. of Kuwait. 21 pp.

Information on Threatened Plants None.

Voluntary Organizations

Ahmadi Natural History and Field Studies Group, c/o Kuwait Oil Co., Ahmadi-103.

Useful Addresses

Kuwait Institute for Scientific Research, P.O. Box 24885, Safat, Kuwait.

Additional References

Clayton, D. and Pilcher, C. (Eds) (1983). *Kuwait's Natural History. An Introduction.* Kuwait Oil Company. 351 pp. (Fully illustrated with colour photographs. See especially chapters by L. Corrall on Vegetation, pp. 24-66, and by C. Pilcher on Conservation, pp. 294-316.)

Halwagy, R. and M. (1974, 1977). Ecological studies on the desert of Kuwait; I: The physical environment. *J. Univ. Kuwait (Science)* 1: 75-86 (1974); II: the vegetation. *Ibid.* 1: 87-95 (1974); III: the vegetation of the coastal salt marshes. *Ibid.* 4: 33-74.

Lakshadweep

Lakshadweep, formerly the Laccadive Islands, are a group of 19 coral atolls north of the Maldives and c. 300 km off the Malabar coast of southern India. They are administered as a Union Territory of the Republic of India. Area 32 sq. km. 10 islands inhabited; population 40,237 (1981 census, *Times Atlas*, 1983).

348 vascular plant species recorded (Raghavan, 1977). The flora is related to that of the Maldives and Pacific Ocean atolls, rather than to that of the west coast of India. According to Prain (1893) and Willis (1901) there are no endemics; many species have pantropical and Indo-Pacific distributions. Apart from planted coconuts the vegetation of most of the islands consists of littoral communities, with *Casuarina*, *Pandanus* and *Terminalia* scrub. 3 islets are open reefs with no vascular plants.

References

Prain, D. (1892, 1893). Botany of the Laccadives. *J. Bombay Nat. Hist. Soc.* 7: 268-295; 7: 460-486. (Introduction in first part; second part includes annotated checklist of 121 species of which 40 indigenous.)

Raghavan, R.S. (1977). Floristic studies in India – the Western Circle. *Bull. Bot. Survey India* 19: 95-108.

Sivadas, P., Narayanan, B. and Sivaprasad, K. (1983). An account of the vegetation of Kavaratti Island, Laccadives. *Atoll Res. Bull.* 266. 9 pp. (Includes checklist of 117 plants on Kavaratti.)

Wadhwa, B.M. (1961). Additions to the flora of Laccadives, Minicoy and Aminidives groups of islands. *Bull. Bot. Survey India* 3: 407-408. (Notes on 11 species in Cyperaceae and Gramineae.)

Willis, J.C. (1901). Note on the flora of Minikoi. *Annals Royal Botanic Gardens Peradeniya* 1: 39-43. (Lists 134 species for Minicoy Island.)

Willis, J.C. and Gardiner, J.S. (1901). The botany of the Maldive Islands. *Annals Royal Botanic Gardens Peradeniya* 1: 45-164. (Includes annotated checklist of 359 species recorded from Chagos Archipelago, Laccadives and Maldives.)

Laos

Area 236,725 sq. km

Population 4,315,000

Floristics No figure for size of flora or number of endemics. Laos, Kampuchea and Viet Nam have c. 600 fern species (Parris, 1985, cited in Appendix 1).

Vegetation 27,000 sq. km of tropical lowland and hill evergreen rain forest, mainly along the Annamite Chain, the Sekong Valley bordering the Bolovens Plateau, and a few patches along the Mekong River; above 1000 m these forests have been extensively converted to grasslands (Myers, 1980, cited in Appendix 1); dry dipterocarp and mixed deciduous forests (with dipterocarps and teak) in south and between Vientiane and Burmese border; 10,000 sq. km of pine forests in the Xieng Khouang region and on sandy soils between 600-1400 m, greatly damaged by military activity; bamboo forests estimated at 6000 sq. km (Myers, 1980).

Estimated rate of deforestation of closed broadleaved forests 1000 sq. km/annum out of a total of 75,600 sq. km (FAO/UNEP, 1981). Myers (1980) quotes a UNDP/UNIDO estimate for "well-stocked forests" of only 46,000 sq. km. Few areas of forest remain undisturbed; much has been converted to grasslands.

Checklists and Floras No national Flora. Laos is included in *Flore du Cambodge, du Laos, et du Vietnam* (1960-), and *Flore Générale de L'Indo-Chine* (1907-1951), both cited in Appendix 1. See also:

Seidenfaden, G. (1972). An enumeration of Laotian orchids. *Bull. Mus. Nat. Hist. Naturelle Bot.* 71: 101-152. (Enumeration of about 316 species.)

Information on Threatened Plants None.

Additional References

Legris, P. (1974). Vegetation and floristic composition of humid tropical continental Asia. In Unesco, *Natural Resources of Humid Tropical Asia*. Natural Resources Research 12. Paris. Pp. 217-238.

Vidal, J. (1934-1960). La végétation du Laos. *Trav. Lab. For. Toulouse* Tome 5, sect. 1, vol. 1. (Part 1 – 103 pp.; part 2 – 582 pp.)

Vidal, J.E. (1979). Outline of ecology and vegetation of the Indochinese Peninsula. In Larsen, K. and Holm-Nielsen, L.B. (Eds), *Tropical Botany*. Academic Press, London. Pp. 109-123.

Lebanon

Area 10,400 sq. km

Population 2,644,000

Floristics No figure for Lebanon, but Syria and Lebanon together have about 3000 species; 11% of the flora of Syria and Lebanon is endemic (Zohary, 1973, cited in Appendix 1). In Lebanon, many endemics are confined to the high mountains of the Mediterranean zone in the west.

Vegetation Steppes and deserts cover most of Lebanon. There is a narrow coastal plain along the Mediterranean Sea, with evergreen maquis; further inland are the Lebanon Mountains, which rise to 3086 m. The western slopes up to 300 m support evergreen maquis, with *Quercus calliprinos*, *Ceratonia* and *Pistacia*; *Pinus halepensis* forest (replaced by *P. brutia* in north) from sea-level to 1200 m, now reduced to remnants; forests with *Cedrus libani* (Cedar of Lebanon), *Pinus nigra* and *Quercus calliprinos*, particularly between 1400-1800 m in the north. The oldest and most famous pure stands of *C. libani* are at Bsharri. The alluvial plains of the Beqaa Valley separate the Lebanon Mountains from the Anti-Lebanon Mountains in the east, which reach 2814 m at Mt Hermon. The Anti-Lebanon Mountains have *Amygdalus/Pistacia* scrub, and fragmented deciduous forests on their western slopes. There are also remnants of steppe/coniferous forests with *Abies cilica*, *Cedrus libani* and *Juniperus excelsa*. Subalpine and alpine communities occur above 2500 m in Lebanon. For detailed description of vegetation see Zohary (1973), cited in Appendix 1.

Checklists and Floras Lebanon will be covered by the *Med-Checklist*, cited in Appendix 1. See also:

Bouloumoy, L. (1930). *Flore du Liban et de la Syrie*, 2 vols. Vigot Freres, Paris. (1 – keys; 2 – plates.)

Mouterde, P. (1966-). *Nouvelle Flore du Liban et de la Syrie*, 3 vols so far. Dar El-Machreq, Beirut. (Vols 1-2 – pteridophytes, gymnosperms, monocotyledons and dicotyledons to Umbelliferae and Cornaceae; 3 – so far 3 fascicles, including Ericaceae, Labiatae, Scrophulariaceae. In addition there are 2 supplementary volumes with line drawings.)

Mouterde, P. (1973). Novitates florae libano-syriacae. *Saussurea* 4: 17-25. (17 new species and 2 varieties described from Lebanon and Syria.)

Thiébaut, J. (1936-1953). *Flore Libano-Syrienne*, 3 vols. Centre National de la Recherche Scientifique, Paris.

Information on Threatened Plants None. The section on Lebanon in the draft list for North Africa and the Middle East produced by IUCN Threatened Plants Committee Secretariat (1980), cited in Appendix 1, contains only 41 endemic species without categories. The list was taken from Mouterde (1966-), cited above.

Additional References
Charpin, A. and Greuter, W. (1975). Données disponibles concernant la flore de la Syrie et du Liban. In CNRS (1975), cited in Appendix 1. Pp. 115-117.
Gómez-Campo, C. (Ed.) (1985), cited in full in Appendix 1.

Lesotho

Area 30,344 sq. km

Population 1,481,000

Floristics 1591 vascular species (Jacot Guillarmod, 1971), predominantly herbaceous; one or two endemic species only.

Flora predominantly Afromontane, but lower altitude land in west in Kalahari-Highveld region.

Vegetation Predominantly montane grassland, with woody montane communities in sheltered valleys and south-facing slopes; communities with ericoid shrubs at highest altitudes. Most available lower altitude land under cultivation.

For vegetation map see White (1983), cited in Appendix 1.

Checklists and Floras Lesotho is included in the incomplete *Flora of Southern Africa*, and in *The Genera of Southern African Flowering Plants* (Dyer, 1975, 1976), both cited in Appendix 1. The national Flora is:

Jacot Guillarmod, A. (1971). *Flora of Lesotho (Basutoland)*. Cramer, Lehre. 474 pp.

Information on Threatened Plants

Hall, A.V. *et al.* (1980), cited in Appendix 1. (List on pp. 85-86 contains one endemic species: *Kniphofia hirsuta*, V, and 6 non-endemic species: V:1 (regional category), R:3, K:2.)

Hedberg, I. (1979), cited in Appendix 1. (List for Lesotho, p. 101, by A. Jacot Guillarmod, contains five species and three genera: E:6, R:1, I:1.)

Talukdar, S. (1983). The conservation of *Aloe polyphylla* endemic to Lesotho. In Killick, D.J.B. (1983), cited in Appendix 1. Pp. 985-989. (Gives details of conservation status and protective legislation.)

Information on *Aloe polyphylla* is included in *The IUCN Plant Red Data Book* (1978).

Laws Protecting Plants Legal Notice No. 36 of 1969 defines the monuments, relics, fauna and flora protected under Act 41 of 1967 (Historical Monuments, Relics, Fauna and Flora Act). The list of protected plants includes all aloes and specifically *A. polyphylla*.

Additional References

Bawden, M.G. and Carroll, D.M. (1968). *The Land Resources of Lesotho*. Land Resource Study 3. Directorate of Overseas Surveys, Tolworth, U.K. 89 pp. (With vegetation map 1:1,000,000.)

Jacot Guillarmod, A. (1968). Lesotho. In Hedberg, I. and O. (1968), cited in Appendix 1. Pp. 253-256.

Werger, M.J.A. (1978), cited in Appendix 1. Citation includes list of relevant chapters.

Liberia

Area 111,370 sq. km

Population 2,123,000

Floristics Size of flora unknown. 59 endemic species and 1 endemic genus (Brenan, 1978, cited in Appendix 1).

Floristic affinities Guinea-Congolian. Mt Nimba, shared with Guinea and Ivory Coast, has an Afromontane element and is especially important floristically, with more than 2000 species.

Vegetation Small areas of mangrove along coast. Coastal strip of lowland rain forest interspersed with secondary grassland and cultivation; transitional rain forest (between lowland and montane) on Mt Nimba. Remainder of country predominantly covered with lowland rain forest.

Estimated rate of deforestation for closed broadleaved forest 460 sq. km/annum out of 20,000 sq. km (FAO/UNEP, 1981). However, Myers (1980, cited in Appendix 1) quotes the following figures: 25,000 sq. km primary forest, plus an additional 23,000 sq. km broken forest; primary forest is degraded by shifting cultivators at 300 sq. km/annum, and by logging at 2000 sq. km/annum.

For vegetation map see White (1983), cited in Appendix 1.

Checklists and Floras Liberia is included in the *Flora of West Tropical Africa*. The Liberian portion of Mt Nimba is included in *Flore Descriptive des Monts Nimba*. Both works are cited in Appendix 1. See also:

Kunkel, G. (1965). *The Trees of Liberia: Field Notes on the More Important Trees of the Liberian Forests and a Field Identification Key*. Report No. 3, German Forestry Mission to Liberia, Munich. 270 pp. (Illustrations, map.)

Voorhoeve, A.G. (1979). *Liberian High Forest Trees*, 2nd Ed. (1st Ed. 1965). Centre for Agricultural Publishing and Documentation, Wageningen. 416 pp. (Extensive notes on the 75 most important or frequent high forest trees; 72 line drawings, 32 black and white photographs.)

Information on Threatened Plants

Hedberg, I. (1979), cited in Appendix 1. (Includes short list of example species and genera, p. 88, by J.M. Thorne.)

IUCN has records of 103 species and infraspecific taxa believed to be endemic, including E:2, V:10, R:5, I:5; the remainder are K.

Useful Addresses

CITES Management Authority: Forestry Development Authority, P.O. Box 3010, Monrovia.

CITES Scientific Authority: University of Liberia, Capitol Hill, Monrovia.

Additional References

Adam, J.-G. (1970). Etat actuel de la végétation des monts Nimba au Libéria et en Guinée. *Adansonia*, Sér. 2, 10: 193-211. (With 10 black and white photographs.)

Cooper, G.P. and Record, S.J. (1931). *The Evergreen Forests of Liberia*. Bulletin 31 of the Yale Univ. School of Forestry, New Haven. 153 pp. (Includes 26 black and white photographs.)

Lamotte, M. (1983). The undermining of Mount Nimba. *Ambio* 12(3-4): 174-179. (Photographs, maps.)

Voorhoeve, A.G. (1968). Liberia. In Hedberg, I. and O. (1968), cited in Appendix 1. Pp. 74-76.

Libya

Area 1,759,540 sq. km

Population 3,471,000

Floristics c. 1600 species of which about 90% (1440) occupy the coastal region, especially Jabal al Akhdar (Boulos, 1975); c. 1800 species (Le Houérou, 1975). Northern Cyrenaica has 134 endemics, of which 109 are endemic to Jabal al Akhdar (Bartolo *et al.*, 1977); IUCN has records of 83 species and infra-specific taxa believed to be endemic.

Floristic affinities Mediterranean and Saharan, although Jabal al Akhdar is the only area with a typical Mediterranean flora. The flora of most of the country is small and has Saharan affinities. Other coastal areas have a flora transitional between the two.

Vegetation Mostly desert with little or no perennial vegetation; the only non-desert vegetation is in a strip along the coast and has been cultivated and overgrazed with the result that very little natural vegetation survives except in a somewhat degraded form in the sclerophyllous forests of Jabal al Akhdar.

For vegetation map see White (1983), cited in Appendix 1.

Checklists and Floras Libya is included in the incomplete *Flore de l'Afrique du Nord*, the computerized *Atlas der Pflanzenwelt des Nordafrikanischen Trockenraumes* (Frankenberg and Klaus, 1980), *Flore du Sahara* (Ozenda, 1977), and is being covered in *Med-Checklist*; all of these are cited in Appendix 1. Below is the recent Flora, and up-to-date checklists:

Ali, S.I., Jafri, S.M.H. and El-Gadi, A. (Eds) (1976-). *Flora of Libya.* Al Faateh University, Tripoli. (86 families published so far: mostly small ones, but including Caryophyllaceae, Chenopodiaceae, Liliaceae and Brassicaceae.)

Boulos, L. (1977-1980). A checklist of the Libyan flora. 1. Introduction and Adiantaceae to Orchidaceae. *Publ. Cairo Univ. Herb.* 7/8: 115-141; 2. Salicaceae to Neuradaceae. *Candollea* 34(1): 21-48; 3. Compositae (by C. Jeffrey). *Ibid.* 34(2): 307-332; corrections (1980). *Ibid.* 35(2): 565-567.

Also published:

Brullo, S. and Furnari, F. (1979). Taxonomic and nomenclatural notes on the Flora of Cyrenaica (Libya). *Webbia* 34(1): 155-174.

Keith, H.G. (1965). *A Preliminary Check List of Libyan Flora*, 2 vols. Ministry of Agriculture and Agrarian Reform, Govt of Libyan Arab Republic. 1047 pp.

Information on Threatened Plants Libya is included in the draft list for North Africa and the Middle East produced by IUCN Threatened Plants Committee Secretariat (1980), cited in Appendix 1.

Boulos, L. (1985). The arid eastern and south-eastern Mediterranean regions. In Gómez-Campo, C. (Ed.), *Plant conservation in the Mediterranean area.*

Latest IUCN statistics: endemic taxa – E:2, V:18, R:18, I:4, K:20, nt:21; non-endemics rare or threatened on a world scale – E:1, V:7, R:6 (world categories).

Botanic Gardens
Sidi Mesri Experiment Station, Tripoli.

Additional References
Bartolo, G., Brullo, S., Guglielmo, A. and Scalia, C. (1977). Considerazioni fitogeografiche sugli endemismi della Cirenaica settentrionale. *Archiv. Bot. Biogeogr. Ital.* 53(3-4): 131-154.

Boulos, L. (1972). Our present knowledge on the flora and vegetation of Libya: bibliography. *Webbia* 26: 365-400.

Boulos, L. (1975). The Mediterranean element in the flora of Egypt and Libya. In CNRS (1975), cited in Appendix 1. Pp. 119-124.

Le Houérou, H.-N. (1975). Etude préliminaire sur la compatibilité des flores nord-africaine et palestinienne. In CNRS (1975), cited in Appendix 1. Pp. 345-350.

Liechtenstein

The principality of Liechtenstein is situated in the European Alps between Austria and Switzerland. One-third of the country lies in the Upper Rhine valley; the rest is mountainous.

Area 160 sq. km

Population 27,000

Floristics Over 1400 native vascular taxa (estimated from Seitter, 1977). Elements: Central European, alpine.

Vegetation About 25% of the country is agricultural; semi-natural and plantation forests occupy c. 34%; alpine pastures c. 16%. Widespread drainage, intensive agriculture and urban expansion responsible for dramatic loss, in recent years, of wetlands, woodlands and alpine pastures (Anon, 1984, cited in Appendix 1 and Broggi, 1977).

Checklists and Floras National Flora:

Seitter, H. (1977). *Die Flora des Fürstentums Liechtenstein*. Botanisch-Zoologische Gesellschaft, Liechtenstein. 573 pp. (In German; no keys; line drawings and colour photographs.)

Regional Floras:

Garcke, A. *et al*. (1972). *Illustrierte Flora, Deutschland und Angrenzende Gebiete*, 23rd Ed. by K. von Weihe. Parey, Berlin. 1607 pp. (Line drawings.)

Hess, H.E., Landolt, E. and Hirzel, R. (1967-). *Flora der Schweiz und angrenzender Gebiete*, 3 vols to date. Birkhäuser, Basel. (Covers all Switzerland and Liechtenstein and parts of Austria, France, Federal Republic of Germany and Italy; 1 – pteridophytes and dicotyledons; 2 and 3 – dicotyledons and monocotyledons; line drawings, and detailed historical and ecological introduction.)

Relevant journal: *Mitteilungen der Botanisch Zoologischen Gesellschaft Liechtenstein*, Sargans Werdenberg.

Field-guides See Grey-Wilson (1979) and Hegi (1935-1979), both cited in Appendix 1.

Information on Threatened Plants A national plant Red Data Book has recently been published (reviewed in *Oryx* 19: 112) identifying 383 rare and threatened flowering plant taxa of which 68 are 'extinct', 102 'endangered', 91 'threatened' and 122 'rare'; about one quarter of these are marshland plants.

Laws Protecting Plants The 1933 Nature Protection Law, revised 1966, (Loi relative à la protection de la nature) provides full protection to 34 plant species and partial protection to 17 additional species, 1 genus and 1 family. For partially protected plants it is prohibited to uproot them, but the picking of their above-ground parts is allowed. Under the Law, it is prohibited to promote, to acquire or to offer for sale, in either a fresh or dry condition, any plants listed. For the list of protected plants see:

Anon (1967). Gesetz vom 21 Dezember 1966, betreffend die Abänderung des Naturschutzgesetzes. *Liechtensteinisches Landesgesetzblatt* 1967, Nr. 5. Pp. 1-4.

Voluntary Organizations

Liechtensteinische Gesellschaft für Umweltschutz (Liechtenstein Society for
Environmental Protection), Heiligkreuz 52, Postfach 53290, 9490 Vaduz.

Useful Addresses

Ministère de l'agriculture et des forêts, Département des forêts, Vaduz.

Additional References

Broggi, M.F. (1977). Nature conservation and landscape management in Liechtenstein.
Parks 2(3): 14-16. (A short descriptive account of the history of nature conservation
in Liechtenstein and habitat degradation.)

Lord Howe Island

Lord Howe Island (31°35'S, 159°05'E) is situated 692 km north-east of Sydney, in the
Tasman Sea. It is a dependency of New South Wales, Australia. Unlike many colonized
islands of similar size, it retains a significant proportion of its native vegetation and flora.
In 1981, 8 sq. km were declared the Lord Howe Island Permanent Park Preserve which has
legislative protection equivalent to a National Park. In 1982, the Lord Howe Island group
(including Ball's Pyramid) was designated a World Heritage Site under the World Heritage
Convention.

Area 13 sq. km

Population 300 (1974)

Floristics 379 vascular plant taxa, of which 219 are native (Rodd and Pickard,
1983). Of the 48 native fern species, 17 endemic; of the 171 flowering plant species, 57
endemic (Rodd and Pickard, 1983). A further 5 flowering plant taxa below the rank of
species are listed as endemic by Rodd and Pickard (1983). Lord Howe has 4 endemic
genera: *Negria* (Gesneriaceae) and the monotypic palm genera *Howea*, *Hedyscepe* and
Lepidorrhachis. Much of the flora has affinities with those of New Zealand and the Pacific
islands.

Vegetation Lowland evergreen rain forest with *Drypetes lasiogyna* var.
australasica and *Cryptocarya triplinervis*, mostly below 460 m in north; lowland evergreen
rain forest with *Cleistocalyx fullageri* and *Chionanthus quadristamineus*, in south below
530 m; palm forest dominated by *Howea*, mostly below 300 m on coral sandstone and
basalt; palm forest dominated by *Hedyscepe* on Mount Gower and Mount Lidgbird, pure
stands mostly above 610 m, but mixed stands as low as 335 m; *Pandanus* forest mostly in
south; mixed montane forest on summit plateau of Mount Gower above 760 m; scrub
vegetation, mostly in south; small areas of grassland on exposed coasts; tiny areas of
mangroves in sheltered creeks. Less than 20% of the vegetation is disturbed, and less than
10% cleared (Pickard, 1983b).

For vegetation maps and more detailed descriptions of vegetation units see (Recher and
Clark, 1974; Pickard, 1983b).

Checklists and Floras Lord Howe will be included in a forthcoming volume of
the *Flora of Australia* (1981-), cited under Australia. The most recent checklist is:

Rodd, A.N. and Pickard, J. (1983). Census of the vascular flora of Lord Howe Island. *Cunninghamia* 1: 267-280.

See also:

Recher, H.F. and Clark, S.S. (Eds) (1974). *Environmental Survey of Lord Howe Island: A Report to the Lord Howe Island Board*. New South Wales Govt Printer, Sydney. 86 pp. (Includes annotated checklist, endemics indicated; chapter on vegetation; vegetation map, scale 2 inches to one mile, prepared by J. Pickard.)

Information on Threatened Plants

Pickard, J. (1983a). Rare or threatened vascular plants of Lord Howe Island. *Biol. Conserv.* 27: 125-139. (Detailed assessment of native and endemic vascular flora of Lord Howe in terms of distribution, abundance and threat.)

A preliminary list of endemic plants with notes on conservation status is given in Leigh *et al.* (1981), cited under Australia. Latest IUCN statistics: endemic taxa – E:2, V:10, R:58, I:3, nt:2, non-endemics rare or threatened worldwide – V:3 (world categories).

Additional References

Pickard, J. (1973). An annotated botanical bibliography of Lord Howe Island. *Contrib. N.S.W. Nat. Herb.* 4: 470-491.

Pickard, J. (1983b). Vegetation of Lord Howe Island. *Cunninghamia* 1: 133-265.

Recher, H.F. and Clark, S.S. (1974). A biological survey of Lord Howe Island with recommendations for the conservation of the island's wildlife. *Biol. Conserv.* 6: 263-273.

Louisiade Archipelago

About 100 islands 200 km south-east of New Guinea and politically part of Papua New Guinea. The largest islands – Tagula, Misima and Rossel – are volcanic and have fringing reefs; however, the majority of islands are coral formations. Population 12,000 (1971, *Encyclopedia Britannica*, 1974).

The Louisiades have tropical rain forest (see the *Vegetation Map of Malaysia* by van Steenis, 1958, cited in Appendix 1); no figure available for current rate of deforestation.

The flora has affinities with that of New Caledonia. No Flora or checklist has been published. No figure for size of flora or number of endemics. No information on threatened plants.

Luxembourg

Area 2586 sq. km

Population 363,000

Floristics About 1200 native and naturalized vascular species (L. Reichling, 1984, *in litt.*). No endemics (IUCN figure).

Vegetation A largely agricultural landscape. Original vegetation cover almost entirely modified except for small forest fragments on steep rocky slopes, covering c. 33% of country, of which beechwoods comprise 38%, oakwoods 28% and conifer plantations 33% (Reichling, *in litt.*). In the Ardennes, near Echternach, is one of Europe's most ancient forests, of oak, beech and hornbeam, now protected as the Deutsch-Luxemburgischer Naturpark (Muller, 1978).

Checklists and Floras Covered by *Flora Europaea* (Tutin *et al.*, 1964-1980), but plant records not distinguished from those for Belgium. No recent Flora except:

De Langhe, J.-E. *et al.* (1983). *Nouvelle Flore de la Belgique, du Grand-Duché de Luxembourg, du Nord de la France et des régions voisines*, 3rd Ed. Jardin Botanique National de Belgique, Meise. 1016 pp. (Ferns and flowering plants.)

For a plant atlas see:

Rompaey, E. van and Delvosalle, L. (1979). *Atlas de la Flore Belge et Luxembourgeoise, Pteridophtyes et Spermatophytes*, 2nd Ed. Jardin Botanique National de Belgique, Meise.

For a floristic bibliography see Hamann and Wagenitz (1977), cited in Appendix 1, and for floristical accounts see:

Reichling, L. (1955-). Notes floristiques. Observations faîtes dans le Grand-Duché de Luxembourg en 1954. *Bull. Soc. Naturalistes Luxembourg*. Vol. 59 onwards.

A computerized floristic databank is to be developed by the Musée d'Histoire Naturelle de Luxembourg (address below) under the direction of the Centre de Recherche Scientifique sur l'Environnement Natural (Anon, 1985, cited in Appendix 1).

Relevant journal: *Bulletin de la Société des Naturalistes Luxembourgeois*.

Information on Threatened Plants A national threatened plant list is in preparation (Reichling, *in litt.*). Luxembourg is included in the European threatened plant list (Threatened Plants Unit, 1983, cited in Appendix 1); latest IUCN statistics, based upon this work: non-endemics rare or threatened worldwide – E:1, R:1 (world categories).

In 1982 IUCN, under contract to the EEC through the UK Nature Conservancy Council, prepared a report (unpublished), *Threatened Plants, Amphibians and Reptiles, and Mammals (excluding Marine Species and Bats) of the European Economic Community*, which includes data sheets on 2 plant species from Luxembourg, both extinct there, 1 Endangered on a world scale, the other of unknown world status.

Laws Protecting Plants The 1967 Grand-Ducal Order (Règlement grand-ducal du 22 décembre 1967 portant protection de certaines espèces végétales) provides 2 main levels of protection: 18 species and 3 genera are given "strict protection", i.e. picking, uprooting, sale and transport are prohibited; a further 24 species and 1 genus are given more limited protection. For details see:

Reichling, L. (1981). *In Luxembourg Geschützte Pflanzen. Ubersicht sowie Anleitung zum Kennenlernen der in Luxemburg geschützten wildwachsenden Pflanzenarten*, 2nd Ed. Natura (Luxemburger Liga für Natur- und Umweltschutz, Luxembourg. 47 pp. (Outlines the law; describes ecology and threats of plants protected; distribution maps; colour photographs.)

Voluntary Organizations

NATURA, 6 bd. Roosevelt, 2450 Luxembourg.
Société des Naturalistes Luxembourgeois, B.P. 327, 2013 Luxembourg.

Useful Addresses

Direction des Eaux et Forêts, Service Conservation de la Nature, 34 av. de la Porte-
   Neuve, 2227 Luxembourg.
Musée d'Histoire Naturelle de Luxembourg, Marché-aux-Poissons, 2345 Luxembourg.
CITES Management Authority: Ministère de l'Agriculture, de la Viticulture et des Eaux
   et Forêts, Administration des Services Techniques de l'Agriculture, Service de la
   Protection des Végétaux, P.O. Box 1904, 16 Route d'Esch, 1019 Luxembourg.

Additional References

Muller, F-C. (1978). One park, two countries. *Naturopa* 30: 24-25.

Macau

Macau, an overseas province of Portugal, consists of the peninsula of the Chinese district
of Fo Shan and two small islands (Taipa and Coloane), 64 km west of Hong Kong. Area 16
sq. km; population 309,000. The highest point is 190 m, on Coloane.

Subtropical evergreen, monsoon forest greatly modified by fuelwood and timber cutting.
Extensive areas of secondary scrub and grassland. No figure for size of native flora or
number of endemics. No information on threatened plants.

References

Nogueira, A.C. de Sá (1984). *Catálogo descritivo de 380 espécies botânicas da Colónia
   de Macau*, 2nd Ed. Seviços Florestais E Agrícolas de Macau, Julho. 181 pp.
   (Describes 380 taxa, mostly introductions, in Portuguese.)

Macquarie Island

Macquarie Island (54°29'S, 158°58'E) is in the South Pacific Ocean, c. 967 km south-west
of New Zealand. It is a dependency of Tasmania, Australia. Area 11 sq. km. No
permanent population, but the Australian National Antarctic Research Expedition
(ANARE) station is manned by about 20 (1981) temporary staff (Clark and Dingwall,
1985, cited in Appendix 1).

36 native vascular plants, of which 3 endemic (all grasses). The vegetation is mainly
tussock grassland and *Pleurophyllum* herbaceous communities; sedges and rushes occupy
wetter areas; 'feldmark' vegetation, consisting of large areas of open ground with cushion-
forming vascular plants, mosses and lichens, on exposed uplands above 200 m. Grazing by
rabbits has reduced *Poa foliosa*, the dominant tussock grass of coastal slopes. Serious
erosion has stripped surface peat to reveal bedrock in places (Costin and Moore, 1960).

Leigh *et al.* (1981), cited under Australia, provides notes on the conservation status of the
endemics. Latest IUCN statistics: endemic taxa – V:1, R:1, K:1.

References

Cheeseman, T.F. (1919). *The Vascular Flora of Macquarie Island*. Scientific Report of the Australian Antarctic Expedition, 1911-1914, Ser. C (Zoology and Botany) 7(3): 63 pp.

Costin, A.B. and Moore, D.M. (1960). The effects of rabbit grazing on the grasslands of Macquarie Island. *J. Ecol*. 48: 729-732.

Greene, S.W. and Walton, D.W.H. (1975). An annotated check list of the sub-antarctic and antarctic vascular flora. *Polar Record* 17(110): 473-484. (Includes tabular list of native vascular plants, distributions indicated.)

Taylor, B.W. (1955). *The Flora, Vegetation and Soils of Macquarie Island*. ANARE Scientific Report, Ser. B, Vol. 2 (Botany). 92 pp.

Madagascar

Area 594,180 sq. km

Population 9,731,000

Floristics Current estimates of flora between 10,000 and 12,000 species (Rauh, 1979; Guillaumet and Mangenot, 1975); more than 80% specific endemism (Rauh, 1979), but this figure probably too high. Seven endemic families.

East and West Malagasy regions. East region much richer, with almost 75% of Madagascar's species, while the West region has 25% (Perrier de la Bathie, 1936). Floristic affinities principally pantropical, African (especially East African) and Asian.

Vegetation North and east: tropical rain forest; west: dry deciduous forest; south: dry xerophytic scrub (spiny desert). All but about 20% of natural vegetation now destroyed; remainder includes 61,500 sq. km rain forest, 25,500 sq. km mountain sclerophyllous and deciduous forest and 29,000 sq. km dry xerophytic scrub (Chauvet in Richard-Vindard and Battistini, 1972). Most of land surface now uniform grassland with chronic problems of erosion, probably caused by man (Rauh, 1979).

Estimated rate of deforestation for closed broadleaved forest 1500 sq. km/annum out of 103,000 sq. km (FAO/UNEP, 1981). However, Myers (1980, cited in Appendix 1) gives a figure of 26,000 sq. km of eastern moist forest, half of which is disrupted by shifting cultivation which accounts for the destruction of 2000-3000 sq. km/annum.

For vegetation map see White (1983), cited in Appendix 1.

For information on Ile de l'Europa (22°20'S 40°20'E) and Juan de Nova (17°02'S 43°42'E), small islands in the Mozambique Channel, see Bosser (1952), Capuron (1966), and Perrier de la Bathie (1921), below.

Checklists and Floras

Humbert, H. (1936-). *Flore de Madagascar et des Comores*. Muséum National d'Histoire Naturelle, Paris. (c. 80% complete, with 132 families written out of 189, the most significant families outstanding being Leguminosae, Rubiaceae and Gramineae; many of the early volumes now out of date.)

Information on Threatened Plants No published lists of rare or threatened plants; IUCN has records of 468 species and infraspecific taxa believed to be endemic –

E:3, V:11, R:23, I:31, K:375, nt:25. By and large these are succulents, information lacking for other life forms.

One species which occurs in Madagascar (*Catharanthus coriaceus*) is included in *The IUCN Plant Red Data Book* (1978).

Index of potentially threatened plants in cultivation:

Threatened Plants Committee Secretariat (1980). *The Botanic Gardens List of Madagascan Succulents 1980*. Botanic Gardens Conservation Co-ordinating Body Report No. 2. IUCN, Kew. 21 pp. (Lists 235 succulents, most endemic to Madagascar, as in cultivation, from a list of 328 species.)

Laws Protecting Plants No plants or seeds may be exported without permission (strongly enforced, and includes botanical collecting), but permission granted for export of thousands of rare succulents.

Botanic Gardens
Jardin Botanique de la DRST Tsimbazaza, B.P. 4096, Antananarivo.

Useful Addresses WWF is represented by Monsieur B. Vaohita, B.P. 4373, Antananarivo.

CITES Management Authority: Direction des Eaux et Forêts et de la Conservation des Sols, Foiben'ny Rano sy Ala, MPAEF, B.P. 243, Antananarivo.
CITES Scientific Authority: Ministère de la Recherche Scientifique et Technologie pour le Développement, Antananarivo.

Additional References
Bosser, J. (1952). Notes sur la végétation des îles Europa et Juan de Nova. *Naturaliste Malg*. 4: 41-42. (Illus.)
Capuron, R. (1966). Rapport succinct sur la végétation et la flore de l'île Europa. *Mém. Mus. Nat. Hist. Nat., Sér. 2/A (Zool.)* 41: 19-21.
Guillaumet, J.-L. and Mangenot, G. (1975). Aspects de la spéciation dans la flore malgache. In Miège, J. and Stork, A.L. (1975, 1976), cited in Appendix 1, pp. 119-123.
Humbert, H. and Cours Darne, G. (1965). *Carte Internationale du Tapis Végétal et des Conditions Ecologiques: "Madagascar"*. Trav. Sect. Sci. Techn. Inst. Franç. Pondichéry, Hors Sér. 6. 162 pp. (Illus., with coloured vegetation map 1:1,000,000.)
IUCN (1972). *Comptes Rendus de la Conférence Internationale sur la Conservation de la Nature et de ses Ressources à Madagascar, 1970*. Publications UICN Nouvelle Série 36. 239 pp. (See especially papers by M. Keraudren-Aymonin, pp. 145-151, on the Didiereaceae thickets of southern Madagascar, and by R. Melville, pp. 139-142, on the floristic significance of Madagascar.)
Keraudren, M. (1968). Madagascar. In Hedberg, I. and O. (1968), cited in Appendix 1. Pp. 261-265.
Koechlin, J., Guillaumet, J.-L. and Morat, P. (1974). *Flore et Végétation de Madagascar*. Cramer, FL-9490, Vaduz, Liechtenstein. 687 pp. (With line drawings and 188 black and white photographs.)
Leroy, J.-F. (1978). Composition, origin, and affinities of the Madagascan vascular flora. *Ann. Missouri Bot. Gard.* 65(2): 535-589.
Paulian, R. *et al.* (1981). *Madagascar, un Sanctuaire de la Nature*. Paris.
Perrier de la Bathie, H. (1921). Note sur la constitution géologique et la flore des îles Chesterfield, Juan-de-Nova, Europa et Nosy-Trozona. *Bull. Ec. Mad.* 170-176.

Perrier de la Bathie, H. (1936). *Biogéographie des Plantes de Madagascar*. Paris. 156 pp., 40 plates.

Rauh, W. Various articles on the succulent flora of Madagascar published in the journal *Kakteen und andere Sukkulenten* between 1961 and 1970.

Rauh, W. (1973). Über die Zonierung und Differenzierung der Vegetation Madagaskars. *Tropische und Subtropische Pfanzenwelt* 1. 146 pp.

Rauh, W. (1979). Problems of biological conservation in Madagascar. In Bramwell, D. (Ed.), *Plants and Islands*. Academic Press, London. Pp. 405-421.

Richard-Vindard, G. and Battistini, R. (Eds) (1972). *Biogeography and Ecology of Madagascar*. Junk, The Hague. 765 pp. (See especially papers by B. Chauvet, pp. 191-199, on the forests, and by J. Koechlin, pp. 145-190, on the flora and vegetation, with 14 black and white photographs.)

The IUCN Conservation Monitoring Centre, at the request of UNEP, has prepared an extensive Environmental Profile of Madagascar, now in press. This provides a comprehensive review of the biota, plant and animal, of Madagascar and of the physical environment. It includes a chapter on vegetation types and an analysis of forest cover and loss.

Madeira Islands

A volcanic archipelago in the North Atlantic Ocean, belonging to Portugal. Comprises Madeira itself and the Desertas to the south-east (uninhabited) and Porto Santo (inhabited) to the north-east. Madeira, itself, is a very precipitous, wooded, volcanic island c. 58 x 23 km. Its backbone is a serrated mountain range reaching the rugged peak of Pico Ruivo (1861 m) and, to the west, the high grassy plateau of Paul da Serra. Deep rugged ravines run to the coast.

Area 796 sq. km

Population 265,100 (1979 estimate, *Times Atlas*, 1983)

Floristics About 760 species of native ferns and flowering plants (Vieira, 1974), of which 131 are endemic (IUCN figures); also c. 380 introduced plants, mostly subtropical and many extensively naturalized.

Vegetation When discovered in 1419, most of the island was covered with forest, now greatly reduced. Sjögren (1972) distinguishes 4 vegetation zones: coastal vegetation of low shrubs, herbs and succulents (e.g. *Aeonium*), much now replaced by cultivated land; laurel forest, a subtropical evergreen cloud forest mainly of *Ilex* and Lauraceae, rich in endemics and with a large ground flora, occurring between 1300 and 1850 m; a transitional zone (700-1250 m) between the previous 2 zones; and above the laurel forest *Erica* scrub.

Checklists and Floras Covered in the *Flora of Macaronesia* checklist (Hansen and Sunding, 1979, cited in Appendix 1). Below is a modern checklist and 2 Floras, both very old and one of them – Lowe – incomplete:

Hansen, A. (1969). Checklist of vascular plants of the Archipelago of Madeira. *Bol. Museu Municipal Funchal* 24: 1-62. (Annotated checklist with extensive bibliography.)

Lowe, R.Th. (1857-1872). *A Manual Flora of Madeira and the Adjacent Islands of Porto Santo and the Desertas*. London.

Menezes, C.A. (1914). *Flora do Archipelago da Madeira*. Funchal. 282 pp. (In Portuguese.)

The British Museum (Natural History), London, are preparing a Flora of Madeira. Hansen has updated his 1969 checklist in a series of papers in *Bocagiana* (Museu Municipal do Funchal, Madeira), namely No. 25 (18 pp., 1970), No. 27 (14 pp., 1971), No. 32 (13 pp., 1973) and No. 36 (37 pp., 1974). See also:

Hansen, A. (1976). A botanical bibliography of the archipelago of Madeira. *Bol. Museu Municipal Funchal* 30: 26-45.

Field-guides

Christensen, T.B., Dalgaard, V. and Hamann, O. (1970). *Oversigt over Madeiras Flora*. Kobenhavens Universitets. 167 pp. (Includes keys; in Danish.)

Delagação de Turismo da Madeira (1976). *Plantas e Flores/Plantes et Fleurs/Plants and Flowers/Pflanzen und Blumen: Madeira*. 151 pp. (Colour photographs of selected species both wild and cultivated.)

Pinto da Silva, A.R. (1975). L'état actuel des connaissances floristiques et taxonomiques du Portugal, de Madère et des Açores, en ce qui concerne les plantes vasculaires. In CNRS, 1975, cited in Appendix 1. Pp. 19-28.

Ramirez (1953). *Flora da Ilha da Madeira, Pteridofitas*. (Not seen.)

Vieira, R. (1974). *Album floristico da Madeira*. Funchal. (Colour photographs of 124 plants, both wild and cultivated; English version available as *Flowers of Madeira*.)

Information on Threatened Plants The only known list is that produced by IUCN Threatened Plants Committee Secretariat (1980) for North Africa and the Middle East, cited in Appendix 1. Latest IUCN statistics, based upon this work: endemics – E:17, V:30, R:39, K:22, nt:23; non-endemics rare or threatened worldwide – E:2, V:17, R:5 (world categories).

Botanic Gardens

Jardím Botânico da Madeira, Quinta do Bom Sucesso-Caminho do Meio, 9000 Funchal.

Jardím Botânico da Ribeiro Frio (maintained by Serviços Florestais, Departamento de Agricultura e Pescas, Avenida do Mar, Funchal).

Useful Addresses

Museu Municipal do Funchal, 9000 Funchal.

Additional References

Bramwell, D., Montelongo, V., Navarro, B. and Ortega, J. (1982). *Informe Sobre la Conservacion de los Bosques y la Flora de la Isla de Madeira*. Report to International Dendrology Society and IUCN, by staff of the Jardín Botánico "Viera y Clavijo", outlining proposals for a protected areas system on Madeira. (In Spanish and Portuguese.)

Bramwell, D. and Synge, H. (1983). A conservation project in Madeira. *Int. Dendrol. Soc. Yb.*, 1982: 73-74. (Summary of Bramwell *et al.*, 1982.)

Malato-Beliz, J. (1977). Considerações sobre a protecção da flora e da vegetação na Madeira. *Natureza e Paisagem* 3: 1-11.

Sjögren, E. (1972). Vascular plant communities of Madeira. *Bol. Museu Municipal Funchal* 26 (114): 45-125.

Sjögren, E. (1973). Conservation of natural plant communities on Madeira and in the Azores. In *Proc. 1 Intern. Congress pro Flora Macaronesica*. Pp. 148-153. (Not seen.)

Tavares, C.N. (1965). *Ilha da Madeira. O meio e a flora*. Lisboa. 174 pp. (In Portuguese.)

Malawi

Area 94,081 sq. km

Population 6,788,000

Floristics c. 3600 species (quoted in Lebrun, 1960, cited in Appendix 1). Endemism generally low, but highest in the mountain areas; Brenan (1978, cited in Appendix 1) estimates 69 endemic species from a sample of *Flora Zambesiaca*. Wild (1964) lists 30 species apparently endemic to Mt Mulanje.

Flora principally Zambezian but with a few islands of Afromontane flora, especially the Misuku forests and Nyika and Viphya Plateaux in the north, and Mt Mulanje and Zomba Plateau in the south.

Vegetation Predominantly more or less open *Brachystegia-Julbernardia* (Miombo) woodland; also considerable areas of Zambezian woodland dominated by species of *Combretum*, *Acacia* and *Piliostigma* around Lilongwe and south of Lake Malawi. Afromontane communities occur at higher altitudes, including small patches of evergreen forest and large expanses of short grassland. Lowland forest occurs on the shores of the northern part of Lake Malawi, on the lower slopes of Mt Mulanje and on the Malawi Hills where they rise from the Shire Valley.

For vegetation maps see Wild and Barbosa (1967, 1968), and White (1983), both cited in Appendix 1.

Checklists and Floras Malawi is included in the incomplete *Flora Zambesiaca* and in *Trees of Central Africa* (Coates Palgrave *et al.*, 1957), both cited in Appendix 1.

Binns, B. (1968). *A First Check List of the Herbaceous Flora of Malawi*. Govt Printer, Zomba. 113 pp.
Burtt Davy, J. and Hoyle, A.C. (Eds) (1958). *Check Lists of the Trees and Shrubs of the Nyasaland Protectorate*, 2nd Ed., revised by P. Topham, 1958. Govt Printer, Zomba. 137 pp. (1st Ed. 1936 as *Check-Lists of the Forest Trees and Shrubs of the British Empire. No. 2: Nyasaland Protectorate*, Oxford.)

Field-guides
Kitchin, A.M. and Pullinger, J.S. (1982). *Trees of Malawi, with Some Shrubs and Climbers*. 229 pp. (Colour paintings of 108 species, mostly by J.S. Pullinger; text by A.M. Kitchin.)
Moriarty, A. (1975). *Wild Flowers of Malawi*. Purnell, Cape Town. 166 pp.

Information on Threatened Plants No published lists of rare or threatened plants; IUCN has records of c. 130 species and infraspecific taxa believed to be endemic, of which roughly half are known to be rare or threatened. Of relevance:

Chapman, J.D. (1981). Conservation of vegetation and its constituent species in Malawi. *Nyala* 6(2): 125-132.

Voluntary Organizations

National Fauna Preservation Society of Malawi, c/o Museums of Malawi, P.O. Box 30360, Blantyre. (Publishes the journal *Nyala*.)

Society of Malawi Historic and Scientific, P.O. Box 125, Blantyre. (Publishes *The Society of Malawi Journal*.)

Useful Addresses

Dept of Forestry, Ministry of Forestry and Natural Resources, Lilongwe 3.

National Herbarium, Chancellor College, P.O. Box 280, Zomba.

CITES Management and Scientific Authority: The Chief Game Warden, Dept of National Parks and Wildlife, P.O. Box 30131, Lilongwe 3.

Additional References

Brass, L.J. (1953). Vegetation of Nyasaland. Report on the Vernay Nyasaland expedition of 1946. *Mem. New York Bot. Gard.* 8: 161-190.

Chapman, J.D. (1962). *The Vegetation of the Mlanje Mountains, Nyasaland*. Govt Printer, Zomba. 78 pp. (With 25 black and white photographs.)

Chapman, J.D. (1968). Malawi. In Hedberg, I. and O. (1968), cited in Appendix 1. Pp. 215-224.

Chapman, J.D. and White, F. (1970). *The Evergreen Forests of Malawi*. Commonwealth Forestry Institute, Univ. of Oxford. 190 pp. (Includes a useful ecological and phytogeographical bibliography; 60 black and white photographs.)

Werger, M.J.A. (1978), cited in Appendix 1. Citation includes list of relevant chapters.

Wild, H. (1964). The endemic species of the Chimanimani Mountains and their significance. *Kirkia* 4: 125-157.

Malaysia

Area 332,669 sq. km

Peninsular Malaysia: 131,587 sq. km; Sabah: 76,115 sq. km; Sarawak: 124,967 sq. km

Population 15,204,000, of which c. 12,000,000 in Peninsular Malaysia

Floristics Peninsular Malaysia has c. 8000 flowering plant species in 1500 genera; c. 500 species of ferns (Keng, 1983). The flora of the Malay Peninsula comprises mainly Malesian elements, with continental Asiatic and some Australian elements at low and medium altitudes. Floristic affinities are discussed by Keng (1970). No figure for number of species in Sabah or Sarawak, but Borneo (whole island) has c. 10,000-11,000 vascular plant species, based on Merrill (1921).

Vegetation Tropical evergreen rain forest is the natural vegetation of most of Malaysia: lowland dipterocarp forest up to 300 m, hill dipterocarp forest at 300-1300 m, montane rain forest above; semi-evergreen rain forest occurs in the far north-west of Peninsular Malaysia; karst limestone supporting rich endemic flora covers 260 sq. km in Peninsular Malaysia (Chin, 1977-); limestone forests at low elevations south of Kuching and at Niah, and at high elevations around Gunung Mulu in Sarawak.

Peninsular Malaysia Lowland forests have been heavily logged; most hill dipterocarp forests selectively logged; only tiny patches of heath forest remaining on east coast; freshwater swamp-forest and c. 1136 sq. km of mangrove forest remaining, mostly in south (Corner, 1978). Estimated rate of deforestation of closed broadleaved forest 900 sq. km/annum out of a total of 75,780 sq. km (FAO/UNEP, 1981). Davison (1982) calculated the area of forest in 1980 to be 53,420 sq. km, of which primary rain forest occupied 27,925 sq. km.

Sabah Lowland and hill dipterocarp forests comprise c. 54% of the total forest cover; montane forests, 14% (FAO/UNEP, 1981). Most remaining forests are 'productive' or 'potentially productive' dipterocarp forests (Myers, 1980, cited in Appendix 1). Upper montane forest (1850-3200 m), subalpine rain forest (3200-4100 m) and alpine scrub occur on Mt Kinabalu. Mangrove forests cover 3500 sq. km (FAO/UNEP, 1981); peat swamp and mangrove forests in Klias Peninsula now being logged. Estimated rate of deforestation of closed broadleaved forest 760 sq. km/annum out of a total of 49,970 sq. km (FAO/UNEP, 1981). According to estimates by the Government of Malaysia (quoted in Myers, 1980), there were 61,488 sq. km still forested in 1977, of which 31,000-34,521 sq. km were undisturbed.

Sarawak Mixed dipterocarp forests cover 78.6% of the forest area; peat swamp forests about 15%; heath forests (kerangas) 3.9%; mangroves 1.8% (FAO/UNEP, 1981). Gunung Mulu National Park contains most of the major vegetation types of Sarawak, including high elevation limestone forest. Estimated rate of deforestation of closed broadleaved forest 890 sq. km/annum out of a total of 84,200 sq. km (FAO/UNEP, 1981). According to estimates by the Government of Malaysia (quoted in Myers, 1980), 97,087 sq. km were still forested in 1977, of which 55,687-62,661 sq. km were undisturbed. Malaysia is included on the *Vegetation Map of Malaysia* (van Steenis, 1958) and on the vegetation map of Malesia (Whitmore, 1984), both covering the *Flora Malesiana* region at scale 1:5,000,000 and cited in Appendix 1. See also:

Thomas, P., Lo, F.K.C. and Hepburn, A.J. (1976). *The Land Capability Classification of Sabah*, 4 vols. Land Resources Study 25. Ministry of Overseas Development, Surbiton, U.K. (Land use and evaluation, includes maps of land capability classification at 1:250,000. Vol. 1 - Tawau Residency; 2 - Sandakan Residency; 3 - West Coast and Kudat Residencies; 4 - Interior Residency and Labuan.)

Wyatt-Smith, J. (1964). A preliminary vegetation map of Malaya with descriptions of the vegetation types. *J. Trop. Geog.* 18: 200-213. (Includes vegetation map of Peninsular Malaysia with notes on vegetation types.)

Checklists and Floras Malaysia is included in the very detailed but incomplete *Flora Malesiana* (1948-), cited in Appendix 1.

Peninsular Malaysia is covered by:

A Revised Flora of Malaya, 3 vols. 1 - *Orchids of Malaya*, by R.E. Holttum. 3rd Ed., 1964. 759 pp. 2 - *Ferns of Malaya* by R.E. Holttum, 2nd Ed., 1966. 653 pp. 3 - *Grasses of Malaya*, by H.B. Gilliland (1971). 319 pp. Govt Printer, Singapore.

Tree Flora of Malaya. Vols 1 and 2 (1972) edited by T.C. Whitmore. Vol. 3 (1978) and 4 (in press) edited by F.S.P. Ng. Longman, Kuala Lumpur and London. (Excludes Dipterocarpaceae, but otherwise complete; keys, descriptions, line drawings of selected taxa. For dipterocarps see *Flora Malesiana* 9(2), 1982.)

Other accounts include:

Anderson, J.A.R. (1980). *A Checklist of the Trees of Sarawak*. Forest Dept, Sarawak. 364 pp. (Over 2500 species enumerated.)

Browne, F.G. (1955). *Forest Trees of Sarawak and Brunei and Their Products*. Govt

Printer, Kuching. 369 pp. (Descriptions of timber trees with notes on distribution and wood properties.)

Cockburn, P.F. (1976, 1980). *Trees of Sabah*, 2 vols so far. Forest Dept, Kuching.

Dransfield, J. (1979). *A Manual of the Rattans of the Malay Peninsula*. Malayan Forestry Records no. 29. Malaysia Forest Dept. 270 pp. (Keys, descriptions, drawings; checklist of 104 species.)

Dransfield, J. (1984). *The Rattans of Sabah*. Forest Dept, Sabah. 182 pp. (Keys, descriptions, drawings; checklist of 82 taxa.)

Fox, J.E.D. (1970). *Preferred Check-list of Sabah Trees*. Sabah Forest Record no. 7. Borneo Literature Bureau, Kuching. 65 pp.

Merrill, E.D. (1921). *A Bibliographic Enumeration of Bornean Plants*. Fraser and Neave, Singapore. 637 pp. (Systematic enumeration with notes on distribution; introduction covers vegetation, history of botanical investigation.)

Ridley, H.N. (1922-1925). *The Flora of the Malay Peninsula*, 5 vols. Reeve, London. (Reprinted 1968; Asher, Amsterdam.)

Whitmore, T.C. (1973). *Palms of Malaya*. Oxford Univ. Press, London. 132 pp.

Wyatt-Smith, J. (1952). *Pocket Check List of Timber Trees*. Malayan Forest Records no. 17. Forest Dept, Peninsular Malaysia. (3rd Ed., 1979, by K.M. Kochummen.)

Field-guides

Corner, E.J.H. (1952). *Wayside Trees of Malaya*, 2nd Ed., 2 vols. Govt Printing Office, Singapore.

Henderson, M.R. (1949, 1954). *Malayan Wild Flowers*, 2 vols. Malayan Nature Soc., Kuala Lumpur. (1 – dicotyledons; 2 – monocotyledons; keys, descriptions of a selection of wildflowers.)

Kurata, S. (1976). *Nepenthes of Kinabalu*. Sabah National Parks, Kota Kinanbalu. 80 pp.

Shivas, R. (1984). *Pitcher Plants of Peninsular Malaysia and Singapore*. Maruzen Asia, Singapore. 58 pp.

Information on Threatened Plants No national list of threatened plants has been published. 4 species are included in *The IUCN Plant Red Data Book* (1978). A preliminary list of endemics from limestone areas, prepared by S.C. Chin in 1984, includes – E:6, V:2, R:72, I:41, nt:30, K:13. IUCN also has a full list of palms, some of which have conservation categories.

Kiew, R. (1983-). Portraits of threatened plants. *Malayan Naturalist* 37(1): 6-7; 37(2): 6-7; 37(4): 4-6; 38(1): 9-10; 38(2): 6. (Data sheets on *Maxburretia rupicola, Ilex praetermissa, Didymocarpus primulinus, Maclurodendron magnificum, Melicope suberosa, Musa gracilis* and *Maingaya malayana*.)

Ng, F.S.P. and Low, C.M. (1982). *Check List of Endemic Trees of the Malay Peninsula*. Forest Research Institute, Kepong. 94 pp. (Lists 654 trees endemic to the Malay peninsula of which 343 'endangered', based on numbers of herbarium specimens.)

Rao, A.N., Keng, H. and Wee, Y.C. (1983). Problems in conservation of plant resources in South East Asia. In Jain, S.K. and Mehra, K.L. (Eds), *Conservation of Tropical Plant Resources*. Botanical Survey of India, Howrah. Pp. 181-204. (Includes list of 90 endemic taxa threatened in Malaysia; useful bibliography.)

Voluntary Organizations

Malayan Nature Society, P.O. Box 10750, Kuala Lumpur, Peninsular Malaysia.

Sabah Society, P.O. Box 547, Kota Kinabalu, Sabah.

WWF-Malaysia, Wisma Damansara, Jalan Semantan, P.O. Box 10769, Kuala Lumpur, Peninsular Malaysia.

Botanic Gardens
Botanic Gardens, Penang, Peninsular Malaysia.

Forest Research Centre (Arboretum and Herbarium), P.O. Box 1407, Sandakan, Sabah.

Forest Research Institute (Arboretum and Herbarium), Kepong, Selangor, Peninsular Malaysia.

Rimba Ilmu Botanic Garden, Department of Botany, University of Malaya, Lembah Pantai, Kuala Lumpur, Peninsular Malaysia.

Sabah Orchid Centre, c/o Cocoa Research Station, P.O. Box 197, Tenom, Sabah.

Semangoh Arboretum, Sarawak Forest Department, Kuching, Sarawak.

Useful Addresses
Sarawak Herbarium, Forest and Department Headquarters, Jalan Badruddin, Kuching, Sarawak.

CITES Management Authority: Wildlife and National Parks, Pejabat-Pejabat Kerajaan, Blok K-19, Jalan Duta, Kuala Lumpur 11-04, Peninsular Malaysia.

CITES Scientific Authority: Secretary General, Ministry of Science, Technology and the Environment, Tingkat 14, Bangunan Oriental Plaza, Jalan Ramli, Kuala Lumpur 04-01, Peninsular Malaysia.

Additional References

Anderson, J.A.R. (1963). The flora of the peat swamp forest of Sarawak and Brunei, including a catalogue of all recorded species of flowering plants, ferns and fern allies. *Gard. Bull. Singapore* 20: 131-228.

Brunig, E.F. (1974). *Ecological Studies in the Kerangas Forests of Sarawak and Brunei.* Borneo Literature Bureau, Kuching. 237 pp.

Burkill, I.H. (1966). *A Dictionary of the Economic Products of the Malay Peninsula,* 2nd Ed., 2 vols. Ministry of Agriculture, Kuala Lumpur. (2432 species, notes on origin, uses, vernacular names.)

Chai, P.K. and Choo, N.C. (1983). Conservation of forest genetic resources in Malaysia with special reference to Sarawak. In Jain, S.K. and Mehra, K.L. (Eds), *Conservation of Tropical Plant Resources.* Botanical Survey of India, Howrah. Pp. 39-47.

Chin, S.C. (1977-). The limestone hill flora of Malaya. *Gard. Bull. Singapore* 30: 165-219; 32: 64-203; 35: 137-190; 36: 31-91. (About 1216 vascular species found on limestone, including 261 endemics; keys, annotated checklist.)

Corner, E.J.H. (1978). *The Freshwater Swamp-forest of South Johore and Singapore.* Gardens Bulletin Supplement 1, Singapore. 266 pp. (Ecology; species lists.)

Davison, G.W.H. (1982). How much forest is there? *Malayan Naturalist* 35: 11-12.

Holttum, R.E. (1954). *Plant Life in Malaya.* Longmans and Green, London. 254 pp. (Useful introduction to the flora.)

Jacobs, M. (1974). Botanical panorama of the Malesian archipelago (vascular plants). In Unesco, *Natural Resources of Humid Tropical Asia.* Natural Resources Research 12. Unesco, Paris. Pp. 263-294.

Keng, H. (1970). Size and affinities of the flora of the Malay Peninsula. *J. Trop. Geog.* 31: 43-56.

Keng, H. (1983). *Orders and Families of Malayan Seed Plants.* Singapore Univ. Press. 441 pp. (Revised edition; keys and brief systematic accounts of 41 orders and 177 families in the Malayan flora.)

Kiew, R. (1983). Conservation of Malaysian plant species. *Malayan Naturalist* 37(1): 2-5. (Conservation problems and priorities.)

Lee, D. (1980). *The Sinking Ark: Environmental Problems in Malaysia and Southeast Asia*. Heinemann, Kuala Lumpur. 85 pp.

Luping, D.M., Wen, C. and Dingley, E.R. (1978). *Kinabalu: Summit of Borneo*. Sabah Society Monograph, Kota Kinabalu. 486 pp. (Covers flora, vegetation, fauna, geology, history of exploitation.)

Shuttleworth, C. (1981). *Malaysia's Green and Timeless World*. Heinemann, Kuala Lumpur. 221 pp. (Covers flora and fauna).

Watson, J.G. (1928). *Mangrove Forests of the Malay Peninsula*. Malayan Forest Records no. 6. Fed. Malay States Govt. 275 pp.

Maldives

The Maldives (298 sq. km) comprise 1201 islands, grouped into 19 coral atolls, extending north-south for about 885 km south-west of Sri Lanka, between latitudes 7°N and 3°S, and longitudes 73-74°E. 202 islands are permanently inhabited; the total population is 173,000.

The islands are mostly below 1.5 m above sea-level, and covered by coconut palms, grassland or scrub. Little native vegetation remains undisturbed.

583 vascular plant species (including cultivated plants). According to Adams (1983), there are 260 "native or naturalized" species, of which about half are likely to have been intentionally introduced. The only recorded endemics are 5 species of *Pandanus* (St John, 1961). There are local restrictions on the cutting of any living plant for firewood, except *Scaevola sericea* (C.D. Adams, 1984, *in litt.*).

References

Adams, C.D. (1983). Report to the Government of the Maldive Islands on Flora Identification. FAO Project RAS 79/123, Rome. 41 pp.

Fosberg, F.R. (1957). The Maldive Islands, Indian Ocean. *Atoll Res. Bull.* 58. 37 pp. (Includes checklist of 4 ferns, one cycad, 322 angiosperms, many of which are introductions.)

Fosberg, F.R., Groves, E.W. and Sigee, D.C. (1966). List of Addu vascular plants. In Stoddart, D.R. (Ed.), Reef studies at Addu Atoll, Maldive Islands. Preliminary results of an expedition to Addu Atoll in 1964. *Atoll Res. Bull.* 116: 75-92. (Checklist of 5 ferns, 2 gymnosperms, 135 angiosperms.)

St John, H. (1961). Revision of the Genus *Pandanus* Stickman, Part 5. *Pandanus* of the Maldive Islands and Seychelles Islands, Indian Ocean. *Pacific Science* 15: 328-346.

Stutz, L.-C. (1982). Herborisation 1981 aux îles Maldives. *Candollea* 37: 599-631. (Lists 123 taxa in Male, Bandos and Thulaagiri; notes on uses, and additional reports on 31 mainly introduced shrubs.)

Willis, J.C. and Gardiner, J.S. (1901). The botany of the Maldive Islands. *Annals Royal Botanic Gardens Peradeniya* 1: 45-164. (Includes annotated list of 359 species recorded from Chagos Archipelago, Laccadives and Maldives; 284 recorded on Maldives, of which c. 90 are native.)

Mali

Area 1,240,142 sq. km

Population 7,825,000

Floristics 1600 species (J.-P. Lebrun, 1984, pers. comm.), with 11 endemic species (Brenan, 1978, cited in Appendix 1). Floristically poor for its enormous size.

Floristic affinities Saharan, Sahelian and Sudanian in north, centre and south of country respectively.

Vegetation Northern half of country desert and semi-desert with little or no perennial vegetation. Southwards: east-west bands of *Acacia* wooded grassland and deciduous bushland, Sudanian woodland without characteristic dominants, and Sudanian woodland with *Isoberlinia*. Also, a large area of swamp grassland with semi-aquatic vegetation in centre of country.

For vegetation map see White (1983), cited in Appendix 1.

Checklists and Floras Mali south of c. 18°N is included in the *Flora of West Tropical Africa*. Mali north of c. 16°N is included in *Flore du Sahara* (Ozenda, 1977), and in the computerized *Atlas der Pflanzenwelt des Nordafrikanischen Trockenraumes* (Frankenberg and Klaus, 1980); these are all cited in Appendix 1.

Boudet, G. and Lebrun, J.-P. (in prep.). *Catalogues des plantes vasculaires du Mali.* To be published by the Institut d'Elevage et de Médecine Vétérinaire des Pays Tropicaux, Maisons-Alfort.

Information on Threatened Plants No published lists of rare or threatened plants; IUCN has records of 11 species and infraspecific taxa believed to be endemic, including V:2, R:3.

Jaeger, P. (1956). Contribution à l'étude des forêts reliques du Soudan occidental. *Bull. IFAN* 18A: 993-1053. (Includes small map of distribution of threatened timber tree *Gilletiodendron glandulosum*.)

Additional References
Jaeger, P. (1968). Mali. In Hedberg, I. and O. (1968), cited in Appendix 1. Pp. 51-53.
Jaeger, P. and Winkoun, D. (1962). Premier contact avec la flore et la végétation du plateau de Bandiagara. *Bull. IFAN* 24A: 69-111.
Rossetti, C. (1962). Observations sur la Végétation au Mali Oriental (1959). Projet Pèlerin, Rapp. No. UNSF/DL/ES/4, FAO, Rome. 68 pp.

Malta

The Republic of Malta includes Malta, Gozo, Comino and 2 uninhabited islands, in the central Mediterranean.

Area 316 sq. km

Population 380,000

Floristics 900 native vascular species (E. Lanfranco, 1984, pers. comm.); 5 endemics (IUCN figures). A Mediterranean flora.

Vegetation Little natural vegetation due to agriculture, building construction and tourism. Most remaining vegetation is semi-natural and confined to inaccessible coastal cliffs, e.g. fragments of garigue and maquis with remnants of Holm Oak (*Quercus ilex*) woodland, now reduced to a few individuals. Inland, on the jagged coralline limestone plateau in the north and west, there is a thin scattered scrub of garigue, with occasional trees in the valleys. Elsewhere garigue is the dominant vegetation cover with *Euphorbia*, *Thymus* and *Teucrium* spp. Little maquis remains.

Priority areas for protection are as follows: "Wardija Ridge, the pool and sand dunes at Ghadira and in Gozo, the dunes at Ramla bay and the coralline plateau and valley between Ta' Cenc and Mgarr ix-Xini. These together with the Wieds contain much of what is left of the semi-natural vegetation of the Islands" (Haslam *et al.*, 1977).

Checklists and Floras Malta is covered by the completed *Flora Europaea* (Tutin *et al.*, 1964-1980), cited in Appendix 1, but plant records are not distinguished from those for Sicily. Malta is also being covered by the *Med-Checklist* (cited in Appendix 1). National Floras:

Borg, J. (1927). *Descriptive Flora of the Maltese Islands*. Government Printing Office, Malta. 846 pp. (Extensive introductory text describes geology, climate, vegetation and botanical exploration; reprinted 1976.)

Haslam, S.M., Sell, P.D. and Wolseley, P.A. (1977). *A Flora of the Maltese Islands*. Malta University Press, Msida. 560 pp. (Introduction outlines history of floristic studies in Malta, plant communities and habitats; line drawings.)

Relevant journal, which includes conservation articles: *The Maltese Naturalist*, Society for the Study and Conservation of Nature (SSCN), address below.

Field-guides
Lanfranco, G.G. (1977). *Field Guide to the Wild Flowers of Malta*, 2nd Ed. Progress Press, Malta. 83 pp. (Illus.)

Information on Threatened Plants No national plant Red Data Book, but see:

Lanfranco, E. (1976). Report on the present situation of the Maltese flora. *The Maltese Naturalist* 2(3): 69-80. (Describes threats to the flora; lists over 300 extinct and endangered taxa in 2 appendices; line drawings of over 50 species.)

Included in the European threatened plant list (Threatened Plants Unit, 1983, cited in Appendix 1); latest IUCN statistics, based upon this work: endemic taxa – V:1, R:2, I:1, nt:1.

Laws Protecting Plants The Antiquities Act of 1933 (Article 3) provides protection for historical trees and those over 200 years old. This includes *Quercus ilex* as well as several cultivated trees. Legislation for the protection of Maltese wildlife has been prepared by the Environment Protection Centre (address below) and now awaits finalization.

Voluntary Organizations
Society for the Study and Conservation of Nature (SSCN), P.O. Box 459, Valetta. (Formerly the Natural History Society of Malta.)

Botanic Gardens

Argotti Botanic Gardens, Floriana.

Useful Addresses

Environment Protection Centre (EPC), Ministry of Health and Environment, Bighi, Malta.

Additional References

Kramer, K.U. *et al.* (1972). Floristic and cytotaxonomic notes on the flora of the Maltese Islands. *Acta Bot. Neerl.* 21(1): 54-66.

Lanfranco, E. (1980). A survey of natural sites in Gozo and the updating of flora and fauna lists. *Gozo Agricultural Study*. Working Paper no. III/i. Unesco and University of Malta. (Not seen.)

Lanfranco, E. (1981). Suggestions on the conservation of the unique flora associated with the Gozo Citadel. *Soc. Stud. Cons. Nat.* 3 pp.

Lanfranco, E. (1982). Maltese succulents and conservation. *Kakti u Sukkulenti Ohra* 24: 13-15.

Mariana Islands

14 islands to the north of Guam, in the Pacific Ocean, and extending in a 925 km arc between latitudes 12-23°N and longitudes 145-150°E. The northern islands are volcanic, some still active; Tinian (102 sq. km) and Rota (86 sq. km) in the south are raised limestone terraces overlying extinct volcanoes. The Marianas are part of the United Nations Trust Territory of the Pacific Islands administered by the United States, but currently form the Commonwealth of the Northern Mariana Islands (Ballendorf, 1984).

Area 477 sq. km

Population 16,780 (1980 census)

Floristics No overall figure for the size of the flora, but 478 dicotyledon taxa, including introductions. Of the 221 native dicotyledons, 78 are endemic (Fosberg, Sachet and Oliver, 1979, cited in Appendix 1). The only native gymnosperm is *Cycas circinalis*, which is non-endemic. There are 64 native fern taxa, of which 3 are endemic (Fosberg, Sachet and Oliver, 1982, cited in Appendix 1). The flora is mostly related to that of S.E. Asia, Melanesia and New Guinea.

Vegetation Pioneer stands of *Casuarina*, broadleaved evergreen thickets, mixed scrub forest, with some *Miscanthus* and *Nephrolepis* herbaceous communities on the northern islands. Broadleaved evergreen forest on old lava flows; *Miscanthus* and tree ferns on ash slopes of those northern islands with dormant volcanoes (Douglas, 1969, cited in Appendix 1). Tinian has mostly secondary forests; Rota has some closed evergreen and limestone forests (Fosberg, 1973, cited in Appendix 1). Small areas of cloud forest occur on the volcanic islands of Saipan, Agrihan, Alamagan and Anatahan (Dahl, 1980, cited in Appendix 1). The lower slopes on many islands have been cleared for cultivation.

Checklists and Floras The Marianas are included in *Flora Micronesica* (Kanehira, 1933), the regional checklists of Fosberg, Sachet and Oliver (1979, 1982), cited in Appendix 1, and will be covered by the *Flora of Micronesia* (1975-), cited in Appendix 1. Separate lists include:

Fosberg, F.R., Falanruw, M.V.C. and Sachet, M.-H. (1975). Vascular flora of the Northern Marianas Islands. *Smithsonian Contrib. Bot.* 22. 45 pp. (Annotated checklist with geographical and ecological data.)

Fosberg, F.R., Falanruw, M.V.C. and Sachet, M.-H. (1977). Additional records of vascular plants from the Northern Mariana Islands. *Micronesica* 13(1): 27-31.

Information on Threatened Plants *Heritiera longipetiolata* and *Serianthes nelsonii* are included in *The IUCN Plant Red Data Book* (1978).

Additional References

Ballendorf, D.A. (1984). American social, political and economic interests in Micronesia. *Ambio* 13(5-6): 294-295.

Marion and Prince Edward Islands

The volcanic islands of Marion and Prince Edward in the Southern Ocean are 22 km apart; the nearest continent is Africa 1800 km NNW. Marion Island (46°55'S, 37°45'E) has a central highland plateau rising to over 1200 m, the top of which is permanently covered with ice. The area of Marion is 300 sq. km; that of Prince Edward is 90 sq. km. There is a permanently manned weather and scientific station on Marion, with up to 12 persons. In 1948 South Africa proclaimed sovereignty of the islands.

Marion has 22 native and 13 introduced vascular species; Prince Edward has 21 native and 1 introduced vascular species. One endemic (*Elaphoglossum randii*). (Gremmen, 1982). Cryptogams show quite a high degree of endemism. There are no trees or shrubs. The vegetation of the coastal areas consists of herbaceous communities dominated by salt-resistant species. Otherwise the islands are mostly covered by various sorts of tundra-type mire in which the important peat-forming plants are bryophytes, closed communities of tussock-forming grasses, cushion-forming flowering plants, and communities with large-leaved perennial species.

References

Greene, S.W. and Walton, D.W.H. (1975). An annotated check list of the sub-antarctic and antarctic vascular flora. *Polar Record* 17(110): 473-484.

Gremmen, N.J.M. (1982). *The Vegetation of the Subantarctic Islands Marion and Prince Edward.* (Geobotany 3.) Junk, The Hague. 149 pp. (With tables of the indigenous vascular plants and their distributions.)

van Zinderen Bakker Sr, E.M., Winterbottom, J.M. and Dyer, R.A. (Eds) (1971). *Marion and Prince Edward Islands: Report on the South African Biological and Geographical Expedition, 1965-1966.* Balkema, Cape Town. 427 pp. (Includes numerous papers on the islands; see especially that of B.J. Huntley, pp. 98-160, on the vegetation.)

Marquesas Islands

The Marquesas are an isolated group of 14 volcanic islands in the central Pacific Ocean, between latitudes 7°50' and 10°35'S, and longitudes 138°25' and 140°50'W. Their nearest neighbours are the atolls of the Tuamotu Archipelago, 483 km to the south. Apart from Ua Pu, each island appears to consist of half an original volcanic peak. The highest point is 1260 m, on Hiva Oa. The Marquesas form an administrative division of French Polynesia.

Area 1275 sq. km

Population 800, most on Tahuata and Fatu Hiva (Douglas, 1969, cited in Appendix 1).

Floristics 76 ferns and 171 native angiosperm taxa (*Flora of Southeastern Polynesia*, 1931-1935, cited in Appendix 1); 103 endemic vascular plant taxa (IUCN figures). 24 species are found only on Nuku Hiva, 13 are confined to Hiva Oa, 6 to Fatu Hiva, 5 to Ua Pu, 2 to Eiao, and one confined to each of Ua Huka and Mohotani (Melville, 1970). *Lebronnecia* and *Cyrtandroidea* are monotypic endemic genera.

Vegetation The natural vegetation included upland rain forest, with *Metrosideros*, *Weinmannia* and tree ferns, above 600 m, in northern and western Nuku Hiva, Fatu Hiva, Ua Huku and Ua Pu, and above 1000 m on Hiva Oa; dry forest, with *Hibiscus, Pandanus, Thespesia* and *Cordia*, on the lower slopes below the cloud line, and originally covering most of Eiao and Fatu Huku (Melville, 1970); and intermediate or 'mesophytic' forest, with *Hibiscus, Piper* and *Cordyline*, on the plateaux to the west and east of Mt Ootua on Hiva Oa, and over most of central Nuku Hiva (Adamson, 1936). *Eragrostis* grassland and xerophytic scrub is still found on the lower, more arid islands such as Hatutu.

All the islands have been devastated by overgrazing by feral and domestic animals. Much of the original dry forest on the lower slopes below 1000 m, has been totally destroyed, or reduced to *Gleichenia* and tussock grassland, and on some islands, such as Eiao, the drier parts of Nuku Hiva, and in north-west Ua Pu, there is no vegetation left at all (Melville, 1970, 1979; Schäfer, 1977). Feral cattle have caused extensive damage to upland rain forests on the larger islands (Melville, 1970).

Checklists and Floras The only complete account is the *Flora of Southeastern Polynesia* (Brown and Brown, 1931-1935), cited in Appendix 1. See also:

Sachet, M.-H. (1975). Flora of the Marquesas, 1: Ericaceae-Convolvulaceae. *Smithsonian Contrib. Bot.* 23. 34 pp.

Information on Threatened Plants *Lebronnecia kokioides* and *Pelagodoxa henryana* are included in *The IUCN Plant Red Data Book* (1978). Latest IUCN statistics: endemic taxa – Ex:1, E:17, V:13, R:7, I:21, K:40, nt:4.

Additional References

Adamson, A.M. (1936). Marquesan insects: environment. *Bull. Bernice P. Bishop Mus.* 139. 73 pp. (Includes description of vegetation.)

Gillett, G.W. Report on botanical research in the Marquesas Islands (1970). *Bull. Soc. Etud. Océanien.* (Not seen.)

Hallé, F. Arbres et forêts de Iles Marquises. *Cah. Pacifiq.* 27. (Not seen.)

Melville, R. (1970). The endemic plants of the Marquesas Islands and their conservation status. (Unpublished Red Data Bulletin material.)

Melville, R. (1979). Endangered island floras. In Bramwell, D. (Ed.), *Plants and Islands*. Academic Press, London. Pp. 361-377.

Sachet, M.H., Schäfer, P.A. and Thibault, J.C. (1975). Mohotani: une île protégée aux Marquises. *Bull. Soc. Etudes Océanien* 16(6): 557-568.

Salvat, B. (1974). Mesures en faveur de la Protection de la Iles Marquises. Unpublished report. (Not seen.)

Schäfer, P.A. (1977). *La Vegetation et L'Influence Humaine aux Iles Marquises*. Academie de Montpellier, Languedoc. 31 pp.

Marshall Islands

The Marshall Islands are the easternmost island group of Micronesia in the western Pacific Ocean, between latitudes 8-12°N and longitudes 162-172°E. There are two island chains: the Ralik Chain (18 atolls) and the Ratak Chain (15 atolls). They form a district of the United Nations Trust Territory of the Pacific Islands administered by the United States. All the atolls are low with numerous islets, some of which enclose a central lagoon. The largest island is Kwajalein (16 sq. km) with 92 islets.

Area 181 sq. km

Population 30,873

Floristics No overall figure for size of flora, but 293 dicotyledon taxa, of which 88 are native (Fosberg, Sachet and Oliver, 1979, cited in Appendix 1); one native cycad (*Cycas circinalis*) and 10 native fern taxa (Fosberg, Sachet and Oliver, 1982, cited in Appendix 1). Most of the atolls are species-poor, the majority of plants having a widespread distribution throughout the Pacific and Indian Oceans. No endemic ferns or gymnosperms; 4 endemic *Pandanus* spp. (St John, 1960). Pokak, in the Ratak Chain, has an endemic grass (*Lepturus gassaparicensis*) (Douglas, 1969, cited in Appendix 1).

Vegetation Small remnants of atoll/beach forest (mostly comprising pan-Pacific species such as *Pisonia grandis*, *Casuarina equisetifolia*, *Pandanus tectorius* and *Scaevola* spp.) on some northern atolls (e.g. Wotho, Ujae and some of the islets of Kwajalein); small areas of mangrove forest on Jaluit, Ailinglapalap and Mejit (Dahl, 1980, cited in Appendix 1). All the Marshall Islands have been greatly modified; most atolls have coconut and breadfruit plantations and some islands have been drastically damaged by the testing of atomic weapons. For an account of the condition and status of the forests see Fosberg (1973), cited in Appendix 1.

Checklists and Floras The Marshall Islands are included in *Flora Micronesica* (Kanehira, 1933), the regional checklists of Fosberg, Sachet and Oliver (1979, 1982), and will be covered by the *Flora of Micronesia* (1975-), all cited in Appendix 1. Separate lists for individual islands include:

Fosberg, F.R. (1955). Northern Marshalls expedition 1951-1952: land biota; vascular plants. *Atoll Res. Bull.* 39. 22 pp. (Annotated list; notes on habitats, distribution. For additions see *ibid.*, 68. 9 pp., 1959.)

Fosberg, F.R. (1956). *Military Geography of the Northern Marshalls*. U.S. Army Engineers and U.S. Geological Survey. 320 pp. (Describes 21 atolls, notes on vegetation, lists about 150 species on 13 atolls.)

Fosberg, F.R. and Sachet, M.-H. (1962). Vascular plants recorded from Jaluit Atoll. *Atoll Res. Bull.* 92. 39 pp.

Hatheway, W.H. (1953). The land vegetation of Arno Atoll, Marshall Islands. Scientific investigations in Micronesia. *Atoll Res. Bull.* 16. 68 pp. (Arno has c. 125 species of which 44 are native; all are wide-ranging species of the Pacific and Indian Oceans.)

Koidzumi, G. (1915). The vegetation of Jaluit Island. *Bot. Mag. (Tokyo)* 29: 242-252. (59 species listed; 40 indigenous, all of widespread distribution.)

Okabe, M. (1941). An enumeration of the plants collected in Marshall Islands. *J. Jap. Forestry Soc.* 23: 261-272.

St John, H. (1951). Plant records from Aur Atoll and Majuro Atoll, Marshall Islands, Micronesia. Pacific Plant Studies 9. *Pacific Science* 5: 279-286. (Annotated list of 78 vascular plant taxa collected on the atolls, 43 indigenous.)

St John, H. (1960). Flora of Eniwetok Atoll. *Pacific Science* 14: 313-336. (95 taxa recorded; 42 indigenous, 4 endemic pandans; includes keys and brief descriptions.)

Taylor, W.R. (1950). *Plants of Bikini and Other Northern Marshall Islands.* Ann Arbor, Univ. of Michigan Press. 227 pp. (Results of investigations carried out before the testing of atomic weapons.)

Information on Threatened Plants None.

Mauritania

Area 1,030,700 sq. km

Population 1,832,000

Floristics 1100 species (quoted in Lebrun, 1976, cited in Appendix 1). Levels of endemism not known, but probably low. Floristic affinities Saharan and Sahelian.

Vegetation Mostly desert and semi-desert, with little or no perennial vegetation. As rainfall increases further south, semi-desert grassland grades into rather low wooded grassland with *Acacia tortilis*, increasing in density and height, reaching 8 m or so high in the extreme south.

For vegetation map see White (1983), cited in Appendix 1.

Checklists and Floras Mauritania is included in *Flore du Sahara* (Ozenda, 1977), and in the computerized *Atlas der Pflanzenwelt des Nordafrikanischen Trockenraumes* (Frankenberg and Klaus, 1980). The tropical, southern part of Mauritania is included in the *Flora of West Tropical Africa*. These are all cited in Appendix 1.

Adam, J.G. (1962). Itinéraires botaniques en Afrique occidentale; flore et végétation d'hiver de la Mauritanie Occidentale. Les pâturages. Inventaire des plantes signalées en Mauritanie. *J. Agric. Trop. Bot. Appl.* 9: 85-200, 297-416. Also reprinted separately by Muséum National d'Histoire Naturelle, Paris, according to Frodin. (With 18 plates of black and white photographs.)

Monod, T. (1939). Phanérogams. In *Contributions à l'Etude du Sahara Occidental*, vol. 2: 55-211. Larose, Paris. (Publications du Comité d'Etudes Historiques et Scientifiques de l'Afrique Occidentale Française, Sér. B, No. 5, according to Frodin.)

Information on Threatened Plants No published lists of rare or threatened plants; IUCN has records of only 7 species and infraspecific taxa believed to be endemic, including R:3.

Additional References

Adam, J.G. (1968). La Mauritanie. In Hedberg, I. and O. (1968), cited in Appendix 1. Pp. 49-51.

Audry, P. and Rossetti, C. (1962). Observations sur les Sols et la Végétation en Mauritanie de Sud-Est et sur la Bordure Adjacente du Mali (1959 et 1961). Projet Pèlerin, Rapp. No. UNSF/DL/ES/3, FAO, Rome. 267 pp. (With 24 black and white photographs.)

Monod, T. (1938). Notes botaniques sur le Sahara occidental et ses confins sahéliens. *Mém. Soc. Biogéogr.* 6: 351-374.

Monod, T. (1952). Contribution à l'étude du peuplement de la Mauritanie. Notes botaniques sur l'Adrar (Sahara Occidental). *Bull. IFAN* 14: 405-449; 16A: 1-48.

Murat, M. (1944). Esquisse phytogéographique du Sahara occidental. Remarques et Commentaires par T. Monod, C. Rungs et C. Sauvage. *Mém. Off. Nat. Anti-acrid.* 1: 1-31.

Naegélé, A. (1958-1960). Contributions à l'étude de la flore et des groupements végétaux de la Mauritanie. *Bull. IFAN* 20A: 293-305, 876-908; 21A: 1195-1204; 22A: 1231-1247. (Most of these have several black and white photographs.)

Roberty, G. (1958). Végétation de la guelta de Soungount (Mauritanie méridionale) en mars 1955. *Bull IFAN* 20A: 869-875.

Rossetti, C. (1963). Observations sur la Végétation: Conclusions sur les Travaux Entrepris en 1959 et 1961. Projet Pèlerin, Rapp. No. UNSF/DL/ES/5, FAO, Rome. 71 pp.

Mauritius

The volcanic island of Mauritius, part of the Mascarenes group, lies some 840 km east of Madagascar. It has very varied topography, with ranges of peaks, plateaux and low-lying plains. The highest point is Piton de la Petite Rivière Noire, at 828 m, near the south-west coast. Round Island is a small island of 1.6 sq. km 24 km north-east of Mauritius.

Area 1865 sq. km

Population 1,031,000 (including Rodrigues, *q.v.*, and other dependencies)

Floristics 800-900 species (W. Strahm, 1984, *in litt.*), including 186 ferns (Lorence, 1978); roughly a third of species endemic; eight endemic genera. Baker (1877, cited in Appendix 1) gives 869 'wild' vascular species.

46 species of ferns and flowering plants recorded from Round Island. 70 species of ferns, fern allies and flowering plants recorded from Gunner's Quoin, 28 of which also occur on Round Island; 20 indigenous species and eight species endemic to the Mascarenes. (Bullock *et al.*, 1984.)

Floristically each island of the Mascarenes is related primarily to the others, but relationships also exist with Madagascar (Melville, 1970, cited in Appendix 1), and, somewhat remotely, with Malesia, India and Sri Lanka (M.J.E. Coode, 1984, pers. comm.).

Vegetation Most of the island used to be covered with dense tropical evergreen forest, with heath and dwarf forest at higher altitudes and palm savannas in the dry eastern regions (Procter and Salm, 1975; Vaughan and Wiehe, 1937). Mauritius is now almost totally devoid of indigenous vegetation. The best examples remaining are the patches of upland forest around the Black River Gorges in the south-west.

More than 60% of the area of the island is under sugar cultivation, and tea and other vegetables are also important. An additional cause of destruction of the indigenous vegetation has been the super-abundance of exotic plants and animals introduced deliberately or by accident, which prevent natural regeneration of the native species.

Round Island is now so badly degraded by introduced goats and rabbits that very little vegetation of any sort remains on the island. Goats have been exterminated, but rabbits continue to be a pest.

Checklists and Floras Mauritius is included in the incomplete *Flore des Mascareignes*, and in the rather dated *Flora of Mauritius and the Seychelles* (Baker, 1877), both cited in Appendix 1.

Johnston, H.H. (1895). Additions to the Flora of Mauritius as recorded in Baker's 'Flora of Mauritius and the Seychelles'. *Trans and Proc. Bot. Soc. Edinburgh* 20: 391-407.

Field-guides
Cadet, L.J.T. (1981). *Fleurs et Plantes de la Réunion et de l'Ile Maurice*. Editions du Pacifique, Tahiti. 131 pp. (Incomplete for indigenous flora.)

Information on Threatened Plants
Hedberg, I. (1979), cited in Appendix 1. (List for Mauritius and Rodrigues, p. 103, by A.W. Owadally, contains 34 species: E:12, V:2, R:18, I:2.)

IUCN has records of 222 species and infraspecific taxa believed to be endemic to Mauritius – Ex:19, E:65, V:35, R:39, I:14, K:11, nt:39. Non-endemic taxa rare or threatened worldwide – Ex:1, E:8, V:15, R:9, I:3 (world categories). (Covers the 74 families in *Flore des Mascareignes* (out of 203 in total), and some others as well, including Rubiaceae and Myrtaceae.)

A Red Data Book for Mauritius is being written by W. Strahm as part of the IUCN/WWF Plants Programme (Project 3149).

Four species which occur in Mauritius are included in *The IUCN Plant Red Data Book* (1978).

Laws Protecting Plants The Forests and Reserves Act (1983) gives general protection to the island's forest and reserves, and specific protection to all indigenous orchids and ferns, species of three genera, and to five additional species.

Voluntary Organizations
Mauritius Wildlife Conservation Society.

Botanic Gardens
Botanic Gardens, Curepipe. (Belongs to Curepipe Municipality, but partly managed by the Forestry Service, address below.)
Royal Botanic Gardens, Pamplemousses. (Mailing address: Chief Agricultural Officer, Reduit.)

Useful Addresses

Curator, Herbarium, Mauritius Sugar Industry Research Institute, Reduit.

CITES Management Authority: The Conservator of Forests, Forestry Service, Curepipe.

Additional References

Bullock, D., North, S. and Greig, S. (Eds) (1984). Round Island Expedition 1982: final report. Unpublished, but available from D. Bullock, Dept of Botany, St Andrews KY16 9AL, Scotland. 123 pp. (Includes annotated ckecklists of plants from Round Island and Gunner's Quoin.)

Cadet, L.J.T. (1984). *Plantes Rares ou Remarquables des Mascareignes*. Agence de Coopération Culturelle et Technique, 13 quai André-Citroën, 75015 Paris. 132 pp. (With 48 photographs.)

Lorence, D. (1978). The pteridophytes of Mauritius (Indian Ocean): ecology and distribution. *Bot. J. Linn. Soc.* 76: 207-247.

Procter, J. and Salm, R. (1975). Conservation in Mauritius 1974. IUCN, Morges, Switzerland. (Cyclostyled.)

Vaughan, R.E. and Wiehe, P.O. (1937). Studies on the vegetation of Mauritius, 1: A preliminary survey of the plant communities. *J. Ecol.* 25: 289-343. (With vegetation map, 20 plates of black and white photographs.)

Vaughan, R.E. and Wiehe, P.O. (1941). Studies on the vegetation of Mauritius, 3: The structure and development of the upland climax forest. *J. Ecol.* 29: 127-160. (With 4 black and white photographs.)

Vaughan, R.E. (1968). Mauritius and Rodriguez. In Hedberg, I. and O. (1968), cited in Appendix 1. Pp. 265-272.

Vinson, J. (1964). Sur la disparition progressive de la flore et de la faune de l'Ile Ronde. *Proc. Roy. Soc. Arts Sci. Mauritius* 2: 247-261.

Mexico

Area 1,972,546 sq. km

Population 77,040,000

Floristics Due to its latitudinal and altitudinal range, Mexico contains a very diverse flora of an estimated 20,000 vascular plant species (Rzedowski, 1978, Lot and Toledo, 1980); 3376 endemic species (Toledo, 1984, pers. comm.); a meeting point of boreal and tropical floras.

Vegetation Tropical and subtropical region (c. one third of Mexico, mainly on the Atlantic and Pacific seaboards south of the tropic of Cancer and east of the Isthmus of Tehuantepec): Rain forests, the northernmost in the Americas, once formed a continuous corridor from Veracruz to Chiapas, covering 6% of Mexico; half of them now destroyed, the largest remaining being the 13,000 sq. km Lacandon Forest along the Guatemala border, now partly protected (Estrada and Coates-Estrada, 1983). Where rainfall is lower and the winter dry season more pronounced, the forest canopy is lower and the percentage of deciduous species increases sharply. Low deciduous forest (Selva Baja Caducifolia), with many broadleaved species to c. 15 m tall, occupies 16% of the area.

Temperate region (one third of Mexico), occupying the main cordilleras: The principal forest is of pines (*Pinus* spp.) and oaks (*Quercus* spp.) in varying proportions and with numerous constituent species. In the higher parts of the cordilleras, to 3300 m, forests of silver fir (*Abies* spp.). In all, these vegetation types occupy about 15% of Mexico.

Semi-arid and arid zone, also about a third of Mexico, mainly in the north and centre (Sonoran and Chihuahuan desert regions and central altiplano): Mostly open shrubland (*matorral*), the principal variants dominated by (i) small-leaved shrubs, (ii) cacti, and (iii) xerophytic monocotyledons (*Agave, Yucca, Dasylirion, Nolina* spp., Bromeliaceae).

Estimated rate of deforestation for closed broadleaved forest 4700 sq. km/annum, out of a total of 265,700 sq. km (FAO/UNEP, 1981).

Checklists and Floras The tropical part of Mexico, principally east of the Isthmus of Tehuantepec, is covered by the *Flora Mesoamericana* Project, described in Appendix 1; the part south of the Tropic of Cancer by the family and generic monographs of *Flora Neotropica* (cited in Appendix 1). State and regional Floras are:

Flora de Veracruz (1978-). (Various authors). Instituto de Investigaciones sobre Recursos Bióticos (INIREB). 39 family fascicles so far. (The output of a substantial project to provide a database on Veracruz flora, described by Gómez-Pompa *et al.*, 1984, cited under 'Additional References', below.)

Flora of Chiapas (1981-). (Various authors). Published by the California Academy of Sciences, two parts completed so far: 1 – introduction and descriptions of vegetation types and their endemics, by D.E. Breedlove (1981, 35 pp.); 2 – ferns, by A. Smith (609 species). (Breedlove, 1981, refers to 8200 vascular plant species recorded from Chiapas; "the number ... will probably climb to between 9000 and 10,000 by the time the entire Flora is published".)

Flora Yucatanense project. Edited by V. Sosa, INIREB, Calle 43 No 506, Apdo Postal 281, CP 97000, Mérida, Yucatán. (2100 species – Toledo, 1985, cited in Appendix 1, quoting Sosa, pers. comm.)

Johnston, M.C., Henrickson, J. *et al.* (in press). *Chihuahuan Desert Flora*. Prepared at Dept of Botany, University of Texas at Austin, Texas, U.S.A. (About 3000 species of vascular plants, from southern New Mexico to San Luís Potosí.)

McVaugh, R. (1974-). *Flora Novo-Galiciana*. University of Michigan. 17 vols planned, by various authors. Gramineae (Vol. 14) published; Compositae (12) to be completed in late 1984, Orchidaceae (16) in 1985, Leguminosae (5) in 1986. (Covers Mexican states of Aguascalientes, Jalisco, Colima, and parts of Nayarit, Durango, Zacatecas, Guanajuato and Michoacan.)

Martínez, M. and Matuda, E. (1953-1972). *Flora del Estado de México*. Many separates, reissued as 3 vols by Biblioteca Enciclopédica del Estado de México, 1979.

Rzedowski, J. and Rzedowski, G.C. de (1979-). *Flora Fanerogámica del Valle de México*. Ed. Continental, México. Vol. 1 (introductory, gymnosperms, dicotyledons Saururaceae to Polygalaceae) published, Vol. 2 in press, Vol. 3 in prep.

Sánchez Sánchez, O. (1968). *La Flora de Valle de México*. Herrero, México. 519 pp.

Shreve, F. and Wiggins, I.L. (1964). *Vegetation and Flora of the Sonoran Desert*, 2 vols. Stanford Univ. Press, Stanford. 1740 pp. (Vegetation types and representative species, vegetation map.)

Wiggins, I.L. (1980). *Flora of Baja California*. Stanford Univ. Press, Stanford. 1025 pp. (2705 species with 686 endemic taxa.)

See also:

Bravo-Hollis, H. (1978-). *Las Cactáceas de México*, Ed. 2. Vol. 1. Univ. Nacional Autónoma de México. 743 pp. Vol. 2 in press.

Cowan, C.P. (1983). *Listados Florísticos de México. I. Flora de Tabasco*. Instituto de Biología, UNAM, México. (Checklist with cited specimens.)

Gentry, H.S. (1942). *Rio Mayo Plants: A Study of the Flora and Vegetation of the Valley of the Rio Mayo, Sonora*. Carnegie Institution Publication 527, Washington, D.C. 328 pp. (Annotated list of 1276 species.)

Gentry, H.S. (1982). *Agaves of Continental North America*. Univ. Arizona Press, Tucson, Arizona. 670 pp.

Lundell, C.L. (1942). Flora of eastern Tabasco and adjacent Mexican areas. *Contrib. Univ. Mich. Herb*. 8: 1-74. (Annotated list of c. 700 species.)

Martínez, M. (1963). *Las Pináceas Mexicanas*, 3rd Ed. Universidad Nacional Autónoma de México. 401 pp.

Pennington, T.D. and Sarukhan, J. (1968). *Los Arboles Tropicales de México*. Instituto Nacional de Investigaciónes Forestales, México and FAO, Rome. 413 pp.

Sousa S., M. and Cabrera C., E.F. (1983). *Listados Florísticos de México. II. Flora de Quintana Roo*. Instituto de Biología, UNAM, México. (Checklist with cited specimens.)

Standley, P.C. (1920-1926). Trees and shrubs of Mexico. *Contrib. U.S. Nat. Herb*. 23(1-5). 1721 pp.

Standley, P.C. (1930). Flora of Yucatan. *Field Mus. Nat. Hist., Bot. Ser*. 3(3): 157-492. (Annotated list of 1263 plants.)

Tellez V., O and Sousa S., M. (1982). *Imagenes de la Flora Quintanarroense*. Centro de Investigaciones de Quintana Roo, Puerto Morelo, Q.R.

Williams, L.O. (1951). The Orchidaceae of Mexico. *Ceiba* 2(1): 1-321. (600 species.)

Selected bibliographies:

Jones, G.N. (1966). *An Annotated Bibliography of Mexican Ferns*. Univ. Illinois Press, Urbana. 297 pp. (1200 author entries.)

Langman, I.K. (1964). *A Selected Guide to the Literature of the Flowering Plants of Mexico*. Univ. Pennsylvania Press, Philadelphia. 1015 pp.

The National Council of the Flora of Mexico, which includes about 40 institutions, is promoting and co-ordinating a catalogue of Mexican plants (Flora de México Project).

Field-guides

Clark, P. (1972). *A Flower Lover's Guide to Mexico*. Minutiae Mexicana, México. 128 pp. (Guide to common species.)

Coyle, J. and Roberts, N.C. (1975). *A Field Guide to the Common and Interesting Plants of Baja California*. Natural History Publishing Co., La Jolla, Calif. 206 pp. (259 plants, endemics indicated.)

Tellez Valdes, O. and Sousa Sanchez, M. (1982). *Imagenes de la flora Quintanarroense*. Puerto Morelos, Centro de Investigaciones de Quintana Roo, A.C. (116 of known 1300 species described, illus.)

Information on Threatened Plants There is no national Red Data Book. The most comprehensive list published so far is that of Vovides (1981), see below. IUCN is preparing a threatened plant list for release in a forthcoming report *The list of rare, threatened and endemic plants of Middle America*. Latest IUCN statistics, based upon this work: endemic taxa – Ex:8, E:72, V:176, R:320, I:66, K:2084, nt:732; non-endemics rare or threatened worldwide – E:3, V:22, R:36, I:4 (world categories).

Threatened plants are mentioned in several papers in:

Prance, G.T. and Elias, T.S. (Eds) (1977), cited in Appendix 1. See in particular J.T.
Mickel on rare and endangered ferns (pp. 323-328), H.E. Moore on endangerment
in palms (pp. 267-282), P. Ravenna on endangered bulbous plants (pp. 257-266),
and A.P. Vovides and A. Gómez-Pompa (cited below).

Other relevant publications:

Anon (1979). Especies en peligro de extinción. *Macpalxochitl, Bol. Bimestral de Soc.
Bot. México* 79: 3-4. (24 taxa listed.)

Howard, T.M. (1981). Current status of some endangered Mexican *Hymenocallis*
species. *Pl. Life* 37(1-4): 157-158.

Hunt, D.R. (1982). The conservation status of Mexican Mammillarias: a preliminary
assessment. *Cact. Succ. J. Great Britain* 44(4): 87-88. (IUCN categories assigned to
each of 233 taxa.)

Perez D., J.F. (1982). Especies amenazadas y en peligro de extinción de la península de
Baja California. *Publ. Espec. Inst. Nacion. Invest. Forest. México* 37: 62-67.

Pina, I. (1980). Rare and threatened Agavaceae and Cactaceae of Mexico. Sociedad
Mexicana Cactología. (Unpublished.)

Rzedowski, J. (1979a). Extinción de especies vegetales. In Rzedowski, J. and G. (Eds),
Flora Fanerogámica del Valle de México: Vol. 1. Cited under Checklists and Floras,
above. Pp. 42-45.

Rzedowski, J. (1979b). Deterioro de la Flora. *Memorias sobre Problemas Ambientales
en México.* Instituto Politécnico Nacional, Escuela de Ciencias Biológicas.
Pp. 51-57.

Toledo, V.M. (1985). Criterios fitogeográficos para la conservación de la flora de
México. In Gómez, L.D. (Ed.), *Memorias del Simposio de Biogeografía de
Mesoamérica.* In press.

Vovides, A.P. (1981). Lista preliminar de plantas Mexicanas raras o en peligro de
extinción. *Biótico* 6(2): 219-228. (Preliminary list of 210 rare, threatened and
endangered species.)

Vovides, A.P. and Gómez-Pompa, A. (1977). The problems of threatened and
endangered plant species of Mexico. In Prance, G.T. and Elias, T.S. (Eds) (1977),
cited in Appendix 1. Pp. 77-88.

Laws Protecting Plants No information. The U.S. Government has determined
Abies guatemalensis (Mexico, Guatemala, El Salvador and Honduras) as 'Threatened'
under the U.S. Endangered Species Act.

Voluntary Organizations

Asociación Mexicana de Orquideología A.C., Apdo Postal 53-123, 11320 México 17,
D.F.

Pronatura A.C., Apdo Postal 20-768, Del. Alvaro Obregón, 01000 México, D.F.

Sociedad Botánica de México, Apto Postal 70-385, México 200, D.F.

Sociedad Mexicana de Cactología A.C., 2a Juárez 42, Col. San Alvaro, Deleg.
Azcapotzalco, 02090 México, D.F.

Botanic Gardens

Jardín Botánico, Centro de Investigación Científico de Yucatán, Mérida, Yucatán.

Jardín Botánico, Centro de Investigaciónes de Quintana Roo, 77500 Puerto Morales,
Quintana Roo.

Jardín Botánico, Escuela Nacional de Enseñaza Profesional, Universidad Nacional
Autonoma de México, Ixtapalapa.

Jardín Botánico, La Estación de Biología Tropical "Los Tuxtlas", Instituto de
  Biología, Universidad Nacional Autónoma de México, Municipio de San Andres
  Tuxtla, Catemaco, Veracruz.
Jardín Botánico "Francisco J. Clavijero", INIREB, Km 2.5 Antigua Carretera A.
  Coatepec, 91000 Xalapa, Veracruz.
Jardín Botánico, INIREB, Km 7, Camino San Cristobal de Las Casas a Comitan, San
  Cristobal de Las Casas, Chiapas.
Jardín Botánico, Tuxtla Gutierrez, Chiapas.
Jardín Botánico, Universidad Autónoma Agraria "Antonio Narro", Buenavista,
  Saltillo, Coahuila.
Jardín Botánico, Departamento de Difusion y Enseñanza, Universidad Nacional
  Autónoma de México, Ciudad Universitaria, Deleg. Coycoacan, 04510 México, D.F.
Jardín Botánico Medicinal, Instituto Nacional de Antropología e História, Matamoros
  200, Colonia Acapanzingo, Cuernavaca, Morelos.

A Union of Mexican Botanical Gardens has recently been formed.

Index of threatened plants in cultivation:

Threatened Plants Unit, IUCN Conservation Monitoring Centre (1985). *The Botanic
  Gardens List of Rare and Threatened Species of Mexican Cacti*. Botanic Gardens
  Conservation Co-ordinating Body, Report No. 13. IUCN, Kew. 25 pp. (Lists all
  but 20 of 301 rare, threatened and insufficiently known taxa reported in cultivation,
  with gardens listed against each.)

Useful Addresses
Dirección General de Flora y Fauna Silvestres, Netzahuackoyotl No. 109, 1° Piso,
  Deleg. Cuauhtemoc, 06080 México, D.F.
Herbario Nacional de México, Universidad Nacional Autónoma de México (UNAM),
  Apdo Postal 70-367, México 20, D.F.
Instituto Nacional de Investigaciones sobre Recursos Bióticos (INIREB), P.O. Box 63,
  Xalapa, Veracruz.

Additional References
Avila, J.A.R., Calderon, G. and Chapa, H. (1961). *Los recursos naturales de México;
  estado actual de las investigaciones de hidrología y pesca*. Instituto Mexicana de
  Recursos Naturales Renovables. 421 pp.
Estrada, A. and Coates-Estrada, R. (1983). Rain forest in Mexico: research and
  conservation at Los Tuxtlas. *Oryx* 17: 201-204.
Flores Mata, G. *et al.* (1971). *Mapa de Tipos de Vegetación de la República Méxicana*.
  Secretaria de Recursos Hidráulicos, México. Map (1:2,000,000), with explanatory
  text.
Gómez-Pompa, A. (1973). Ecology of the vegetation of Veracruz. In Graham, A. (Ed.)
  (1973), cited in Appendix 1. Pp. 73-148.
Gómez-Pompa, A., Moreno, N.P., Gama, L., Sosa, V. and Allkin, R. (1984). Flora of
  Veracruz: Progress and prospects. In Allkin, R. and Bisby, F.A. (Eds), *Databases in
  Systematics*. Academic Press, London. Pp. 165-174. (Systematics Assoc. Special
  Vol. No. 26.)
Hagsater, E. (1976). Orchids and conservation in Mexico. *Orchid Review* 84: 39-42.
Lot, A. and Toledo, V.M. (1980). Hacia una Flora de México: vamos por buen
  camino. *Macpalxochitl* 88/89: 1-31.

McCullough, R. (1981). Mexico and its orchids. In Stewart, J. and van der Merwe, C.N. (Eds), *Proceedings of the 10th World Orchid Conference*. South African Orchid Council, Johannesburg. Pp. 111-114.

Miranda, F. and Hernandez, E. (1963). Los tipos de vegetación de México y su clasificación. *Bol. Soc. Bot. Méx.* 28: 29-179.

Pesman, M.W. (1962). *Meet Flora Mexicana*. Northland Press, Flagstaff, Arizona. 278 pp. (2nd Ed. by R. Bye and E. Linares Mazari in press.)

Rzedowski, J. (1966). *Vegetación de Estado de San Luís Potosí*. Universidad Autónoma de San Luís Potosí, México. 291 pp. (Vegetation zones and representative species.)

Rzedowski, J. (1978). *Vegetación de México*. Editorial Limusa, México. 432 pp.

Midway Islands

Midway (5 sq. km), an unincorporated territory of the United States, lies 1850 km north-west of the Hawaiian Islands, in the central Pacific Ocean, at latitude 28°12'N, longitude 177°24'W. It is an atoll with 2 islets, Eastern Island (135 ha) and Sand Island (384 ha) surrounding a lagoon. The population is over 2220 (1970). The vegetation includes extensive *Casuarina* plantations, *Scaevola* and *Boerhavia* scrub. 90 vascular plant species, most of which have been recently introduced (Neff and DuMont, 1955). Military activity and the construction of air and submarine bases has greatly modified the vegetation.

References
Neff, J.A. and DuMont, P.A. (1955). A partial list of the plants of the Midway Islands. *Atoll Res. Bull.* 45. 11 pp.

Minami-Tori-Shima

Minami-Tori-Shima (Marcus Island) is a raised coral atoll with a fringing reef, of area 300 ha, situated 965 km east south-east of the Ogasawara Islands in the north-west Pacific at 24°14'N and 154°E. It is a Japanese dependency. Following extensive levelling, the highest point on the island is 7 m. The vegetation, which has been greatly modified by war damage and construction works, consists mainly of *Tournefortia* and *Pisonia* scrub. Papayas and bananas have been introduced (Douglas, 1969, cited in Appendix 1).

The flora consists of widespread angiosperms, including 18 dicotyledon taxa, of which 9 are indigenous (Fosberg, Sachet and Oliver, 1979, cited in Appendix 1); 4 species of monocotyledons (Sakagami, 1961). There are no endemics.

References
Sakagami, S.F. (1961). An ecological perspective of Marcus Island, with special reference to land animals. *Pacific Science* 15: 82-104. (Includes plant list, notes on vegetation.)

Mongolia

Area 1,565,000 sq. km

Population 1,851,000

Floristics 2272 vascular plant species; of these 229 endemic and a further 143 species restricted to Mongolia and the adjacent territories of Inner Mongolia, Altai and Tuva in the U.S.S.R., and Dzungaria in China (V.I. Grubov, 1984, *in litt.*).

Vegetation Almost 90% grassland, semi-desert and desert; c. 10% forested, mainly of larch, cedar and pine. In the south, the vast Gobi Desert covers c. 1,300,000 sq. km, and supports sparse scrub with *Artemisia*, *Ephedra* and *Haloxylon*; in the west, the vegetation cover is less than 5% and is mainly *Nitraria* scrub; on dunes above 10 m there is no plant life at all. The only natural forests of the Gobi are in the west, around Ala Shan, where *Populus diversifolia* and *Tamarix* spp. are found along river banks. Northern Mongolia has semi-deserts and grass steppes.

Checklists and Floras

Grubov, V.I. (1955). *Konspeckt Flory Mongol'skoi Narodnoi Respubliki*. Mongolian Commission. 307 pp. (Annotated checklist of 1875 species.)

Grubov, V.I. (1972). Additions and corrections to the "Concised Flora of the Mongolian People's Republic". *Novitates Syst. Plantarum Vascularium* 9: 275-305. (Enumeration of 133 species described since Grubov, 1955.)

Grubov, V.I. (1982). *Key to the Vascular Plants of Mongolia (with an Atlas)*. Academy of Sciences of the U.S.S.R., Leningrad. 441 pp. (In Russian.)

Mongolia is also covered in Grubov (1963-) and by the *Flora of the Mongolian Steppe and Desert Areas* (Norlindh, 1949), cited in full in Appendix 1. See also:

Inner Mongolia Botanical Records Compiling Group (1977-1982). *Flora Intramongolica*, 6 vols. Typis Intramongolicae Popularis, Huhhot. (In Chinese.)

Information on Threatened Plants IUCN has a preliminary list, compiled by V.I. Grubov, which includes 11 threatened plants, of which one is endemic to Mongolia, and a further 10 species are also found in Inner Mongolia (China).

Gubanov, I.A. (1982). Zametki o redkikh rasteniyakh Mongolii (Notices on rare plants of Mongolia). *Byull. Most. Obshch. Ispyt. Prir. Biol.* 87(1): 122-129. (In Russian.)

Botanic Gardens

Botanic Garden, The Academy of Sciences of the MPR Institute of Botany, Ulan Bator.

Additional References

Printz, H. (1921). *The Vegetation of the Siberian-Mongolian Frontiers (The Sayansk Region)*. Det Kongelige Norske Videnskabers Selskab. 458 pp. (Includes enumeration of plants in region.)

Walker, E.H. (1941). Plants collected by R.C. Ching in southern Mongolia and Kansu Province, China. *Contrib. U.S. Nat. Herb.* 28(4): 563-675.

Montserrat

Montserrat is a Dependent Territory of the United Kingdom, in the Leeward Islands of the Eastern Caribbean, 43.5 km north-west of Antigua. It is a small island of 104 sq. km and with a population of 13,000. It consists of a serrated range of volcanic peaks; the Soufrière is still active. It has among the best natural vegetation in the Leewards: high forest practically non-existent due to cultivation to near summits and hurricane activity but secondary rain forest to summit peaks; palm brake and elfin woodland along ridges; secondary thickets of young trees and dry scrub woodland below; north slopes of hills better wooded than south slopes due to favourable moist conditions; 40% forested according to FAO (1974, cited in Appendix 1). For botanical information, see the account for Antigua and Barbuda.

R.A. Howard is preparing a checklist of the flora for Montserrat National Trust, Plymouth, Montserrat.

Morocco

Area 659,970 sq. km

Population 22,848,000

Floristics 3500 species (Le Houérou, 1975); 3600 species (Lebrun, 1976, cited in Appendix 1); 3700 species (Sauvage, 1975). 600-650 endemic species estimated, of which c. 170 are from the high Moroccan Atlas (Quézel, 1978, cited in Appendix 1); IUCN figures, from existing Floras, record 537 endemic taxa.

Flora in north and centre of Morocco with Mediterranean affinities; Saharan flora along southern border; transition zone between the two.

Vegetation Desert along southern border, with little or no perennial vegetation. Semi-desert and transition from Mediterranean scrubland to succulent semi-desert shrubland along west coast and in east-central part of country. Mediterranean sclerophyllous forest in band along north coast and at lower altitudes on the Atlas mountains. Mediterranean montane forest, altimontane shrubland and *Cedrus* forests on the Atlas mountains.

For vegetation map see White (1983), cited in Appendix 1.

Checklists and Floras Morocco is included in the incomplete *Flore de l'Afrique du Nord*, the computerized *Atlas der Pflanzenwelt des Nordafrikanischen Trockenraumes* (Frankenberg and Klaus, 1980), *Flore du Sahara* (Ozenda, 1977), and is being covered in *Med-Checklist*. These are all cited in Appendix 1.

Jahandiez, E. and Maire, R. (1931-1941). *Catalogue des Plantes du Maroc*, 4 vols. Alger. (Annotated checklist; 4th vol. by M.L. Emberger and R. Maire. For additions see Sauvage, C. and Vindt, J. (1949-1956), 4 papers in *Bull. Soc. Sci. Nat. Maroc* 29: 131-162, 32: 27-51, 34: 217-234, 36: 185-222.)

Nègre, R. (1961, 1962). *Petite Flore des Régions Arides du Maroc Occidental*, 2 vols. CNRS, Paris. 413, 566 pp. (Covers only west-central Morocco; keys, descriptions, distributions, line drawings, and several colour photographs.)

Sauvage, C. (1961). Flore des subéraies marocaines: catalogue des cryptogames vasculaires et des phanérogames. *Trav. Inst. Sci. Chérif., Sér. Bot.* 22. 252 pp.

Sauvage, C. and Vindt, J. (1952, 1954). Flore du Maroc, analytique, descriptive et illustrée. *Trav. Inst. Sci. Chérif.* 4 and *Ibid., Sér. Bot.* 3. (Incomplete, covering only Ericaceae to Boraginaceae.)

Field-guides

Emberger, L. (1938). *Les Arbres du Maroc et Comment Les Reconnaître.* Larose, Paris. 317 pp.

Information on Threatened Plants Morocco is included in the draft list for North Africa and the Middle East produced by IUCN Threatened Plants Committee Secretariat (1980), cited in Appendix 1.

Mathez, J., Quézel, P. and Raynard, C. (1985). The Maghrib countries. In Gómez-Campo, C. (Ed.), *Plant Conservation in the Mediterranean Area.*

Sauvage, C. (1959). Au sujet de quelques plantes rares et menacées de la flore du Maroc. In *Animaux et Végétaux Rares de la Région Méditerranéenne.* Proceedings of the IUCN 7th Technical Meeting, 11-19 September 1958, Athens, vol. 5. IUCN, Brussels. Pp. 156-158.

Latest IUCN statistics: endemic taxa – E:1, V:3, R:162, I:23, K:54, nt:294. Non-endemic taxa rare or threatened worldwide – V:2 (world category).

Botanic Gardens

Institut Scientifique Chérifien, Laboratoire de Phanérogamie, Avenue Moulay Chérif, Rabat.

Jardins Exotiques de Rabat-Sale, km 13 Route No. 2 par Sale, Rabat.

Useful Addresses

CITES Management Authority: Comité national de l'Environnement, Division de l'environnement, Direction de l'aménagement du territoire Ministère de l'habitat et de l'aménagement du territoire, B.P. 600, Rabat.

Correspondence to:

Administration des Eaux et Forêts et de la Conservation des Sols, Division de la Protection de la Nature, Ministère de l'Agriculture et de la Réforme Agraire, Rabat.

Additional References

Braun-Blanquet, J. and Maire, R. (1924). Etudes sur la végétation et la flore marocaines. *Mém. Soc. Sci. Nat. Maroc* 8(1). 244 pp. (20 black and white photographs.)

Emberger, L. (1939). Aperçu général sur la végétation du Maroc. In Rübel, E. and Lüdi, W. (Eds), Ergebnisse der internationalen pflanzengeographischen Exkursion durch Marokko und Westalgerian 1936. *Veröff. Geobot. Inst. Zürich* 14: 40-157. (With coloured vegetation map 1:1,500,000.) (Published also as an out-of-series number of Mém. Soc. Sci. Nat. Maroc.)

Frödin, J. (1923). Recherches sur la végétation du Haut Atlas. *Lunds Univ. Arsskr.,* N.F., Avd. 2, 19(4): 1-24.

Ionesco, T. and Sauvage, C. (1962). Les types de végétation du Maroc. Essai de nomenclature et de définition. *Rev. Géogr. Maroc.* 1-2: 75-83.

Le Houérou, H.-N. (1975). Etude préliminaire sur la compatibilité des flores nord-africaine et palestinienne. In CNRS (1975), cited in Appendix 1. Pp. 345-350.

Maire, R. (1924). Etudes sur la végétation et la flore du Grand Atlas et du Moyen Atlas marocains. *Mém. Soc. Sci. Nat. Maroc* 7. 220 pp. (32 black and white photographs.)

Mathez, J. (1973). Nouveaux matériaux pour la Flore du Maroc. Fasc. 2. Contribution à l'étude de la flore de la région d'Ifni. *Trav. RCP* 249(1): 105-120. CNRS, Paris.

Nègre, R. (1959). Recherches phytogéographiques sur l'étage de végétation méditérranéen aride (sous-étage chaud) au Maroc occidental. *Trav. Inst. Sci. Chérif., Sér. Bot.* 13. 385 pp. (With coloured vegetation map 1:500,000; 16 black and white photographs.)

Sauvage, C. (1975). L'état actuel de nos connaissances sur la flore du Maroc. In CNRS (1975), cited in Appendix 1. Pp. 131-139.

Mozambique

Area 784,754 sq. km

Population 13,693,000

Floristics 5500 species (quoted in Lebrun, 1960, cited in Appendix 1). Brenan (1978, cited in Appendix 1) estimates 219 endemic species, from a sample of *Flora Zambesiaca*. Northern part of coast especially rich in local endemics because of extension of coastal mosaic south from Tanzania.

Inland flora predominantly Zambezian, with Afromontane elements on high ground. The flora of a broad band along the coast is part of the so-called Zanzibar-Inhambane region, which extends from southern Mozambique to southern Somalia; it has substantial floristic affinities with the Guinea-Congolian region of central and western tropical Africa.

Vegetation Predominantly dry *Brachystegia-Julbernardia* (Miombo) woodland, but wetter Miombo in the north and large areas of *Colophospermum mopane* (Mopane) woodland along the Zambezi and Limpopo valleys in the north-west and south. Also woodland without characteristic dominants in extreme south and in centre of country. Coastal strip occupied by East African coastal mosaic consisting of a rather dry woodland with abundant *Adansonia*, *Acacia* and *Commiphora*; also abundant mangrove forests. Montane communities confined to the border with eastern Zimbabwe. Estimated rate of deforestation for closed broadleaved forest 100 sq. km/annum out of 9350 sq. km (FAO/UNEP, 1981).

For vegetation maps see Wild and Barbosa (1967, 1968), and White (1983), both cited in Appendix 1.

Checklists and Floras Mozambique is included in the incomplete *Flora Zambesiaca*, cited in Appendix 1.

Fernandes, A. and Mendes, E.J. (Eds) (1969-). *Flora de Moçambique*. Junta de Investigações Científicas do Ultramar, Lisboa. (Incomplete: 64 families plus Pteridophytes published, c. 55% of it, so far.)

Gomes e Sousa, A. (1966, 1967). *Dendrologia de Moçambique*, 2 vols. Instituto de Investigação Agronómica de Moçambique. 822 pp. (Numerous black and white photographs and line drawings.)

Information on Threatened Plants No published lists of rare or threatened plants; IUCN has records of 195 species and infraspecific taxa believed to be endemic, including E:6, V:5, R:59, I:15, nt:19.

Botanic Gardens

Departamento de Botanica, Universidade Eduardo Mondlane/Biologia, C.P. 257, Maputo.

Jardim Municipal, Camara Municipal, Lourenco Marques.

Useful Addresses

CITES Management Authority (Plants): Unidad de Direcçao de Florestal, Maputo.

CITES Scientific Authority (Plants): Instituto Nacional de Investigação Agronómica, P.O. Box 3656, Maputo.

Additional References

Barbosa, L.A. Grandvaux (1968). Moçambique. In Hedberg, I. and O. (1968), cited in Appendix 1. Pp. 224-232.

Bruton, M.N. (1981). Major threat to the coastal dune forest in Maputoland. *The Naturalist (South Africa)* 25(1): 26-27. (Discusses invasion by Bardados Gooseberry.)

Mendonça, F.A. (1952/1955). The vegetation of Mozambique. *Lejeunia* 16: 127-135.

Pedro, J. Gomes and Barbosa, L.A. Grandvaux (1955). A vegetação. In *Esboço do Reconhecimento Ecológico-Agrícola de Moçambique, Mems Trab. Cent. Invest. Cient. Algod.* 23(2): 67-224. (With coloured vegetation map 1:2,000,000.)

Werger, M.J.A. (1978), cited in Appendix 1. Citation includes list of relevant chapters.

Namibia

Area 824,293 sq. km

Population 1,507,000

Floristics 3159 species (Merxmüller, 1966-1972). Unknown levels of endemism, but 11 taxa endemic to the Brandberg (Nordenstam, 1974).

The flora of the north-eastern part bordering Angola has Zambezian affinities. The flora of the Namib desert along the coast is related to the flora of the Karoo further south. Most of the centre of the country has a flora transitional between the two, the so-called Kalahari-Highveld transition zone, with affinities with the Kalahari flora.

Vegetation Vegetation predominantly of a dry type. Rainfall decreases from the north-east to the coastal Namib desert and to the south. In the north-east corner mosaic of dry deciduous forest (rich in species) and transition from woodland without characteristic dominants to *Acacia* deciduous bushland and wooded grassland. Large areas of Kalahari *Acacia* wooded grassland and deciduous bushland, sand dunes with sparse grassland/wooded grassland, *Colophospermum mopane* woodland, scrub woodland (including the swampy Etosha pan), and shrubland. Parallel with the coast: band of bushy shrubland and, along the coast, the Namib desert. This is almost devoid of vegetation, but includes the desert gymnosperm *Welwitschia mirabilis*.

For vegetation map see White (1983), and for vegetation map of Caprivi Strip only see Wild and Barbosa (1967, 1968). Both are cited in Appendix 1.

Checklists and Floras Namibia is included in the incomplete *Flora of Southern Africa,* and in *The Genera of Southern African Flowering Plants* (Dyer, 1975, 1976), both cited in Appendix 1. The Caprivi Strip is included in *Flora Zambesiaca,* cited in Appendix 1. See also:

Merxmüller, H. (1966-1972). *Prodromus einer Flora von Südwest-afrika,* 35 fasc. Cramer, Lehre. (Keys, descriptions, distributions, specimens. For additions see Roessler, H. and Merxmüller, H. (1976). Nachträge zum Prodromus einer Flora von Südwestafrika. *Mitt. Bot. Staatssamml. München* 12: 361-373.)

Nordenstam, B. (1970). Notes on the flora and vegetation of Etosha Pan, South West Africa. *Dinteria* 5: 3-18. (Includes list of 134 species.)

Nordenstam, B. (1974). The flora of the Brandberg. *Dinteria* 11: 3-67. (Annotated checklist of 337 species.)

Information on Threatened Plants

Hall, A.V. *et al*. (1980), cited in Appendix 1. (List for Namibia, p. 78, contains 12 endemic: R:4, I:3, K:5 and 44 non-endemic: V:2 (regional category), R:17, I:3, K:22 species and infraspecific taxa.)

IUCN has records of 31 species and infraspecific taxa believed to be endemic; most are succulents. (R:4, I:4, K:23.)

Laws Protecting Plants 49 taxa (mostly whole genera but including all orchids) are specifically protected under Ordinance No. 4 of 1975 (Nature Conservation Ordinance). This also prohibits the picking of any indigenous plant without written permission from the owner of the land.

Useful Addresses

Dept of Agriculture and Nature Conservation, Private Bag x13306, Windhoek 9000.

Additional References

Giess, W. (1962). Some notes on the vegetation of the Namib Desert. *Cimbebasia* 2: 1-35. (Includes annotated list of plants; black and white photographs throughout.)

Giess, W. (1971). A preliminary vegetation map of South West Africa. *Dinteria* 4: 5-114. (Includes 70 black and white photographs and coloured vegetation map 1:3,000,000.)

Giess, W. and Tinley, K.L. (1968). South West Africa. In Hedberg, I. and O. (1968), cited in Appendix 1. Pp. 250-253.

Werger, M.J.A. (1978), cited in Appendix 1. Citation includes list of relevant chapters.

Nauru

A raised limestone island of 20.7 sq. km in the west-central Pacific Ocean at 0°31'S, 160°56'E. Population 8000. The highest point is 71 m surrounded by a terrace and fringing reef. Vegetation of mixed plateau forest, dominated by *Calophyllum*; a few remaining areas of atoll forest, with *Pandanus* and *Cocos* (Douglas, 1969, cited in Appendix 1). About two-thirds of the island has been mined for phosphates.

4 native fern species, no gymnosperms (Fosberg, Sachet and Oliver, 1982, cited in Appendix 1); no figure for monocotyledons but 87 dicotyledon taxa, of which 35 native

(Fosberg, Sachet and Oliver, 1979, cited in Appendix 1). One endemic, an undescribed *Phyllanthus*.

Nauru will be covered by the *Flora of Micronesia* (1975-), cited in Appendix 1.

Navassa Island

A 3.5 sq. km islet, belonging to U.S.A., at 18°25'N, 75°00'W, 50 km west of the western extremity of Hispaniola in the West Indies. Uninhabited except for lighthouse staff and a large introduced population of goats; no streams or rivers.

102 species of vascular plants, 44 possibly indigenous to the island and only 4 species of trees (Ekman, 1929).

The island rises abruptly from the sea to a table-land. Towards the margin of the table-land forest of low stunted trees; in centre, grass savanna; on lower terraces, similar but more stunted savanna, with cacti and shrubs, usually less than 30 cm (Ekman, 1929).

Ekman, E.L. (1929). Plants of Navassa Island, West Indies. *Arkiv för Botanik* 22A(16): 1-12. Plates.

Nepal

Area 141,414 sq. km

Population 16,107,000

Floristics An estimated 6500 species of flowering plants of which c. 315 endemic; 30 species of gymnosperms, and c. 450 species of ferns (Hara *et al.*, 1978). Many endemics in Western Himalaya do not extend into the wetter Eastern Himalaya (Stainton, 1972). Sino-Japanese floristic elements in east and centre; western Himalayan and Mediterranean elements in west; central Asiatic elements north to Himalayan foothills; Indo-Gangetic elements in southern Himalayan foothills and in the plains (Terai).

Vegetation Tropical moist deciduous or Sal (*Shorea robusta*) forest in northern Terai and valleys of Churia hills below 1000 m, little remaining; tropical evergreen rain forest along river valleys below 1000 m, the richest forests being those in the east; subtropical mixed broadleaved forest (1000-2000 m) with *Schima-Castanopsis* in east, dry oak forest in centre, and Chir Pine (*Pinus roxburghii*) forest in west; moist temperate broadleaved forest, with laurel, evergreen oak and rhododendron at 1500-3000 m, in east and centre; mixed coniferous forests on Churia hills, Mahabharat range (1000-1800 m) and southern Himalayas (above 2450 m). Subalpine forests occur around 3500 m; alpine scrub dominated by birch and rhododendron, and alpine meadows at 4000-4500 m; alpine steppes north of Dhaulagiri-Annapurna massif (Stainton, 1972). Estimated rate of deforestation of closed broadleaved forest 800 sq. km/annum out of a total of 16,100 sq. km (FAO/UNEP, 1981).

Checklists and Floras No modern Flora, but see the *Flora of British India* (Hooker, 1872-1897), cited in Appendix 1. For ferns see Beddome (1892) and, Nayar and Kaur (1972), cited in Appendix 1. Recent checklists of the flora are:

Flora of Eastern Himalaya (1966-1975), 3 vols, by H. Hara (vols 1-2) and H. Ohashi (vol. 3), cited in full in Appendix 1.

Hara, H. *et al.* (1978-1982). *An Enumeration of the Flowering Plants of Nepal*, 3 vols. British Museum (Natural History), London. (1 – gymnosperms, monocotyledons, including keys and notes on distribution; 2-3 – dicotyledons. Vols 1 and 2 by H. Hara, W.T. Stearn and L.H.J. Williams; vol. 3 by H. Hara, A.O. Chater and L.H.J. Williams.)

Other relevant literature:

Banerji, M.L. (1965). Contributions to the Flora of East Nepal. *Rec. Bot. Survey India* 19(2). 90 pp. (Enumeration of 583 dicotyledons; introductory notes on vegetation.)

Kitamura, S. (1955). Flowering plants and ferns. In Kihara, H. (Ed.), *Fauna and Flora of Nepal Himalaya: Scientific Results of the Japanese Expeditions to Nepal Himalaya 1952-1953*, 1. Fauna and Flora Research Society, Kyoto. Pp. 73-290. (Annotated checklist of 34 ferns, 14 gymnosperms and 910 angiosperms; notes on vegetation.)

Malla, S.B., Shrestha, A.B., Rajbhandari, S.B., Shrestha, T.B., Adhikari, P.M. and Adhikari, S.R. (Eds) (1976). *Flora of Langtang and Cross Section Vegetation Survey (Central Zone)*. Bull. Dept of Medicinal Plants no. 6, Kathmandu. 269 pp. (Enumeration of 911 vascular species; northern half of area covered by Langtang National Park; detailed analysis of vegetation types.)

An earlier list, covering about half the flora is:

Malla, S.B., Shrestha, A.B., Rajbhandari, S.B., Shrestha, T.B., Adhikari, P.M., and Adhikari, S.R. (1976). *Catalogue of Nepalese Vascular Plants*. Bull. Dept Medicinal Plants no. 7, Kathmandu. 211 pp. (Lists 308 ferns and fern allies, 24 gymnosperms and 3121 angiosperm species; based mainly on collections by the Dept of Medicinal Plants, address below.)

Field-guides
Polunin, O. and Stainton, J.D.A. (1984). *Flowers of the Himalaya*. Oxford University Press. 580 pp.

Storrs, A. and J. (1984). *Discovering Trees in Nepal and the Himalayas*. Sahayogi Press, Kathmandu. 366 pp. (Descriptions and photographs of nearly 200 species.)

Information on Threatened Plants
Sahni, K.C. (1979). Endemic, relict, primitive and spectacular taxa in eastern Himalayan flora and strategies for their conservation. *Indian J. Forestry* 2(2): 181-190. (Mentions 30 taxa rare or threatened in the Himalayan region, including Nepal; notes on vegetation.)

IUCN/WWF are sponsoring an inventory of endemic and threatened plants, to result in a Nepalese Plant Red Data Book, as part of their Plants Programme.

Botanic Gardens
Royal Botanical Garden, Department of Medicinal Plants, Ministry of Forests, Godawari, Lalitpur.

Useful Addresses

Department of Medicinal Plants, Thapathali, Kamaladi, Kathmandu.

Royal Nepal Academy, Kamaladi, Kathmandu.

The King Mahendra Trust for Nature Conservation, P.O. Box 3712, National Parks Building, Babar Mahal, Kathmandu.

CITES Management Authority: The Director General, Dept of Botany, Thapathali, Kathmandu.

Additional References

Dobremez, J.F. *et al.* (1969-1975). *Cart Ecologique du Nepal.* Documents de Cartographie Ecologique 15: 1-7. Grenoble. (Vegetation maps covering central and eastern Nepal, at 1:50,000 and 1:250,000.)

Hara, H. (1968). *Photo-Album of Plants of Eastern Himalaya.* Inoue, Tokyo. 89 pp. (249 plates with notes on vegetation; in Japanese.)

Khadka, R.B. (1983). Mountain flora and their conservation in Nepal. In Jain, S.K. and Mehra, K.L. (Eds), *Conservation of Tropical Plant Resources.* Botanical Survey of India, Howrah. Pp. 132-141. (Includes outline of vegetation and human impact on mountain flora.)

Majupuria, T.C. (Ed.) (1984). *Nepal – Nature's Paradise (Insight into Diverse Facets of Topography, Flora and Ecology).* White Lotus, Bangkok. 476 pp. (Chapters cover vegetation; checklists of ferns in Nepal; orchids of Kathmandu Valley; economic plants; man and the environment.)

McNeely, J.A. (1985). Man and nature in the Himalaya: what can be done to ensure that both can prosper. 14 pp. (Paper presented to the International Workshop on the Management of National Parks and Protected Areas in the Hindukush, Himalaya. Kathmandu, Nepal, 6-11 May 1985.)

Nakao, S. (1964). *Living Himalayan Flowers.* Mainichi Newspapers, Tokyo. 194 pp. (253 colour plates with chapters covering vegetation and major plant families; introduction to Himalayan plants by S. Kitamura.)

Numata, M. (Ed.) (1983). *Biota and Ecology of Eastern Nepal.* Chiba University, Japan. (Includes plant lists.)

Stainton, J.D.A. (1972). *Forests of Nepal.* Murray, London. 181 pp.

For useful background to the Himalayas see Lall and Moddie (1981), cited in Appendix 1.

A Prospectus for a National Conservation Strategy was prepared in 1983 by His Majesty's Government of Nepal and IUCN as a first step toward the formulation of a complete National Conservation Strategy.

Netherlands

Area 41,160 sq. km

Population 14,339,551 (1983 estimate)

Floristics 1400-1600 native vascular species, estimated by D.A. Webb (1978, cited in Appendix 1) from *Flora Europaea*; 1436 native and naturalized species (Meijden *et al.*, 1983); no endemics. Floristic element: predominantly Atlantic, although the rocky terrain of the far south (Limburg district) supports an isolated central European flora.

Vegetation Natural vegetation grossly modified by agriculture, forestry and urban development; c. 40% of land-surface is man-made, the result of reclamation from the sea. Despite the drainage of the large marsh and peat bog region (the Polders) in the west, a valuable wetland flora still remains in places. The original acid oak woodland of the higher parts of the east and south, and oak/beech woodland with birch, was cleared in the middle of the 19th century. Remaining areas of floristic interest: the Wadden Sea area, dunes along the North Sea, especially the Isle of Voorne, relict heathlands of the Veluwe and the Biesbos delta (J. Mennema, 1984, *in litt.*).

Checklists and Floras Included in the completed *Flora Europaea* (Tutin *et al.*, 1964-1980) cited in Appendix 1. National Floras include:

Heimans, E., Heinsius, H.W. and Thijsse, J.P. (1983). *Geïllustreerde Flora van Nederland*, 22nd Ed. Versluys, Amsterdam. 1242 pp. (Line drawings.)

Heukels, H. and Meijden, R. van der (Ed.) (1983). *Flora van Nederland*, 20th Ed. Wolters-Noordhoff, Groningen. 583 pp. (Line drawings.)

Weevers, T. *et al.* (Eds) (1948-). *Flora Neerlandica: Flora van Nederland*. De Koninklijke Nederlands Botanische Vereeniging, Amsterdam. 9 parts to date. (Line drawings.)

For a detailed checklist see:

Meijden, R. van der, Arnolds, E.J.M., Adema, F., Weeda, E.J. and Plate, C.L. (1983). *Standaardlijst van de Nederlandse Flora 1983*. Rijksherbarium, Leiden. 32 pp.

The Central Bureau of Statistics (CBS – address below) has a data-bank on plant distributions, using a 5 km square grid system (Anon, 1985, cited in Appendix 1).

Field-guides The popular field-guide in English by Fitter, Fitter and Blamey (1974), cited in Appendix 1, has been translated into Dutch and revised by H. Korthof and J. Mennema (1984) (*Elseviers Nieuwe Bloemengids*, Elsevier, Amsterdam). See also:

Heukels, H. and Ooststroom, S.J. van (1968). *Beknopte School-En Excursieflora voor Nederland*, 12th Ed. by S.J. van Oostroom. Wolters-Noordhoff, Groningen. 425 pp.

Information on Threatened Plants A national plant Red Data Book is in preparation (J. Mennema, 1984, *in litt.*). The first 2 volumes of the plant atlas by Mennema *et al.* (1980-) are devoted to extinct, threatened and rare species:

Mennema, J., Quené-Boterenbrood, A.J. and Plate, C.L. (Eds) (1980-). *Atlas van de Nederlandse Flora*, 1 vol. so far, by Kosmos, Amsterdam. English edition by Junk, The Hague. 226 pp. 3 vols planned. (1 – Uitgestorven en zeer zeldzame planten (Extinct and very rare species); contains conservation data and maps for over 300 vascular plant species (native and introduced); ecological and phytogeographical descriptions. 2 (in press) – zeldzame en vrij zeldzame planten (Rare and rather rare species); includes a chapter, by E.J. Weeda, about the changes in the occurrence of vascular plants in the Netherlands; Bohn, Scheltema and Holkema, Utrecht. 3 (in prep.) – Vrij algemene en algemene planten (Rather common and common species); a threatened plant list will be included in the introduction.)

See also:

Leeuwen, C.G. van and Westhoff, V. (1961). De nivellering van flora en vegetatie. *Natura* 58: 132-140.

Mennema, J. (1973). La régression des espèces vègètales en Hollande, basée sur les premiers résultats de l'atlas de la flore néerlandaise en préparation. Rijksherbarium, Leiden. 9 pp. (Mimeo.)

Mennema, J. (1975a). Threatened and protected plants in the Netherlands. *Naturopa* 22: 10-13.

Mennema, J. (1975b). Zeldzame planten tellen (Census of rare plants). *Levende Nat.* 78(2): 29-31.

Quené-Boterenbrood A.J. (1974). Een 'tussenrapport' over zeldzame Nederlandse plantesoorten (An interim report of rare Dutch plant species). *Natuur en Landschap* 28: 297-308.

Westhoff, V. (1956). De verarming van flora en vegetatie (The impoverishment of the flora and vegetation). *In Gedenkboek 50 jaar Natuurmonumenten.* Pp. 151-184. (Not seen.)

Westhoff, V. (1976). Die Verarmung der Niederländischen Gefässpflanzenflora in den letzten 50 Jahren und ihre Teilweise Erhaltung in Naturreservaten (The decline of the Dutch vascular plant flora during the past 50 years and the contribution of nature reserves to its conservation). *Schr.-R. Vegetationskunde* 10: 63-73.

Westhoff, V. (1979). Bedrohung und Erhaltung seltener Pflanzengesellschaften in den Niederlanden. In Wilmans, O. and Tüxen, R. (Eds), *Werden und Vergehen von Pflanzengesellschaften*, Vaduz. Pp. 285-313.

Westhoff, V. and Weeda, E.J. (1984). De achteruitgang van de Nederlandse flora sinds het begin van deze eeuw. (The decline of the Dutch flora since the beginning of the first century). *Natuur en Milieu* 8(8): 8-17.

Wijnands, D.O. (1981). Bedreigde Nederlandse Waterplanten (Threatened Dutch water plants). *Bull. Arbor. Waasland* 4(1): 38-42. (English translation pp. 48-50; describes over 40 species.)

See also a series of papers written by many authors (S.L. van Oostroom, J. Mennema and Th.J. Reichgelt *et al.*) entitled 'Nieuwe vondsten van zeldzame planten in Nederland' (New discoveries of rare plants in the Netherlands) in *Gorteria* from 1964 onwards.

Included in the European threatened plant list (Threatened Plants Unit, 1983, cited in Appendix 1); latest IUCN statistics, based upon this work: non-endemics rare or threatened worldwide – V:5, R:1, I:1 (world categories).

In 1982 IUCN, under contract to the EEC through the U.K. Nature Conservancy Council, prepared a report (unpublished), *Threatened Plants, Amphibians and Reptiles, and Mammals (excluding Marine Species and Bats) of the European Economic Community*, which included a data sheet on 1 Dutch plant, now extinct in the country.

Laws Protecting Plants The Besluit of 6 August 1973 specifies 31 plant species and 5 genera as being absolutely protected. It is prohibited to uproot or take any part of these plants. In addition, it is forbidden to possess these plants, or to offer them for sale, unless they have originated from propagated stock in a nursery or garden. In some provinces and municipals there are local regulations forbidding the collection of certain plants, for example *Eryngium maritimum*.

Voluntary Organizations

Christian Youth Organization for Nature Study (ACJN), Driebergseweg 16, 3708 7B Zeist.

Koninklijke Nederlandse Botanische Vereeniging (KNBV) (Royal Botanical Society of the Netherlands), Lange Nieuwstraat 106, 3512 PN Utrecht.

Netherlands Youth Organization for Nature Study (NJN), Noordereinde 60, 1243 77
's-Graveland.

Royal Naturalists' Organization for the Netherlands (KNNV), Burg. Hoogenboomlaan
24, 1718 B7 Hoogwoud.

Vereniging tot behoud van Natuurmonumenten in Nederland (Society for Nature
Preservation in the Netherlands), Schaep en Burgh, Noordereinde 60, 1243 JJ
's-Graveland.

WWF-Netherlands (Wereld Natuur Fonds), P.O. 7, 3700 AA Zeist.

Botanic Gardens Numerous botanic gardens, as listed in Henderson (1983), cited
in Appendix 1. Only subscribers to the Botanic Gardens Conservation Co-ordinating Body
are listed below:

Arboretum Trompenburg, Groene Wetering 46, 3062 PC Rotterdam.

Botanical Gardens of the State University, Harvardlaan 2, Postbus 80-162, 3508 TC
Utrecht.

Botanische Tuinen en Belmonte Arboretum Wageningen, Generaal Foulkesweg 70, 6703
BL Wageningen.

Botanische Tuin I.V.N.-Elsloo, Op den Berg 7, Elsloo.

Botanische Tuin "Jochum-Hof", Maashoek 2b, Steyl, Gem. Tegelen.

Hortus Botanicus der Katholieke Universiteit Nijmegen, Toernooiveld, 6525 ED
Nijmegen.

Hortus Botanicus der Rijksuniversiteit Leiden, Nonnensteeg 3, 2311 VJ Leiden.

Hortus Botanicus Vrije Universiteit, Postbus 7161, 1007 MC Amsterdam.

University of Amsterdam Botanic Garden, Plantage Middenlaan 2, 1018 DD
Amsterdam.

Useful Addresses

Central Bureau of Statistics (CBS), Department of the Natural Environment, P.O. Box
959, 2270 AZ Voorburg.

Institute for the Investigation of the Vegetation in the Netherlands (IVON),
Schelpenkade 6, 2313 ZT Leiden.

Natuur en Milieu (Foundation for Nature Conservation and Environmental Protection),
Donkerstraat 17, 3511 KB Leiden.

Natuurbeschermingsraad (Nature Conservancy Council), Maliebaan 12, 3581 CN
Utrecht.

Research Institute for Nature Management (RIN), Kasteel Broekhuizen, 3956 ZR
Leersum.

Rijksherbarium, Schelpenkade 6, 2313 ZT Leiden.

Staatsbosbeheer (Government Nature Conservancy Service), P.O. 20020, 3505 CA
Utrecht.

CITES Management Authority: Hoofd van de Directie Natuur-en-
Landschapsbescherming, Ministerie van Landbouw en Visserij, Prins Clauslaan 6,
P.O. 20401, 2500 EK 's-Gravenhage.

CITES Scientific Authority: Adviescommissie wet bedreigde uitheemse diersoorten,
Prins Clauslaan 6, P.O. 20401, 2500 EK 's-Gravenhage.

TRAFFIC (Nederland), Muur 10, 1422 Uithoorn.

Additional References

Bakker, P.A. (1979). Vegetation science and nature conservation. In Werger, M.J.A.
(Ed.), *The Study of Vegetation*. Junk, Den Haag. Pp. 249-288. (Historical and
theoretical account of nature conservation; maps and diagrams.)

Donselaar, J. van (1970). De Nederlandse natuurbescherming gezien in internationaal verband-Botanie (Dutch nature conservation in the context of international botany). In J.C. van de Kramer *et al.*, *Het Veerstoorde Evenwicht*. Oosthoek, Utrecht. Pp. 231-244. (Describes important botanical areas in international context; in Dutch.)

Leeuw, W.C. de (1935). *The Netherlands as an Environment for Plant Life*. E.J. Brill, Leiden. 19 pp. (Describes edaphic, climatic and biotic factors; maps.)

Ministry of Cultural Affairs, Recreation and Social Welfare (1981). *Conservation in the Netherlands: Factsheet on the Netherlands*. 7 pp. (History of growth of nature conservation in the Netherlands, including plants; statistics.)

Ooststroom, S.J. van (1975). Floristic literature published in the Netherlands mainly between 1962 and 1972. *Mem. Soc. Brot.* 24(2): 747-763.

Westhoff, V., Bakker, P.A., Leeuwen, C.G. van and Voo, E.E. van der (1970-1973). *Wilde Planten - Flora en Vegetatie in Onze Natuurgebieden* (Wild Plants - Flora and Vegetation in our Nature Areas), 3 vols. Vereniging tot Behoud van Natuurmonumenten in Nederland. 320 pp, 303 pp, 359 pp. (1 - Algemene inleiding, duinen, zilte gronden; 2 - Het lage land; 3 - De hogere gronden.)

Westhoff, V. and Den Held, A.J. (1975). *Planten Gemeenschappen in Nederland* (Plant communities in the Netherlands). W.J. Thieme and CIE-Zutphen. 324 pp.

Netherlands Antilles

The Netherlands Antilles, two widely separated groups of islands of the Lesser Antilles in the Caribbean, are an integral part of the Kingdom of the Netherlands. The southern group, igneous with coral reefs, comprises Curaçao, Aruba and Bonaire and are less than 100 km off the coast of Venezuela. The northern group, volcanic and within the Leeward Islands, comprise St Eustatius, Saba and the southern part of St Martin (see also under Guadeloupe and Martinique).

Area 993 sq. km

Population 260,000

Floristics Accounts of Flora of the region are incomplete but the study of published Floras revealed 7 species endemic to the southerly group and 12 doubtfully endemic.

Vegetation On the southern group of Curaçao, Aruba and Bonaire xerophytic vegetation of thorny shrubs and cacti; on St Eustatius, Saba and St Martin, where the climate is more humid, vegetation of *Croton* shrubs and some woodland; mostly modified by man.

Checklists and Floras St Eustatius, St Martin and Saba are covered by the *Flora of the Lesser Antilles, Leeward and Windward Islands* (only monocotyledons and ferns published so far, Howard, 1974- , cited in Appendix 1) and by the family and generic monographs of *Flora Neotropica* (cited in Appendix 1). See also:

Arnoldo, M. (A.N. Broeders) (1967). *Handleiding tot het gebruik van inheemse en ingevoerde planten op Aruba, Bonaire en Curaçao*. Uitgare: Boekhandel 'St. Augustinus', Curaçao. 257 pp. (In Dutch, with keys and black and white photographs.)

Arnoldo, M. (A.N. Broeders) (1971). *Gekweekte en Nuttige Planten van de Nederlandse Antillen*. Utigaven van de Natuurwetenschappelijke Werkgroep Nederlandse Antillen, Curaçao no. 20. 279 pp. (In Dutch, with keys and black and white photographs.)

Stoffers, A.L. *et al.* (1963, 1966). *Flora of the Netherlands Antilles*, Uitgaven 'Natuurwetenschappelijke, studierkring voor Suriname en de Nederlandse Antillen', Utrecht. 3 parts. (Covers ferns and 25 angiosperm families.)

Field-guides

Arnoldo, M. (A.N. Broeders) (1964). *Zakflora, wat in het wild groeit en bloeit op Curaçao, Aruba en Bonaire* (Pocket Flora of Curaçao, Aruba & Bonaire.) Uitgaven van de Natuurwetenschappelijke Werkgroep Nederlandse Antillen, Caraçao no. 16. 2nd Ed. 232 pp. (68 plates; in Dutch, with keys.)

Information on Threatened Plants None.

New Caledonia

The French Overseas Territory of New Caledonia, 1200 km east of Australia in the southwest Pacific Ocean, includes the main island of New Caledonia (16,750 sq. km), the Loyalty Islands (2227 sq. km), the Isle of Pines (134 sq. km) and the uninhabited Huon Islands. In addition, Hunter (40.5 ha), Matthew (12 ha) and Walpole (125.5 ha) are 550 km east of the main island, whereas the Chesterfield Islands are 450 km to the west. The highest point, Mt Panié (1649 m), is on the main island. The Loyalty and Huon Islands are low coral; Walpole is a raised limestone island; Hunter and Matthew are active volcanoes.

Population 152,000

Floristics c. 3250 vascular plant species (Morat, *et al.*, 1984), including c. 300 fern species (Parris, 1985, cited in Appendix 1). 2474 endemic vascular plant species (Morat *et al.*, 1984), including all conifers (44 spp.), Cunoniaceae (70-80 spp.), Proteaceae (43 spp.) and palms. 5 endemic families – Amborellaceae (1 sp.), Oncothecaceae (2 spp.), Paracryphiaceae (1-2 spp.), Phellineaceae (10 spp.), Strasburgeriaceae (1 sp.). Pantropical and Indo-Australian genera represent 45% of the rain forest flora, Malesian genera 9.6% (Morat *et al.*, 1984). Lowland rain forests, and maquis scrub on ultrabasic rocks (especially serpentine) have a large number of primitive relict species.

Vegetation Tropical evergreen rain forest up to 1000 m; tropical montane rain forest above 1000 m; a variant of evergreen rain forest, sometimes with *Araucaria columnaris*, dominant near coast on raised coral, especially on Loyalty Islands and Isle of Pines; dry sclerophyllous forest on western slopes; various types of maquis scrub on acidic and ultrabasic rocks (e.g. peridotites and serpentinites), covering about 30% of the land area; mangroves along western coasts. About 50% of the land area covered by secondary forests, savanna and grasslands, due to clearance for mining, logging and agriculture. Hunter has some grassland with occasional trees; Walpole is covered by dense scrub (Douglas, 1969, cited in Appendix 1); Matthew has almost no vegetation.

According to figures of the Forestry Department (quoted in Myers, 1980, cited in Appendix 1), forests of all types cover 16,000 sq. km; however, Thomson and Adloff (1971) estimated that relatively undisturbed rain forest covered only 10% of the territory, and that "the high forest resource will be exhausted in 30-40 years". For a more detailed account of vegetation and maps, see Morat, Jaffré, Veillon and MacKee (1981). See also Schmid (1978).

Checklists and Floras

Aubréville, A., Leroy, J.-F. and MacKee, H.S. (Eds) (1967-). *Flore de la Nouvelle-Calédonie et Dépendances*. Muséum National d'Histoire Naturelle. (13 fascicles so far, covering ferns, gymnosperms, and 25 flowering plant families, including Apocynaceae, Lauraceae, Myrtaceae, Orchidaceae, and Proteaceae.)

Guillaumin, A. (1911). Catalogue des plantes phanérogames de la Nouvelle-Calédonie et Dépendances. *Ann. Mus. Col. Marseille* 19. 86 pp. (Includes checklists; rather dated and incomplete.)

Guillaumin, A. (1948). *Flore Analytique et Synoptique de la Nouvelle-Calédonie – Phanérogames*. Office de la Recherche Scientifique Coloniale, Paris. 369 pp. (Keys to families, genera, species; rather dated.)

Morat, Ph., Veillon, J.-M. and MacKee, H.S. (1984). Floristic relationships of New Caledonian rain forest phanerogams. In Radovsky, F.J., Raven, P. and Sohmer, S.H. (Eds), *Biogeography of the Tropical Pacific*. Bernice P. Bishop Mus. Special Publ. no. 72. Honolulu. Pp. 71-128. (Includes checklist of c. 1000 rain forest species; endemics indicated.)

Sarasin, F. and Roux, J. (Eds) (1914-1921). *Nova Caledonia – Recherches Scientifiques en Nouvelle-Calédonie et aux Iles Loyalty*. Kreidel, Berlin. 311 pp. (Checklists of lower plants, ferns and some flowering plants; chapters on plant geography.)

Separate lists for Hunter, Matthew, Walpole, Chesterfield, Loyalty·and the Huon Islands include:

Cochic, F. (1959). Report on a visit to the Chesterfield Islands, September 1957. *Atoll Res. Bull.* 63. 11 pp. (Lists 20 vascular plant species; notes on vegetation.)

Guillaumin, A. (1973). Contributions à la flore de la Nouvelle-Calédonie, 130: plantes des îles Walpole et Matthew. *Bull. Mus. National d'Histoire Naturelle (Paris), sér. 3*, 192 (Bot., no. 12): 180-183. (Lists 45 species from Walpole, 10 from Matthew.)

Guillaumin, A. and Veillon, J.M. (1969). Plantes des archipels Huon et Chesterfield. *Bull. Mus. National d'Histoire Naturelle (Paris), sér. 2*, 41: 606-607. (Lists 10 species.)

Information on Threatened Plants No published list of threatened plants. 2 palms, *Burretiokentia hapala* and *Cyphophoenix nucele*, are included in *The IUCN Plant Red Data Book* (1978). Latest IUCN statistics: endemic taxa – Ex:1, E:14, V:24, R:108, I:21, nt:48 (mainly covering gymnosperms, Lauraceae, Myrtaceae, Palmae).

Voluntary Organizations It is reported that a nature protection association has recently been formed.

Additional References

Jaffré, T. (1980). *Végétation des Roches Ultabasiques en Nouvelle Calédonie*. Traveaux et Documents no. 124. ORSTOM, Nouméa. Pp. 228. (Includes map.)

Morat, Ph., Jaffré, T., Veillon, J.M. and MacKee, H.S. (1981). *Les Formations Végétales, Carte no. 15 Atlas de la Nouvelle-Calédonie*. ORSTOM, Nouméa. (Scale 1:1,000,000.)

Sarlin, P. (1954). *Bois et Forêts de la Nouvelle-Calédonie*. Centre Technique Forestier Tropical, Nogent-sur-Marne, France. 303 pp. (Includes treatments of principal forest trees.)

Schmid, M. (1978). The Melanesian forest ecosystems (New Caledonia, New Hebrides, Fiji Islands and Solomon Islands). In Unesco/UNEP/FAO (1978), cited in Appendix 1. Pp. 654-683.

Schmid, M. (1981). *Fleurs et Plantes de Nouvelle-Calédonie*. Les éditions du Pacifique. Papeete, Tahiti. 164 pp. (181 taxa with notes on distribution, ecology, and vegetation; many colour photographs.)

Thomson, V. and Adloff, R. (1971). *The French Pacific Islands: French Polynesia and New Caledonia*. Univ. Press, Berkeley, California.

Thorne, R.F. (1965). Floristic relationships of New Caledonia. *Univ. Iowa Stud. Nat. Hist.* 20(7): 1-14.

Virot, R. (1956). La végétation Canaque. *Mem. Mus. Nat. Paris (Bot.)* 7. 398 pp.

New Zealand

Area 268,704 sq. km

Population 3,264,000

Floristics c. 2000 species of flowering plants and ferns; about 81% endemic (Given, 1981a), reaching over 90% in the alpine flora. Over 200 species are shared with Australia. There are also subantarctic and palaeotropical elements (*Flora of New Zealand*, 1961).

Vegetation Kauri (*Agathis australis*) forests in the warmer parts of North Island, north of latitude 38°S; lowland podocarp and mixed podocarp/beech (*Nothofagus*)/hardwood forests along west coast of South Island; beech forests over much of South Island and south of latitude 39°S in North Island, and in montane and subalpine regions (Molloy, 1984); remnants of swamp-forest in west South Island. The forested area is reduced from 80% (1200 years ago) to 26% today, of which 23% consists of montane remnants of the indigenous forests, and 3% plantations of exotic softwoods (Molloy, 1984). Scrubland, wetland and coastal communities have also been seriously depleted.

Checklists and Floras The Flora is:

Flora of New Zealand (1961, 1970, 1980). Vol. 1 by H.H. Allan. Owen, Wellington. 1085 pp. (Ferns, fern allies, gymnosperms, dicotyledons; bibliography.) Vol. 2 by L.B. Moore and E. Edgar. Shearer, Wellington. 354 pp. (Monocotyledons except Gramineae; bibliography.) Vol. 3 by A.J. Healy and E. Edgar. Hasselberg, Wellington. 220 pp. (Adventive monocotyledons; covers 168 introduced species.)

See also:

Eagle, A. (1982). *Eagle's Trees and Shrubs of New Zealand: Second Series*. Collins, Auckland. 382 pp. (405 botanical paintings, notes on distribution, short descriptions.)

Poole, A.L. and Adams, N.M. (1963). *Trees and Shrubs of New Zealand*. Owen, Wellington. 250 pp. (Complete coverage; line drawings of 400 species.)

Field-guides

Cooper, D. (1981). *A Field Guide to New Zealand Native Orchids*. Price Milburn, Wellington. 103 pp.

Mark, A.F. and Adams, N.M. (1979). *New Zealand Alpine Plants*, 2nd Ed. Reed, Wellington. 262 pp.

Moore, L.B. and Adams, N.M. (1963). *Plants of the New Zealand Coast*. Paul's, Auckland and Hamilton. 113 pp.

Richards, E.C. (1956). *Our New Zealand Trees and Flowers*, 3rd Ed. Simpson and Williams, Christchurch. 297 pp.

Salmon, J.T. (1963). *New Zealand Flowers and Plants in Colour*. Reed, Wellington. 203 pp. (Colour photographs and short descriptions of over 500 species arranged according to habitats.)

Salmon, J.T. (1968). *Field Guide to the Alpine Plants of New Zealand*. Reed, Wellington. 326 pp.

Wilson, H.D. (1978). *Field Guide: Wild Plants of Mount Cook National Park*. Field Guide Publications, Christchurch. 294 pp.

Wilson, H.D. (1982). *Field Guide: Stewart Island Plants*. Field Guide Publication, Christchurch. 528 pp.

Information on Threatened Plants New Zealand is covered by a technical loose-leaf Red Data Book (Given, 1976, 1977, 1978), an official RDB (Williams and Given, 1981) and a popular account of threatened plants (Given, 1981a).

Given, D.R. (1976, 1977, 1978). *Threatened Plants of New Zealand: A Register of Rare and Endangered Plants of the New Zealand Botanical Region*. DSIR, Christchurch. (Loose-leaf series of detailed double-paged sheets on 50 selected threatened species.)

Given, D.R. (1976). A register of rare and endangered indigenous plants in New Zealand. *N.Z. J. Bot.* 14(2): 135-149. (Lists 314 taxa under consideration for threatened status.)

Given, D.R. (1981a). *Rare and Endangered Plants of New Zealand*. Reed, Wellington. 154 pp. (Descriptive text, chapters on each threat, with examples, introductory chapters on vegetation; lists 279 taxa, the majority of which are rare and threatened endemics; includes 'Code of Conduct for conservation of wild plants'.)

Williams, G.R. and Given, D.R. (1981). *The Red Data Book of New Zealand: Rare and Endangered Species of Endemic Terrestrial Vertebrates and Vascular Plants*. Nature Conservation Council, Wellington. 175 pp. (Includes data sheets on 66 selected threatened plants.)

For a more comprehensive bibliography of publications and papers on the conservation of New Zealand's flora, see Given (1981a). 11 species from New Zealand, including *Xeronema callistemon* from the Poor Knights, and Hen and Chicken Islands, are included in *The IUCN Plant Red Data Book* (1978). Latest IUCN statistics: endemic taxa – Ex:4, E:41, V:5, R:86, I:23.

Laws Protecting Plants The Native Plants Protection Act (1934) gives limited protection to native plants growing on any Crown Land, or in any State Forest or public reserve, or roads. Under the provisions of the Act it is an offence to take native plants from such land without the consent of the owner or occupier. A few serious weeds are exempt from the Act. A Supreme Court ruling in 1973 decided that the Act does not apply to trees, and does not recognize degrees of endangerment with provision for various levels and types of protection. An extensive revision of the Act is proposed following discussion and public submissions (D. Given, 1984, *in litt.*).

Other legislation giving various degrees and types of protection to threatened plants include:

The Forest Act (1949): makes it illegal to take, destroy or injure without lawful authority, forest produce in, on or from any State Forest land.

The National Parks Act (1980): gives similar protection to plants in National Parks and Reserves administered by the Department of Land and Survey.

The Town and Country Planning Act: has provision for preservation of "trees, bushes, plants, or landscape of scientific, wildlife, or historic interest or visual appeal".

Provisions in the Land Act (1961) make it an offence to interfere with forest, wood or timber, or to remove bark and flax from Crown lands without permission.

Voluntary Organizations
Auckland Botanical Society, c/o Secretary, 14 Park Road, Titirangi, Auckland 7.
Canterbury Botanical Society, P.O. Box 8212, Christchurch.
Waipahihi Botanical Society (Inc.), c/o Secretary, 45 Ingle Avenue, Taupo.
Wellington Botanical Society, c/o Secretary, 116 Korokora Road, Petone.
WWF-New Zealand, 110-116 Courtenay Place, P.O. Box 6237, Wellington.

Botanic Gardens
Auckland City Council Botanic Garden, Private Bag, Wellesley Street, Auckland.
Botany Division Experimental Gardens, DSIR, Private Bag, P.O. Box 237, Christchurch.
Christchurch Botanic Gardens, Parks and Recreation Dept, City Council, P.O. Box 237, Christchurch.
Dunedin Botanic Garden, Parks and Recreation Dept, City Council, P.O. Box 5045, Dunedin.
Massey University Botanic Garden, Palmerston North.
Otari Open-Air Native Plant Museum, Wilton, P.O. Box 2199, Wellington.
Pukeiti Rhododendron Trust (Inc.), P.O. Box 385, New Plymouth.
Pukekura Park, Parks and Recreations Dept, City Council, Private Bag, New Plymouth.
Timaru Botanic Garden, Parks and Recreation Dept, City Council, P.O. Box 522, Timaru.

Index of threatened plants in cultivation:

Threatened Plants Unit, IUCN Conservation Monitoring Centre (1983). *The Botanic Gardens List of New Zealand Threatened Species*. Botanic Gardens Conservation Co-ordinating Body, Report No. 8. IUCN, Kew. 11 pp. (Lists 96 rare and threatened endemic taxa reported in cultivation, with gardens listed against each.)

Useful Addresses
Botany Division, DSIR, Private Bag, Christchurch.
Nature Conservation Council, Box 12/200, Wellington North.

Additional References
Given, D.R. (1981b). Threatened plants of New Zealand: documentation in a series of islands. In Synge, H. (Ed.), *The Biological Aspects of Rare Plant Conservation*. Wiley, Chichester. Pp. 67-80.
Given, D.R. (Ed.) (1983). *Conservation of Plant Species and Habitats*. Nature Conservation Council, Wellington. 128 pp. (Symposium proceedings of 15th Pacific Science Congress, Dunedin, February 1983. See in particular D.R. Given on monitoring and strategies for threatened plant conservation in New Zealand, pp. 83-101; K. Thompson on the status of New Zealand's wetlands, pp. 103-116.)

Molloy, L.F. (1984). The reservation of commercially important lowland forests in New
Zealand. In McNeely, J.A. and Miller, K.R. (Eds), *National Parks, Conservation,
and Development: the Role of Protected Areas in Sustaining Society*. Smithsonian
Institution Press, Washington, D.C. Pp. 394-401. (Proceedings of the World
Congress on National Parks, Bali, Indonesia, 11-22 October 1982.)

Nicaragua

Area 148,000 sq. km

Population 3,162,000

Floristics Not explored botanically in great detail; an estimated 5000 species of
vascular plants (Gentry, 1978, cited in Appendix 1); 57 endemic species known so far
(IUCN figures).

Vegetation In the Mosquitia region tropical moist forest (believed to be the
largest remaining tract in Central America, about 3600 sq. km still undisturbed);
coniferous forest (c. 1300 sq. km); at the upper reaches of the cerros and cordilleras moist
cloud forests; in some summit areas elfin forest, some undisturbed and rich in new species.
Estimated rate of deforestation for closed broadleaved forest 1050 sq. km/annum out of
41,700 sq. km (FAO/UNEP, 1981).

Checklists and Floras Nicaragua is covered by the *Flora Mesoamericana* Project,
described in Appendix 1, as well as by the family and generic monographs of *Flora
Neotropica* (cited in Appendix 1). Also "most plants" of Nicaragua are included in the
completed *Flora of Guatemala* and related articles in *Fieldiana*, outlined under
Guatemala. Country accounts are:

Hamer, F. (1983). Orchids of Nicaragua. Part 2, 3. *Icones Plantarum Tropicarum* 8, 9:
701-900. (Descriptions, illustrations and dot maps.)

Seymour, F. (1980). A check list of the vascular plants of Nicaragua. *Phytologia
Memoirs* 1: 1-314. (List of species based on collections made by the author,
1968-1976.)

A 10-year project to prepare a 2-volume *Flora of Nicaragua Manual* (in Spanish) was
begun in 1977 under the aegis of the Missouri Botanical Garden and the Herbario Nacional
de Nicaragua, Universidad Centroamericana.

Information on Threatened Plants There is no national Red Data Book. IUCN is
preparing a threatened plant list for release in a forthcoming report *The list of rare,
threatened and endemic plants of Middle America*. Latest IUCN statistics, based upon this
work: endemic taxa – Ex:1, R:6, K:48, nt:2; non-endemics rare or threatened worldwide –
V:7, R:6, I:1 (world categories).

7 species are listed as threatened in the Annex to the Convention on Nature Protection and
Wildlife Preservation in the Western Hemisphere (1940). Threatened plants are mentioned
in several papers in:

Prance, G.T. and Elias, T.S. (Eds) (1977), cited in Appendix 1. See in particular W.G.
D'Arcy on endangered landscapes in the region (pp. 89-104), J.T. Mickel on rare
and endangered ferns (pp. 323-328).

Useful Addresses

Herbario Nacional de Nicaragua, Universidad Centroamericana, Apdo 69, Managua.

CITES Management and Scientific Authorities: Departamento de Regulación y Control, Instituto Nicaragüense de Recursos Naturales y de Ambiente (IRENA), Km 12 1/2 Carretera Norte, Apdo Postal 5123, Managua.

Additional References

Ashton, J. (1945). On the plant resources and flora of Nicaragua. In Verdoorn, F. (Ed.), cited in Appendix 1. Pp. 60-64.

Holdridge, L.R. (1962). *Mapa Ecológico de Nicaragua*. Agencia para el Desarollo Internacional de Gobierno de los Estados Unidos de America, Managua.

Niger

Area 1,186,408 sq. km

Population 5,940,000

Floristics 1178 species (Lebrun *et al.*, 1983), 2 dubiously endemic (Brenan, 1978, cited in Appendix 1).

Flora north of c. 16°N with Saharan affinities; flora of central and southern parts, including the Aïr and Ténéré area with Sahelian affinities. In extreme south flora has Sudanian affinities. The Aïr and Ténéré area is especially rich floristically, with even Mediterranean and Afromontane elements.

Vegetation Mostly desert and semi-desert. As rainfall increases further south, semi-desert grassland grades into low wooded grassland with *Acacia tortilis*, increasing in density and height, and grading into Sudanian woodland without characteristic dominants in extreme south. Saharamontane vegetation, including woody shrubland and grassland communities, occurs on the northernmost peaks of Aïr.

For vegetation map see White (1983), cited in Appendix 1.

Checklists and Floras Niger south of c. 18°N is included in the *Flora of West Tropical Africa*. Niger north of c. 16°N is included in *Flore du Sahara* (Ozenda, 1977), and in the computerized *Atlas der Pflanzenwelt des Nordafrikanischen Trockenraumes* (Frankenberg and Klaus, 1980). These are all cited in Appendix 1.

Boudouresque, E., Kaghan, S. and Lebrun, J.-P. (1978). Premier supplément au "Catalogue des plantes vasculaires du Niger". *Adansonia, Sér. 2*, 18(3): 377-390.

Lebrun, J.-P., Boudouresque, E., Dulieu, D., Garba, M., Saadou, M. and Roussel, B. (1983). Second supplément au "Catalogue des plantes vasculaires du Niger". *Bull. Soc. Bot. Fr. 130 (Lettres Bot.)* 1983(3): 249-256.

Peyre de Fabrègues, B. (1979). *Lexique de Noms Vernaculaires de Plantes du Niger*, 2nd Ed. Institut d'Elevage et de Médecine Vétérinaire des Pays Tropicaux, Maisons-Alfort. 156 pp.

Peyre de Fabrègues, B. and Lebrun, J.-P. (1976). *Catalogue des plantes vasculaires du Niger*. Institut d'Elevage et de Médecine Vétérinaire des Pays Tropicaux, Maisons-Alfort, France. 433 pp. (Annotated checklist with botanical bibliography of Niger.)

Information on Threatened Plants

Gillet, H. and Peyre de Fabrègues, B. (1982). Quelques arbres utiles, en voie de disparition, dans le centre-est du Niger. *Rev. Ecol. (Terre Vie)* 36(3): 465-470. (Includes *Khaya senegalensis, Terminalia avicennioides*.)

No published lists of rare or threatened plants; IUCN has records of 4 species and infraspecific taxa believed to be endemic; no categories available.

One species which occurs in Niger (*Olea laperrinei*) is included in *The IUCN Plant Red Data Book* (1978).

Useful Addresses

CITES Management Authority: Ministère de l'hydraulique et de l'environnement, B.P. 241, Niamey.

Additional References

Dundas, J. (1938). Vegetation types of the Colonie du Niger. *Inst. Pap. Imp. For. Inst.* 15. 10 pp. (With small-scale vegetation map.)

Fairbairn, W.A. (1943). Classification and description of the vegetation types of the Niger Colony, French West Africa. *Inst. Pap. Imp. For. Inst.* 23. 38 pp. (With small-scale vegetation map.)

Nigeria

Area 923,850 sq. km

Population 92,037,000

Floristics 4614 species (quoted in Lebrun, 1976, cited in Appendix 1); northern region (as defined in the *Flora of West Tropical Africa*, cited in Appendix 1) with mainly Sudanian (but also Guinea-Congolian) affinities, 39 endemic species; western and central region (38 endemic species) and eastern region (128 endemic species) with Guinea-Congolian affinities. Eastern region especially rich round Oban (Brenan, 1978, cited in Appendix 1).

Vegetation Large areas of mangrove and swamp forest round the Niger River delta. Inland, lowland rain forest, changing gradually to Guinea Savanna of *Isoberlinia doka* woodland, and, in the most northerly regions of Nigeria, Sudanian woodland without characteristic dominants. Montane communities including forest and grassland are found on the Jos Plateau, and in places on high ground near the south and eastern border (Vogel Peak massif, Mambilla Plateau complex, and Obudu Plateau).

Estimated rate of deforestation for closed broadleaved forest 3000 sq. km/annum out of 59,500 sq. km (FAO/UNEP, 1981). However, Myers (1980, cited in Appendix 1) gives the following figures: 45,000 sq. km moist forests remaining, of which 25,495 sq. km are worth classifying as forest reserves and of which 16,000 sq. km are sufficiently stocked to warrant further timber exploitation.

For vegetation map see White (1983), cited in Appendix 1.

Checklists and Floras Nigeria is included in the *Flora of West Tropical Africa*, cited in Appendix 1.

Gbile, Z.O. (1981). Dichotomous key to the Nigerian species of ferns and fern-allies. *Nigerian J. For.* 11(1,2): 33-48.

Keay, R.W.J., Onochie, C.F.A. and Stanfield, D.P. (1960, 1964). *Nigerian trees*, 2 vols. Dept of Forest Research, Ibadan. 334, 495 pp. (Keys, including multi-access key to genera; descriptions, specimens, distributions; line drawings.)

Lowe, J. and Stanfield, D.P. (Eds) (1970-). *The Flora of Nigeria*. Ibadan University Press, Ibadan. (Published in fascicles; only 2 produced so far: Grasses + illustrations, 1970, 118 pp., 58 plates, by D.P. Stanfield; Sedges, 1974, 144 pp., by J. Lowe and D.P. Stanfield. Multi-access keys included.)

Information on Threatened Plants

Chapman, J.D. (1982). Conservation of Afromontane forest: Ngel Nyaki Forest Reserve. *Nigerian Field* 47(1-3): 133.

Gbile, Z.O., Ola-Adams, B.A. and Soladoye, M.O. (1978). Endangered species of the Nigerian flora. *Nigerian J. For.* 8(1,2): 14-20.

Gbile, Z.O., Ola-Adams, B.A. and Soladoye, M.O. (1981). List of rare species of the Nigerian flora. *Research Paper (Forest Series)* 47. Forest Research Institute of Nigeria, Ibadan.

Hedberg, I. (1979), cited in Appendix 1. (Only three species reported as known to be threatened, p. 92.)

Kinako, P.D.S. (1977). Conserving the mangrove forest of the Niger Delta. *Biol. Conserv.* 11(1): 35-39. (Includes map.)

IUCN has records of 282 species and infraspecific taxa believed to be endemic, including E:8; no other categories assigned.

Laws Protecting Plants There are forest laws restricting harvesting of timber trees without permission.

Botanic Gardens

Biological Gardens, University of Ife, Ife-Ife, Oyo State.

Botanical Garden, University of Ibadan, Ibadan, Oyo State.

Dept of Biological Sciences, University of Zaria, Zaria, North Central State.

Useful Addresses

Forestry Research Institute of Nigeria, P.M.B. 5054, Ibadan.

CITES Management Authority: Federal Department of Forestry, Federal Ministry of Agriculture, P.M.B. No. 12613, 6, Ijeh Village, Obalende, Lagos.

Additional References

Charter, J.R. (1968). Nigeria. In Hedberg, I. and O. (1968), cited in Appendix 1. Pp. 91-94.

Keay, R.W.J. (1948-1959). *An Outline of Nigerian Vegetation*. Govt Printer, Lagos. 52 pp. with coloured vegetation map 1:3,000,000 (1948). 2nd Ed., 55 pp. (1953). 3rd Ed. (minor corrections only), 1959.

Ola-Adams, B.A. (1977). Conservation of genetic resources of indigenous forest tree species in Nigeria: possibilities and limitations. *Forest Genetic Resources Inf.* 7: 1-9.

Ola-Adams, B.A. and Iyamabo, D.E. (1977). Conservation of natural vegetation in Nigeria. *Envir. Conserv.* 4(3): 217-226. (With two black and white photographs.)

Niue

Niue (169°55'W, 19°2'S), a self-governing territory associated with New Zealand, is a raised coral plateau 480 km east of Tonga, in the south-west Pacific Ocean. Area 259 sq. km; population 4000. Mutalau Reef reaches 61 m, Alofi Terrace 25 m. Soils are shallow and porous, and on Mutalau Reef only present in pockets. Settlements are found on the coast; much of the interior is uninhabited.

629 vascular plant taxa, of which c. 175 indigenous (Sykes, 1970). Most species are also found on Tonga and the Samoan Archipelago; many are widespread throughout the Pacific. No information on threatened plants.

Niue was originally covered by tropical rain forest, now found only in the centre, east and south-east; coastal forests on the terraces; large areas of secondary forest and scrub in central basin. Shifting cultivation has greatly modified the vegetation over much of the island.

There was an Environmental Protection Ordinance with conservation provisions under consideration in 1975; current status not known.

References
Sykes, W.R. (1970). *Contributions to the Flora of Niue*. DSIR, Bull. no. 100. Christchurch, N.Z. 321 pp.

Yuncker, T.G. (1943). The flora of Niue Island. *Bull. Bernice P. Bishop Mus*. 178. 126 pp.

Norfolk Island

The Norfolk Island complex, an External Territory of Australia, is an isolated volcanic outcrop, c. 800 km north-west of New Zealand, in the south-west Pacific Ocean, at latitude 29°S and longitude 168°E. It comprises Norfolk (36 sq. km), Philip (2.5 sq. km), Nepean and satellite islands.

Area 39 sq. km

Population 1700 (1980 estimate, *Times Atlas*, 1983)

Floristics Norfolk Island has 174 native vascular plant species (Turner *et al.*, 1968), of which 48 are endemic. Philip Island has 3 endemic species (Melville, 1969).

Vegetation On Norfolk Island, forests greatly reduced by clearance for agriculture and settlement, and disturbed by timber exploitation. There are remnants of coniferous, mixed hardwood, palm/hardwood and palm/tree fern forest, particularly in the Norfolk Island National Park (formerly the Mt Pitt Reserve, of area 460 ha), which includes 100 native plants and is the best plant site remaining on the island (Australian National Parks and Wildlife Service, 1984).

Philip Island originally supported a dense forest but has been devastated by introduced pigs, goats and rabbits. Today little vegetation remains; WWF-Australia are sponsoring a rescue project for the endemic *Hibiscus insularis* and the Australian National Parks and Wildlife Service are controlling the rabbits, as described by Coyne (1983).

Checklists and Floras Norfolk Island will be included in a forthcoming volume of the *Flora of Australia* (1981-), cited under Australia.

Turner, J.S., Smithers, C.N. and Hoogland, R.D. (1968). *Conservation of Norfolk Island.* Australian Conservation Foundation Special Publ. no. 1. 41 pp. (Includes checklist of plants on the islands with notes on local distribution, frequency, habitats; chapters on conservation problems and recommendations.)

Information on Threatened Plants 49 vascular plants of Norfolk and Philip islands, with notes on conservation status are listed in Leigh *et al.* (1981), cited under Australia. *Hibiscus insularis* and *Streblorrhiza speciosa* are included in *The IUCN Plant Red Data Book* (1978). See also:

Melville, R. (1969). The endemics of Phillip Island. *Biol. Conserv.* 1: 170-172. (Of the 3 endemics, *Agropyron kingianum* was last seen in 1912, *Streblorrhiza speciosa* is Extinct and *Hibiscus insularis* is Endangered.)

Latest IUCN statistics: endemic taxa – Ex:5, E:11, V:29, I:1; non-endemic taxa rare or threatened worldwide – V:3 (world categories).

Botanic Gardens In 1984 the Norfolk Island Government changed the status of Mt Pitt Reserve to a National Park and decided to establish a Botanic Garden for native species (P. Coyne, 1984, *in litt.*).

Useful Addresses
Australian National Parks and Wildlife Service, P.O. Box 310, Norfolk Island.

Additional References
Australian National Parks and Wildlife Service (1984). *Plan of Management Norfolk Island National Park and Plan of Management Norfolk Island Botanic Garden.* ANPWS, Canberra. 112 pp.
Coyne, P. (1983). Revegetation attempt on Philip Island, South Pacific. *Threatened Plants Newsletter* 12:14.

Norway

Area 323,895 sq. km

Population 4,140,000

Floristics Based on *Flora Europaea*, D.A. Webb (1978, cited in Appendix 1) estimates a flora of 1600-1800 native vascular species; 1 endemic species and 1 endemic subspecies (IUCN figures). Elements: Arctic/alpine, Boreal and Atlantic.

Vegetation Large tracts of vegetation still untouched. Forests, mostly coniferous, occupy c. 30% of country. Species diversity highest in south-east with deciduous forest of oak, elm and lime up to 550 m, replaced by widespread pine and spruce at higher altitudes. Atlantic influence felt only in extreme south-west. Along west coast, forests of birch, oak and alder predominate together with blanket bogs and mires. On the central, longitudinal mountains at 1200-1600 m, alpine flora with dwarf shrubs, while at higher levels plant communities become dominated by cryptogams. In the extreme north and north-east pine and birch gives way to Arctic/alpine vegetation with lichen-tundra accompanied by dwarf shrubs, grasses and rushes.

Checklists and Floras Included in the completed *Flora Europaea* (Tutin *et al.*, 1964-1980) and Lindman's *Nordens Flora* (1964), both cited in Appendix 1. Below are recent national and regional Floras:

Hylander, N. (1953, 1966). *Nordisk Kärlväxtflora*, 2 vols. Almqvist and Wiksell, Stockholm. (Line drawings.)

Lid, J. (1974). *Norsk og Svensk Flora*, 4th Ed. Norske Samlaget, Oslo. 808 pp. (Covers Norway and Sweden; line drawings.)

Nordhagen, R. (1940-1970). *Norsk Flora*, 2 vols. Aschehoug, Oslo. 766 pp., 638 plates. (1 - text with keys; 2 - illustrations of pteridophytes and angiosperms.)

For plant atlases see Hultén (1971), cited in Appendix 1, and:

Faegri, K. *et al.* (Eds) (1960). *Maps of Distribution of Norwegian Vascular Plants. Vol. 1. Coast plants.* Oslo Univ. Press, Oslo. 134 pp. (Describes geography of Norway and ecology of coast plants; 156 taxa mapped.)

Relevant journals: *Blyttia*, Journal of the Norwegian Botanical Society (Norsk Botanisk Forening); *Nordic Journal of* Botany, Copenhagen; *Norsk Natur*, Journal of the Norwegian Society for Nature Conservation (Norges Naturvernforbund).

Information on Threatened Plants The National Council for Nature Conservation (Statens Naturvernrad), address below, publishes the list of threatened Norwegian plant species and undertakes a regular revision of the list every 4 years (Norderhaug, 1984). There is no national plant Red Data Book but see:

Gjerlaug, H.C. (1975). Liste over truede oglleller sjeldne planter i Norge, karsporeplanter og froplanter. Oslo. (Unpublished; includes a list of rare and threatened plants; not seen.)

Gjerlaug, H.C. (1977). Liste over antatt utdødde, truete, sarbare og sjeldne plantearter i Norge. (Unpublished.) 7 pp.

Halvorsen, R. and Fagernaes, K.E. (1980-). Sjeldne og sarbare plantearter i Sør-Norge (Rare and threatened plant species in South Norway). *Blyttia* 38(1): 3-8; 38(3): 127-132; 38(4): 171-179; 40(2): 85-93 (by T. Schumacher, E. Bendiksen and R. Halvorsen); 40(3): 163-173. (English summaries.)

Norderhaug, M. (Ed.) (1984). *Truete Planter og Dyr i Norge* (Threatened Plants and Animals in Norway). Statens Naturvernrad. 24 pp. (A popular booklet prepared by the National Council for Nature Conservation (address below) outlining general problems related to threatened species and their conservation; lists 126 threatened vascular plants with IUCN categories; in Norwegian with English summary; colour illus.)

Norway is included in the Nordic Council of Ministers' threatened plant list (Ovesen *et al.*, 1978, 1982, cited in Appendix 1) and in the European threatened plant list (Threatened Plants Unit, 1983, cited in Appendix 1); latest IUCN statistics, based upon this latter work: endemic taxa – E:1, nt:1; doubtful endemics – V:1; non-endemics rare or threatened worldwide – V:6, R:8, I:1 (world categories).

Laws Protecting Plants Section 13 of the Nature Conservation Act 1970 states that "The King may decide that wild-growing plant species or plant colonies which are rare or are in danger of extinction, shall be protected in the whole country or in specified areas". Only 4 plant species are nationally protected: *Viscum album*, *Aster sibiricus*, *Oxytropis deflexa* subsp. *norvegica* and *Braya purpurascens*. The following taxa have a lower grade of protection: *Saxifraga paniculata*, *Cladium mariscus*, *Papaver radicatum* subsp. *relictum*, *Carex scirpoidea*, *Polemonium boreale* and *Onopordium acanthium*. For details see Koester (1980, cited in Appendix 1).

Voluntary Organizations

Norges Naturvernforbund (The Norwegian Society for Conservation of Nature), Box 8268, Hammersborg, Oslo 1.

Norsk Botanisk Forening (Norwegian Botanical Society), Botanical Museum, University of Oslo, Trondheimsveien 23 B, 0560 Oslo 5.

Nyttevekstforeningen, Botanical Museum, University of Oslo, Trondheimsveien 23 B, 0560 Oslo 5.

WWF-Norway (Verdens Villmarksfond – Norge), Rosenkrantzgt. 22, 0160 Oslo 1.

Botanic Gardens

Botanical Garden, University of Oslo, Trondheimsveien 23 B, 0560 Oslo 5.

Botanisk Hage, P.O. Box 12, 5014 Bergen.

Milde Arboretum, P.O. Box 41, 5067 Store Milde.

Ringve Botaniske Hage, University of Trondheim, 7000 Trondheim.

Useful Addresses

Statens Naturvernrad (National Council for Nature Conservation), Postboks 266, 3101 Tønsberg.

The Norwegian NGO Secretariat on Acid Rain, c/o The Norwegian Society for Conservation of Nature, address as above.

CITES Management Authority: Miljøverndepartementet, Ministry of Environmental Affairs, Postboks 8013 Dep., Oslo 1.

Additional References

Holmboe, J. (1924-1925). *Einige Grundzüge von der Pflanzengeographie Norwegens.* Bergens Museums Aarbok. 54 pp.

Kleppa, P. (1973, 1979). *Norsk Botanisk Bibliografi 1814-1964, and 1964-1975*, 2 vols. Universitetsforlaget, Oslo.

Miljøverndepartementet (1976). Oversikt over omrader og forekomster i Norge som er fredet eller vernet etter naturvernloven, samt omrader og forekomster som er administrativt fredet (List of areas and objects in Norway protected by the Nature Conservation Act, and areas and objects protected by administrative regulations). Norwegian Ministry of Environmental Affairs.

Wielgolaski, F.E. (1971). IBP Ecosystems studies in Norway. *Biol. Conserv.* 4(1): 71-72.

Ogasawara-Gunto

72 small volcanic islands, formerly called the Bonin Islands, 966 km south of Tokyo. Area 73 sq. km; population 1798 (1984 official report). The islands are a dependency of Japan.

483 vascular plant species of which 369 native (Kobayashi, 1978). Of the 70 species of pteridophytes, 25 are endemic; of the 298 angiosperm species, 126 endemic. The only native gymnosperm (*Juniperus taxifolia*) is endemic. The Ogasawaras are floristically distinct from the adjacent Mariana Islands (van Balgooy, 1971, cited in Appendix 1). There are Asian and tropical Pacific elements represented in both broadleaved evergreen and broadleaved deciduous forests (Tuyama, 1972). Many areas cleared for agriculture, grazing and settlements.

Checklists and Floras

Kobayashi, S. (1978). A list of the vascular plants occurring in the Ogasawara (Bonin) Islands. *Ogasawara Research* 1: 1-33. (Annotated checklist, endemics indicated, notes on distribution and habitats.)

A number of species from Ogasawara-Gunto are included in the Atlas by Horikawa (1972, 1976), cited under Japan.

Information on Threatened Plants

Woolliams, K.R. (1978, 1979). Observations on the flora of the Ogasawara Islands. *Notes from Waimea Arboretum* 5(2): 2-10; 6(1): 6-14. (Reports on 19 species, most of which are threatened.)

Woolliams, K.R. (1983). Ogasawara Islands: news from Hahajima. *Notes from Waimea Arboretum* 10(1): 4-5. (Notes on 4 rare or threatened plants.)

Yoshida, A. and Tannawa, T. (1977). Endangered plant species of the Ogasawara Islands. *Notes from Waimea Arboretum* 3(2): 8-12. (Tentative list of 31 'endangered'; 17 'rare' and 6 'depleted' taxa.)

Additional References

Toyota, T. (1981). *Flora of Bonin Island*. Abochsha, Kamakura. 396 pp. (Covers 236 taxa in Japanese, colour photographs.)

Tuyama, T. (1972). The status of the Bonin Islands flora in the Pacific. In Graham, A. (Ed.), *Floristics and Palaeofloristics of Asia and Eastern North America*. Elsevier, Amsterdam. Pp. 79-81. (Discussion of floristic affinities.)

Tuyama, T. and Asami, S. (1970). *The Nature in the Bonin Islands*, 2 vols. Hirokawa Shoten, Tokyo. (1 – Notes on flora and fauna, in Japanese; 2 – coloured illustrations.)

Wilson, E.H. (1919). The Bonin Islands and their ligneous vegetation. *J. Arnold Arb.* 1: 97-115. (Descriptive account with lists of important trees, shrubs and climbers.)

Oman

Area 271,950 sq. km

Population 1,181,000

Floristics A very provisional estimate is 1100 species (Edmondson, 1980; A.G. Miller, 1984, *in litt.*), with up to 50 endemic species (Miller, *in litt.*). Most endemics concentrated in the southern part of Dhofar.

Floristic affinities of the south (southern Dhofar) Sudano-Deccanian, with Africa, Yemen and southern India. Affinities of northern Dhofar and the edge of the Empty Quarter Saharo-Sindian, with the Sahara and north-west India. Affinities of northern Oman: low altitudes Irano-Turanian, with Iran; at higher altitudes, affinities with western Himalayas.

Vegetation Most of country desert to semi-desert with patches of extreme verdure on mountain masses and along the coastal strip. In extreme south Oman (part of Dhofar), on a coastal strip 100 miles long and up to 20 miles deep, the vegetation is rather atypical, consisting of grassland, thick scrub and forest with a very low canopy. On the mountains there is extensive cover of grassland and thicket, and a low juniper "forest" on Jabal

Akhdar in the north. The Musandam peninsula in the north consists mostly of stony mountains with small patches of alluvium producing rich grassy pastures.

Checklists and Floras Works relating to the Arabian peninsula as a whole are outlined under Saudi Arabia. See also:

Mandaville, J.P. (1977). Plants. In Harrison, D.L. *et al.* (Eds), *Scientific Results of the Oman Flora and Fauna Survey 1975*. J. Oman Studies, Special Report, Ministry of Education and Culture. Pp. 229-267. (Includes list of plant species collected from the mountains of northern Oman.)

Radcliffe-Smith, A. (1979). Flora. In *Interim Report on the Results of the Oman Flora and Fauna Survey, Dhofar, 1977*. Sultanate of Oman. Pp. 41-48. (Selected species only; brief description of vegetation, with colour plates.)

A new checklist of the flora of Oman by A.G. Miller and R. Whitcombe is in preparation which they hope will be published in 1985.

Field-guides

Mandaville, J.P. (1978). *Wild Flowers of Northern Oman*. Bartholomew Books, London. 64 pp. (85 species, illustrated in colour by D. Bovey; available in Arabic and English.)

Information on Threatened Plants No published lists of rare or threatened plants. Two species from Oman are included in *The IUCN Plant Red Data Book* (1978): *Dionysia mira* and *Ceratonia* sp. nov., now *Ceratonia oreothauma*.

Voluntary Organizations There is a Natural History Group, based at the Natural History Museum.

Useful Addresses

Conservation Adviser, Diwan of Royal Court Affairs, The Palace, Muscat.

Natural History Museum, Ministry of National Heritage and Culture, P.O. Box 668, Muscat.

Additional References

Edmondson, J.R. (1980). Botanical collections from Oman, February-March 1980. Unpublished report. 16 pp.

Radcliffe-Smith, A. (1980). The vegetation of Dhofar. In *Scientific Results of the Oman Flora and Fauna Survey 1977 (Dhofar)*. J. Oman Studies, Special Report No. 2. Pp. 59-86. (Includes systematic plant list.)

A group of IUCN consultants will be in Oman in 1985 to investigate setting up reserves, conservation laws, etc.

Pakistan

Area 803,941 sq. km

Population 98,971,000

Floristics 5500-6000 vascular plant species (M.N. Chaudhri, 1984, *in litt.*); c. 300 of these are endemic (Ali, 1978). Centres of endemism are in western and northern mountain regions, over 1200 m (Ali, 1978).

Vegetation Little natural vegetation left due to agricultural encroachment, overgrazing and urbanization. In the south (Sind, Baluchistan), extensive areas of semi-desert and desert, with tropical thorn scrub of *Prosopis*, *Capparis* and *Acacia*; subtropical dry evergreen scrub to 1000 m, greatly modified by grazing and fuelwood harvesting; small patches of broadleaved forests with *Quercus* and *Juglans* on mountains above 1500 m; Chir Pine (*Pinus roxburghii*) forest between 900-1650 m; temperate coniferous forest between 1650-3000 m; "juniper tracts" in Baluchistan at 2000-3000 m with *Juniperus macropoda*, *Pistacia* and *Fraxinus*; alpine scrub between 2850-3600 m with *Abies*, *Betula* and *Rhododendron*; riverine forest along the Indus; mangrove swamps along the Sind coast (Stewart, 1982). Estimated rate of deforestation of closed broadleaved forest 10 sq. km/annum out of a total of 8600 sq. km (FAO/UNEP, 1981).

Checklists and Floras The national Flora is:

Nasir, E. and Ali, S.I. (Eds) (1970-1979). *Flora of West Pakistan*, continued as *Flora of Pakistan* (1980-). Pakistan Agricultural Research Council, Islamabad. (Presently 157 fascicles, covering c. 160 families.)

See also:

Stewart, R.R. (1972). *An Annotated Catalogue of the Vascular Plants of West Pakistan and Kashmir*. Islamabad. 1028 pp. (Part of Nasir and Ali, 1970-1979, *Flora of West Pakistan*, cited above; enumeration of 128 ferns, 23 gymnosperms, c. 1140 monocotyledons and c. 4500 dicotyledon taxa, including introductions.)

Pakistan is included in the *Flora of British India* (Hooker, 1872-1897), cited in Appendix 1, and for ferns in Beddome (1892) and the companion volume by Nayar and Kaur (1972), both of which are cited in Appendix 1. The North-West Frontier Province is covered by the incomplete *Flora Iranica* (1963-), cited in Appendix 1. Other relevant works include:

Bamber, C.J. (1916). *Plants of the Punjab: A Descriptive Key to the Flora of the Punjab, North-West Frontier Province and Kashmir*. Govt Printing Press, Lahore. 652 pp.

Jafri, S.M.H. (1966). *The Flora of Karachi (Coastal West Pakistan)*. Book Corporation, Karachi. 375 pp. (Covers 403 native vascular plant species.)

Information on Threatened Plants On 18 March 1984 WWF-Pakistan launched a Plant Conservation Programme, which includes the identification of threatened plants. A preliminary list of c. 500 species, suspected to be rare or threatened, has been compiled by K.H. Sheikh, based on accounts of c. 160 families in Nasir and Ali (1978-).

IUCN also has a preliminary list by S.I. Ali which includes 8 threatened medicinal plants and 10 other rare species.

Voluntary Organizations
WWF-Pakistan, P.O. Box 1312, Lahore.

Botanic Gardens
Botanical Garden, Karachi.
Botanical Gardens, Punjab University, New Campus, Lahore.
Botanical Gardens, University of Agriculture, Faisalabad.
Government College Botanical Garden, Lahore.
Pakistan Forest Institute Botanical Garden, Peshawar.

Useful Addresses

National Herbarium, Pakistan Agricultural Research Council, House 97G, Street 1, F-7/4, Islamabad.

CITES Management and Scientific Authority: National Council for Conservation of Wildlife in Pakistan, Ministry of Agriculture and Cooperatives, Government of Pakistan, 4-G, Street No. 51 F. 6/4, Islamabad.

Additional References

Ali, S.I. (1978). The flora of Pakistan: some general and analytical remarks. *Notes Roy. Bot. Gard. Edinburgh* 36: 427-439. (Describes 'Flora of Pakistan' project; notes on endemism.)

Kitamura, S. (Ed.) (1964). *Plants of West Pakistan and Afghanistan*. Kyoto Univ. 283 pp. (Results of Kyoto Univ. expedition to Karakoram and Hindukush, 1955.)

Stewart, R.R. (1982). History and exploration of plants in Pakistan and adjoining areas. In Nasir, E. and Ali, S.I. (Eds), *Flora of Pakistan*. Pakistan Agricultural Research Council, Islamabad. 186 pp. (Published as a separate fascicle of the Flora.)

Panama

Area 78,513 sq. km

Population 2,134,000

Floristics An estimated 8000-9000 species of vascular plants (Gentry, 1982); 1226 endemic taxa (IUCN figures). Areas high in endemism are Santa Rita Ridge, El Valle de Anton and Cerros Azul, Pirre, Campana, Jefe and Pilon.

Vegetation Principally tropical forest; under the Holdridge system, Panama has the following Life Zones: Tropical Moist Forest (extensive areas along the Caribbean seaboard), Tropical Dry Forest (mainly in the south), Subtropical Moist Forest (small strips surrounding Tropical Dry Forest), Montane Wet Forest and Lower Montane Wet Forest (along the Cordillera Central east from Costa Rica, the former small in extent), surrounded by Subtropical Dry Forest on either side; and two Transition Zones: Tropical Moist Forest (Transition) (the major part of the east of Panama, from the Darién to the Pacific coast), and Tropical Dry Forest (Transition) (smaller areas also in the east) (Holdridge and Budowski, 1959; Porter, 1973). Also extensive mangrove, especially on Pacific coast. Panama's largest and most species-rich forest is in Darién Province, partly protected but much threatened.

According to FAO/UNEP (1981), estimated rate of deforestation for closed broadleaved forest 360 sq. km/annum out of 41,650 sq. km; according to Myers (1980, cited in Appendix 1), out of 40,816 sq. km officially classified as forests, 38,873 sq. km is lowland rain forest and 1736 sq. km moist montane forest. But he states that reputedly at least 10,000 sq. km of these have been seriously disrupted by slash-and-burn agriculture, especially on the Panama Canal watersheds.

Checklists and Floras Panama is covered by the *Flora Mesoamericana* Project, described in Appendix 1, as well as by the family and generic monographs of *Flora Neotropica* (cited in Appendix 1). The country Flora is:

Woodson, R.E. *et al.* (1943-1980). Flora of Panama. *Ann. Missouri Bot. Gard.* 30-67. (All families covered, but numerous new species and new records discovered subsequently.)

Staff at the Missouri Botanical Garden are making final revisions to the Flora of Panama Database in preparation for the first *Flora of Panama Checklist*. The following cover parts of Panama:

Croat, T. (1978). *Flora of Barro Colorado Island*. Stanford Univ. Press, California. 943 pp. (Includes account of vegetation and map; covers 1369 taxa of vascular plants for a 15.6 sq. km island in Gatun Lake.)

Johnston, I.M. (1949). Flora of San Jose Island. *Sargentia* 8: 1-306.

Standley, P.C. (1928). Flora of the Panama Canal Zone. *Contrib. U.S. Nat. Herb.* 27: 1-416.

Information on Threatened Plants There is no national Red Data Book. IUCN is preparing a threatened plant list for release in a forthcoming report *The List of rare, threatened and endemic plants of Middle America*. Latest IUCN statistics, based upon this work: endemic taxa – Ex:2, E:19, V:35, R:92, I:51, K:830, nt:197; non-endemics rare or threatened worldwide – E:5, V:31, R:16, I:3 (world categories).

Threatened plants are mentioned in several papers in:

Prance, G.T. and Elias, T.S. (Eds) (1977), cited in Appendix 1. See in particular W.G. D'Arcy on endangered landscapes in the region (pp. 89-104), J.T. Mickel on rare and endangered ferns (pp. 323-328), H.E. Moore on endangerment in palms (pp. 267-282).

Botanic Gardens

Summit Gardens, Smithsonian Tropical Research Institute, address below.

Useful Addresses

Smithsonian Tropical Research Institute, P.O. Box 2072, Balboa, Panama; and APO, Miami, Florida 34002, U.S.A.

CITES Management Authority: Dirección Nacional de Recursos Naturales Renovables (RENARE), Ministerio de Desarrollo Agropecuario, Apdo 2016-Paraíso, Corregimiento de Ancón, Panamá 5.

Additional References

Croat, T.B. and Busey, P. (1975). Geographical affinities of the Barro Colorado Island Flora. *Brittonia* 27: 127-135.

Gentry, A.H. (1982). Phytogeographic patterns as evidence for a Chocó Refuge. In Prance, G.T. (Ed.) (1982), cited in Appendix 1. Pp. 112-136.

Holdridge, L.R. and Budowski, G. (1959). *Mapa Ecológica de Panama*. Instituto Interamericano Ciencias, Agrícolas, Turrialba, Costa Rica. (Life Zone map.)

Porter, D.M. (1973). The vegetation of Panama: a review. In Graham, A. (Ed.) (1973), cited in Appendix 1. Pp. 168-201. (Review of knowledge and papers on Panamanian vegetation.)

RARE (1983). *Draft plan for the development of a private sector initiative in natural resource and environment programs in the Republic of Panama*. RARE, c/o WWF-U.S. (address under U.S.A.) 42 pp.

Tosi, J.A., Jr. (1970). *Mapa Ecológico de Panama*. Programa de las Naciones Unida para Desarollo, Rome.

Papua New Guinea

Area 462,840 sq. km

Population 3,601,000

Floristics Good (1960) estimates that the island of New Guinea, of which Papua New Guinea is the eastern portion, has c. 9000 angiosperm species, of which 90% endemic, and Parris (1985, cited in Appendix 1) estimates that it has c. 2000 fern species; R. Johns (1984, *in litt.*) believes there to be more than 11,000 vascular plant species. There are 1465 genera in New Guinea, of which 124 are endemic (van Balgooy, in Paijmans, 1976). According to Gressitt (1982), 55% of the vascular flora of Papua New Guinea is endemic. The flora of the lowland forests is mainly related to that of Malesia, whereas the montane flora is mainly related to that of Australasia (Paijmans, 1976).

Vegetation c. 85% has some form of forest cover, including subclimax secondary forest. Tropical lowland evergreen and semi-evergreen rain forests on the coastal plains; these forests differ markedly from those elsewhere in Malesia with dipterocarps poorly represented; extensive limestone rain forests, especially in west (Whitmore, 1984). Extensive mangroves, brackish-water forest with the salt-water palm (*Nypa fruticans*), freshwater swamp-forest with sago palm (*Metroxylon sagu*), particularly along the south coasts around the Gulf of Papua and along the Sepik and Fly rivers; *Saccharum/Imperata* grassland on alluvial plains; dry evergreen forests and mixed savanna woodlands, with *Acacia* and Proteaceae in south-west. Much of the country is mountainous, rising to over 4000 m in the central highlands; montane forests above 1200 m with *Araucaria*, *Castanopsis*, *Lithocarpus*, *Nothofagus*; alpine vegetation above 4000 m (e.g. on Mt Wilhelm). Extensive *Imperata* and *Themeda* grasslands due to fires, and forest clearance for shifting and permanent agriculture (Paijmans, 1976; Whitmore, 1975b, cited in Appendix 1).

Estimated rate of deforestation of closed broadleaved forest 220 sq. km/annum out of a total of 337,100 sq. km (FAO/UNEP, 1981); however, c. 40,000 sq. km of forestland have been either denuded or rendered unproductive, and another 2500 sq. km, including at least 250 sq. km of primary forest, are cleared each year for shifting cultivation (figures quoted in Myers, 1980, cited in Appendix 1). Over 1000 sq. km of primary forest were cleared in 1978 as a result of commercial logging (figures quoted in Myers, 1980).

For vegetation maps see:

Haantjens, H.A. (Ed.) (1964-1965). *CSIRO Land Research Series*. Melbourne. (Land use and evaluations; includes maps of land use and forest types at 1:250,000. Vol. 10 – Buna-Kokoda area; 12 – Wanigela-Cape Vogel area; 14 – Port Moresby-Kairuku area.)

Paijmans, K. (1975). *Vegetation Map of Papua New Guinea* (1:1,000,000) and *Explanatory Notes to the Vegetation Map of Papua New Guinea*. Land Research Series no. 35, Melbourne.

Papua New Guinea is included on the *Vegetation Map of Malaysia* (van Steenis, 1958), and on the vegetation map of Malesia (Whitmore, 1984), both covering the *Flora Malesiana* region at scale 1:5,000,000 and cited in Appendix 1.

Checklists and Floras Papua New Guinea is included in the incomplete but very detailed *Flora Malesiana* (1948-), cited in Appendix 1. National accounts include:

Handbooks of the Flora of Papua New Guinea (1978-). Melbourne Univ. Press.
(2 vols so far. Vol. 1, 1978, edited by J.S. Womersley, includes introductory
chapter on vegetation; keys, treatments of Combretaceae, Magnoliaceae, Meliaceae
and many smaller families. Vol. 2, 1981, edited by E.E. Henty, covers
Elaeocarpaceae, Juglandaceae, Loranthaceae and other families.)

Johns, R.J. and Stevens, P.F. (1971). Mount Wilhelm flora: a checklist of the species.
Bot. Bull. Dept Forests Papua New Guinea 6. 60 pp. (Checklist of high-altitude
flora.)

Royen, P. van (1959). *Compilation of Keys to the Families and Genera of Angiosperms
and Gymnosperms in New Guinea*, 3 vols. Rijksherbarium, Leiden.

Royen, P. van (Ed.) (1980-1983). *The Alpine Flora of New Guinea*, 4 vols. Cramer,
FL-9490, Vaduz, Liechtenstein. (1 – Comprehensive account of physical and
biological features; 2 – keys and descriptions of gymnosperms, monocotyledons; 3,4
– angiosperms.)

Streimann, H. (1983). *The Plants of the Upper Watut Watershed of Papua New
Guinea*. National Botanic Garden, Canberra. 209 pp. (2114 vascular plant taxa,
including 325 fern taxa, in checklist of collections with notes on ecology.)

See also the botanical results of the Archbold expeditions, including treatments of new
species and revisions of some families and genera, based principally on expeditions made
between 1933 and 1939. Published as:

Merill, E.D. and Perry, L.M. (Eds) (1939-1949). Plantae Papuanae Archboldianae, 1-8.
J. Arnold Arbor. 20-30.

Perry, L.M. (1949-1953). Plantae Archboldianae, 9-13. *Ibid.*, 30: 139-165; 32: 369-389;
34: 191-257.

Smith, A.C. (1941-1944). Studies of Papuasian plants, 1-6. *Ibid.*, 22: 60-80; 22:
231-252; 22: 497-528; 23: 417-443; 25: 104-298.

Contributions to the flora and vegetation of New Guinea have been published in the
journal *Nova Guinea* (Contributions to the anthropology, botany, geology and zoology of
the Papuan region).

Field-guides

Havel, J.J. (1975). *Training Manual for the Forestry College 3(2): Botanical
Taxonomy*. PNG Forestry Dept, Port Moresby. 317 pp. (Identification manual for
forest botanists.)

Johns, R.J. (1975-1976). *Common Forest Trees of Papua New Guinea*, 12 parts.
Forestry College, Bulolo. (Keys, descriptions and drawings; 1-3 revised 1983,
Forestry Dept, PNG Univ., Lae.)

Johns, R.J. (1979-). *The Ferns and Fern Allies of Papua New Guinea*, 12 parts so far.
(Parts 1-5: Forestry College, Bululo; parts 6-12: PNG University of Technology,
Lae.)

Royen, P. van (1964-1970). *Manual of the Forest Trees of Papua and New Guinea*,
9 parts. Forestry Dept, Port Moresby. (Descriptions, keys, line drawings of selected
trees. Part 1 – Combretaceae, revised 1969 by M.J.E. Coode; Forestry Dept, Lae.
Other parts include Anacardiaceae, Dipterocarpaceae, Sapindaceae and
Sterculiaceae.)

Information on Threatened Plants None, but lists of rare and threatened species
are to be compiled by a recently established 'Flora Committee' (J.R. Croft, 1985, *in litt.*).

See also:

Kores, P. (1977). Papua New Guinea's orchids: an exploited resource. *Science in New Guinea* 5: 51-66. (Paper presented to the PNG Botanical Society, Wau Ecology Institute, October 1977; describes the threats to orchids.)

Specht, R.L., Roe, E.M. and Boughton, V.H. (Eds) (1974). *Conservation of Major Plant Communities in Australia and Papua New Guinea*. Australian J. Bot. Supp. Series 7. 667 pp. (Detailed assessment of conservation status of major plant communities.)

Botanic Gardens

Gardens of the University, Box 4820, University, Port Moresby.

National Botanic Garden and Herbarium, Office of Forests, Division of Botany, P.O. Box 314, Lae.

Useful Addresses

Department of Forestry, University of Lae, Institute of Technology (Unitec), Lae.

Wildlife Branch, Office of Environment and Conservation, P.O. Box 6601, Boroko.

CITES Management Authority: Director of Forestry, Department of Primary Industry, Frangipani Street, P.O. Box 5055, Boroko.

Additional References

Gressitt, J.L. (Ed.) (1982). *Biogeography and Ecology of New Guinea*, 2 vols. Junk, Hague. (1 – General and physical background, man's impact, vegetation and flora; 2 – fauna, conservation.)

Paijmans, K. (Ed.) (1976). *New Guinea Vegetation*. Elsevier, Amsterdam. 213 pp. (Includes lists of medicinal and other useful species.)

Paraguay

Area 406,750 sq. km

Population 3,576,000

Floristics An estimated 7000-8000 plant species (R. Spichiger, pers. comm., quoted in Toledo, 1985, cited in Appendix 1). One of the least known countries botanically in South America.

Vegetation West of the Río Paraguay, which bisects the country from north to south, the Gran Chaco, a plain covered with savannas and xerophytic scrub vegetation. In the centre, around the Río Paraguay, seasonally inundated swamp (the Pantanal). East of the river, where 96% of the population live, is palm savanna with fertile grasslands and wooded hills in the south; in the north subtropical seasonal evergreen forest, exploited commercially (FAO/UNEP, 1981). On the eastern border, along the Río Parana, where rainfall is heaviest, the only tropical rain forest in Paraguay, 25,000-40,000 ha of it affected by the Itapú Dam (FAO/UNEP, 1981). Further west, as rainfall decreases, the vegetation changes to subtropical seasonal evergreen lowland forest.

Estimated rate of deforestation for closed broadleaved forest 1900 sq. km/annum out of 40,700 sq. km (FAO/UNEP, 1981).

Checklists and Floras About half of Paraguay, that part north of the Tropic of Capricorn, is covered in the family and generic monographs of *Flora Neotropica*, described in Appendix 1. The country Floras are:

Chodat, R. and Hassler, E. (1898-1907). *Plantae Hasslerianae*. Bull. Herb. Boissier, Geneva. Many parts, in Vols 3-5 and 7, each cited in Bertoni, Mascherpa and Spichiger (1982), below. (Most complete checklist available.)

Lopez, J. (1979). *Arboles de la Región Oriental del Paraguay*. Asunción. 227 pp.

Michalowski, M. (1954). *Catálogo sistemático de las malezas del Alto Paraguay*. Servicio Técnico Interamericano de Cooperación Agrícola, Boletín No. 169, Asunción. 158 pp. Mimeo. (Annotated list.)

Spichiger, R. and Bocquet, G. (Eds.) (1983-). *Flora del Paraguay*. 10 year project co-ordinated by the Geneva Herbarium, Switzerland, to produce a multi-part Flora, in Spanish, to be printed by the Missouri Botanical Garden. Includes a computerized database. 2 vols so far, including Annonaceae, by R. Spichiger and J.-M. Mascherpa.

Teague, G.W. (1965). Plants of central Paraguay. *Anales del Museo de Historia Natural, Serie 2*, 7(4): 1-55. Montevideo. (List of species collected, with economic and medicinal values.)

Boletín del Inventario Biológico (Biological Inventory News), published quarterly by the Museo Nacional de Historia Natural del Paraguay, Servicio Forestal Nacional, outlines present and future activities on the inventory of plants and animals and on the establishment of a natural history museum.

Information on Threatened Plants None.

Voluntary Organizations

Sociedad de Botánica y Zoología del Paraguay, Caja de Correo 811, Asunción.

Botanic Gardens

Jardín Botánico y Zoológico, Parque y Museo de Historia Natural, Asunción.

Useful Addresses

Departamento de Manejo de Bosques, Parques Nacionales y Vida Silvestre, Ministerio de Agricultura y Ganadería, Edificio Patria, Tacuary 443-4° piso, Asunción.

Herbarium, Conservatoire et Jardin Botaniques de la Ville de Genève, Case Postal 60, 1292 Chambésy, Switzerland.

Museo Nacional de Historia Natural del Paraguay, Projecto de Inventario Biológico Nacional, Edificio Patria, Piso 6, Tacuary 443, Asunción.

CITES Management Authority: Director del Departamento de Control Agrícola Forestal y Conservación de Recursos Naturales, Ministerio de Agricultura y Ganadería, Calle Pte. Franco 472, Asunción.

CITES Scientific Authority: Jefe del Departamento de Manejo de Bosques, Parques Nacionales y Vida Silvestre, Servicio Forestal Nacional, Ministerio de Agricultura y Ganadería, Edificio Patria, Tacuary 443-4° piso, Asunción.

Additional References

Arenas, P. (1981). *Ethnobotánica Lengua-Maskoy*. Fundación para la Educación, la Ciencia y la Cultura. Buenos Aires, Argentina. 358 pp. (Flora and ethnobotany of the Chaco.)

Bertoni, B.S., Mascherpa, J.-M. and Spichiger, R. (1982). Datos bibliográficos para el estudio de la vegetación y de la flora del Paraguay. *Candollea* 37: 277-313. (Bibliography on Paraguay vegetation; in French and Spanish.)

Cabrera, A.L. (1970). La vegetación del Paraguay en el cuadro fitogeográfico de América del Sur. *Bol. Soc. Arg. Bot.* 2, suppl. (Not seen.)

Chodat, R. and Vischer, W. (1916-1920?). La végétation du Paraguay. *Bull. Soc. Bot. Genève*. 14 parts in vol. 8 onwards. Reprinted in 1977 as one book by Cramer,

FL-9490, Vaduz, Liechtenstein. Includes many line drawings of individual plants and some species descriptions, but is not a Flora in the normal sense.

Esser, G. (1982). *Vegetationsgliederung Und Kakteenvegetation Von Paraguay*. Akad. d. Wiss. u. d. Literatur, Mainz, Germany. 113 pp. (Vegetation map.)

Peru

Area 1,285,215 sq. km

Population 19,197,000

Floristics Gentry (1980) suggested that "well over" 20,000 vascular plant species will eventually be found in Peru. 14,000 species will be included in the published Flora (A. Gentry, 1984, pers. comm.). Over 900 orchid species (Schweinfurth, 1958, 1970).

Vegetation Along the coast, extending inland up to 100 km, is a desert region that includes the northern part of the Atacama Desert. Due to precipitation from fog, vegetation is luxuriant (the Loma Formation), consisting of annuals, shrubs and scattered trees, with many succulents and high in endemics (Ferreyra, 1953, 1977). In extreme northwest is woodland (the Algarrobal Formation) (Ferreyra, 1977), much degraded from agriculture and grazing (Unesco, 1981, cited in Appendix 1). Further inland and into the Andes to 4000 m are deciduous thickets and montane deciduous scrub. Above 4000 m, on both slopes, are the páramo (open herbaceous/grass communities, mainly north of 8°S.) and the puna, which is drier, colder and includes dwarf shrubs (mainly 12-16°S.). The eastern slopes of the Andes range from moist to wet depending on elevation: at 2000-4000 m subhumid montane forests and scattered thorn forests; below 2000 m evergreen submontane and semi-deciduous forests. In eastern Peru tropical rain forest constitutes over 610,000 sq. km (8.7%) of the Amazonian forest (Unesco, 1981); in southern Amazon dry forest similar to the Cerrado of Brazil.

Estimated rate of deforestation for closed broadleaved forest 2600 sq. km/annum out of 693,100 sq. km (FAO/UNEP, 1981). Areas protected include the vast Manu National Park (15,328 sq. km), which spans Andean and Amazonian vegetation and may contain more plant species than any other protected area in the world.

Checklists and Floras Peru is covered by the family and generic monographs of *Flora Neotropica*, as described in Appendix 1. Floristic knowledge of the Tumbes (Pacific) region is summarized by Gentry (1978), cited in Appendix 1. The country Floras are:

Flora of Peru (various authors) (1936-). *Field Mus. Nat. Hist., Bot. Ser.* 13 (1-5C), and *Fieldiana, Bot. New Series* 5, 7, 9, 10 and 11. About 3/4 completed, recently re-activated as joint project of Field Museum and Missouri Botanical Garden, in collaboration with Universidad Major de San Marcos and Universidad Nacional de Amazonía Peruana. Includes an ecological inventory and a search for economically useful plants.

Schweinfurth, C. (1958-1961). Orchids of Peru. *Fieldiana, Bot.* 30(1-4).

Schweinfurth, C. (1970). First supplement to the orchids of Peru. *Fieldiana, Bot.* 33: 1-80.

Tryon, R. (1964). The ferns of Peru: Polypodiaceae (Dennstaedtieae to Oleandreae). *Contr. Gray Herb.* 194: 1-253. (176 species, which, according to the author, is about 1/4 of the Peruvian fern flora.)

See also:

Cerrate de Ferreyra, E. (1979). *Vegetación del Valle de Chiquian: Provincia de Bolognesi, Departamento de Ancha.* Los Pinus, Lima. 65 pp.

Vargas Calderón, C. (1974). *La Flora del Departamento de Madre de Dios (Perú).* Universidad Nacional Mayor de San Marcos, Lima. 93 pp. (Species lists.)

Williams, L. (1936). Woods of northeastern Peru. *Field Mus. Nat. Hist., Bot. Ser.* 15: 1-587. (Descriptions of trees, their habitats, local uses, physical properties and wood structures; plant associations in N.E. Peru.)

Information on Threatened Plants There is no national Red Data Book. See:

Ferreyra, R. (1977). Endangered plant communities in Andean and Coastal Peru. In Prance, G.T. and Elias, T.S. (Eds) (1977), cited in Appendix 1. Pp. 150-157. (Includes several endangered species lists for different vegetation types.)

Gentry, A.H. (1977). Endangered plant species and habitats of Ecuador and Amazonian Peru. In Prance, G.T. and Elias, T.S. (Eds) (1977), cited in Appendix 1. Pp. 136-149.

Individual threatened plants are mentioned in several other papers in:

Prance, G.T. and Elias, T.S. (Eds) (1977), cited in Appendix 1. See in particular J.T. Mickel on endangered and rare ferns (pp. 323-328), H.E. Moore Jr. on endangerment in palms (pp. 267-282), and P. Ravenna on endangered bulbous plants (pp. 257-266).

Other references:

Dourojeanni, M.T. (1968). Estado actual de la conservación de la flora y la fauna en el Peru. *Ciencia Interamericana* 9(106): 51-64.

Laws Protecting Plants Ley Forestal y de Fauna Silvestre, Decreto Ley No. 21147 of 13 May 1975, assigns the Ministerio de Agricultura jurisdiction over plants (and animals). Decreto Supremo No. 158-177-AG of 13 March 1977 directs the Ministerio to assign plants (and animals) to specified threatened categories, and thereby afford them legal protection (Fuller and Swift, 1984, cited in Appendix 1).

Voluntary Organizations

Asociación Peruana para la Conservación (APECO), Atahualpa 335, Lima 18.

Pro Defensa de la Naturaleza (PRODENA), Avenida Nicolas de Pierola, 742, Of. 703, Edificio Internacional, Lima.

A relevant reference:

Lieberman, G.A. and Swift, B. (1984). *The development of private voluntary organizations dealing with natural resources and environmental management in Peru and a strategy for enhancement of their programs.* RARE, c/o WWF-US, 1601 Connecticut Avenue N.W., Washington, D.C. 20009, U.S.A. (RARE have also produced a list of these organizations, 1983.)

Botanic Gardens

Jardín Botánico de la Universidad Nacional Agraria, Apto 456 La Molina, Lima.

Jardín Botánico de la Universidad Nacional de Huanuco "Hermilio Valdizan", Cuidad Universitaria, Cayhuayna.

Jardín Botánico de la Universidad Nacional Mayor de San Marcos, Jiron Puno 1002, Lima.

Useful Addresses

Centro de Datos para la Conservación, Departamento de Manejo Forestal, Universidad
Agraria La Molina, Apdo 456, Lima.

Museo de Historia Natural "Javier Prado", Ave. Arenales 1256, Apto Postal 11010,
Lima 14.

CITES Management and Scientific Authorities: Dirección General Forestal y de Fauna,
Ministerio de Agricultura, Jiron Natalio Sanchez 220, 3° Piso, Jesus Maria, Lima.

Additional References

Ellenberg, H. (1959). Typen tropischer urwältder in Peru. *Schweiz. Ziets. Forstw.* 110:
109-187.

Ferreyra, R.H. (1953). Communidades vegetales de algunas Lomas Costaneras del
Peru. *Estación Experimental Agrícola de "La Molina"*, Bol. No. 53. 88 pp.

Ferreyra, R.H. (1960). Algunas aspectos fitogeográficos de Perú. *Rev. Inst. Geogr.* 6:
41-88.

Gentry, A.H. (1980). The Flora of Peru: A conspectus. *Fieldiana, Bot. New Series* 5:
1-73.

Mapa Ecologico del Peru: Guia Explicativa. Oficina Nacional de Evaluación de
Recursos Naturales (ONERN), 1976.

Svenson, H.K. (1945). Vegetation of the coast of Ecuador and Peru and its relation to
the Galapagos Islands. *Am. J. Bot.* 33: 394-498.

Weberbauer, A. (1936). Phytogeography of the Peruvian Andes. *Field Mus. Nat. Hist.,
Bot. Ser.* 13: 13-81.

Weberbauer, A. (1945). *El mundo vegetal de los Andes Peruanos*. Ministerio de
Agricultura, Lima.

Philippines

The Philippines comprise 7100 islands of which 800 are inhabited; the largest are Luzon
(104,688 sq. km), Mindanao (94,630 sq. km) and Visayas (50,000 sq. km). Nearly all the
larger islands have interior mountain ranges; the highest point is 2954 m at Mt Apo on
Mindanao.

Area 300,000 sq. km

Population 53,395,000

Floristics c. 8000 flowering plant species of which 3500 are endemic (Madulid,
1982); c. 900 fern species (Parris, 1985, cited in Appendix 1). Floristic affinities with
Borneo, Malaysia, the Sino-Himalayan region and Australia. The island of Palawan has
c. 1500 flowering plant species; while species endemism was once thought to be as high as
15%, recent research based on *Flora Malesiana* suggests it is nearer 5% (A. Podzorski,
1984, pers. comm.).

Vegetation Tropical forests originally covered most of the Philippines; now
extensively deforested. Most remaining forests modified by shifting cultivation ('kaingin');
the only extensive areas of forest outside national parks are on Palawan. Where native
vegetation survives, it consists of mixed dipterocarp forest up to 800 m; tropical montane
and subalpine (mossy) forests; small areas of 'molave' forest on limestone, with
Pterocarpus and *Vitex*; seasonally dry monsoon forest on the western coastal strip; pine

forests (with *Pinus insularis* and *P. merkusii*) in uplands of north and west Luzon and Mindanao; extensive kogon grassland *(Imperata cylindrica)*, scrub and secondary forest. Mangroves cover 2450 sq. km (FAO/UNEP, 1981).

According to FAO/UNEP (1981), estimated rate of deforestation of closed broadleaved forest 900 sq. km/annum out of a total of 93,200 sq. km (only 31% of the country); according to government surveys using Landsat imagery for 1972-1976 and aerial photographs, forests covered an estimated 114,616 sq. km, of which 60,119 sq. km were "full-canopy forests" (figures quoted in Myers, 1980, cited in Appendix 1). Between 800 and 1400 sq. km of forest (including previously logged forest) are converted to agriculture by 'kaingineros' each year; Landsat surveys reveal that between 1971 and 1976 forests were converted to other land uses at an average rate of 3000 sq. km/annum (figures quoted in Myers, 1980).

The Philippines are included on the *Vegetation Map of Malaysia* (van Steenis, 1958), and on the vegetation map of Malesia (Whitmore, 1984), both covering the *Flora Malesiana* region at scale 1:5,000,000 and cited in Appendix 1.

Checklists and Floras The Philippines are included in the incomplete, but very detailed *Flora Malesiana* (1948-), cited in Appendix 1. The standard checklist is:

Merrill, E.D. (1923-1926). *An Enumeration of Philippine Flowering Plants*, 4 vols. Bureau of Printing, Manila. (Annotated list including 5532 endemic vascular plant taxa.)

Other accounts include:

Brown, W.H. (Ed.) (1920). *Minor Forest Products of Philippine Forests*, 3 vols. Bureau of Forestry, Manila. (Comprehensive account of Philippine plants with economic uses.)
Brown, W.H. (1951-1958). *Useful Plants of the Philippines*, 3 vols. Bureau of Forestry, Manila. (Reprinted edition.)
Copeland, E.B. (1958-1960). *Fern Flora of the Philippines*, 3 parts. Monogr. Philippine Inst. Sci. Tech., 6. Manila.
Merrill, E.D. (1912). *Flora of Manila*. Bureau of Science, Manila. 490 pp. (Reprinted 1974, Bookmark, Manila.)
Pancho, J.V. (1983-). *Vascular Flora of Mount Makiling and Vicinity (Luzon; Philippines)*. Kalikasan Suppl., New Mercury, Quezon. (Vol. 1 – Introduction to flora, keys, treatments of all gymnosperms and c. 60 angiosperm families including Dipterocarpaceae, Leguminosae and Moraceae. 476 pp. Vols 2-4 – in prep.)

Descriptions of new species are often published in the *Philippine J. Science*.

Information on Threatened Plants The main list is:

Gutierrez, H.G. (1974). The endemic flowering plant species of the Philippines. Bound manuscript, 242 pp. (List of 5221 endemic taxa, assigned to earlier IUCN numerical system, 0-4, to indicate degree of threat. Taxonomy rather dated and degree of threat based mainly on literature and number of herbarium specimens.)

Threatened plants are also listed in:

Madulid, D.A. (1982). Plants in peril. *Filipinas Journal* 3: 8-16. (Mentions 20 threatened plants, lists 2 plants on Appendix I and 5 on Appendix II of CITES.)
Quisumbing, E. (1967). Philippine species of plants facing extinction. *Araneta J. Agric.* 14: 135-162. (Lists about 100 taxa at risk, including non-endemics.)

Laws Protecting Plants Act No. 3983 provides protection for the flora and prescribes conditions under which plants may be collected, kept, sold, exported and for other purposes. The Bureau of Forestry is charged with enforcing this legislation.

Presidential Decree No. 1152 establishes specific environment management policies and prescribes environmental quality standards. Among the provisions are "conserving threatened flora as well as increasing their rate of propagation".

Presidential Decree No. 1586 requires the submission of Environmental Impact Statements on projects in critical areas among which are "those which constitute the habitats of any endangered or threatened species of indigenous Philippine wildlife (flora and fauna)".

Voluntary Organizations

Association of Systematic Biologists of the Philippines, c/o Botany Division, National Museum, P.O. Box 2659, Manila.

Philippine Wildlife Conservation Foundation, Bancora Building, Amorsolo Street, Legaspi Village, Makati, Rizal, Luzon.

Botanic Gardens

Makiling Botanic Gardens, University of the Philippines, Los Baños, Laguna 3720.

Manila Zoological and Botanical Garden, Harrison Park, Malate, Manila.

Philippine National Botanic Garden, Real, Quezon.

The Hortorium, Museum of Natural History, U.P. at Los Baños, College, Laguna.

Useful Addresses

Bureau of Plant Industry, San Adres, Manila.

Forest Research Institute, College, Laguna 3720.

CITES Management Authority: Director Bureau of Forest Development, Ministry of Natural Resources, Visayas Avenue, Diliman, Quezon City.

CITES Scientific Authority: Forest Research Institute, U.P.L.B., College of Forestry, Laguna.

Additional References

Brown, W.H. (1919). *Vegetation of Philippine Mountains*. Manila. 434 pp. (Useful data on physical environment, forest types; many black and white photographs.)

Madulid, D.A. (in prep.). *A Dictionary of Philippine Plant Names*, 3 vols. (Alphabetical list of 35,000 local names, cross-referenced to scientific names.)

Nemenzo, C.A. (1969). The flora and fauna of the Philippines, 1851-1966: an annotated bibliography. Part 1: Plants. *Nat. Appl. Sci. Bull. Univ. Philipp.* 21. 307 pp. (1493 entries; not seen, citation from Frodin.)

Whitford, H.N. (1911). *The Forests of the Philippines*, 2 parts. Philippine Bureau of Forestry Bull. 10(1) (94 pp.) and 10(2) (113 pp.). Manila. (1 – Forest types and products; 2 – descriptions of forest types.)

Pitcairn Islands

The Pitcairn Island District is a British Dependent Territory, situated in the middle of the South Pacific Ocean, to the west of the Society Islands, between latitudes 23-26°S and longitudes 125-128°W. It consists of the volcanic Pitcairn Island, Henderson Island (an elevated limestone island), Ducie Island and Oeno Atoll.

In 1982 world attention was focused on Henderson Island following a proposal to the U.K. Government for the construction of a settlement and airstrip there. This prompted 2 reviews of the biological importance of Henderson as one of the very few elevated atolls with vegetation intact; these are Fosberg *et al.* (1983) and Serpell *et al.* (1983). A popular account is given by Serpell (1983). The U.K. Government refused the application. Various conservation groups are now pressing for all or part of the Pitcairn Islands to be nominated for inscription as a World Heritage Site under the Convention Concerning the Protection of the World Cultural and Natural Heritage, adopted in Paris, 1972 (otherwise known as the World Heritage Convention).

Area 43.5 sq. km

Pitcairn: 5 sq. km; Henderson: 37 sq. km; Ducie: 0.7 sq. km; Oeno: 0.8 sq. km

Population 63, all on Pitcairn Island (1981 census, *Times Atlas*, 1983)

Floristics By far the richest flora is on Henderson, which has 9 ferns, all of which are widespread, and 54 native angiosperm taxa, 10 of which are endemic (Fosberg, Sachet and Stoddart, 1983). Oeno has 2 ferns and 15 angiosperms including 2 endemic taxa (St John and Philipson, 1960). Ducie has only 3 species; no endemics (St John and Philipson, 1962). No figure for number of angiosperms on Pitcairn Island, but Brownlie (1961) lists 20 fern species, of which 2 are endemic.

Vegetation Henderson is one of the least disturbed Pacific islands. It has dense scrub forest c. 5-10 m tall, with *Pandanus tectorius*; the central part of the island is more sparsely vegetated (St John and Philipson, 1962). Pitcairn Island has remnants of rain forest, scrub and grassland, but the vegetation has been greatly modified, particularly in the south and centre.

Checklists and Floras The most recent list is that of Fosberg, Sachet and Stoddart (1983), cited below. The Pitcairn Islands are also included in the *Flora of Southeastern Polynesia* (Brown and Brown, 1931-1935), cited in Appendix 1. See also:

Brownlie, G. (1961). Studies on Pacific ferns, 4. The pteridophyte flora of Pitcairn Island. *Pacific Science* 15(2): 297-300. (Annotated list of 20 ferns.)

St John, H. and Philipson, W.R. (1960). List of the flora of Oeno Atoll, Tuamotu Archipelago, south-central Pacific Ocean. *Trans. R. Soc. N.Z.* 88(3): 401-403. (Checklist of ferns and angiosperms.)

St John, H. and Philipson, W.R. (1962). An account of the flora of Henderson Island, South Pacific Ocean. *Trans. R. Soc. N.Z. Bot.* 1(14): 175-194. (Brief description of 8 ferns, 55 angiosperms.)

Information on Threatened Plants For Henderson Island, the reviews by Fosberg *et al.* (1983) and Serpell *et al.* (1983) outline the known status of the flora; *Bidens hendersonensis*, one of the endemics, was included in *The IUCN Plant Red Data Book* (1978); for the other islands no information.

Additional References

Fosberg, F.R., Sachet, M.-H. and Stoddart, D.R. (1983). Henderson Island (Southeastern Polynesia): summary of current knowledge. *Atoll Res. Bull.* 272. 47 pp. (Physical geography, history of exploration, vegetation, bibliography; includes revised list of vascular plants of Henderson.)

Rehder, H.A. and Randall, J.E. (1975). Ducie Atoll: its history, physiography and biota. *Atoll Res. Bull.* 183. 40 pp.

Serpell, J. (1983). Desert island risk. *New Scientist* 1356: 320. (Describes the
importance of Henderson Island and threats to its flora and fauna; reprinted in
Threatened Plants Newsletter 11: 14, 1983.)

Serpell, J., Collar, N., Davis, S. and Wells, S. (1983). Submission to the Foreign and
Commonwealth Office on the future conservation of Henderson Island in the
Pitcairn Group. WWF-UK, IUCN, ICBP. 27 pp. Mimeo.

Poland

Area 312,683 sq. km

Population 37,228,000

Floristics 2250-2450 native vascular species, estimated by D.A. Webb (1978,
cited in Appendix 1) from *Flora Europaea*; 3 endemics (IUCN figures). Elements:
Atlantic, Central European, southern Boreal, Arctic/alpine. Many species reach their
western or eastern distributional limit in Poland.

Vegetation Principally an agricultural landscape, especially in the north and
central lowlands; less than 25% of the country has natural or semi-natural vegetation.
Extensive re-afforestation with pine and spruce, particularly in the foothills of the
Carpathians. In the north-east, fragments of natural forest with oak, lime and hornbeam,
including alder, ash and conifers, notably in the Bialowiéza National Park; in the south,
patches of beech/fir woodland, subalpine spruce and alpine pine forests. Original
extensive cover of steppe grassland confined to poor soils and steep slopes, but swamp and
peat bogs still widespread (Szafer and Zarzycki, 1972).

Checklists and Floras Poland is covered by the completed *Flora Europaea* (Tutin
et al., 1964-1980, cited in Appendix 1). For national Floras see:

Raciborski, M., Szafer, W., Pawlowski, B. and Jasiewicz, A. (Eds) (1919-1980). *Flora
Polska: Rosliny Naczyniowe Polski i Ziem Osciennych*, 14 vols. Polska Akademia
Umiejetnosci, Cracow (vols 1-6); Panstwowe Wydawnictwo Naukowe, Warsaw (vols
7-14).

Szafer, W., Kulczynski, S. and Pawlowski, B. (1967). *Rosliny Polskie* (Flora of
Poland). Panstwowe Wydawnictwo Naukowe, Warsaw. 1020 pp. (Covers post-1945
Poland; illus.)

Atlases:

Bialobok, S. *et al.* (Ed.) (1963-). *Atlas rozmieszczenia drzew i krzewów w Polsce (Atlas
of distribution of trees and shrubs in Poland)*. Panstwowe Wydawnictwo Naukowe,
Warsaw. (Large-scale dot maps, accompanied by notes in Polish, Russian and
English.)

Kulczynski, S. and Madalski, J. (Eds) (1930-). *Atlas Flory Polskiej*. Polska Akademia
Umiejetnosci (1930-1936), Cracow, then Polska Akademia Nauk (1954-), Warsaw.
(From 1954 entitled *Atlas Flory Polskiej i Ziem Osiennych*; atlas with descriptive
commentary in Polish and Latin; illus.; 21 vols planned.)

Field-guides

Fabiszewski, J. (1971). *Rosliny Sudetow*, Atlas (Plant atlas of the Sudety Mts).
Warsaw. 160 pp. (Not seen.)

Zarzycki, K. and Zwolinska, Z. (1984). *Rosliny Tatr Polskich* (Plants of the Polish Tatra Mts). Warsaw. 160 pp. (Not seen.)

Information on Threatened Plants A 5-year Red Data Book programme was begun in 1981, under the co-ordination of A. Jasiewicz, Cracow. The first phase of the programme involved extensive field study to identify threatened species throughout the country and record their localities. As a result, a Red List was published (Jasiewicz, 1981). The second phase, the production of a Red Data Book, is underway; ecological and phytosociological data will be provided for each species, together with dot maps, threats and conservation proposals. A more popular Red Data Book is also in preparation by K. Zarzycki, Cracow, which will include species data on approximately 200 threatened plants.

Jasiewicz, A. (1981). Wykaz gatunków rzadkich i zagrozonych flory polskiej – List of rare and endangered plants from the Polish flora. *Fragmenta Floristica et Geobotanica* 27(3): 401-414. (English abstract and summary; more than 400 threatened taxa listed.)

Other relevant publications:

Falinski, J.B. (Ed.) (1976). Synantropizacja szaty roslinnej VI. Wymieranie skladników flory polskiej i jego przyczyny (Synanthropization of plant cover VI. The decline and extinction of native plant species in Poland). *Phytocoenosis* 5(3/4): 157-409. (22 research papers in Polish with English summaries.)

Jasnowska, J. and Jasnowski, M. (1977). Zagrozone gatunki flory torfowisk (Endangered plant species in the flora of peatbogs). *Chron. Przyr Ojczysta* 33(4): 5-14. (Includes 3 lists of endangered species and 1 of protected species; English summary.)

Kornas, J. (1971a). Changements récents de la Flore polonaise. *Biol. Conserv.* 4(1): 43-47.

Kornas, J. (1971b). Uwagi o wspólczesnym wymieraniu niektórych gatunkow roslin synantropijnych w Polsce (Recent decline of some synanthropic plant species in Poland). *Mater. Zakt. Fitosoc. Stos. U.W.* 27: 51-64.

Kornas, J. (1982). Man's impact upon the flora: processes and effects. *Memorabilia Zool.* 37: 11-30.

Michalik, S. (1979). Zagadnienia ochrony zagrozonych gatunków roslin w Polsce (Some problems on the conservation of threatened plant species in Poland). *Ochr. Przyr.* 42: 11-28. (English summary; not seen.)

Molski, B.A. (1979). The relationship between the national reserves and the activities of botanic gardens in plant genetic resource conservation. In Synge and Townsend (1979), cited in Appendix 1. Pp. 53-62. (Lists endemic and protected species.)

Included in the European threatened plant list (Threatened Plants Unit, 1983, cited in Appendix 1); latest IUCN statistics, based upon this work: endemic taxa – E:2, R:1; non-endemics rare or threatened worldwide – V:11, R:7, I:2 (world categories).

Laws Protecting Plants The 1983 Order of the Ministry of Forestry and Timber Industry (Dz. U. No. 27, item 134), based on Article 15 of the Law on the Preservation of Nature (7 April, 1949; Dz. U. No. 25, item 180), grants complete protection for over 190 vascular plant species and 10 taxa of fungi, and partial protection for 25 vascular plants, 1 genus and 2 species of lichen. For fully protected species it is prohibited to collect, damage, destroy or sell the entire or any part of the plants. Klosowski (1984-) describes the species protected by this Order and Olaczek (1983) examines its effectiveness.

Klosowski, S. (1984-). Rosliny nowo objete ochrona gatunkowa. *Przyroda Polska* 3: 22-25; 5: 25-26; 6: 23-25. (Describes selected plant species protected under the 1983 Order; colour illus.)

Relevant literature:

Kostyniuk, M. and Mraczek, E. (1961). *Nasze rosliny chronione* (Plant species protected in Poland). Wroclawskie Towarzystwo Naukowe, Wroclaw. 202 pp. (In Polish with short English summary; describes biology, economic potential and distribution of 37 protected species, 5 protected genera and 1 protected family (Orchidaceae); illus.)

Olaczek, R. (1983). O wspólczesnym rozumieniu ochrony gatunkowej roslin (On modern comprehension of plant species protection). *Przyroda Polska* 11: 3-6. (Colour photographs; English summary p. 47.)

Pawlowska, S. (1953). Rosliny endemiczne w Polsce i ich ochrona (Les espèces endémiques en Pologne et leur protection). *Ochr. Przyr.* 21: 1-33. (In Polish, with French summary; maps; illus.)

Szafer, W. (1958). Chronione w Polsce gatunki roslin (Plant species protected in Poland). *Zakl. Ochrony Przyrody PAN, Wyd.* 14: 1-108. (Illus.)

Voluntary Organizations No specific plant conservation societies but see:

Liga Ochrony Przyrody (League for Nature Protection in Poland), c/o Ministerstwo Lesnictwa i Przemyslu Drzewnego (Ministry of Forestry and Timber Industry), ul. Reja 3/5, 00-922 Warsaw.

Botanic Gardens

Arboretum, Lesny Zaklad Doswiadczalny SGGW-AR, Experimental Forests, Warsaw Agricultural University, 96-135 Rogow.

Botanic Garden of Adam Mickiewicz University, ul. Dabrowskiego 165, 60-594 Poznan.

Botanic Garden of Warsaw University, A1 Ujazdowskie 4, 00-478 Warsaw.

Botanic Garden, Universitatis Jagellonicae, ul. Kopernika 27, 31-501 Cracow.

Botanical Garden of Marie Curie-Sklodowska University, ul. Akademicka 19, 20-033 Lublin.

Hortus Botanicus, Instituti Plantarum Medicinalium, ul. Libelta 27, 61-707 Poznan.

Instytut Hodowlii Aklimatyzacji Roslin, Ogrod Botaniczny, Plac Weyssenhoffa nr 11, 85-950 Bydgoszcz P 618.

Instytut Technologii i Analizy Leku Akademii Medycznej w Gdansku, Ogrod Roslin Leczniczych, ul. K. Marks 107, 80-416 Gdansk-Wrzeszcz.

Kornik Arboretum of the Institute of Dendrology, Polish Academy of Sciences, ul. Parkowa 5, 62-035 Kornik.

Ogrod Botaniczny, ul. Krzemieniecka nr 36/38, 94-303 Lodz 28.

University Botanical Garden, ul. Sienkiewicza 21, 50-335 Wroclaw.

Wyzsza Szkola Rolnicza w Poznaniu, Arboretum Goluchow, Poczta Golochow, Powiat Pleszew.

Useful Addresses

Departament Ochrony Przyrody, Ministerstwo Lesnictwa i Przemyslu Drzewnego, (Department for Nature Conservation, Ministry of Forestry and Timber Industry), 02-067 Warsaw, ul. Wawelska 52/54. (Responsible for management of protected areas, ecosystems, and all practical aspects of species conservation, including plants.)

Instytut Botaniki, Polska Akademia Nauk, ul. Lubicz 46, 31-512 Cracow.

Komitet Ochrony Przyrody PAN (Committee for Nature Conservation of the Polish Academy of Sciences), 31-512 Cracow, Lubicz 46. (A committee of voluntary biological experts, from universities and other scientific institutions.)

Straz Ochrony Przyrody (The Guardian of Nature), 02-053 Warsaw, ul. Reja 3/5. (A voluntary body of experts responsible for monitoring the effectiveness of protected areas and other sites of environmental concern.)

Zaklad Ochrony Przyrody i Zasobów Naturalnych Polskiej Akademii Nauk (Research Centre for the Conservation of Nature and Natural Resources), 31-512 Cracow, ul. Lubicz 46. (Main centre for scientific documentation of protected landscapes and species.)

Additional References

Szafer, W. and Zarzycki, K. (Eds) (1972). *Szata Roslinna Polski*, 2nd Ed., 2 vols. Panstwowe Wydawnictwo Naukowe, Warsaw. 615 pp. and 347 pp. (Descriptive account of the vegetation of Poland; English translation of 1st Ed. (1966) by W.H. Paryski, Pergamon, Oxford.)

Portugal

Area 91,631 sq. km

Population 10,008,000

Floristics 2400-2600 native vascular plant species, estimated by D.A. Webb (1978, cited in Appendix 1) from *Flora Europaea*; c. 3000 native vascular species according to A.R. Pinto da Silva (1975). 99 endemic taxa (IUCN figures); 127 according to Parker (1981). Elements: Mediterranean, Atlantic and alpine.

Areas of floristic interest: Serra de Estrêla, Serra de Gerês, Serra de Arrábida, Braganca (serpentine flora), Serra da Penêda, Serra do Bussaco and the Algarve (A.R. Pinto da Silva, 1984, *in litt.*; Polunin and Smythies, 1973, cited in Appendix 1).

Vegetation Original cover of temperate deciduous and evergreen forest largely cleared for agriculture or converted to maquis and garigue. In the north, some original oak forest remains, with ericaceous heathland in montane regions where forest removed; in central Portugal (near the Tagus river) transitional Mediterranean/Atlantic vegetation with evergreen oaks dominant; true Mediterranean influence felt in the south (e.g. *Ceratonia siliqua* and *Chamaerops humilis*). Cork Oak (*Quercus suber*) still abundant as natural stands up to 1600 m and in extensive plantations.

Checklists and Floras Included in the completed *Flora Europaea* (Tutin *et al.*, 1964-1980) and *Med-Checklist*, both cited in Appendix 1. National Floras:

Coutinho, A.X.P. (1939). *A Flora de Portugal (Plantas Vasculares)*, 2nd Ed. Irmãos, Lisboa. 933 pp. (Native and naturalized vascular plants; brief ecological notes; reprinted 1974.)

Franco, J.A. (1971-). *Nova Flora de Portugal (Continente e Açores)*. Sociedade Astoria, Lisboa. 647 pp. (Incomplete, 1 vol. to date: Lycopodiaceae to Umbelliferae; includes the Azores.)

Oliveira Feijão, R. (1960-1963). *Elucidário Fitológico. Plantas vulgares de Portugal continental, insular e ultramarino*, 3 vols. 1328 pp. (Not seen.)

Sampaio, G. (1947). *Flora Portuguesa*, 2nd Ed. Impressa Moderna, Porto. 792 pp. (Covers native and naturalized vascular plants; illus.)

See also:

Parker, P.F. (1981). The endemic plants of metropolitan Portugal, a survey. *Bol. Soc. Brot.*, sér. 2, 53(2): 943-994. (Checklist of Portuguese endemics.)

For a summary of floristic literature see:

Fernandes, A. (1955). Progrès récents dans l'étude de la flore vasculaire du Portugal. *Ann. Soc. Broteriana* 21: 6-25.

Pinto da Silva, A.R. (1963). L'étude de la flore vasculaire du Portugal continental et des Açores les dernières années (1955-1961). *Webbia* 18: 397-412.

Field-guides

Fernandez-Casas, J. and Ceballos, A. (1982). *Plantas Sylvestres de la Peninsula Iberica (Rupicoles)*. H. Blume, Madrid. 430 pp.

Gonzalez, G.L. (1982). *La Guia de Incafo de los Arboles y Arbustos de la Peninsula Iberica*. Incafo, Madrid.

Taylor, A.W. (1972). *Wild Flowers of Spain and Portugal*. Chatto and Windus, London. 103 pp. (Colour and black and white photographs.)

Also see Polunin and Smythies (1973), cited in Appendix 1.

Information on Threatened Plants A list of threatened vascular plants for Portugal is being compiled as part of a data-base project, by the Serviço Nacional de Parques, Reservas e Conservação da Natureza (address below).

See also:

Malato-Beliz, J. (n.d.). Plantes vasculaires à protéger au Portugal. 480 pp. (Unpublished; contains distribution maps for over 450 species of threatened plants.)

Tavares, C.N. (1959). Protection of the flora and plant communities in Portugal. In *Animaux et Végétaux Rares de la Région Méditerranéenne*, Proceedings of the IUCN 7th Technical Meeting, 11-19 September, 1958, Athens, vol. 5. IUCN, Brussels. Pp. 86-94.

Included in the European threatened plant list (Threatened Plants Unit, 1983, cited in Appendix 1); latest IUCN statistics, based upon this work: endemic taxa – Ex:2, E:11, V:12, R:31, I:2, K:31, nt:10; doubtful-endemics – R:2, nt:1; non-endemics rare or threatened worldwide – E:2, V:13, R:8 (world categories).

Laws Protecting Plants The Ministério da Habitação e Obras Publicas (address below) is preparing a list of plants for protection.

Voluntary Organizations

Liga Portuguesa para a Protecção da Natureza (The Portuguese League for the Protection of Nature), Estrada do Calhariz a Benfica, 187 1500 Lisboa.

Botanic Gardens

Instituto de Botânica "Dr Gonçalo Sampaio", Universidade do Porto, Rua do Campo Alegre 1191, 4100 Porto.

Jardím Botânico da Ajuda, Calçada da Ajuda, c/o Instituto Superior de Agronomia, Tapada da Ajuda, 1300 Lisboa.

Jardím Botânico da Universidade, Acres de Jardím, 3049 Coimbra.

Jardím Botânico, Universidade de Lisboa, Rua da Escola Politécnica, 1200 Lisboa.

Jardím e Museu Agrícola Tropical, Calçada do Galvão, Belém, 1400 Lisboa.

Useful Addresses Relevant government departments:

Direcção-Geral das Florestas, Av. João Crisóstomo no. 26, 1000 Lisboa.

Direccão de Serviços de Conservação da Natureza, Rua Filipe Folque no. 46-2, 1000 Lisboa.

Estado do Urbanismo e Ambiente, Ministério da Habitação e Obras Publicas, Rua da Lapa no. 73, 1200 Lisboa.

Ministério da Qualidade de Vida y Secretaria de Estado do Ambiente, Rua do Século no. 51, 1200 Lisboa.

Serviço Nacional de Parques, Reservas e Conservação da Natureza, Rua da Lapa no. 73, 1200 Lisboa.

CITES Management Authority: Serviço de Estudos do Ambiente, Rua Barata Salgueiro 37-50, 1100 Lisboa.

Additional References

Font Quer, P. (1953). *Geografía Botánica de la Península Ibérica.* Muntaner y Simón, Barcelona. 271 pp. (Map; illus.)

Gómez-Campo, C. *et al.* (1984). Endemism in the Iberian Peninsula and Balearic Islands. *Webbia* 38: 709-714.

Gómez-Campo, C. and Malato-Beliz, J. (1985). The Iberian Peninsula. In Gómez-Campo, C. (Ed.) (1985), cited in Appendix 1. Pp. 47-70. (Includes case-studies on selected threatened species.)

Malato-Beliz, J. (1976). Relations entre agriculture et conservation de la végétation naturelle dans la region Méditerranéene. *Collana Verde* 39: 269-290. (Not seen.)

Pinto da Silva, A.R. (Ed.) (1946-1980). De Flora Lusitana Commentarii. *Agron. Lusit.* 8-40. (22 fascicles of taxonomy and floristics.)

Pinto da Silva, A.R. (1975). L'état actuel des connaissances floristiques et taxonomiques du Portugal, de Madère et des Açores, en ce qui concerne les plantes vasculaires. In CNRS, 1975, cited in Appendix 1. Pp. 19-28. (English abstract.)

Rivas-Goday, S. (1955). Los grados de vegetación de la Península Ibérica. *Anal. Inst. Bot. Cavanilles* 13: 269-331.

Vasconcellos, J. de C. (1950). Protecção à Flora do Gerês. *Agron. Lusit.* 12(4): 611-617. (French summary.)

Puerto Rico

The smallest and easternmost of the Greater Antillean Islands, 178.6 km west to east by 58 km north to south, mostly mountainous. Puerto Rico is an autonomous political entity in voluntary association with the United States. Includes Mona Island c. 60 km to the west.

Area 8959 sq. km

Population 3,404,000

Floristics Slightly over 3000 vascular plant species recorded as native or introduced to Puerto Rico and its adjacent islands. 234 (7.7%) of them considered endemic.

Mona Island 393 species of vascular plants, of which 4 are endemic. An estimated 11% of the flora is rare or endangered. (Woodbury, 1973.)

Vegetation Much disturbed montane rain forest; palm brake on slopes and elfin woodland on summits; seasonal forest in the central highlands; dry evergreen vegetation on limestone on inland lowlands; some mangrove swamp on parts of the coast and freshwater swamp at river mouths (Beard, 1949, cited in Appendix 1). Strikingly diverse, Ewel and Whitmore (1973) describe 6 subtropical life zones (in the Holdridge system) and Figueroa and Schmidt (1983) 20 geo-climatic zones. 31.5% forested (Birdseye and Weaver, 1982); 14.2% forested (FAO, 1974, cited in Appendix 1).

Mona Island Dry forest in the uplands covers 74% of the island; tall moist forest covers 6%, only in the lowlands; low cactus growth near cliffs; shrub forest found on the NE and east sides of the island; vegetation greatly modified by man.

Checklists and Floras

Britton, N. and Wilson, P. (1923-1930). Botany of Porto Rico and the Virgin Islands. *Scientific survey of Porto Rico and the Virgin Islands*, 5 (626 pp.) and 6 (663 pp.). New York Academy of Sciences, New York. (Keys, descriptions, general ranges and distributions by island.)

Liogier, A.H. and Martorell, L.F. (1982). *Flora of Puerto Rico and adjacent islands: A systematic synopsis*. Edit. Univ. Puerto Rico, Río Piedras. 342 pp. (Updated checklist with synonyms; general distributions and world ranges.)

See also:

Liogier, A.H. (1965). Nomenclatural changes and additions to Britton and Wilson's "Flora of Porto Rico and the Virgin Islands". *Rhodora* 67(772): 315-361.

Liogier, A.H. (1967). Further changes and additions to the Flora of Porto Rico and the Virgin Islands. *Rhodora* 69(779): 372-376.

Little, E.L., Jr. and Wadsworth, F.H. (1964). *Common trees of Puerto Rico and the Virgin Islands*. Agriculture Handbook No. 249, U.S.D.A. Forest Service, Washington, D.C. 548 pp. (Keys, mainly to families; descriptions, illus., distributions.) Spanish edition by authors and J. Marrero, Editorial UPR, Puerto Rico, 1967.

Little, E.L., Jr. et al. (1974). *Trees of Puerto Rico and the Virgin Islands, Second volume*. Agriculture Handbook No. 449, U.S.D.A. Forest Service, Washington, D.C. 1024 pp. (2nd vol. to Little and Wadsworth, 1964, above; includes endemic, rare and endangered tree species.)

Little, E.L., Jr., and Woodbury, R.O. (1976). *Trees of the Caribbean National Forest, Puerto Rico*. Forest Service Research Paper ITF-20. Institute of Tropical Forestry, Puerto Rico. 27 pp. (Includes a checklist of the trees of the Caribbean National Forest in the Luquillo Mts.)

Proctor, G.R. (1984). *Annotated checklist of Puerto Rican Ferns*; a preliminary report on *Ferns of Puerto Rico* Book Project. Department of Natural Resources, Commonwealth of Puerto Rico. 16 pp.

Stahl, A. (1936). *Estudios Sobre la Flora de Puerto Rico*, 2nd edition, 3 vols. San Juan de Puerto Rico.

Woodbury, R.O., Martorell, L.F. and Garcia Tuduri, J.C. (1971). The Flora of Desecho Island, Puerto Rico. *J. Agric. Univ. Puerto Rico* 55(4): 478-505.

Woodbury, R.O. et al. (1977). *The Flora of Mona and Monito Islands, P.R.* Bulletin 252, Agricultural Experiment Station, Univ. Puerto Rico. 60 pp.

Information on Threatened Plants

Ayensu, E.S. and DeFilipps, R.A. (1978). *Endangered and Threatened Plants of the United States*. Smithsonian Institution and WWF-U.S., Washington, D.C. Pp. 225-232. (Lists 102 'Endangered' and 'Threatened' taxa from Puerto Rico and the Virgin Islands, both U.S. and British, with bibliography.)

Figueroa-Colón, J.C. *et al.* (1984). Directices para la evaluación de áreas naturales en Puerto Rico. Department of Natural Resources, Commonwealth of Puerto Rico. 69 pp. (In Spanish; the most recent report of the most critical elements of the flora and fauna of Puerto Rico and its adjacent islands and the natural areas containing those elements; contains the most recent list of endangered and rare plants of Puerto Rico, brought up to date by the Puerto Rico Natural Heritage Program.)

Little, E.L., Jr. and Woodbury, R.O. (1980). *Rare and Endemic Trees of Puerto Rico and the Virgin Islands*. Conservation Research Report No. 27, U.S.D.A. Forest Service, Washington, D.C. 26 pp.

Woodbury, R.O. *et al.* (1975). *Rare and endangered plants of Puerto Rico: a committee report*. U.S.D.A. Soil Conservation Service and Dept of Natural Resources, Commonwealth of Puerto Rico. 85 pp. (Lists 515 rare and endangered species of endemic and non-endemic plants, with their habitat, distribution and threat.)

The IUCN Plant Red Data Book has one data sheet for Puerto Rico, on *Calyptronoma rivalis*. Threatened plant conservation is discussed in:

Howard, R.A. (1977). Conservation and the endangered species of plants in the Caribbean islands. In Prance, G.T. and Elias, T.S. (Eds) (1977), cited in Appendix 1. Pp. 105-114.

Laws Protecting Plants Included under the U.S. Endangered Species Act of 1973, see under the United States. About 111 plant species are candidates for listing under the Act and a few of them have been formally proposed, but none designated so far.

Voluntary Organizations

Conservation Trust of Puerto Rico, P.O. Box 4747, San Juan, Puerto Rico 00905.

Natural History Society of Puerto Rico, P.O. Box 3936, Hato Rey 00936.

Botanic Gardens

Arboretum and Casa Maria Gardens of the Inter-American University, San German, Puerto Rico.

Jardín Botánico de la Universidad de Puerto Rico, Apartado 4984-G Correo General, San Juan, Puerto Rico 00936.

Jardín Botánico, Estación Experimental Agrícola, Río Piedras, Puerto Rico 00928.

Palmas Botanical Gardens, 130 Candelero Abejo, Humacao, Puerto Rico 00661.

Plant Collection, Mayagüez Institute of Tropical Agriculture, P.O. Box 70, Mayagüez, Puerto Rico 00708.

Useful Addresses

Caribbean Islands Field Station, Fish and Wildlife Service, U.S. Dept of the Interior, P.O. Box 3005, Marina Station, Mayagüez, Puerto Rico 00709.

Estación Experimental Agrícola, Río Piedras, Puerto Rico 00928.

Institute of Tropical Forestry, P.O. Box AQ, Río Piedras, Puerto Rico 00928.

Programa Pro-Patrimonio Natural de Puerto Rico (Puerto Rico Natural Heritage Program), DRN, Puerta de Tierra, Box 5887, Puerto Rico 00906.

Additional References

Birdseye, R.A. and Weaver, P.L. (1982). *The Forest Resources of Puerto Rico*. U.S. Forest Service Southern, Forest Experiment Station Resources Bulletin, New Orleans, La.

Dansereau, P. (1966). Description and integration of the plant communities. In Institute of Caribbean Science, Special Publication No. 1, *Studies in the Vegetation of Puerto Rico*. University of Puerto Rico (Mayagüez). Pp. 3-45.

Ewel, J.J. and Whitmore, J.L. (1973). *The ecological life zones of Puerto Rico and the U.S. Virgin Islands*. Research Paper ITF-18. Institute of Tropical Forestry, Puerto Rico.

Figueroa, J.C. and Schmidt, R. (1983). *Species Diversity and Forest Structure in Lower Montane Wet Serpentine Forest in Puerto Rico*. 7th Symposium of Natural Resources, Dept of Natural Resources of Puerto Rico.

Little, E.L., Jr. (1955). Trees of Mona Island. *Caribbean Forester* 16: 36-53.

Martorell, L.F. *et al.* (1981). *Catálogo de los nombres vulgares y científicos de las Plantas de Puerto Rico*. Boletín 263, Univ. de Puerto Rico, Estación Experimental Agrícola, Río Piedras. 231 pp. (In Spanish; useful bibliography.)

Woodbury, R.O. (1973). The Vegetation of Mona Island. In *Las Islas de Mona y Monito, una evaluacion de sus recursos naturales e historicos*. Vol 2. Estado Libre Asociado de Puerto Rico Oficina de Gobernador Junta de Calidad Ambiental. 2 pgs: Apéndice G.

Qatar

Area 11,437 sq. km

Population 291,000

Floristics 301 species of vascular plants (Batanouny, 1981). Few if any endemics. Affinities with the flora of Iraq.

Vegetation A series of desert communities, often of low shrubs (e.g. *Acacia tortilis*), other perennials, grasses (e.g. *Panicum turgidum*) and salt-tolerant plants, as outlined by Batanouny (1981). Disturbance from cultivation, grazing and oil exploitation.

Checklists and Floras Works relating to the Arabian peninsula as a whole are outlined under Saudi Arabia. Works specifically on Qatar are:

Batanouny, K.H. (1981). *Ecology and Flora of Qatar*. Univ. of Qatar, P.O. Box 2713, Doha. 245 pp. (Keys to genera and species; many species illustrated in colour; includes chapter on vegetation.)

Boulos, L. (1978). Materials for a flora of Qatar. *Webbia* 32(2): 369-396. (List of 260 species and 3 varieties, collected by the author in 1977.)

Information on Threatened Plants None.

Additional References

Batanouny, K.H. and Turki, A.A. (1983). Vegetation of south-western Qatar. *Arab Gulf J. Scient. Res.* 1(1): 5-19.

Réunion

Réunion, a French overseas *département*, is the largest of the volcanic Mascarene islands, situated 780 km east of Madagascar and 200 km south-west of Mauritius, 20°51'-21°22'S, 55°15'-55°54'E. Much of the island is mountainous, so there is less pressure on the native plant communities than on Mauritius and Rodrigues.

Area 2510 sq. km

Population 555,000

Floristics c. 500 species indigenous seed plants (including c. 120 orchids); c. 220 indigenous ferns and fern allies. 30% specific endemism (seed plants); 12% for ferns and fern allies. 63% generic endemism. (Cadet, 1980.) Jacob de Cordemoy (1895) gives 1156 species.

Vegetation Littoral vegetation badly degraded; none left in north-west, but some communities survive on exposed cliffs in south and south-west. Low-altitude (0-600 m) forest in west has almost disappeared. Moist low-altitude mixed evergreen forest (up to 1000 m) exists as fragments in the St-Philippe and Grand-Brûlé area and on cliffs flanking the valleys in the east. Mid-altitude forest (1000-1800 m) better preserved and occurs as several different types: principally mixed evergreen forest called 'bois de couleurs', *Acacia heterophylla* forest, low *Cyathea* forest, swampy dwarf cloud forest with *Pandanus* and *Acanthophoenix*. High-altitude ericoid vegetation (1900-3000 m) extensive, but flora being degraded by tourism, fire and cultivation.

Checklists and Floras Réunion is included in the incomplete *Flore des Mascareignes*, cited in Appendix 1, and in *Les Fougères des Mascareignes et des Seychelles* (Tardieu-Blot, 1960), cited under the Seychelles.

Jacob de Cordemoy, E. (1895). *Flore de l'Ile de la Réunion*. Klincksieck, Paris.
  Reprinted 1972 by Cramer, Lehre, according to Frodin.

Field-guides
Cadet, L.J.T. (1981). *Fleurs et Plantes de la Réunion et de l'Ile Maurice*. Editions du Pacifique, Tahiti. 131 pp. (Incomplete for indigenous flora.)
Cadet, L.J.T. (1984). *Plantes Rares ou Remarquables des Mascareignes*. Agence de Coopération Culturelle et Technique, 13 quai André-Citroën, 75015 Paris. 132 pp. (With 48 photographs.)

Information on Threatened Plants No published lists of rare or threatened plants; latest IUCN statistics: endemic taxa (principally for families covered by the *Flore des Mascareignes*) – Ex:2, E:13, V:9, R:18, I:7, K:8, nt:46; non-endemics rare or threatened worldwide – E:8, V:12, R:9 (world categories).

Two species which occur in Réunion are included in *The IUCN Plant Red Data Book* (1978).

Laws Protecting Plants Exploitation of *Cyathea* (for making pots for horticulture) and *Acanthophoenix* (for palm-cabbage) subject to permission from the Office National des Forêts. The illegal exploitation of *Acanthophoenix* is very severe, and the plant has been wiped out in all accessible areas.

Voluntary Organizations
Société Réunionnaise pour l'Etude et la Protection de l'environnement, B.P. 1012, 97481 St-Denis. (This society publishes an annual journal *Info-Nature, Ile Réunion*,

and has produced a list of 58 species which require legal protection. Recently the society has set up a plant section.)

Botanic Gardens There is a botanic garden at St-Denis in the old French East India Company's garden; mostly exotic trees. The Office National des Forêts have a small arboretum near St-Denis with exotic species and some Mascarene endemics.

Useful Addresses

Herbier de la Réunion, Campus Universitaire Ste-Clotilde.
Office National des Forêts, Domaine de la Providence, St-Denis.

Additional References

Badré, F. and Cadet, T. (1978). The pteridophytes of Réunion Island. *Fern Gaz.* 11(6): 349-365. (With 11 black and white photographs; gives figure of c. 240 species.)

Cadet, L.J.T. (1980). *La Végétation de l'Ile de la Réunion: Etude Phytoécologique et Phytosociologique.* Imprimerie Cazal, St-Denis. 312 pp. (Thesis, with small vegetation map and 8 plates of black and white photographs.)

Rivals, P. (1952). Etudes sur la végétation naturelle de l'Ile de la Réunion. *Trav. Lab. For. Toulouse,* T.5: Géographie forestière du Monde, sect. 3, L'Afrique, 1(2). 214 pp. (With 20 plates of black and white photographs.)

Rivals, P. (1968). La Réunion. In Hedberg, I. and O. (1968), cited in Appendix 1. Pp. 272-275.

Rodrigues

Rodrigues is a dependency of the western Indian Ocean state of Mauritius. It is the smallest and most remote of the three Mascarene islands, 574 km east of Mauritius, 19°43'S, 63°26'E. It is generally hilly, rising to 393 m on Mt Limon.

Area 104 sq. km

Population 35,594 (Official estimate, W. Strahm, 1984, pers. comm.)

Floristics 145 indigenous species (White, 1983, cited in Appendix 1) including 40 endemic species (Strahm, 1983); 26 indigenous fern species, including three endemics; four endemic genera (W. Strahm, 1984, *in litt.*). Cadet (1971) (as reported in Frodin) gives over 375 vascular species, but this includes over 100 introductions since 1874.

Floristic affinities primarily with Mauritius and Réunion, but also with Madagascar (Melville, 1970, cited in Appendix 1).

Vegetation The original vegetation over most of Rodrigues was a low forest only 10-15 m high, with palm stands with *Pandanus* on the drier east coast. Today only pockets of degraded native vegetation occur in a few of the more inaccessible valleys and hillsides.

Checklists and Floras Rodrigues is included in the incomplete *Flore des Mascareignes*, and also in *Flora of Mauritius and the Seychelles* (Baker, 1877), both cited in Appendix 1.

Balfour, I.B., Mitten, W., Crombie, J.M. and Dickie, G. (1879). Botany of Rodriguez. Transit of Venus expedition. *Phil. Trans. Roy. Soc. Lond.* 168 (extra volume): 302-419.

Friedmann, F. and Guého, J. (1977). Guide des principales plantes indigènes de l'Ile Rodrigues. *Rev. Agric. Sucr.* 56(1): 1-19. (Annotated checklist with short diagnoses and indication of degree of rarity.)

Lorence, D. (1976). The pteridophytes of Rodrigues Island. *Bot. J. Linn. Soc.* 72(4): 269-283. (Checklist with notes on distributions and collections.)

Field-guides

Cadet, L.J.T. (1984). *Plantes Rares ou Remarquables des Mascareignes*. Agence de Coopération Culturelle et Technique, 13 quai André-Citroën, 75015 Paris. 132 pp. (With 48 photographs.)

Information on Threatened Plants

Hedberg, I. (1979), cited in Appendix 1. (List for Mauritius and Rodrigues by A.W. Owadally, p. 103, contains 34 species – E:12, V:2, R:18, I:2.)

Strahm, W. (in prep.) *Red Data Book for Rodrigues*. To be published by the IUCN Conservation Monitoring Centre, Cambridge, U.K. (Includes detailed accounts of 75 species and infraspecific taxa: Ex:18, E:32, V:13, R:10, nt:2. 48 of these are endemic or extinct elsewhere. The author is presently completing a similar Red Data Book for Mauritian plants.)

Tirvengadum, D.D. (1980). On the possible extinction of *"Randia" heterophylla*, a Rubiaceae of great taxonomic interest, from Rodrigues island. *Bull. Mauritius Inst.* 9(1): 1-21 (7 plates). (Includes observations on the conservation status of 36 endemic species; reviewed in *Threatened Plants Committee – Newsletter*, No. 7: 5, 1981.)

IUCN has records of 48 species and infraspecific taxa believed to be endemic to Rodrigues – Ex:10, E:20, V:8, R:8, nt:2; non-endemic rare or threatened taxa – Ex:8, E:12, V:5, R:2 (Rodrigues status).

One species which occurs in Rodrigues (*Zanthoxylum paniculatum*) is included in *The IUCN Plant Red Data Book* (1978).

Laws Protecting Plants The Forests and Reserves Act 1983 (Parliament of Mauritius) also applies to Rodrigues, but of the list of protected taxa given, the only ones to occur on Rodrigues are orchids, ferns and *Diospyros*.

Useful Addresses

Curator, Herbarium, Mauritius Sugar Industry Research Institute, Reduit, Mauritius.
CITES Management Authority: The Conservator of Forests, Forestry Service, Curepipe, Mauritius.

Additional References

Cadet, L.J.T. (1971). Flore de l'Ile Rodrigues: espèces spontanées introduites depuis Balfour (1874). *Bull. Mauritius Inst.* 7: 1-12.

Cadet, L.J.T. (1975). Contribution à l'étude de la végétation de l'Ile Rodrigue (Ocean Indien). *Cah. Centre Univ. Réunion* 6: 5-29.

Strahm, W. (1983). Rodrigues: can its flora be saved? *Oryx* 17(3): 122-125.

Vaughan, R.E. (1968). Mauritius and Rodriguez. In Hedberg, I. and O. (1968), cited in Appendix 1. Pp. 265-272.

Wiehe, P.O. (1949). The vegetation of Rodrigues Island. *Bull. Mauritius Inst.* 2: 279-304.

Romania

Area 237,500 sq. km

Population 22,048,305 (1979 census)

Floristics 3300-3400 native vascular species, estimated by D.A. Webb (1978, cited in Appendix 1) from *Flora Europaea*, including 41 endemic species and 5 endemic subspecies (IUCN figures). According to Beldie (1977-1979), the flora contains 3063 species and 504 subspecies. Elements: Circumpolar and alpine (14%), Eurasiatic (29%), Mediterranean (6%), south and south-eastern European (18%) and Atlantic (3%).

Areas of floristic interest include the Black Sea coast, with sand-dune flora, and south-east Romania with continental sands and salt-marsh plants.

Vegetation 3 main latitudinal vegetation zones present: 1 – steppe, in the southeast (today largely an agricultural landscape); 2 – forest-steppe, in central Romania, supporting *Quercus pedunculiflora* and *Q. pubescens*; 3 – forest, mainly oak species (*Q. robur*, *Q. cerris* and *Q. frainetto*) especially in the Carpathians. In the Carpathian forests, 4 altitudinal belts can be distinguished: at 300-1400 m oaks and beech; at 1300-1700 m (Boreal zone) spruce forests; at 1700-1850 m (subalpine zone) spruce, Mountain Pine (*Pinus mugo*) and junipers; and at 2000-2500 m (alpine zone) dwarf shrubs dominated by *Salix* species (G. Dihoru, 1984, *in litt.*).

Checklists and Floras Romania is covered by the completed *Flora Europaea* (Tutin *et al.*, 1964-1980, cited in Appendix 1).

Beldie, A. (1977-1979). *Flora României. Determinator ilustrat al plantelor vasculare*, 2 vols. Academiei Republicii Socialiste România, Bucharest. (Systematic checklist; illus.)
Borza, A. (1947-1949). *Conspectus florae Romaniae*, 2 vols. Instituti Botanici Universitatis Clusiensis, Cluj. (Checklist.)
Savulescu, T. (Ed.) (1952-1976). *Flora Republicii Socialiste România*, 13 vols. Academiei Republicii Socialiste România, Bucharest. (Includes extensive notes on phytosociology and habitats; line drawings.)
Simionescu, I. (1960). *Flora Rominiei*. Combinatul Poligrafic, Bucharest. 356 pp.

Field-guides
Pauca, A. and Roman, S. (1959). *Flora alpina si montana*. Editura stiintifică, Bucharest. (Not seen.)
Prodan, I. and Buia, A. (1966). *Flora mica ilustrata a României*, 5th Ed. Agro-Silvica, Bucharest. 676 pp. (Keys; line drawings.)
Tarnavschi, I. and Andrei, M. (1971). *Determinator de plante superioare*. Editura didactica si pedagogica, Bucharest. 443 pp. (Not seen.)
Todor, I. (1968). *Mic atlas de plante din flora Republicii Socialiste România*. Editura didactica si pedagogica, Bucharest. 277 pp. (175 colour plates; not seen.)

Information on Threatened Plants A national threatened plant list has been published (Péterfi *et al.*, 1977) and a Red Data Book is in preparation together with a mapping programme of threatened species (G. Dihoru, 1984, *in litt.*).

Péterfi, S. *et al.* (1977). Noi initiative ale Comisiei Monumentelor Naturii pentru conservarea genofondului României. *Ocrotirea naturii maramuresene*. Pp. 7-48. (Not seen.)

Plants in Danger: What do we know?

Few of the following have been seen by the present authors:

Ardelean, A. (1977). *Ocrotirea naturii în Judetul Arad*. Arad.

Cilievici, E. (1975). *Ocrotirea naturii în Judetul Sibiu*. Sibiu. 116 pp.

Heltman, H. (1968). *Seltene Pflanzen Rumäniens* (Rare Romanian Plants). Editura tineretului, Bucharest. 95 pp.

Mohan, G. (1971). *Pionier, cunoaste si ocroteste monumentele Naturii*. Editura Didactica si Pedagogica, Bucharest. 204 pp.

Morariu, I. *et al.* (1971). *Ce ocrotim din natura judetului Brasov*, 2nd Ed. Brasov. 221 pp.

Mosneaga, M. (1969). Monumente ale naturii din bazinele Ialomitei. *Prahovei si Buzăului*. 36 pp. Editura Meridiane, Bucharest.

Nadasan, I. *et al.* (1976). *Monumente ale naturii din Maramures*. Editura sport-turism, Bucharest.

Opris, T. (1954-1984). *Ocrotirea Naturii (si a Mediului Inconjurator)*, 28 vols. Editura Academiei Republicii Socialiste România, Bucharest.

Opris, T. (1972). *Aceste uimitoare plante*. Editura Albatros, Bucharest.

Pop, E. and Sălăgeanu, N. (Eds), (1965). *Monumente ale naturii din România*. Editura Meridiane, Bucharest.

Ratiu, F. and Gergely, I. (1974). New and rare plant associations in Romania. *Stud. Univ. Babes-Bolyai. Ser. Biol.* 19(2): 7-15.

Resmerita, I. (1971). Statium noi cu plante rare din România (Status of some rare plants in Romania). *Stud. Cerc. Biol., Ser. Bot.* 23(6): 491-493. (12 rare taxa described.)

Resmerita, I. (1983). *Conservarea dinamica a naturii*. Editura stiintifică si Enciclopedica, Bucharest.

Schrott, L. (1979). Necesitatea ocrotirii unor taxoni rari, pe cale de disparitie in flore judetului Timis (The need to protect some rare taxa, due to their disappearance in the flora of the province of Timis). *Tisbiscus Stiinte Nat.* 11-16.

Stoiculescu, C. (1978). Arbori seculari si de mari dimensiuni din Oltenia, propusi pentru ocrotire. *Ocrotirea Naturii* 1: 55-58.

Included in the European threatened plant list (Threatened Plants Unit, 1983, cited in Appendix 1); latest IUCN statistics, based upon this work: endemic taxa – Ex:1, E:6, V:12, R:17, K:3, nt:7; non-endemics rare or threatened worldwide – E:2, V:23, R:13, I:12 (world categories).

Laws Protecting Plants There is a law for the protection of nature (1950) and a Decree (1950, nr. 237) concerning protection of natural monuments. There is a national programme, begun in 1976, to protect and develop forest resources. For further details contact the Comisia Monumentelor Naturii (CMN), address below.

Voluntary Organizations

Societatea de Stiinte Biologice (SSB) din România, Aleea Portocalior nr. 1.

Botanic Gardens

Gradina Agrobotanica, Institutul Agronomic "Dr Petru Groza", Str. Mănăstur 3, 3400 Cluj-Napoca.

Gradina Botanica a Universitatii, "Al.I. Cuza", Str. Dumbrava Rosie 9, Iasi.

Gradina Botanica a Universitatii Bucaresti, Soseaua Cotroceni 32, 76258 Bucharest.

Gradina Botanica a Universitatii din Cluj-Napoca, Str. Republicii 42, 3400 Cluj-Napoca.

Gradina Botanica, Str. Comuna din Paris 26, Craiova.

Useful reference:

Milhailescu-Firea, S. (1979). Plante rare din flora R.S.R., cultivate în Gradina Botanica din Bucuresti (Rare plants of the Romanian flora cultivated at Bucharest Botanic Garden). *Culegere de studii si articole de Biologie.* Gradina Botanica Iasi, 1: 189-194. (Describes 70 rare taxa in cultivation.)

Useful Addresses

Comisia Monumentelor Naturii (CMN), Academia Republicii Socialiste România, Calea Victoriei nr. 125, Bucharest.

Institutul Central de Biologie, Splaiul Independentei 296, 77748 Bucharest VI.

Subcomisia Monumentelor Naturii Cluj-Napoca, Str. Republicii 9, 3400 Cluj-Napoca.

Subcomisia Monumentelor Naturii Iasi, Str. 23 August nr. 11, Iasi.

Subcomisia Monumentelor Naturii Timisoara, Str. Mihai Viteazu nr. 24, 1900 Timisoara.

Additional References

Baicu, T. (1976). Citeva probleme ecologice ale protoctiei plantelor (Some ecological problems of plant protection). *Prod. Veg. Cereale Plante Teh* 28(12): 33-38. (Not seen.)

Borza, A. (1931). *Die Vegetation und Flora Rumäniens. Guide de la sixième excursion phytogéografique internationale. Roumaine, 1931. Part 1.* Institut de Literatura si Tipografia 'Minerva' S.A., Cluj. 54 pp. (In German.)

Borza, A. (1941). Die Pflanzenwelt Rumäniens und ihr Schutz (The flora of Romania and its protection). *Ber. Deutsch. Bot. Ges.* 59(5): 153-168.

Borza, A. and Pop, E. (1921-1947). Bibliographia Botanica Romaniae, 387 parts. *Bull. Grad. Mus. Bot. Univ. Cluj.* 1-27.

Boscaiu, N. (1975a). La protection de la flore dans les Carpates Roumains. In Jordanov, D. *et al.* (Eds), *Problems of Balkan flora and vegetation.* Proceedings of the 1st International Symposium on Balkan Flora and Vegetation, 7-14 June 1973, Varna. Bulgarian Academy of Sciences, Sofia. Pp. 428-430. (In French.)

Boscaiu, N. (1975b). Probleme le conservanii vegetatiei alpine si subalpine. *Ocrotirea Naturii* 19(1): 17-23. (Not seen.)

Sanda, V. *et al.* (1983). *Caracterizarea Ecologica si Fitocenologica a Speciilor Spontane din Flora României.* Stud. si Com. – Stiinte Naturale 25, Supliment, Muzeul Brukenthal, Sibiu. (Not seen.)

Rwanda

Area 26,330 sq. km

Population 5,903,000

Floristics 2150 species (Troupin, 1971). Levels of endemism unknown but unlikely to be high. Brenan (1978, cited in Appendix 1) gives a figure of 26 species endemic to Rwanda and Burundi out of a c. 39% sample of the *Flore du Congo Belge et du Ruanda-Urundi.*

Floristic affinities with Lake Victoria and Afromontane regions.

Vegetation Mostly mosaic of East African evergreen bushland and secondary *Acacia* wooded grassland. Large areas of Afromontane communities of montane forest and grassland in the west. Afroalpine communities on the top of the volcanic mountains Muhavura and Karisimbi on the border with Zaïre. Estimated rate of deforestation for closed broadleaved forest 28 sq. km/annum out of 1010 sq. km (FAO/UNEP, 1981).

For vegetation map see White (1983), cited in Appendix 1.

Checklists and Floras Rwanda is included in the incomplete *Flore du Congo Belge et du Ruanda-Urundi* (cited in Appendix 1), continued since 1972 as *Flore d'Afrique Centrale (Zaïre - Rwanda - Burundi)*. Rwanda's plants of high altitudes are listed in *Afroalpine Vascular Plants* (Hedberg, 1957), cited in Appendix 1.

Troupin, G. (1971). *Syllabus de la Flore du Rwanda, Spermatophytes*. Ann. Mus. Roy. Afrique Centr., Sér. in-8°, Sci. Econ. 7. (Key to families; family and genus descriptions, species diagnoses, line drawings.)

Troupin, G. (1978-). *Flore du Rwanda, Spermatophytes*. Ann. Mus. Roy. Afrique Centr., Sér. in-8°, Sci. Econ. 9, 13, 15. (3 vols so far out of expected 4; last vol. due end-1986. Descriptive keys, distributions, specimens, line drawings.)

Troupin, G. (1982). *Flore des Plantes Ligneuses du Rwanda*. Ann. Mus. Roy. Afrique Centr., Sér. in-8°, Sci. Econ. 12. 747 pp. (Keys, descriptions, distributions, specimens, line drawings.)

Field-guides

Combe, J. (1977). *Guide des Principales Essences de la Forêt de Montagne du Rwanda*. Projet Pilote Forestier, Kibuye. 241 pp.

Troupin, G. and Girardin, N. (1975). Plantes ligneuses du Parc National de l'Akagera et des savanes orientales du Rwanda. Clés de détermination scientifique. *Bull. Jard. Bot. Nat. Belg.* 45: 1-96, and Publication No. 13, Inst. Nat. Rech., Sci. Butare. 96 pp.

Information on Threatened Plants No published lists of rare or threatened plants; IUCN has records of 23 species and infraspecific taxa believed to be endemic; no categories assigned.

Laws Protecting Plants The only laws protecting plants apply to National Parks.

Useful Addresses

CITES Management Authority: Office Rwandais du Tourisme et des Parcs Nationaux (ORTPN), B.P. 905, Kigali.

Additional References

Deuse, P. (1968). Rwanda. In Hedberg, I. and O. (1968), cited in Appendix 1. Pp. 125-127.

Devred, R. (1958). La végétation forestière du Congo belge et du Ruanda-Urundi. *Bull. Soc. R. For. Belg.* 65: 409-468. (With vegetation map.)

Lebrun, J. (1955). *Esquisse de la Végétation du Parc National de la Kagera*. Fasc. 2 of Exploration du Parc National de la Kagera Mission J. Lebrun (1937-1938), Publ. Inst. Parcs Nat. Congo Belge, Bruxelles. 89 pp. (With 12 black and white photographs.)

Lebrun, J. (1956). La végétation et les territoires botaniques du Ruanda-Urundi. *Natural. Belges* 37: 230-256.

Troupin, G. (1966). *Etude Phytocénologique du Parc National de l'Akagera et du Rwanda Oriental*. Publication No. 2, Inst. Nat. Rech., Sci. Butare. 293 pp.

There is a series of vegetation and soil maps covering Zaïre, Rwanda and Burundi in c. 25 parts, published between 1954 and c. 1970 by the Institut National pour l'Etude Agronomique du Congo (INEAC); each is accompanied by a descriptive memoir, and several of the maps are to different scales. The series is called *Carte des Sols et de la Végétation du Congo Belge et du Ruanda-Urundi*, or, more recently, ... *du Congo, du Rwanda et du Burundi*.

Ryukyu Retto

More than 100 islands and islets in a 1100 km arc south of Japan, of which they are a dependency. Total land area of 2196 sq. km, maximum elevation 500 m. Population 1,250,000 (1980 census), mostly on Okinawa, the largest island (1176 sq. km).

Over 2000 vascular plant species, but this includes introductions (Walker, 1976). Subtropical and warm temperate broadleaved forests are the climax vegetation in the lowlands, with *Castanopsis* forests above 300 m (Numata, 1974, cited under Japan). There are also coastal mangroves. Most of the natural vegetation has been cleared for agriculture, settlements, and plantations of pine and pineapple. Degraded areas support *Miscanthus* grassland and secondary forests (Walker, 1952). No information on threatened plants.

References
Hatusima, S. (1971). *Flora of the Ryukyus (including Amami Islands, Okinawa Islands and Sakishima Archipelago)*. Study Group of Biological Education, Okinawa. 940 pp. (In Japanese, revised 1975.)

Hatusima, S. and Nackejima, C. (1979). *Flowers of the Ryukyu Islands*. Japan. 368 pp. (In Japanese.)

Nakajima, C. (1971). *An Enumeration of the Orchids of the Ryukyus (A Preliminary List)*, 2 parts. Okinawa. (1 – Okinawa Islands; 2 – Amami, Miyako and Yaeyama Islands. Descriptions in Chinese, short summaries in English.)

Walker, E.H. (1952). A botanical mission to Okinawa and the southern Ryukyus. *A. Gray Bull.* 1(3): 225-244. (Describes vegetation.)

Walker, E.H. (1954). *Important Trees of the Ryukyu Islands*. U.S. Civil Administration Special Bulletin no. 3. Okinawa. 350 pp. (Short descriptions, keys and line drawings of selected trees; notes on habitats, uses; in Japanese and English.)

Walker, E.H. (1976). *Flora of Okinawa and the Southern Ryukyu Islands*. Smithsonian Institution, Washington, D.C. 1159 pp. (Describes 2080 species, including introductions, with keys.)

A number of species from the Ryukyus are included in the Atlas by Horikawa (1972, 1976), cited under Japan.

St Helena

St Helena is a British Dependent Territory in the South Atlantic Ocean, 1850 km from the west coast of Africa, 15°58'S 5°43'W. It is rugged and volcanic, with a unique but devastated flora. Recently the London-based Flora and Fauna Preservation Society (FFPS), and the Royal Botanic Gardens, Kew, with the help of IUCN and grant-aided by WWF-UK, have co-ordinated and sponsored a plant rescue programme.

Area 121 sq. km

Population 6000 (including Tristan da Cunha and Ascension)

Floristics J.D. Hooker (in Melliss, 1875) predicted that there were once about 100 endemic species in this "wonderfully curious little flora": most of the destruction occurred before a botanist saw the island. Today the known flora consists of c. 320 species of pteridophytes and flowering plants; of these 50 are endemic, c. 10 native but not endemic, and c. 260 naturalized aliens (Q.C.B. Cronk, 1984, pers. comm.); of the endemics, 7 are Extinct, a declining total as recent searching, especially by G. Benjamin, has refound plants not seen for generations.

Vegetation When discovered in 1502, said to contain "fine woods" in early Portuguese accounts. Since the introduction of goats (early 16th Century) and other animals, natural forest became reduced to a small montane area at the top of the central ridge ('The Peaks'), rich in endemics but now invaded by exotics; below, at 450-750 m, pasture and plantations of New Zealand Flax (*Phormium tenax*), now abandoned; at 300-450 m mostly pasture and replanted woodland; over the remaining two-thirds of the island, dry lowland of rocky desert with sparse vegetation (Kerr, 1971).

Checklists and Floras Q.C.B. Cronk (Corpus Christi College, Cambridge) is preparing a Flora of St Helena and has completed a manuscript entitled "A provisional check-list of St Helena pteridophytes". The only published account is still:

Melliss, J.C. (1875). *St Helena*. London. 426 pp. (Illus.)

Field-guides S. Goodenough (Royal Botanic Gardens, Kew) is preparing a field-guide.

Information on Threatened Plants
Kerr, N. (1971). Report on a preliminary nature conservation project, Island of
    St Helena; July-August 1970. IBP/4(71). Mimeo. (Documents changes in the
    vegetation; assesses conservation status and distribution of each of the endemics.)

There are various reviews of progress in rescuing the endemic flora, with news of the species rediscovered, in *Threatened Plants Newsletter* 8: 12-13 (1981); 12: 10-11 (1983); and 13: 9-10 (1984). Latest IUCN statistics: endemics – Ex:7, E:23, R:17, K:2.

Voluntary Organizations
Fauna and Flora Preservation Society, c/o Zoological Society of London, Regent's
    Park, London NW1 4RY. (Has a St Helena Sub-committee – Chairman: Sir P.
    Watkin Williams; Secretary: R.S.R. Fitter.)
St Helena Heritage Society, c/o The Bishop of St Helena, Bishopholm, St Helena.

Botanic Gardens The Agriculture and Forestry Department have a small arboretum at Cason's, where endemics have been propagated and planted. G. Benjamin and the Forestry Department have established a small Ebony propagation plot at Pounceys, and undertake further propagation at their headquarters at Scotland, St Helena.

Useful Addresses

Agriculture and Forestry Department, Scotland, St Helena.

Additional References

Brown, L.C. (1982). *The Flora and Fauna of St Helena*. Project Record 59, S.HEL-01-12/REC-59/82, Overseas Development Administration, Land Resources Centre, Surbiton, Surrey, U.K. 88 pp. (Includes plant checklist and description of vegetation.)

Cronk, Q.C.B. (1981). *Senecio redivivus* and its successful conservation in St Helena. *Envir. Conserv.* 8(2): 125-126.

Goodenough, S. (1983). Saint Helena: A Plant Propagation Project and Recommendations for the Conservation of the Endemic Flora of the Island. Royal Botanic Gardens, Kew. Typescript.

Turrill, W.B. (1948). On the flora of St. Helena. *Kew Bull.* 3: 358-362.

St Kitts-Nevis

Two islands at the northern end of the Eastern Caribbean in the Leeward Islands, 117 km west of Antigua: St Kitts, with a volcanic mountainous ridge down the centre, Mt Misery the highest point (1156 m); and Nevis, surrounded by coral reefs, 3.2 km south-east of St Kitts, consisting almost entirely of the mountain Nevis Peak (985 m).

Area St Kitts: 168.4 sq. km; Nevis: 93 sq. km

Population St Kitts: c. 35,000; Nevis: c. 11,200

Floristics No information available

Vegetation Both islands are volcanic and are cultivated or covered with secondary scrub woodland and thorn bush; good rain forest on sheltered NW face above Jessops Mt (Nevis) and in two small areas of St Kitts, bordering on dry evergreen forest reaching to the sea; palm brake to the summits and covering all mountains of St Kitts; montane thicket between the rain forest and palm brake; elfin woodland caps the summits and gives way to herbaceous growth on Misery Peak; some coastal mangrove in the south. (Beard, 1949, cited in Appendix 1.) FAO (1974, cited in Appendix 1) recorded 16.7% forest cover, but this included Anguilla, then part of St Kitts-Nevis.

Checklists and Floras Covered by the *Flora of the Lesser Antilles, Leeward and Windward Islands* (only monocotyledons and ferns published so far, Howard, 1974- , cited in Appendix 1) and by the family and generic monographs of *Flora Neotropica* (cited in Appendix 1).

Box, H.E. and Alston, A.H.G. (1937). Pteridophyta of St Kitts. *J. Bot.* 75: 241-258.

Burdon, K.J. (1920). *A Handbook of St Kitts-Nevis*. Crown Agents. 247 pp. (Illus., maps, notes on flora.)

See also:

Beard, J.S. (1944). Provisional list of trees and shrubs of the Lesser Antilles. *Caribbean Forester* 5(2): 48-67. (428 species assigned in a table to individual islands.)

Stehlé, H. and Stehlé, M. (1947). Liste complémentaire des arbres et arbustes des
    petites Antilles. *Caribbean Forester* 8: 91-123. (A further 328 species to Beard, 1944,
    in similar format.)

Information on Threatened Plants None.

St Lucia

A mountainous island in the Windward group of the Lesser Antilles, 43.5 km long by
22.5 km wide. Agriculture occupies 227 sq. km.

Area 616 sq. km

Population 126,000

Floristics No information available.

Vegetation Some untouched primary rain forest and lower montane rain forest in
the interior, surrounded by cut-over areas of secondary rain forest; thorn bush in the low
dry north; a belt of dry scrub woodlands with *Pterocarpus* swamp around the coast and
with small areas of mangrove on the east (windward) side; interior forests separated from
the coastal woodlands by land devoid of trees, used for cultivation. (Beard, 1949, cited in
Appendix 1.) 21% forested according to FAO (1974, cited in Appendix 1).

Checklists and Floras Covered by the *Flora of the Lesser Antilles, Leeward and
Windward Islands* (only monocotyledons and ferns published so far, Howard, 1974- ,
cited in Appendix 1) and by the family and generic monographs of *Flora Neotropica* (cited
in Appendix 1). See also:

Beard, J.S. (1944). Provisional list of trees and shrubs of the Lesser Antilles.
    Caribbean Forester 5(2): 48-67. (428 species assigned in a table to individual
    islands.)
Lang, W.G. (1955). *Forest Trees of St Lucia*. Government Printer, Castries. 21 pp. (27
    trees, 23 local and 4 introduced and cultivated, described, with local and/or
    botanical names.)
Stehlé, H. and Stehlé, M. (1947). Liste complémentaire des arbres et arbustes des
    petites Antilles. *Caribbean Forester* 8: 91-123. (A further 328 species to Beard, 1944,
    in similar format.)

C.D. Adams, at British Museum (Natural History), London, has prepared a tentative
checklist of vascular plants of St Lucia from literature and collections seen.

Information on Threatened Plants None.

Useful Addresses
St Lucia CITES Management Authority: Forestry Division, Ministry of Agriculture and
    Lands, Castries.

St-Pierre and Miquelon

An archipelago c. 15 km off the southern coast of Newfoundland, Canada, comprising two small groups of islands, 241 sq. km in size and with a population of c. 6000. The only remaining overseas territory of France in North America. Arsène (1927) recorded 391 native species and 96 introduced species of vascular plants that he had found there. Rouleau (1978) gives a more up-to-date list but without any accompanying data on each taxon other than its occurrence on the islands. Ponds, swamps, marshes and bogs cover more than half the area and at least 50% of the flora is wetland or aquatic. Some of the valleys are wooded. (Arsène, 1927.)

Arsène, L. (1927). Contribution to the Flora of the islands of St-Pierre et Miquelon. *Rhodora* 29: 117-221.

Rouleau, E. (1978). *List of the Vascular Plants of the Province of Newfoundland (Canada)*. Oxen Pond Botanic Park, Newfoundland.

St Vincent

A very fertile and densely populated island in the Windward group of the Lesser Antilles, located 34 km south of St Lucia, 161 km west of Barbados; 29 km long by 18 km wide, with active volcanoes. The state of St Vincent includes some of the 600 Grenadines, a chain of small islands stretching 64 km south of St Vincent to Grenada. The larger include Bequia, Canouan, Mayreau, Mustique, Union Island, Petit St Vincent and Prune Island.

Area 389 sq. km

Population 104,000

Floristics Little published information; a study of the St Vincent checklist (Anon, 1893) and of other works cited below found about 12 endemics, but this could radically alter when the *Flora of the Lesser Antilles* is completed.

Vegetation Natural vegetation remains on uncultivable steep slopes of montane areas; palm brake on central mountains; elfin woodland along the summits; good stands of rain forest, mostly between 300 and 490 m but reaching 580 m; secondary growth only, on Soufrière Mt, all previous vegetation destroyed in 1902 when the volcano erupted. The littoral forest in the western part of the island, described by Beard (1949), has disappeared and virtually all the forest in the east (windward) side of the island, much of the north and extensive parts of the north-west, appears secondary. (Lambert, 1983; Beard, 1949, cited in Appendix 1.) 41.2% forested according to FAO (1974, cited in Appendix 1).

On the Grenadines predominantly deciduous and semi-deciduous forests; dry evergreen littoral stunted vegetation on windward slopes (Howard, 1952).

Checklists and Floras Covered by the *Flora of the Lesser Antilles, Leeward and Windward Islands* (only monocotyledons and ferns published so far, Howard, 1974- , cited in Appendix 1) and by the family and generic monographs of *Flora Neotropica* (cited in Appendix 1). The only checklist is:

Anon (1893). Flora of St Vincent and adjacent islets. *Bull. Misc. Inf., Kew* 81: 231-296. (Lists 1150 species, now very out of date, for St Vincent and the

Grenadines, Bequia, Canouan, Mustique, Union; probably prepared by Rolfe –
C.D. Adams, pers. comm.)

See also:

Beard, J.S. (1944). Provisional list of trees and shrubs of the Lesser Antilles.
   Caribbean Forester 5(2): 48-67. (428 species assigned in a table to individual
   islands.)
Stehlé, H. and Stehlé, M. (1947). Liste complémentaire des arbres et arbustes des
   petites Antilles. *Caribbean Forester* 8: 91-123. (A further 328 species to Beard, 1944,
   in similar format.)

   Information on Threatened Plants None. Howard, in the following reference,
outlines the status of *Spachea perforata*, widely publicized as an extinct St Vincent
endemic, but believed by Howard to be an introduction from South America.

Howard, R.A. (1977). Conservation and endangered species of plants in the Caribbean
   Islands. In Prance, G.T. and Elias, T.S., cited in Appendix 1. Pp. 105-114.

   Botanic Gardens
St Vincent and the Grenadines Botanic Gardens, Kingstown.

   Additional References
Howard, R.A. (1952). The vegetation of the Grenadines, Windward Islands, British
   West Indies. *Contr. Gray Herb. Harv. Univ.* 174: 1-129, 29 plates.
Lambert, F. (1983). Report on an expedition to survey the status of the St Vincent
   Parrot *Amazona guildingii*. International Council for Bird Preservation and
   University of East Anglia.

Salvage Islands

A group of 3 small, uninhabited desolate islands between the Canary Islands and Madeira,
closer to the Canaries but administratively part of Madeira; also known as Ilhas Selvagens.
The largest, Selvagem Grande, is 1-2 km across, with 2 low hills reaching around 100 m.
All 3 are difficult of access due to reefs and rocks. The vegetation is low, stunted and
dominated by *Mesembryanthemum*, *Suaeda* and *Nicotiana glauca*. Rabbits are common
on Selvagem Grande only, which is devoid of trees. Goats were introduced but are now
extinct (Pickering and Hansen, 1969). Selvagem Pequena has apparently suffered much
less and has a more diverse flora (R. Press, 1985, pers. comm.).

The flora consists of 2 species of ferns and 92 of flowering plants (Museo de Ciencias
Naturales del Cabildo Insular de Santa Cruz de Tenerife, 1978), most of which are invasive
weeds. Many are succulents. Of them, 4 species and 4 varieties of flowering plants are
endemic; of the endemic species, IUCN list 2 as E, 1 as V and 1 as R; the endemic varieties
are all listed as I. The endemic species are listed, without categories, in IUCN Threatened
Plants Committee Secretariat (1980), cited in Appendix 1. This report also lists 5 regionally
threatened non-endemics.

The islands are covered by the *Flora of Macaronesia* checklist (Hansen and Sunding, 1979,
cited in Appendix 1). For recent, annotated checklists of the flora, see:

Museo de Ciencias Naturales del Cabildo Insular de Santa Cruz de Tenerife (1978). *Contribución al estudio de la historia natural de las Islas Salvajes.* Resultados de la expedición científica "Agamenon 76" (23 February-3 March 1976). Aula de Cultura de Tenerife, Selecciones Graficas (Ediciones), Paseo de la Dirección, 52-Madrid-29. 209 pp. Multi-authored work, including chapters on flora and vegetation, and a revised checklist of ferns and flowering plants by P.L. Perez de Paz and Acebes Ginoves (pp. 79-105).

Pickering, C.H.C. and Hansen, A. (1969). Scientific expedition to the Salvage Islands July 1963: IX. List of higher plants and cryptogams known from the Salvage Islands. *Bol. Museu Municipal Funchal* 24: 63-72. (Includes bibliography; map.)

São Tomé and Príncipe

The volcanic archipelago of São Tomé and Príncipe is situated in the Gulf of Guinea, off the west coast of Africa. The main islands São Tomé (0°25'N 6°35'E) and Príncipe (1°37'N 7°27'E) are respectively 220 and 210 km from the nearest points on the mainland. Príncipe is about 210 km SSW of Bioko (Fernando Po). The highest point is 2024 m on São Tomé.

Area 964 sq. km. of which São Tomé comprises 854 sq. km

Population 94,000

Floristics

São Tomé 601 species (Exell, 1973), with 108 endemic (out of a total of 556, Exell, 1944).

Príncipe 314 species (Exell, 1973), with 35 endemic (out of a total of 276, Exell, 1944).

Floristic affinities with neighbouring Pagalu and Bioko and also with the countries of the African mainland round the Gulf of Guinea.

Vegetation Original cover almost exclusively lowland rain forest up to c. 800 m, with submontane evergreen forest and mist forest above that (Exell, 1968). Mangrove vegetation round coasts. Most of forest now destroyed and replaced by plantations, mainly of cocoa.

Checklists and Floras

Exell, A.W. (1944). *Catalogue of the Vascular Plants of S. Tomé (with Principe and Annobon).* British Museum (Natural History), London. 428 pp. (Annotated checklist; line drawings.)

Exell, A.W. (1956). *Supplement to the Catalogue of the Vascular Plants of S. Tomé (with Principe and Annobon).* British Museum (Natural History), London. 58 pp.

Exell, A.W. (1959). Additions to the Flora of S. Tomé and Principe. *Bull. IFAN* 21(2)A: 439-476.

Exell, A.W. (1973). Angiosperms of the islands of the Gulf of Guinea (Fernando Po, Príncipe, S. Tomé and Annobon). *Bull. Brit. Mus. (Nat. Hist.) Bot.* 4(8): 325-411. London. (Checklist with distributions.)

Liberato, M.C. and Santo, J. do Espirito (1972-). *Flora de S. Tomé e Príncipe.* Junta de Investigações Científicas do Ultramar; Jardim e Museu Agrícola do Ultramar; Ministério do Ultramar, Lisboa. (6 fascicles so far, each containing one family; keys, descriptions, distributions, specimens.)

Information on Threatened Plants No published lists of rare or threatened plants of the islands in the Gulf of Guinea, but IUCN has records of 95 species and infraspecific taxa believed to be endemic to São Tomé, and 33 to Príncipe; no categories assigned.

Useful Addresses
Ministério de Agricultura e Pecuária, São Tomé.

Additional References

Chevalier, A. (1938). La végétation de l'île de San-Thomé. *Bol. Soc. Brot., Sér. 2* 13: 101-116.

Exell, A.W. (1952/1955). The vegetation of the islands of the gulf of Guinea. *Lejeunia* 16: 57-66.

Exell, A.W. (1968). Príncipe, S. Tomé and Annobon. In Hedberg, I. and O. (1968), cited in Appendix 1. Pp. 132-134. (Includes lists of examples of endemic species for each of the three islands.)

Lains e Silva, H. (1958). *São Tomé e Principe e a Cultura do Café*. Lisbon. 499 pp. Coloured vegetation maps 1:100,000 (São Tomé), 1:200,000 (Príncipe). (Mems Junta Invest. Ultram., vol. 1.)

Monod, T. (1960). Notes botaniques sur les îles de São Tomé et de Príncipe. *Bull. IFAN* 22A: 19-94. (With 11 plates of black and white photographs.)

Saudi Arabia

Area 2,401,554 sq. km

Population 10,824,000

Floristics For the Arabian Peninsula 4500-5000 species of vascular plants (T.A. Cope and G.E. Wickens, 1985, pers. comm.), of which very approximately 3500 are likely to occur in Saudi Arabia. Initial studies identified 23 taxa as endemic to Saudi Arabia, most of them tropical (reported in Brooks *et al.*, 1982).

The flora of most of Saudi Arabia is Saharo-Sindian (i.e. made up of widespread species ranging from north-west Africa to the Sind); the flora of the mountains of south-west Saudi Arabia has links with that of neighbouring Africa.

Vegetation Principally desert, about one third of mobile sand, mainly in the Empty Quarter of the south and west; in northern and central Arabia a mosaic of steppe and dwarf shrub communities, as outlined by Vesey-Fitzgerald (1957b); in the mountains of the south-west, the least arid part of the country, extensive savanna and woodland, often of *Juniperus procera* at altitudes over 1600 m, of mixed composition including Acacias lower down; small extent of mangrove on the southern parts of the coasts of the Red Sea and Arabian Gulf.

Checklists and Floras Works specifically on the flora of Saudi Arabia are:

Chaudhary, S.A. and Cope, T.A. (1983). A checklist of grasses of Saudi Arabia (Studies in Flora of Arabia VI). *Arab Gulf J. Scient. Res.* 1(2): 313-354. (269 taxa.)

Collenette, S. (1985). *An Illustrated Guide to the Flowers of Saudi Arabia*. Scorpion, Essex, U.K. (Colour photos of c. 1500 species.)

Mandaville, J.P. (1973). *A contribution to the flora of Asir, Southwestern Arabia*. Field Research Publications, Coconut Grove, Miami, Florida 33133. 13 pp.

Migahid, A.M. (1978). *Migahid and Hammouda's Flora of Saudi Arabia*, 2 vols, 2nd Ed., revised and illustrated. Riyadh Univ. Publication. 939 pp. (With keys, drawings and colour photos. 3rd Ed. to come out soon.)

There is no modern definitive Flora of the Arabian Peninsula, a natural floristic unit, although much preparatory work has been done, especially at the British Museum (Natural History), the Royal Botanic Gardens, Kew, and the Royal Botanic Garden, Edinburgh. Recent studies towards a checklist for the Arabian Peninsula include:

Cope, T.A. (1985). A key to the grasses of the Arabian Peninsula (Studies in the Flora of Arabia XV). *Arab Gulf J. Scient. Res*, Special Publication No. 1. 82 pp.

Cribb, P.J. (1979). The Orchids of Arabia. *Kew Bull.* 33(4): 651-678. (18 species.)

Hedge, I.C. and King, R.A. (1983). The Cruciferae of the Arabian Peninsula: A checklist of species and a key to genera (Studies in the flora of Arabia IV). *Arab Gulf J. Scient. Res.* 1(1): 41-66. (97 species.)

King, R.A. and Kay, K.J. (in press). The Caryophyllaceae of the Arabian peninsula: a checklist and key to taxa (Studies in the Flora of Arabia XII). *Arab Gulf J. Scient. Res.* 2. (83 species.)

Also relevant to the whole peninsula:

Blatter, E. (1919-1936). Flora Arabica. *Rec. Bot. Surv. India* 8. 519 pp. (Annotated checklist; the only extant checklist covering the whole of the Arabian peninsula.)

Khattab, A. and El-Hadidi, M.N. (1971). Results of a botanic expedition to Arabia in 1944-1945. *Publ. Cairo Univ. Herb.* 4. 95 pp. (700 species listed.)

Miller, A.G., Hedge, I.C. and King, R.A. (1982). Studies in the flora of Arabia: I. A botanical bibliography of the Arabian Peninsula. *Notes R. Bot. Gard. Edinb.* 40(1): 43-61.

Schwartz, O. (1939). *Flora des tropischen Arabien*. Mitteilungen aus dem Institut für allgemeine Botanik in Hamburg 10. 393 pp. (Annotated checklist with distributions for tropical Arabia.)

Wickens, G.E. (1982). A biographical index of plant collectors in the Arabian peninsula (including Socotra) (Studies in the Flora of Arabia III). *Notes R. Bot. Gard. Edinb.* 40(2): 301-330.

Zohary, M. (1957). A contribution to the flora of Saudi Arabia. *J. Linn. Soc. Bot.* 55: 632-643. (Plants collected by the anitilocust unit 1942-1945 from Saudi Arabia and other parts of the Arabian peninsula, including Oman.)

Field-guides

Lipscombe Vincett, B.A. (1977). *Wild flowers of central Saudi Arabia*. Milan. 114 pp. (Colour photos and short descriptions of 80 species.)

Information on Threatened Plants Brooks *et al.* (1982) mention one endangered plant (p. 15). More recently MEPA have prepared a provisional list of 62 native plants that may be threatened.

Voluntary Organizations

Saudi Biological Society; holds a biennial symposium.

Useful Addresses

King Abdulaziz University, P.O. Box 1540, Jeddah.

Meteorology and Environment Protection Administration (MEPA), Ministry of Defence and Aviation, P.O. Box 1358, Jeddah.

National Herbarium, P.O. Box 17285, Riyadh.

Additional References

Abulfatih, H.A. (1979). Vegetation of higher elevations of Asir, Saudi Arabia. *Proc. Saudi Biol. Soc.* 3: 139-148.

Allred, B.W. (1968). *Woodlands in Saudi Arabia*. Ministry of Agriculture and Water, Riyadh; FAO, Rome. 17 pp. (Mimeo.)

Batanouny, K.H. (1978). *Natural History of Saudi Arabia: A bibliography*. Publications of King Abdulaziz University, Biology 1. 113 pp., Appendix. (Includes section on flora and vegetation.)

Batanouny, K.H. and Baeshin, N.A. (1978). Wild plants in Jiddah. *J. Saudi Arabian Nat. Hist. Soc.* 23: 19-40.

Brooks, W.H. and IUCN Conservation Monitoring Centre (1982). Conservation and sustainable use of natural resources: Part I Terrestrial (Draft). One of a series of papers prepared by IUCN for MEPA (see above), for the Expert Meeting of the Gulf Co-ordinating Council; includes short sections on flora and vegetation of Arabian Peninsula.

Vesey-Fitzgerald, D.F. (1955). Vegetation of the Red Sea coast south of Jedda, Saudi Arabia. *J. Ecol.* 43: 477-489. (With 14 black and white photographs.)

Vesey-Fitzgerald, D.F. (1957a). The vegetation of the Red Sea coast north of Jedda, Saudi Arabia. *J. Ecol.* 45: 547-562. (With 8 black and white photographs.)

Vesey-Fitzgerald, D.F. (1957b). The vegetation of central and eastern Arabia. *J. Ecol.* 45: 779-798. (With 4 black and white photographs and small-scale vegetation map.)

Zahran, M.A. (1982). *Vegetation types of Saudi Arabia*. King Abdulaziz University, Jeddah. 63 pp.

Senegal

Area 196,722 sq. km

Population 6,352,000

Floristics 2100 species (Lebrun, 1976, cited in Appendix 1); 26 endemics (Brenan, 1978, cited in Appendix 1), mostly woodland and grassland species.

Floristic affinities Saharan in the north, Sahelian in the centre, and Sudanian in the south of the country.

Vegetation Dry *Acacia* wooded grassland and deciduous bushland in northern quarter of the country; Sahelian woodland in a central band, covering about two-thirds of country, and lowland rain forest and secondary grassland derived from it in south-west corner and along southern border.

For vegetation map see White (1983), cited in Appendix 1.

Checklists and Floras Senegal is included in the *Flora of West Tropical Africa*, cited in Appendix 1. See also:

Berhaut, J. (1967). *Flore du Sénégal*, 2nd Ed. (1st Ed. 1954). Editions Clairafrique, Dakar. 485 pp. (Descriptive keys, colour photographs, line drawings.)

Berhaut, J. (1971-). *Flore Illustrée du Sénégal*. Direction des Eaux et Forêts, Dakar. (6 vols out of a probable 11 published so far; descriptions, distributions, specimen citations; line drawings throughout.)

Lebrun, J.-P. (1973). *Enumération des Plantes Vasculaires du Sénégal*. Etude
   Botanique No. 2, Institut d'Elevage et de Médecine Vétérinaire des Pays Tropicaux,
   Maisons-Alfort. 209 pp. (Checklist, with botanical bibliography of Senegal since
   1930.)

   Information on Threatened Plants No published lists of rare or threatened
plants; IUCN has records of 32 species and infraspecific taxa believed to be endemic –
R:19, I:3, K:9, nt:1.

   Useful Addresses
CITES Management and Scientific Authority: Direction des eaux, fôrets et chasses,
   Parc Forestier de Hann, B.P. 1831, Dakar.

   Botanic Gardens
Institut Fondamental d'Afrique Noire, B.P. 206, Dakar.
Jardin Botanique de la Faculté des Sciences, Dakar-Fann.
Parc Forestier et Zoologique de Hann, B.P. 1831, Dakar.

   Additional References
Adam, J.G. (1965). Généralités sur la flore et la végétation du Sénégal. *Etud. Sénégal.*
   9(3): 155-214.
Adam, J.G. (1968). Sénégal. In Hedberg, I. and O. (1968), cited in Appendix 1.
   Pp. 65-69.

Seychelles

The Seychelles archipelago consists of some 37 granitic and 52 coralline islands, plus
numerous rocks and small cays, lying along about 1000 km roughly south-west to north-
east in the western Indian Ocean. The two broad groups are usually treated separately in
the literature.

The Granitic Islands are grouped at the north-east end of the archipelago (4-5°S 55-56°E)
and represent fragments of the original land-mass of Gondwanaland. The topography is
rugged and the islands are of great scenic beauty. Only the four largest have any noteable
permanent population.

The Coralline Islands are spread for about 1000 km south-west from the Granitic Islands,
and are grouped into the Aldabra Islands (at the south-west end of the archipelago and
including Aldabra, Assumption, Cosmoledo and Astove), the Amirante Islands (which are
near the Granitic Islands and include African Banks, Remire, Desroches and Poivre), the
Farquhar Islands (Farquhar and Providence), and several more or less isolated islands that
include Coëtivy. They are divided into sand cays and low limestone reefs. Many are
unpopulated. Aldabra atoll was declared a World Heritage Site in 1982.

   Area (Total land) 404 sq. km

   Granitic Islands (Land) Mahé: 142.4 sq. km; Silhouette: 19.8 sq. km; Praslin:
39.3 sq. km; La Digue: 10.1 sq. km (Procter, 1979).

   Coralline Islands (Land) Aldabra: 155 sq. km; Assumption: 10.5 sq. km;
Coëtivy: 9.2 sq. km; Farquhar: 7.5 sq. km; Cosmoledo: 5.2 sq. km; Astove: 4.3 sq. km;
Desroches: 3.2 sq. km (Stoddart, 1970).

Population 74,000

Granitic Islands Mahé: 45,200, Silhouette: 400; Praslin: 4200; La Digue: 2000 (1971 census).

Floristics 1139 species of seed plant in 669 genera, including exotics and cultivated plants; c. 250 indigenous species, including endemics (Robertson, in press).

Granitic Islands Several different figures available:

Flora of Granitic Islands, including exotic and naturalized species: 520 species (Procter, 1979); 480 species (Renvoize, 1979). Indigenous flora, including endemics: 233 species (Summerhayes, 1931; Renvoize, 1979); 225 species, including 3 ferns and fern allies (Procter, 1979). Endemic species: 90 (Renvoize, 1979); 72 species of flowering plants (Procter, 1974); 72 species of vascular plants, including 3 ferns and fern allies (Procter, 1979).

Individual island counts for species of angiosperms (Procter, 1979): Mahé: 173 indigenous, inc. 59 endemic to the group; 135 naturalized (the monotypic family Medusagynaceae is confined to Mahé). Praslin: 92 indigenous, inc. 30 endemic to the group; 64 naturalized. Silhouette: 138 indigenous, inc. 51 endemic to the group; 90 naturalized. La Digue: 79 indigenous, inc. 11 endemic to the group; 79 naturalized. Curieuse: 83 indigenous, inc. 17 endemic to the group; 42 naturalized.

Floristic relationships with Africa (particularly East Africa), Madagascar and Indomalaysia.

Coralline Islands Aldabra Islands: 274 vascular plant species/infraspecific taxa, including 185 angiosperms, of which 43 are endemic to the Aldabra Islands; 17 endemic to Aldabra, one to Assumption, one to Cosmoledo (Fosberg and Renvoize, 1980). Amirante Islands: 97 species total, including 72 indigenous (Renvoize, 1979); no endemics. Farquhar Islands: 59 species total (Renvoize, 1979), including 46 indigenous; Farquhar 44 species, St Pierre 24, Providence 14 (Renvoize, 1979); no endemics. Coëtivy: 65 species, including 49 indigenous (Renvoize, 1979); no endemics.

The Aldabra group is exceptional amongst coral islands in that the land surface is more elevated, reaching 8 m in altitude. This means that, whereas the low coral islands have floras mostly composed of pantropical and Indo-Pacific species, the flora of the Aldabra Islands has in addition African, Madagascan, Seychelles (Granitic Islands) and endemic species.

Vegetation

Granitic Islands The vegetation of all the Granitic Islands is essentially similar except that the higher altitude formations are not represented on the lower and smaller islands. The coastal formations include mangroves and littoral communities of typical Indian Ocean species. Inland, and with increasing altitude, rain forest predominates, zoned into lowland rain forest, intermediate forest, and, above 550 m, mountain moss forest, which has luxuriant growth of epiphytic bryophytes. In the drier parts of the three largest islands these forests are replaced by palm-rich communities dominated by the endemic *Dillenia ferruginea*. Little natural vegetation remains, mostly replaced by cultivation of coconuts, cinnamon and vanilla. There are also large areas of secondary communities dominated by introduced species.

Coralline Islands The sand cays are mostly worked as coconut estates, with a narrow encircling littoral community. The limestone islands still retain some of the natural vegetation, but most are also worked as coconut estates and for bird guano. Aldabra is the best-preserved of these; its natural vegetation consists mostly of evergreen or semi-deciduous scrub or scrub-forest including sclerophyllous scrub, broad-leaved evergreen and semi-deciduous scrub and scrub-forest types, mangrove scrub and forest. There is a single patch of taller forest on a small area of porous limestone at Takamaka. Peripheral sand dunes and ridges have grass and patches of scrub-forest of strand trees. Groves of introduced coconuts and *Casuarina* have become established on sand flats in several places, forming tall forests. There are also patches of short grassy turf grazed by the resident tortoise population. (Fosberg, 1971.)

Checklists and Floras The whole nation is included in:

Robertson, S.A. (in press). The flowering plants of Seychelles, an annotated checklist with line drawings. (To be published by Missouri Botanical Garden in *Monographs in Systematic Botany*.)

Granitic Islands:
Bailey, D. (1961). *List of the Flowering Plants and Ferns of the Seychelles*, 2nd Ed. Imprimerie Saint-Fidele, Seychelles. 39 pp.
Summerhayes, V.S. (1931). An enumeration of the angiosperms of the Seychelles archipelago. *Trans. Linn. Soc. Lond., 2nd Ser., Zool.* 19(2): 261-299.
Tardieu-Blot, M.L. (1960). *Les Fougères des Mascareignes et des Seychelles*. Notulae Systematicae 16(1-2): 151-201. (Checklist with distributions and specimen citations.)

Coralline Islands:
Fosberg, F.R. and Renvoize, S.A. (1980). *The Flora of Aldabra and Neighbouring Islands*. Kew Bull., Add. Series 7. 358 pp. (Keys, descriptions, line drawings throughout.)

Information on Threatened Plants
Granitic Islands:
Procter, J. (1974). The endemic flowering plants of the Seychelles: an annotated list. *Candollea* 29(2): 345-387. (Lists 72 endemic species with the previous IUCN numerical system (0-4) to indicate degree of threat.)
Swabey, C. (1970). The endemic flora of the Seychelle Islands and its conservation. *Biol. Conserv.* 2(3): 171-177.

Four species which occur in the Granitic Islands are included in *The IUCN Plant Red Data Book* (1978).

Latest IUCN statistics: endemic species and infraspecific taxa – E:21, V:35, R:15, I:2

Coralline Islands No published lists of rare or threatened plants. One species from Aldabra (*Peponium sublitorale*) is included in *The IUCN Plant Red Data Book* (1978).

Botanic Gardens
Victoria Botanic Garden, c/o Dept of Agriculture, Victoria, Mahé.

Useful Addresses
CITES Management Authority: Conservation Officer, Ministry of National Development, Independence House, Mahé.
Seychelles Islands Foundation, c/o The Royal Society, 6 Carlton Terrace, London SW1 5AG, U.K.

Additional References

Gwynne, M.D. and Wood, D. (1969). Plants collected on islands in the western Indian Ocean during a cruise of the M.F.R.V. 'Manihine', Sept.-Oct. 1967. *Atoll Res. Bull.* 134. 15 pp.

Granitic Islands:

Jeffrey, C. (1962). The botany of the Seychelles. Report by the visiting botanist of the Seychelles Botanical Survey, 1961-62. Dept of Technical Co-operation, London. 69 pp., 5 maps. (Mimeo.)

Jeffrey, C. (1968). Seychelles. In Hedberg, I. and O. (1968), cited in Appendix 1. Pp. 275-279.

Procter, J. (1979?). Floristics of the Granitic Islands. (Unpublished manuscript.)

Renvoize, S.A. (1979). The origins of Indian Ocean island floras. In Bramwell, D. (Ed.), *Plants and Islands*. Academic Press, London. Pp. 107-129.

Sauer, J.D. (1967). *Plants and Man on the Seychelles Coast*. Univ. of Wisconsin Press, Madison, U.S.A. 132 pp., 20 plates, 10 pp. of references.

Vesey-Fitzgerald, D. (1940). On the vegetation of the Seychelles. *J. Ecol.* 28: 465-483.

Vesey-Fitzgerald, D. (1942). Further studies of the vegetation on islands in the Indian Ocean. *J. Ecol.* 30(1): 1-16.

Coralline Islands:

Fosberg, F.R. (1971). Preliminary survey of Aldabra vegetation. *Phil. Trans. Roy. Soc. Lond., Ser. B* 260: 215-225. (The whole of vol. 260 is given over to a discussion on the results of the Royal Society Expedition to Aldabra 1967-68.)

Renvoize, S.A. (1971). The origin and distribution of the flora of Aldabra. *Phil. Trans. Roy. Soc. Lond., Ser. B* 260: 227-236. (See note under Fosberg, 1971.)

Renvoize, S.A. (1975). A floristic analysis of the western Indian Ocean coral islands. *Kew Bull.* 30(1): 133-152.

Stoddart, D.R. (Ed.) (1967). Ecology of Aldabra Atoll, Indian Ocean. *Atoll Res. Bull.* 118. 141 pp. (Includes 41 black and white photographs, list of endemic plant species, bibliography of Aldabra.)

Stoddart, D.R. (1968). The conservation of Aldabra. *Geog. Journ.* 134(4): 471-486. (6 black and white photographs.)

Stoddart, D.R. (Ed.) (1970). Coral Islands of the western Indian Ocean. *Atoll Res. Bull.* 136. 224 pp. (Includes species inventories of several islands by F.R. Fosberg and S.A. Renvoize.)

Sierra Leone

Area 72,326 sq. km

Population 3,536,000

Floristics 2480 species (Gledhill, 1962); 2393 species (quoted in Lebrun, 1960, cited in Appendix 1); 1685 species (quoted in Lebrun, 1976, cited in Appendix 1). 74 endemic species, 1 endemic genus (Brenan, 1978, cited in Appendix 1). 1576 species and infraspecific taxa on the Loma Mountains and surrounding foothill country (Jaeger, 1983).

Floristic affinities Guinea-Congolian. Especially important are the Gola High Forests shared with Liberia, and the Afromontane forests and grasslands of the Loma Mountains (Bintumane at 1947 m) and Tingi Hills (Sankan Biriwa at 1860 m).

Vegetation Most of country once covered by moist evergreen and semi-deciduous closed forests but human activities have reduced high forests to 5% of country as forest reserves and protected forests. At present, most of lowland areas (up to 170 m) covered by closed secondary forests and forest regrowth in various stages of succession, interrupted by shifting cultivation and bush fallowing. Closed savanna woodland with tall grasses (2-3 m high) predominate in the north-eastern slopes (200-600 m); mangrove/swamp forest fringe the coastal mudflats; and pockets of Afromontane vegetation (gallery forests, shrub and grass savanna) cover the Loma Mountains and the Tingi Hills in the north-eastern plateau region (600-1000 m).

Estimated rate of deforestation for closed broadleaved forest 60 sq. km/annum out of 7400 sq. km (FAO/UNEP, 1981). According to Myers (1980, cited in Appendix 1), remaining primary moist forest amounts to 2900 sq. km.

For vegetation map see White (1983), cited in Appendix 1.

Checklists and Floras Sierra Leone is included in the *Flora of West Tropical Africa*, cited in Appendix 1.

Gledhill, D. (1962). *Check List of the Flowering Plants of Sierra Leone*. Bunumba Press, Bo. 38 pp. Published privately. (Includes 2480 species.)

Lane-Poole, C.E. (1916). *A List of the Trees, Shrubs, Herbs and Climbers of Sierra Leone*. Govt Printer, Freetown. 159 pp. (Checklist with short notes on usage and flowering time.)

Field-guides

Cole, N.H. Ayodele (1968). *The Vegetation of Sierra Leone (Incorporating a Field Guide to Common Plants)*. Njala Univ. College Press. 198 pp.

Savill, P.S. and Fox, J.E.D. (1967). *Trees of Sierra Leone*. Published privately in Omagh, Northern Ireland. 316 pp. (Includes 24 plates of black and white photographs, coloured map of forest reserves and protected forests, 114 line drawings.)

Information on Threatened Plants No published lists of rare or threatened plants; IUCN has records of 86 species and infraspecific taxa believed to be endemic, including R:10; the remainder are K.

Laws Protecting Plants No plants specifically protected, but there is legislation regarding the conservation and exploitation of forest and forest products. The Wildlife Conservation Act No. 27 (1972), although mainly concerned with fauna, also deals with the protection of their habitats.

Voluntary Organizations

Sierra Leone Wildlife and Conservation Society, Freetown.

Botanic Gardens

Fourah Bay College Botanical Gardens and Forest Reserve, Botany Department, University of Sierra Leone, Freetown.

Additional References

Clarke, J.I. (Ed.) (1966). *Sierra Leone in Maps*. University of London Press. 119 pp.

Cole, N.H. Ayodele (1974). Climate, life forms and species distribution on the Loma Montane grassland, Sierra Leone. *Bot. J. Linn. Soc.* 69: 197-210.

Cole, N.H. Ayodele (1980). The Gola Forest in Sierra Leone: a remnant primary tropical rain-forest in need of conservation. *Envir. Conserv.* 7(1): 33-40. (Includes small-scale map.)

Deighton, F.C. (1957). *Vernacular Botanical Vocabulary for Sierra Leone.* Crown Agents, London. 175 pp.

Jaeger, P. (1976). Le massif des monts Loma (Sierra Leone) – son importance phytogéographique. In Miège, J. and Stork, A.L. (1975, 1976), cited in Appendix 1. Pp. 473-475.

Jaeger, P. (1983), Le recensement des plantes vasculaires et les originalités du peuplement végétal des Mont Loma en Sierra Leone (Afrique Occidentale). In Killick, D.J.B. (1983), cited in Appendix 1. Pp. 539-542.

Jaeger, P. and Adam, J.G. (1972). Contribution à l'étude de la végétation des Monts Loma (Sierra Leone). *Compt. Rend. Somm. Séanc. Soc. Biogéogr.* 424: 77-103.

Morton, J.K. (1968). Sierra Leone. In Hedberg, I. and O. (1968), cited in Appendix 1. Pp. 72-74.

Singapore

Area 616 sq. km

Population 2,540,000

Floristics Ridley (1900) estimates over 2030 vascular plant species including over 130 ferns. The flora is Malayan; many species found in neighbouring Johore State (Malaysia) and in Sumatra.

Vegetation Singapore was originally covered by lowland tropical rain forest, mangrove and swamp forest. Most has been cleared for cultivation and for urban and industrial uses. Only about 19.5 sq. km of protected forest remains, most of it secondary; Bukit Timah Nature Reserve (75 ha) is the last remnant of primary rain forest (Wee, 1964); c. 8 sq. km of mangroves, mostly consisting of disturbed or scattered patches along the northern coast with relatively undisturbed mangrove forest on some offshore islands (H. Keng, 1984, *in litt.*).

Checklists and Floras Singapore is included in the incomplete, but very detailed *Flora Malesiana* (1948-), cited in Appendix 1; *A Revised Flora of Malaya* (1964, 1966, 1971), cited under Malaysia, and the rather dated *Flora of the Malay Peninsula* (Ridley, 1922-1925), cited under Malaysia. No recent national Flora has been published. Notes and new additions to the flora are published occasionally in *Gardens Bulletin Singapore*. In particular see:

Keng, H. (1973-). Annotated list of seed plants of Singapore. *Gard. Bull. Singapore* 26: 233-237 (gymnosperms); 27: 67-83; 27: 247-266; 28: 237-258; 31: 84-113; 33: 329-367, 35: 83-103; 36: 103-124. (8 parts to date, native species indicated, notes on distribution.)

Ridley, H.N. (1900). The flora of Singapore. *J. Roy. Asiatic Soc. Straits Branch* 33: 27-196. (Annotated checklist, endemics indicated, taxonomy dated.)

Sinclair, J. (1953, 1956). Additions to the flora of Singapore and new localities in Singapore for some plants thought to be extinct. *Gard. Bull. Singapore* 14: 30-39; 15: 22-30. (Part 1 lists 37 species new to the flora of which 12 records new for Malay Peninsula; part 2 lists 18 new species for Singapore of which 5 new to Malay Peninsula.)

Field-guides See Corner (1952), Henderson (1949, 1954) and Shivas (1984), cited under Malaysia.

Information on Threatened Plants No national list of threatened plants has been published, but see:

Ng, F.S.P. and Low, C.M. (1982). *Check List of Endemic Trees of Malay Peninsula.* Forest Research Institute, Kepong. 94 pp. (Lists 654 trees endemic to the Malay Peninsula, some of which occur in Singapore. 'Endangered' species indicated; degree of threat based on numbers of herbarium collections.)

Laws Protecting Plants Permits are required for collecting plants in Nature Reserves and water catchment areas.

Voluntary Organizations
Malay Nature Society (Singapore Branch), c/o Botany Department, National University of Singapore, Kent Ridge, Singapore 0511.

Botanic Gardens
Singapore Botanic Gardens, Cluny Road, Singapore 1025.

Additional References
Corner, E.J.H. (1978). *The Freshwater Swamp-forest of South Johore and Singapore.* Botanic Gardens, Singapore. 266 pp. (Ecology; species lists.)
Wee, Y.C. (1964). A note on the vegetation of Singapore Island. *Malay Forester* 27: 257-266.

Society Islands

10 high volcanic islands and 5 coral atolls, situated 3220 km east of Fiji in the Pacific Ocean, between latitudes 15-20°S and longitudes 147-153°W. The highest point is 2237 m, on Tahiti. The total land area is 1614 sq. km. Population 67,034 (Douglas, 1969, cited in Appendix 1), of which c. 50,000 live on Tahiti, the largest island (1000 sq. km). The Society Islands form part of French Polynesia.

c. 700 vascular plant species (Grant *et al.*, 1974), including c. 200 fern species (Parris, 1985, cited in Appendix 1). 12 endemic vascular plant genera (van Balgooy, 1971, cited in Appendix 1). Floristic affinities with the Indo-Malesian region. No information on threatened plants.

The vegetation consists of strand forest, with *Cocos* and *Pandanus*; lowland rain forest (greatly modified); extensive montane rain forest, with cloud forests on mountain peaks. *Gleichenia* scrub on lower ridges (Grant *et al.*, 1974). All the low atolls of the group have been largely converted to coconut plantations.

References

Copeland, E.B. (1932). Pteridophytes of the Society Islands. *Bull. Bernice P. Bishop Mus.* 93. 86 pp. (Descriptions and keys; notes on distribution; includes 146 fern species of which 30% considered endemic.)

Grant, M.L., Fosberg, F.R. and Smith, H.M. (1974). Partial Flora of the Society Islands: Ericaceae to Apocynaceae. *Smithsonian Contrib. Bot.* 17. 85 pp.

Nadeaud, J. (1873). *Enumération des Plantes Indigènes de l'Ile de Tahita.* Paris. 86 pp. (Descriptions of 127 ferns, 508 angiosperms.)

Papy, H.R. (1951-1954). La végétation des Iles de la Société et de Makatéa (Océanie Française). *Trav. Lab. For. Toulouse* Tome 5, vol. 1, art. 3. 386 pp.

Sachet, M.-H. (1983). Botanique de l'Ile de Tupai, Iles de la Société. *Atoll Res. Bull.* 276. 26 pp. (Includes annotated checklist of 35 native fern species and 95 angiosperm taxa, of which 42 native.)

Sachet, M.-H. (1983). Natural history of Mopelia Atoll, Society Islands. *Atoll Res. Bull.* 274. 37 pp. (Includes annotated checklist of 3 fern species and 81 angiosperm taxa, including 45 introductions.)

Sachet, M.-H. and Fosberg, F.R. (1983). An ecological reconnaissance of Tetiaroa Atoll, Society Islands. *Atoll Res. Bull.* 275. 67 pp. (Includes annotated checklist of 4 native fern species and 91 angiosperm taxa, including 44 introductions.)

The Society Islands are included in *Flore de la Polynésie Française* (Drake del Castillo, 1893, cited in Appendix 1), which forms the only complete Flora.

Socotra

A predominantly limestone island lying in the Indian Ocean (12°30'N 54°00'E) south-east of South Yemen, to which it belongs. The dissected limestone plateau is surrounded by wide alluvial plains of recent origin. In the north-east of the island is a mountainous granitic outcrop (the Hagghier massif) reaching 1500 m. Three small islands (Abd al Kuri and The Brothers) lie between Socotra and mainland Africa.

Area 3625 sq. km

Population No information

Floristics 680 species (quoted in Lebrun, 1976, cited in Appendix 1). 215 species endemic to Socotra; about eight endemic to Abd al Kuri (IUCN figures). Limestone plateaux on both islands especially rich.

Floristic affinities with African mainland, and especially the Somalia-Masai region; but also strong links with drier areas in Dhofar (Oman) and South Yemen; also large endemic element. Climate predominantly dry, so succulent life-forms especially well-represented.

Vegetation Coastal plains: mostly semi-desert dwarf shrubland and grassland, with small areas of mangrove and herbaceous halophytic communities. Limestone Plateau: variable vegetation, including grassland thickets, succulent shrubland and semi-evergreen bushland dominated by *Dracaena cinnabari*. Granite massif of Hagghier: evergreen bushland and thicket, grassland. Most of the original vegetation has been destroyed by overgrazing caused by introduced goats. The vegetation of the coastal plains is affected most; inaccessibility protects many of the endemics in the highland areas.

The vegetation of Abd al Kuri is much more sparse, and much of the island is semi-desert. None of the tree-succulents, so characteristic of Socotra, are found; few of Socotra's endemics are represented.

Checklists and Floras Socotra is included in *Adumbratio Florae Aethiopicae*, cited in Appendix 1. See also:

Balfour, I.B. (1888). Botany of Socotra. *Trans. Roy. Soc. Edinb.* 31. 446 pp. (Over 100 species illustrated.)
Popov, G.B. (1957). The vegetation of Socotra. *J. Linn. Soc. Bot.* 55: 706-720. (Includes species list by J.B. Gillett; 18 black and white photographs.)
Royal Botanic Gardens, Kew (1971). New or noteworthy species from Socotra and Abd al Kuri. *Hooker's Icones Plantarum, Ser. 5*, 7(4): t. 3673-3700.
Vierhapper, F. (1907). *Beiträge zur Kenntnis der Flora Südarabiens und der Inseln Sokótra, Sémha und 'Abd el Kûri*. Denkschr. K. Akad. Wiss. Wien, Math.-Naturwiss. 71: 321-490. (Descriptions; no keys.)

Information on Threatened Plants No published lists of rare or threatened plants; IUCN has records of 215 species and infraspecific taxa believed to be endemic – Ex:1, E:84, V:17, R:29, I:1, K:2, nt:81.

Seven species which occur in Socotra and one from Abd al Kuri are included in *The IUCN Plant Red Data Book* (1978).

Additional References

Forbes, H.O. (Ed.) (1903). *The Natural History of Sokotra and Abd-el-Kuri*. Liverpool. 598 pp. (Report of the results of an expedition by W.R. Ogilvie-Grant and H.O. Forbes 1898-1899; 'Botany of Sokotra and Abd-el-Kuri' by I.B. Balfour, pp. 445-570.)
Gwynne, M.D. (1968). Socotra. In Hedberg, I. and O. (1968), cited in Appendix 1. Pp. 179-185.
Wettstein, R. (1906). Sokotra. In Karsten, G. and Schenck, H., *Vegetationsbilder* 3(5): t. 25-30.

Solomon Islands

By T.C. Whitmore

The Solomon Islands, including the Santa Cruz Islands, are a scattered archipelago, mainly of volcanic islands and several low-lying coral atolls, c. 2000 km east of Queensland. The 6 largest islands – Choiseul, New Georgia, Santa Isabel, Guadalcanal, Malaita and Makira (San Cristobal) – have precipitous, thickly-forested mountain ranges. Mt Popomaneseu (2331 m) on Guadalcanal is the highest point. Bougainville, politically part of Papua New Guinea and covered in a separate sheet, lies to the north-western end of the Solomons.

Area 29,790 sq. km

Population 269,000

Floristics c. 1750 flowering plant species; c. 400 fern species (Parris, 1985, cited in Appendix 1). c. 230 orchid species recorded, but this figure is likely to double as the islands become better known (P. Cribb, 1984, pers. comm.). Endemism is low; no reliable figure for species, and only 3 endemic genera of flowering plants. The tree flora is not as rich as that of Malesia, to which the flora is mostly related. The flora of the Santa Cruz Islands is more related to that of New Caledonia and Vanuatu.

Vegetation Lowland rain forests, the most distinctive of which are the forests on ultrabasic rocks, dominated by *Casuarina papuana*; riverine gallery forests; no distinct lower montane forest but upper montane 'mossy' forest well-developed; extensive grasslands, particularly in rain-shadow areas; on the coast, mangroves, which reach greater stature than in much of Malesia; beach forest well-developed in places. Extensive areas of lowland forests have been felled for timber, notably in parts of the New Georgia Group. For a description of forest types see Schmid (1978).

Checklists and Floras

Thorne, A. and Cribb, P. (1984). Orchids of the Solomon Islands and Bougainville: a preliminary checklist. Royal Botanic Gardens, Kew. 33 pp. Mimeo. (Compiled from herbarium and literature records at Kew.)

Whitmore, T.C. (1966). *Guide to the Forests of British Solomon Islands*. Oxford Univ. Press, London. 208 pp. (Includes checklist, with short descriptions of trees.)

Information on Threatened Plants None.

Botanic Gardens

Honiara Botanic Garden, Forest Department, Ministry of Primary Resources, Honiara.

Useful Addresses

Conservation Division, Ministry of Primary Resources, P.O. Box G 26, Honiara.
Forest Department, Ministry of Primary Resources, P.O. Box G 26, Honiara.
Rainforest Information Centre, P.O. Box 368, Lismore, N.S.W. 2480, Australia.

Additional References

Corner, E.J.H. (Ed.) (1969). A discussion on the results of the Royal Society Expedition to the British Solomon Islands Protectorate, 1965. *Phil. Trans. Royal Society*, B 225: 185-631. (See T.C. Whitmore on vegetation, pp. 259-270, and plant geography, pp. 549-568.)

Hansell, J.R.F. and Wall, J.R.D. (1976). *Land Resources of the Solomon Islands*, 8 vols. Land Resources Study 18. Ministry of Overseas Development, London. (Land use and evaluation; includes maps of forest types at 1:150,000. 1 – Introduction and recommendations; 2 – Guadalcanal and the Florida Islands; 3 – Malaita and Ulawa; 4 – New Georgia Group and the Russell Islands; 5 – Santa Isabel; 6 – Choiseul and the Shortland Islands; 7 – San Cristobal and adjacent islands; 8 – Outer islands.)

Schmid, M. (1978). The Melanesian forest ecosystems (New Caledonia, New Hebrides, Fiji Islands and Solomon Islands). In Unesco/UNEP/FAO (1978), cited in Appendix 1. Pp. 654-683.

Somalia

Area 630,000 sq. km

Population 5,423,000

Floristics 3000 species (J.B. Gillett, 1984, pers. comm.); c. 2610 species (Glover, 1947). c. 518 endemic species, estimated from a sample of Cufodontis' *Enumeratio* (Brenan, 1978, cited in Appendix 1). Especially important are the species of the succulent scrub, threatened by chronic desertification.

Floristic affinities of most of the country with Somalia-Masai region which has c. 50% specific endemism, but is dissected by country boundaries. The northern mountains have Afromontane affinities, and the extreme south has affinities with East African coastal semi-evergreen bushland.

Vegetation Mostly *Acacia-Commiphora* deciduous bushland and thicket, particularly in the southern leg. Large areas of semi-desert grassland and deciduous shrubland in the north and extending a long way south along the coast. Smaller areas of evergreen and semi-evergreen bushland and thicket surrounding small patches of *Juniperus* forest on the tops of the escarpment, especially near Erigavo. Semi-evergreen bushland in extreme south resembles the Kenya coastal belt. Extensive areas of vegetation are being lost through desertification.

Checklists and Floras Somalia is included in *Enumeratio Plantarum Aethiopiae Spermatophyta* (Cufodontis, 1953-1972), and in *Adumbratio Florae Aethiopicae*, both cited in Appendix 1. See also:

Chiovenda, E. (1916). Le collezioni botaniche della Missione Stefanini-Paoli nella Somalia italiana. *Pubbl. R. Istituto Studi Superiori di Firenze*. Firenze.

Chiovenda, E. (1929-1936). *Flora Somala*, 3 vols. Vol. 1 published in Rome; vol. 2 published in Modena; vol. 3 published in *Atti. Ist. Bot. Pavia (Series 4)* 7: 117-160. (Checklist with full descriptions of new species; specimen citations. These works complement his 1916 publication, since records published in that are not included in later works.)

Glover, P.E. (1947). *A Provisional Checklist of British and Italian Somaliland Trees, Shrubs and Herbs*. Published for Govt of Somaliland, by Crown Agents, London. 446 pp. (Checklist with synonyms; 13 black and white plates.)

Information on Threatened Plants No published lists of rare or threatened plants; IUCN has records of 171 species and infraspecific taxa believed to be endemic, including E:7, V:18, R:8, I:7, nt:3. (By and large, categories only assigned to succulent species so far.)

Four species occuring in Somalia are included in *The IUCN Plant Red Data Book* (1978).

Useful Addresses

National Range Agency (Herbarium), Box 1759, Mogadishu.

Additional References

Bally, P.R.O. (1968). Somali Republic South. In Hedberg, I. and O. (1968), cited in Appendix 1. Pp. 145-148.

Bally, P.R.O. (1976). Notes on the vegetation of Somalia. In Miège, J. and Stork, A.L. (1975, 1976), cited in Appendix 1, pp. 447-450.

Hemming, C.F. (1966). The vegetation of the northern region of the Somali Republic. *Proc. Linn. Soc. Lond.* 177: 173-250.

Hemming, C.F. (1968). Somali Republic North. In Hedberg, I. and O. (1968), cited in Appendix 1. Pp. 141-145.

Pichi-Sermolli, R.E.G. (1957). Una carta geobotanica dell'Africa orientale (Eritrea, Ethiopia, Somalia). *Webbia* 13: 15-132. (Includes map 1:5,000,000.)

Senni, L. (1935). *Gli Alberi e le Formazioni Legnose della Somalia.* Firenze. 305 pp. (References to illustrations, very short diagnoses, illustrations. Rather old, but summarizes all Chiovenda's knowledge of woody plants.)

South Africa

Area 1,184,827 sq. km

Population 31,586,000

Floristics c. 23,000 species (H.P. Linder, 1984, pers. comm.); southern Africa (South Africa, Lesotho, Swaziland, Namibia and Botswana) has 23,201 species of vascular plants (Gibbs Russell, 1984), of which c. 80% are endemic, Natal c. 4800 species, unknown levels of endemism; Cape Peninsula (470 sq. km extreme south-west of Cape Province) 2256 species, of which 157 (7%) endemic (all from Goldblatt, 1978). The Cape Floristic Region (the south-west corner of the Cape Province) c. 8579 vascular species, of which c. 68% endemic (Bond and Goldblatt, 1984).

Floristic affinities many and varied, ranging from Zambezian in the north-east, South African coastal in the east, South African Mediterranean in the south (sometimes elevated to "Cape Floristic Kingdom"); most of centre of country with floristic affinities with Kalahari-Highveld region, and Karoo-Namib region in the west; flora of high ground on Drakensberg with Afromontane affinities.

Vegetation Complicated mosaic of vegetation types due to great diversity of topography and climate. Large areas of savanna and open shrubby woodland in Transvaal and Natal. Large area of Highveld grassland in Orange Free State and Transvaal. Most of Cape Province (northern and central parts) occupied by arid and semi-arid Karoo shrubland and grassy shrubland; smaller areas of Afromontane communities (forest and mountain grasslands) on the Drakensberg along the eastern escarpment; also Afroalpine communities near Lesotho. Along the east coast: East African coastal mosaic of grassland, savanna and forest patches. South-western Cape occupied by bushy Karoo (succulent) and Fynbos (sclerophyll) shrubland. Fynbos now greatly reduced in area due to agriculture and development, and to invasion of exotics.

For vegetation maps see White (1983), cited in Appendix 1, and Acocks (1975), cited below.

Checklists and Floras South Africa is included in the incomplete *Flora of Southern Africa, Trees of Southern Africa* (Palmer and Pitman, 1972), and in *The Genera of Southern African Flowering Plants* (Dyer, 1975, 1976), all cited in Appendix 1. See also:

de Winter, B., Vahrmeijer, J. and von Breitenbach, F. (1978). *The National List of Trees*, 2nd Ed. (1st Ed. 1972). Van Schaik, Pretoria.

Gibbs Russell, G.E. *et al.* (1984). List of species of Southern African plants. *Mem. Bot. Surv. S. Afr.* 48. 144 pp.

Harvey, W.H. and Sonder, O.W. (1859-1865). *Flora Capensis*. Vols 1-3. Hodges and Smith, Dublin. (See also Thistleton Dyer, 1896-1933, below.)

Marloth, R. (1913-1932). *Flora of South Africa*, 4 vols. Darter, Cape Town. (Fully illustrated with line drawings, colour paintings, and black and white photographs.)

Schelpe, E.A.C.L.E. (1969). A revised check-list of the Pteridophyta of Southern Africa. *J. S. Afr. Bot.* 35: 127-140.

Thiselton Dyer, W.T. (1896-1933). *Flora Capensis*. Vols 4-7. Reeve, London. (Frodin says vols 4 and 5 reprinted 1973 by Cramer, Lehre. Covers South Africa and the parts of Namibia, Botswana and Mozambique south of the Tropic of Capricorn.)

Cape Province:

Adamson, R.S. and Salter, T.M. (Eds) (1950). *Flora of the Cape Peninsula*. Juta, Cape Town. 889 pp. (Keys, descriptions, distributions; covers 2622 vascular species including 57 pteridophytes.)

Bond, P. and Goldblatt, P. (1984). *Plants of the Cape Flora. A Descriptive Catalogue*. J. S. Afr. Bot. Suppl. Vol. No. 13, National Botanic Gardens of South Africa. 455 pp. (With 12 colour photographs. Covers c. 8579 vascular species including c. 75 pteridophytes, 45% of the flora of southern Africa.)

Duthie, A.V. (1930). List of vascular cryptogams and flowering plants of the Stellenbosch flats. *Ann. Univ. Stellenbosch, Ser. A* 8(4). 52 pp.

Fourcade, H.G. (1941). Check-list of the flowering plants of the divisions of George, Knysna, Humansdorp and Uniondale. *Mem. Bot. Surv. S. Afr.* 20. 127 pp. (Gives 2969 species of native seed plants.)

Leistner, O.A. (1967). The plant ecology of the southern Kalahari. *Mem. Bot. Surv. S. Afr.* 38. 172 pp. (With 51 black and white photographs, and a checklist of species on pp. 115-133.)

Martin, A.R.H. and Noel, A.R.A. (1960). *The Flora of Albany and Bathurst*. Rhodes University, Grahamstown. 128 pp. (Checklist, with brief distributions and diagnoses; 2390 vascular species.)

Schonland, S. (1919). Phanerogamic flora of the divisions of Uitenhage and Port Elizabeth. *Mem. Bot. Surv. S. Afr.* 1. 118 pp. (2312 species of seed plants.)

Wilman, M. (1946). *Preliminary Checklist of the Flowering Plants and Ferns of Griqualand West*. Deighton Bell, Cambridge. 381 pp.

Natal:

Edwards, D. (1967). A plant ecological survey of the Tugela river basin, Natal. *Mem. Bot. Surv. S. Afr.* 36. 285 pp. (With a checklist of the flora on pp. 258-277.)

Ross, J.H. (1973). The Flora of Natal. *Mem. Bot. Surv. S. Afr.* 39. 418 pp. (Keys to families and genera; 1 specimen and distribution of each species; covers 4818 species of seed plants.)

Transvaal and Orange Free State:

Burtt Davy, J. (1926-1932). *A Manual of the Flowering Plants and Ferns of the Transvaal with Swaziland, South Africa*, 2 parts. Longmans and Green, London. 529 pp. (Never completed; pteridophytes, gymnosperms and dicotyledons up to end of Umbelliferae; remainder in manuscript or proof.)

Cohen, S. (1939). A preliminary checklist of the flowering plants and ferns of parts of the Transvaal and the Orange Free State. Unpublished M.Sc. Thesis, Univ. of Witwatersrand.

Schijff, H.P. van der (1969). *A Check List of the Vascular Plants of the Kruger National Park*. Publication No. 53 of the Univ. of Pretoria. 100 pp. (Annotated with distributions, specimens; 12 black and white photographs; 1838 vascular species.)

Wyk, P.van (1972-1974). *Trees of the Kruger National Park*, 2 vols. Purnell, Cape Town. 597 pp. (Colour photographs and distribution maps throughout.)

Field-guides There are far too many field-guides published to include them all; this is just a selection of the more useful ones:

Coates Palgrave, K. (1977). *Trees of Southern Africa*. Struik, Cape Town. 959 pp. (Keys, descriptions, distribution maps, line drawings, colour paintings, colour and black and white photographs. Also includes Botswana, Lesotho, Mozambique south of the Zambezi River, Namibia, Swaziland and Zimbabwe.)

Palmer, E. (1977). *A Field-Guide to the Trees of Southern Africa*. Collins Field-Guide Series, Collins, London. 352 pp. (Line drawings throughout; 32 plates of colour paintings; also covers Swaziland, Namibia, Lesotho and Botswana.)

Cape Province:

Batten, A. and Bokelmann, H. (1966). *Wild Flowers of the Eastern Cape Province*. Books of Africa, Cape Town. 185 pp.

Gledhill, E. (1981). *Eastern Cape Veld Flowers*, 2nd Ed. Cape Provincial Dept of Nature Conservation, Cape Town. 276 pp. (Line drawings throughout.)

Kidd, M.M. (1983). *Cape Peninsula*, 3rd Ed. (First published 1950 as *Wild Flowers of the Cape Peninsula*.) South African Wild Flower Guide No. 3, Botanical Society of South Africa, Cape Town. 239 pp. (With colour illustrations of 814 characteristic species.)

le Roux, A. and Schelpe, E.A.C.L.E. (1981). *Namaqualand and Clanwilliam*. South African Wild Flower Guide No. 1, Botanical Society of South Africa, Cape Town. 173 pp. (Colour photographs throughout.)

Levyns, M.R. (1966). *Guide to the Flora of the Cape Peninsula*, 2nd Ed. Juta, Cape Town. 310 pp. (Keys, descriptions, line drawings.)

Mason, H. (1972). *Western Cape Sandveld Flowers*. Struik, Cape Town. 203 pp. (With 84 pages of colour paintings.)

Moriarty, A. (1982). *Outeniqua, Tsitsikamma and Eastern Little Karoo*. South African Wild Flower Guide No. 2, Botanical Society of South Africa, Cape Town.

Natal:

Gibson, J.M. (1975). *Wild Flowers of Natal Coastal Region*. Durban. 136 pp. (With c. 800 colour paintings.)

Irwin, P., Ackhurst, J. and Irwin, D. (1980). *A Field Guide to the Natal Drakensberg*. Wild Life Society of Southern Africa, Durban.

Moll, E.J. (1981). *Trees of Natal. A Comprehensive Field-Guide to over 700 Indigenous and Naturalized Species*. Univ. of Cape Town Eco-Lab Trust Fund, Cape Town. 567 pp. (With line drawings and a distribution map for each species.)

Wright, W.G. (1963). *The Wild Flowers of Southern Africa: Natal. A Rambler's Pocket Guide*. Nelson, London. 168 pp. (With colour paintings and line drawings throughout. A work of the same name, but not seen, by J. Adams, 1976, may be a later edition.)

Transvaal:

Letty, C. (1962). *Wild Flowers of the Transvaal*. Wild Flowers of the Transvaal Book Fund; Dept of Agriculture, Pretoria. 362 pp. (Colour paintings by C. Letty; text by R.A. Dyer, I.C. Verdoorn and L.E. Codd.)

Tree Society of Southern Africa (1969). *Trees and Shrubs of the Witwatersrand*, 2nd Ed., illus. by B. Jeppe. Witwatersrand Univ. Press, Johannesburg. 309 pp. (Illustrated in colour and black and white.)

Orange Free State:

Venter, H.J.T. (1976). *Bome en Struike van die Oranje-Vrystaat* (Trees and shrubs of the Orange Free State). de Villiers, Bloemfontein. 240 pp. (Illus.)

Information on Threatened Plants

Geldenhuys, R. (1977). *Threatened Species: Habitat Conservation and their Survival.* Dept of Nature and Environmental Conservation, Provincial Administration of the Cape of Good Hope. 48 pp. (Illus.)

Hall, A.V. (1983). Threatened plants at the south-western corner of Africa. In Killick, D.J.B. (1983), cited in Appendix 1. Pp. 981-984. (The species statistics given here are also included in Hall *et al.*, 1984, cited below.)

Hall, A.V. *et al.* (1980), cited in Appendix 1. (Gives lists of endemic and non-endemic threatened plant taxa under the following area headings: Transvaal: 105 endemic: E:4, V:10, R:59, I:1, K:31; 112 non-endemic: E:1, V:4, I:11, K:37. Bophuthatswana: 3 endemic: E:2, R:1; 9 non-endemic: V:2, R:6, K:1. Orange Free State: 4 endemic: V:1, R:1, K:2; 19 non-endemic: V:3, R:8, I:1, K:7. Natal: 73 endemic: Ex:3, E:1, V:13, R:30, I:6, K:20; 95 non-endemic: V:8, R:54, I:12, K:21. Transkei: 21 endemic: E:1, V:4, R:4, I:2, K:10; 53 non-endemic: V:8, R:26, I:6, K:13. Cape Province: 1480 endemic: Ex:37, E:95, V:117, R:317, I:230, K:683; 46 non-endemic: V:8, R:19, I:8, K:11. Regional categories are given for non-endemics.) This provisional list, already out-of-date, is being revised regionally (e.g. Hall and Ashton, 1983).

Hall, A.V. and Ashton, E.R. (1983). *Threatened Plants of the Cape Peninsula.* Threatened-Plants Research Group, Bolus Herbarium, address below. 26 pp. (Lists 174 Cape Peninsula plants, endemic and non-endemic (world categories): Ex:5, E:25, V:28, R:50, I:40, K:26.)

Hall, A.V., de Winter, B., Fourie, S.P. and Arnold, T.H. (1984). Threatened plants in southern Africa. *Biol. Conserv.* 28(1): 5-20. (This contains the latest statistics: southern Africa as a whole, i.e. south of Mozambique, Zimbabwe, Zambia and Angola: 2373 threatened plants, Ex:39, E:110, V:223, R:700, I:393, K:908; Cape Floristic Kingdom: 1621 threatened plants, Ex:36, E:98, V:137, R:383, I:279, K:688; Natal: 215 threatened plants, Ex:3, E:1, V:27, R:90, I:32, K:62; Transvaal: 286 threatened plants, E:8, V:31, R:131, I:32, K:84.)

12 species which occur in South Africa are included in *The IUCN Plant Red Data Book* (1978). General groups which are receiving particular attention are those threatened by trade and horticulture. These include many types of succulents, flowering shrubs from the Fynbos (e.g. Proteaceae), cycads, some forest epiphytes, etc.

A.V. Hall is the editor of the *Rare Plants Gazette*, a periodical series giving up-to-date information about the Cape's most critically endangered plants. (Available from the Bolus Herbarium, address below.)

Laws Protecting Plants For a general overview of plant conservation legislation in South Africa, see:

Fuggle, R.F. and Rabie, M.A. (1983). *Environmental Concerns in South Africa.* Juta, Cape Town. Pp. 173-183.

See also:

Anon. (1967). *Some Protected Wild Flowers of the Cape Province*. Dept of Nature Conservation of the Cape Provincial Administration, Cape Town. Unpaginated. (With colour paintings of 244 protected species.)

Gill, K. (1982). Legal protection for trees: Forest Amendment Act 1982. *Trees S. Afr.* 33(4): 100-101. (Reproduced from a newsletter, Society for the Protection of the Environment.)

The following regional legislation applies:

Cape Province: Nature Conservation Ordinance No. 26 (1965): it is an offence to pluck any protected indigenous plant without written permission. (Several thousand species listed.)

Natal: Nature Conservation Ordinance No. 15 (1974): provides protection for 28 species, 7 genera, and Proteaceae.

Transvaal: Nature Conservation Ordinance No. 17 (1967): gives specific protection to 34 species, 47 genera, all Pteridophyta (except bracken), and Orchidaceae.

Orange Free State: Nature Conservation Ordinance No. 8 (1969): gives specific protection to 29 genera, 15 species and 2 families.

Voluntary Organizations There is a large number of national and local societies with an interest in plants; a small selection here only:

Botanical Society of South Africa, Kirstenbosch Botanic Gardens, address below.
Southern African Nature Foundation (WWF-South Africa), P.O. Box 456, Stellenbosch 7600, Cape Province.
Wildlife Society of South Africa, P.O. Box 44189, Linden 2104, Transvaal.

Botanic Gardens The country's main gardens:

National Botanic Gardens of South Africa, Kirstenbosch Botanic Gardens, Private Bag X7, Claremont 7735, Cape Province.

Important regional gardens:

Harold Porter Botanic Garden, P.O. Box 35, Betty's Bay 7141, Cape Province.
Karoo Botanic Garden, P.O. Box 152, Worcester 6850, Cape Province.
Lowveld Botanic Garden, P.O. Box 1024, Nelspruit 1200, Transvaal.
Natal Botanic Garden, Mayor's Walk, Pietermaritzburg 3201, Natal.
Orange Free State Botanic Garden, P.O. Box 1536, Bloemfontein 9300.
Pretoria National Botanic Garden, Botanical Research Institute, Department of Agriculture and Fisheries, Private Bag X101, Pretoria 0001.
The Drakensberg Botanic Garden, P.O. Box 157, Harrismith 9880, Orange Free State.
Transvaal Botanic Garden, P.O. Box 2194, Wilropark 1731.

Index of threatened species in cultivation:

Threatened Plants Committee Secretariat (1982). *The Botanic Gardens List of Threatened Species for Southern Africa 1982 First Draft*. Botanic Gardens Conservation Co-ordinating Body Report No. 5. IUCN, Kew. 26 pp. (Lists 411 rare and threatened taxa from South Africa as in cultivation, with gardens listed alongside each entry, out of 1045 taxa overall.)

Useful Addresses
Bolus Herbarium, University of Cape Town, Rondebosch 7700, Cape Province.
Foundation for Research Development (formerly Cooperative Scientific Programmes),

Council for Scientific and Industrial Research, P.O. Box 395, Pretoria 0001.

CITES Management Authority: Dept of Environment Affairs, Environmental Conservation Branch, Private Bag X447, Pretoria 0001. (This is also the contact address for the Council for the Environment, the National Plan for Nature Conservation, the South African Natural Heritage Programme, and the National Atlas for Critical Environmental Components.)

Other Management Authorities who grant permits:

Bophuthatswana Parks Board, Private Bag X2078, Mafiking 8670, Bophuthatswana.

Chief Director, National Parks Board, P.O. Box 787, Pretoria 0001.

Dept of Agriculture and Forestry, Private Bag X2247, Sibasa 0970, Republic of Venda.

Dept of Agriculture and Forestry, Private Bag 5002, Umtata, Transkei.

Dept of Agriculture (and Nature Conservation), Private Bag X501, Zwelitsha 5600, Ciskei.

Director of Nature Conservation, Transvaal, Private Bag X209, Pretoria 0001.

Director of Nature Conservation, P.O. Box 517, Bloemfontein 9300, Orange Free State.

Director of Nature and Environmental Conservation, Private Bag X9086, Cape Town 8000.

Director, Natal Parks, Game and Fish Preservation Board, P.O. Box 662, Pietermaritzburg 3200, Natal.

Directorate of Forestry, Dept of Environment Affairs (see above).

Kwazulu Bureau for Natural Resources, Private Bag X05, Ulundi 3838, Natal.

Additional References

Acocks, J.P.H. (1975). Veld types of South Africa, 2nd ed. *Mem. Bot. Surv. S. Afr.* 40. 128 pp. (With coloured vegetation map 1:1,500,000.)

Adamson, R.S. (1934). The vegetation and flora of Robben Island. *Trans. R. Soc. S. Afr.* 22(4): 279-296. (Lists 116 native flowering plants.)

Anon. (1983). *Official Year Book of the Republic of South Africa.* van Rensburg Publications, Johannesburg.

Bayer, A.W., Bigalke, R.C. and Crass, R.S. (1968). Natal. In Hedberg, I. and O. (1968), cited in Appendix 1. Pp. 243-247.

Goldblatt, P. (1978). An analysis of the flora of Southern Africa: its characteristics, relationships and origins. *Ann. Missouri Bot. Gard.* 65(2): 369-436.

Killick, D.J.B. (1968). Transvaal. In Hedberg, I. and O. (1968), cited in Appendix 1. Pp. 239-243.

Lighton, C. (1973). *Cape Floral Kingdom*, 3rd Ed. (1st Ed. 1960). Juta, Cape Town. 221 pp. (Good general text with colour photographs.)

Oliver, E.J.H., Linder, H.P. and Rourke, J.P. (1983). Geographical distribution of present-day Cape taxa and their phytogeographical significance. In Killick, D.J.B. (1983), cited in Appendix 1. Pp. 427-440.

Roberts, B.R. (1968). The Orange Free State. In Hedberg, I. and O. (1968), cited in Appendix 1. Pp. 247-250.

Rycroft, H.B. (1968). Cape Province. In Hedberg, I. and O. (1968), cited in Appendix 1. Pp. 235-239.

Scheepers, J.C. (1983a). Progress with vegetation studies in South Africa. In Killick, D.J.B. (1983), cited in Appendix 1. Pp. 683-690. (Includes a good bibliography.)

Scheepers, J.C. (1983b). The present status of vegetation conservation in South Africa. In Killick, D.J.B. (1983), cited in Appendix 1. Pp. 991-995.

Scheepers, J.C. (1984). The status of conservation in South Africa. *J. S. Afr. Biol. Soc.* 23: 64-71.

Smith, C.A. (1966). Common names of South African plants. *Mem. Bot. Surv. S. Afr.* 35. 642 pp.

Werger, M.J.A. (1978), cited in Appendix 1. Citation includes list of relevant chapters.

Spain

Area 504,879 sq. km

Population 38,717,000

Floristics 4750-4900 native vascular species, estimated by D.A. Webb (1978, cited in Appendix 1) from *Flora Europaea*; 720 endemic vascular taxa (Barreno *et al.*, 1984). Figures are for peninsular Spain only. One of the richest floras in Europe. Elements: predominantly Atlantic and Mediterranean, but also Central European, North African and alpine.

Areas of diversity: Cantabrian Mts, Pyrenees, and the following Sierras: Nevada, de Montserrat, de Montseny, Cuenca, de Teruel, de Cazorla, Tejeda, de Almijara, Morena, Serrania de Ronda and the Central Sierras, together with the gypsum-soil areas in central Spain. The most species-rich area is Sierra Nevada, in the south-east, supporting 177 Iberian and 66 Spanish endemics (C. Gómez-Campo, 1984, pers. comm.; Polunin and Smythies, 1973, cited in Appendix 1).

Vegetation Original sclerophyllous forest has long been removed or degraded into stunted woods and maquis scrub. Today about 20% of Spain is forested, including conifer plantations (Gómez-Campo, pers. comm.). In the wetter north-west, from Galicia to the Pyrenees, including the coastal strip of Cantabria, scattered areas of mixed woodland (Central European type) are present with oak, beech and Scots Pine.

Within the Mediterranean zone, 2 main vegetation types present: original coastal vegetation of Aleppo Pine (*Pinus halepensis*), Stone Pine (*P. pinea*), Holm and Kermes Oaks (*Quercus ilex* ssp. *rotundifolia* and *Q. coccifera* respectively), grossly modified in many areas, where it forms secondary scrub (maquis) with many introduced exotics, mostly palms and cacti. Inland, Holm Oak, Pyrenian Oak (*Q. pyrenaica*) and Cork Oak (*Q. suber*) become more widespread. On the 2 large, dry, central tablelands, original Holm Oak woodland now largely degraded to *Q. coccifera* scrub and garigue. Large expanses of dry grassland survive on southern slopes of Pyrenees and Cantabrian Mts and to lesser extent on sub-Mediterranean calcareous mountains in eastern Spain (C. Gómez-Campo, 1984, *in litt.*; Ozenda, 1979, cited in Appendix 1).

Checklists and Floras Spain is covered by the completed *Flora Europaea* (Tutin *et al.*, 1964-1980), and will also be covered in the *Med-Checklist* (both cited in Appendix 1). A national Flora is in preparation under the co-ordination of the Botanic Garden of Madrid. Vol. 1 (in press) contains pteridophytes, gymnosperms and angiosperms up to Papaveraceae. For earlier national Floras and checklists see:

Caballero, A. (1940-). *Flora Analítica de España*, 1 vol. Sociedad Anánima Española de Traductores y Autores, Madrid. 617 pp. (Incomplete; covers widespread species and those economically or medically important; illus.; reprinted 1983.)

Garcia-Rollán, G.M. (1981-1983). *Claves de Flora de España (Península y Baleares)*, 2 vols. Mundi-Prensa, Madrid.

Guinea López, E. and Ceballos-Jiménez, A. (1974). *Elenco de la Flora Vascular Española (Península y Baleares)*. ICONA, Madrid. 403 pp. (Checklist; includes habitat details.)

Heywood, V.H. (Ed.) (1961-). *Catalogus Plantarum Vascularium Hispaniae*, 1 fasc. Instituto Botánico 'A.J. Cavanilles', Madrid. 58 pp. (Incomplete; pteridophytes, gymnosperms, dicotyledons – Ranunculaceae to Cruciferae.)

Lázaro e Ibiza, B. (1920-1921). *Compendio de la Flora Española: Estudio de las Plantas*, 3rd Ed., 3 vols. Madrid. (Includes lower plants; appendix covers vegetation and phytogeography; illus.)

Sáinz-Ollero, H. and Hernández-Bermejo, J.E. (1981). *Síntesis Corológica de las Dicotiledóneas Endémicas de la Península Ibérica e Islas Baleares*. Instituto Nacional de Investigaciones Agrarias, Madrid. 111 pp. (A checklist of the endemic dicotyledons of the Iberian Peninsula and the Balearic Islands with IUCN conservation categories for taxa considered rare and threatened; English summary.)

Smythies, B.E. (1984). Flora of Spain and the Balearic Islands. Checklist of Vascular Plants. *Englera* 3(1), 212 pp. and 3(2), 486 pp. (2 vols. to date: 1 – pteridophytes, gymnosperms, Acanthaceae-Crassulaceae; 2 – Cruciferae-Rutaceae.)

Willkomm, H.M. and Lange, J. (1861-1880). *Prodromus Florae Hispanicae*, 3 vols and supplement. Schweizerbart, Stuttgart.

Regional Floras are too numerous to list, but for summaries of floristic literature see:

Galiano, E.F. (1975). Données disponibles et lacunes de la connaissance floristique de l'Espagne. CNRS (1975), cited in Appendix 1. Pp. 29-39. (Lists local Floras, monographs and ecological works; centres of floristic studies throughout country.)

Galiano, E.F. and Valdés, B. (1971). Botanical research in Spain, 1962-1969. *Boissiera* 19: 23-60.

Galiano, E.F. and Valdés, B. (1974-1979). Bibliografia Botánica Española (plantas vasculares). *Mem. Soc. Brot.* 24(1): 377-394; *Lagascalia* 4(2): 239-258; 7(1): 83-120; 9(1): 3-28.

Heywood, V.H. and Ball, P.W. (1963). Taxonomic and floristic research in Spain 1940-1962. *Webbia* 18: 445-472.

Field-guides

Ceballos, A., Fernández-Casas, J. and Muñoz Garmendia, F. (1980). *Plantas Silvestres de la Península Ibérica (Rupícolas)*. H. Blume, Madrid. 448 pp.

González, G.L. (1982). *La Guía de Incafo de los Arboles y Arbustos de la Península Ibérica*. Incafo, Madrid. 866 pp. (Keys; colour photographs; line drawings.)

Ruíz, de la Torre, J. and Ceballos-Jimenez, L. (1971). *Arboles y Arbustos de la España Peninsular*. IFIE and ETSI Montes, Madrid. 512 pp. (Tree atlas; includes details on ecology, disease, forestry; illus.)

Taylor, A.W. (1972). *Wild Flowers of Spain and Portugal*. Chatto and Windus, London. 103 pp. (Colour and black and white photographs.)

See also Polunin and Smythies (1973), cited in Appendix 1.

Information on Threatened Plants The Dirección General del Medio Ambiente has recently co-ordinated the publication of a threatened plant list for Spain, based upon the Council of Europe 'List of rare, threatened and endemic plants in Europe' (Threatened Plants Unit, 1983, cited in Appendix 1):

Barreno, E. *et al.* (Eds) (1984). Listado de Plantas Endemicas, Raras o Amenazadas de España. *Informacion Ambiental. Conservacionismo en España.* No. 3. 7 pp. (Includes 563 threatened taxa; compiled with the agreement of numerous authoritative Spanish botanists, it is now the definitive list.)

The Universidad Politécnica in Madrid is undertaking a long-term programme to gather data on threatened plants and produce a Red Data Book. Computerized lists of endemics to support this project and the Artemis project (see below) are now in an advanced stage (Anon, 1985, cited in Appendix 1; Gómez-Campo, pers. comm.).

In May 1984, ADENA (WWF-Spain) launched a national Plants Campaign in the Real Jardín Botánico, Madrid, as part of their contribution to the IUCN/WWF Plants Programme 1984-85. For details contact ADENA (address below) and for summary of Spanish Campaign projects see *Threatened Plants Newsletter* 14: 12-14, 1985 and:

Anon (1984). ADENA presenta una campaña internacional de protección de las plantas. *Quercus* 14: 40. (Outlines 6 project proposals; in Spanish.)

For additional threatened plant literature see:

Costa, M. *et al.* (1984). *Estado Actual de la Flora y Fauna Marinas en el Litoral de la Comunidad Valenciana.* Excelentísimo Ayuntamiento, Castellón de la Plana. 209 pp. (Describes plant and animal communities of the marine and littoral zones in Valencia province, assesses their conservation status and lists individual threatened plant species and plant communities; similar studies, by the same authors, are being undertaken for the entire Mediterranean coastline of Spain.)

Gómez-Campo, C. and Malato-Beliz, J. (1985). The Iberian Peninsula. In Gómez-Campo, C. (Ed.), cited in Appendix 1. (Includes case studies on selected threatened species.)

Rivas-Goday, S. (1959). Algunas especies raras o relictícas que deben protegerse en la España Mediterránea. In *Animaux et Végétaux Rares de la Région Méditerranéenne*, Proceedings of the IUCN 7th Technical Meeting, 11-19 September, 1958, Athens, vol. 5. IUCN, Brussels. Pp. 95-101. (Briefly describes 38 threatened plant species.)

Sáinz-Ollero, H. and Hernández-Bermejo, J.E. (1979). Experimental reintroductions of endangered plant species in their natural habitats in Spain. *Biol. Conserv.* 16(3): 195-206.

Sáinz-Ollero, H. and Hernández-Bermejo, J.E. (1981), cited in full above.

Spain is included in the European threatened plant list (Threatened Plants Unit, 1983, cited in Appendix 1); latest IUCN statistics, from Barreno *et al.* (1984): endemic taxa – Ex:1, E:31, V:76, R:237, I:27, K:9, nt:339; doubtfully endemic taxa – nt:2; non-endemics rare or threatened worldwide – Ex:1, E:17, V:65, R:100, I:8 (world categories). *The IUCN Plant Red Data Book* (1978) includes 7 species for Spain.

For a relevant environmental journal see *Quercus: Observacion, Estudio y Defensa de la Naturaleza*, Madrid.

Laws Protecting Plants A Royal Decree of 1982 (Article 1) declares complete protection for 7 plant species. This list can be extended by the Ministère de l'agriculture, de la pêche et de l'alimentation on the advice of ICONA, as necessary. An earlier decree (1962, No. 485) provided more limited protection for 33 species from peninsular Spain and 26 species from the Canary Islands. The species protected by the 1982 Decree correspond to the Spanish taxa listed as 'Strictly Protected Flora Species' in the 1979 Berne Convention.

Voluntary Organizations

ADENEX (Asociación para la Defensa de la Naturaleza y los Recursos de Extremadura), Larra 50, Mérida, Badajoz.

AEPDEN (Asociación para el estudio, defensa y protección de la Naturaleza), calle de Campomanes n. 13, Madrid 13.

ANDALUS (Asociación para la Supervivencia de la Naturaleza y el Medio Ambiente de Andalucia), Marqués de Pereda 18, P.O. 143, Seville.

DEPANA (Lliga per a la Defensa del Patrimoni Natural), Aragón 28, P.O. Box 2809, Barcelona 9.

WWF-Spain (Asociación para la Defensa de la Naturaleza – ADENA), Santa Engracia 6, 28010 Madrid.

Botanic Gardens

Jardín Botánico de Barcelona, Avenida de Montanans, Parc de Montjuic, 08004 Barcelona 4.

Jardín Botánico de Córdoba, Apdo 348, 14005 Córdoba.

Jardín Botánico de la Universidad Valencia, Cuarte 96, 46008 Valencia.

Jardín Botánico Marimurta, Fundación Carlos Faust, Estación Int. de Biología Mediterránea, Blanes, Prov. Gerona.

Jardín de Aclimatación "Pinya de Rosa", Blanes, Gerona.

Real Jardín Botánico de Madrid, Plaza de Murillo 2, 28014 Madrid.

A seed-banking scheme, known as the Artemis Project, is in progress, under the direction of C. Gómez-Campo in the Universidad Politécnica, Madrid (address below). Its purpose is to develop as a seed bank for Iberian and other Mediterranean endemics. For further details see:

Gómez-Campo, C. (1981). The 'Artemis' Project for seed collection of Mediterranean endemics. *Threatened Plants Committee – Newsletter* 8: 9-11.

Useful Addresses

Artemis Project, Departamento de Organografía y Fisiología Vegetal, Escuela T.S. Ing. Agrónomos, Universidad Politécnica, 28040 Madrid.

Dirección General del Medio Ambiente (Directorate General of the Environment), Ministerio de Obras Públicas y Urbanismo, Paseo de la Castellana 67, 28046 Madrid.

Instituto Nacional para la Conservación de la Naturaleza (ICONA), Gran Vía de San Francisco 35, 28005 Madrid.

Additional References

Bellot, F. (1978). *El Tápiz Vegetal de la Península Ibérica*. H. Blume, Madrid. 423 pp.

De Viedma, M.G. (1976). Nature conservation in Spain: A brief account. *Biol. Conserv.* 9(3): 181-190.

Font Quer, P. (1953). *Geografía Botánica de la Península Ibérica*. Muntaner y Simón, Barcelona. 271 pp. (Maps; illus.)

Font Quer, P. (1954). La vegetación. *Geog. de España y Portugal*. Barcelona.

Galiano, E.F. (1971). Problèmes de la conservation de la végétation et de la flore en Espagne. *Boissiera* 19: 81-86.

Gómez-Campo, C. *et al.* (1984). Endemism in the Iberian Peninsula and Balearic Islands. *Webbia* 38: 709-714.

Rivas-Martínez, S. (1971). Bases ecológicas para la conservación de la vegetación. *La Revista las Ciencias* 36(2), 6 pp.

Rivas-Martínez, S. (1981). Les étages bioclimatiques de la végétation de la Península Ibérique. *Anal. Jardín Botánico de Madrid* 37(2): 251-268.

Tamames, R. *et al.* (Eds), (1984). *El Pais: El Libro de la Naturaleza.* El Pais, Madrid. 304 pp. (A handbook of environmental issues in Spain, including many articles about flora and vegetation, its destruction and protection.)

Spain: Balearic Islands

(The islands of Mallorca, Menorca and Ibiza)

Area 5,014 sq. km

Population 621,925

Floristics 1250-1450 native vascular species, estimated by D.A. Webb (1978, cited in Appendix 1) from *Flora Europaea*; 94 endemics (IUCN figures). A Mediterranean flora. Due to its varied geology, the flora of Mallorca is the most diverse of the 3 islands.

Vegetation Little natural vegetation in lowland areas. Remaining vegetation comprises Aleppo Pine (*Pinus halepensis*) and maquis, especially on Mallorca, where it survives in the mountains and along the coast. Ibiza is probably the most wooded of the islands, with extensive Aleppo Pine forests at higher levels. On Menorca, floristic diversity high in sheltered gorges and coastal cliffs, especially those of the southern limestone areas. Increasing tourist pressure on the islands continues to cause degradation and loss of vegetation.

Checklists and Floras Covered by *Flora Europaea* (Tutin *et al.*, 1964-1980) and the *Med-Checklist*, both cited in Appendix 1. See also:

Bonafè Barceló, F. (1977-). *Flora de Mallorca*, 4 vols to date. Moll, Mallorca. (Incomplete; 1 – pteridophytes, gymnosperms, angiosperms (Typhaceae to Iridaceae; 2 – Orchidaceae to Leguminosae; 3 – Leguminosae to Boraginaceae; 4 – Verbenaceae to Compositae; illus; in Mallorcan.)

Bonner, A. (1982). *Plantes de les Baleares*, 6th Ed. Moll, Palma de Mallorca. 150 pp. (Translated from the Catalan by P. Mathews; contains a descriptive account by habitat; valuable bibliography; line drawings and colour photographs.)

Duvigneaud, J. (1979). *Catalogue Provisoire de la Flore des Baléares*, 2nd Ed. Société pour l'Echange des Plantes Vasculaires de l'Europe Occidentale et du Bassin Méditerranéen, Liège. 43 pp. (A systematic checklist.)

Knoche, H. (1921-1923). *Flora Balearica*, 4 vols. Roumégouis et Déhan, Montpellier. (1 – algae, fungi, pteridophytes, monocotyledons, dicotyledons (Salicaceae to Ranunculaceae); 2 – Lauraceae to Compositae; 3 - history, climate, geology, floristics, phytosociology; 4 – tables, maps, illus.; in French, without keys; reprinted in 1974.)

Rodriguez Femenias, J.J. (1904). *Flórula de Menorca*. Mahon. 198 pp.

The Balearics are also covered by the checklists of Sáinz-Ollero and Hernández-Bermejo (1981) and of Smythies (1984-) and by the keys of Garcia-Rollán (1981-1983), all cited under Spain above.

Field-guides See under Spain.

Information on Threatened Plants The recently published threatened plant list for Spain by Barreno *et al.* (1984), cited under Spain, above, includes a separate list for the Balearic Islands. This covers 136 threatened plant taxa with IUCN conservation categories; agreed by many renowned Spanish botanists, it is now the definitive list for the Balearic Islands.

IUCN statistics: endemic taxa – Ex:1, E:12, V:8, R:40, nt:33; non-endemics rare or threatened worldwide – Ex:1, E:13, V:16, R:45, I:2 (world categories).

See also the earlier checklist covering threatened plants by Sáinz Ollero and Hernández-Bermejo (1981), above. Although the Balearic endemics are well known, little appears to have been published on their conservation other than the above.

Additional References

Bolós, O. de and Molinier, R. (1969). Vue d'ensemble de la végétation des Iles Baléares. *Vegetatio* 17: 251-270. (Phytosociological account.)

Canigueral, J. (1953). Algunos datos sobre la flora de Mallorca. *Collect. Bot. Barcinone* 3(1): 309-323.

Cardona, M.A. (n.d.). *Estudi de les zones d'interès botanic i ecologic de Menorca*. Consell Insular de Menorca. 44 pp. (Includes ecological and conservation data.)

Goldsmith, F.B. (1974). An assessment of the nature conservation value of Majorca. *Biol. Conserv.* 6(2): 79-83.

Gómez-Campo, C. *et al.* (1984). Endemism in the Iberian Peninsula and Balearic Islands. *Webbia* 38: 709-714.

Palau, P. (1955). Plantas de Baleares. *Collect. Bot. Barcelona* 4: 207-215.

Sri Lanka

Area 65,610 sq. km

Population 16,076,000

Floristics c. 2900 native flowering plant species; c. 850 endemic (Abeywickrama, 1983). 314 native pteridophytes, 57 endemic (Sledge, 1982). The endemic flora is concentrated in the wet zone in south-west Sri Lanka. In particular, c. 60% of the tree flora of the Sinharaja Forest Reserve is endemic (Gunatilleke, 1978). There are 12 endemic genera in Sri Lanka, all but one of which are confined to the primary rain forests of the south-west.

Vegetation Extensive broadleaved savanna woodlands and tropical thorn scrub in north-west and south-east; moist deciduous forests up to 600 m on the plains to north, east and south of central highlands; montane semi-evergreen forests at 900-1500 m; montane wet evergreen forests above 1500 m, mostly replaced by 'patana' grassland above 1800 m; tropical lowland evergreen rain forest in south-west to 900 m, only c. 1590 sq. km remaining of which the Sinharaja Forest Reserve (88 sq. km) is the only extensive undisturbed patch.

Estimated rate of deforestation of closed broadleaved forest 580 sq. km/annum out of a total of 16,590 sq. km (FAO/UNEP, 1981). According to Myers (1980) 24,400 sq. km are classified as forestlands, but "adequately stocked forests" amount to only 17,000 sq. km.

335

Checklists and Floras

Abeywickrama, B.A. (1959). A provisional check list of the flowering plants of Ceylon. *Ceylon J. Sci. (Biol. Sci.)* 2: 119-240.

Dassanayake, M.D. and Fosberg, F.R. (Eds) (1980-). *A Revised Handbook to the Flora of Ceylon*, 4 vols to date, 6 planned. Amerind Publ. Co., New Delhi. (Families treated include Clusiaceae, Compositae, Dipterocarpaceae, Moraceae, Myrtaceae and Orchidaceae.)

Sledge, W.A. (1982). An annotated check-list of the Pteridophyta of Ceylon. *Bot. J. Linn. Soc.* 84: 1-30. (Separate lists of endemics and species confined to Sri Lanka and southern India.)

Trimen, H. and Hooker, J.D. (1893-1900). *A Hand-book to the Flora of Ceylon*, 5 vols. Dulau, London. (Vol. 6 by A.H.G. Alston was added in 1931; descriptions, notes on distribution and status; 326 species considered to be endemic to the lowland wet zone of which 195 noted as 'rare' or ;'very rare'. Reprinted 1974, Bishen Singh Mahendra Pal Singh, Dehra Dun.)

For ferns see Beddome (1892) and Nayar and Kaur (1972), both of which are cited in Appendix 1.

Field-guides

Bond, T.E.T. (1953). *Wild Flowers of the Ceylon Hills.* Oxford Univ. Press, London. 240 pp.

De Thabrew, W.H. and W.V. (1983). *Water Plants of Sri Lanka.* Suhada Press, New Malden. 126 pp.

Fernando, D. (1954). *Wild Flowers of Ceylon.* West Brothers, Mitcham. 86 pp. (Illustrations and short descriptions of 173 plants, endemics indicated.)

Information on Threatened Plants *Areca concinna* is included in *The IUCN Plant Red Data Book* (1978). A preliminary IUCN list, covering mainly dipterocarps and palms, includes: endemic taxa – E:8, V:4, R:12, I:14, nt:31. Useful articles are:

Abeywickrama, B.A. (1983). Threatened or endangered plants of Sri Lanka and the status of their conservation measures. In Jain, S.K. and Mehra, K.L. (Eds), *Conservation of Tropical Plant Resources*. Botanical Survey of India, Howrah. Pp. 11-18. (Reports on c. 190 'very rare' endemic taxa and 110-160 rare or threatened non-endemic taxa, but only lists 24 rare or threatened endemics.)

Gunatilleke, I.A.U.N. and C.V.S. (1984). Distribution of endemics in the tree flora of a lowland hill forest in Sri Lanka. *Biol. Conserv.* 28(3): 275-285. (Ecological studies at Hinidumkanda; mentions 2 rare endemic trees.)

Laws Protecting Plants The Fauna and Flora Protection Advisory Committee was appointed for the first time in 1938, under Section 70 of The Fauna and Flora Protection Ordinance (1937). 9 species (including 6 orchids) are totally protected by the Fauna and Flora Protection Act (Crusz, 1973).

Voluntary Organizations

March for Conservation, c/o Department of English, University of Colombo, Munidasa Kumaratunga Mawatha, Colombo 3.

Wildlife and Nature Protection Society of Sri Lanka (Ceylon), Chaitiya Road, Fort, Colombo 1.

Botanic Gardens

Botanic Garden, Heneratogoda, Gampaha.

Botanic Gardens, Hakgala.

Royal Botanic Gardens, Peradeniya.

Useful Addresses

Forest Research Institute, University of Sri Lanka, Colombo.

CITES Management Authority: Department of Wildlife Conservation, P.O. Box 06, Anagarika Dharmapala Mawatta, Dehiwala.

CITES Scientific Authority: Department of Wildlife Conservation, P.O. Box 06, Anagarika Dharmapala Mawatta, Dehiwala.

Additional References

Crusz, H. (1973). Nature conservation in Sri Lanka (Ceylon). *Biol. Conserv.* 5: 199-208.

Gaussen, H. *et al.* (1965). *Notice de la Feuille Ceylon – Carte International du Tapis Vegetal et des Conditions Ecologiques à 1:1,000,000.* Hors serie no. 5. Pondichery, France.

Gunatilleke, C.V.S. (1978). Sinharaja today. *Sri Lankan Forester* 13: 57-61.

Gunatilleke, C.V.S. and Wijesundara, D.S.A. (1982). *Ex-situ* conservation of woody plant species in Sri Lanka. *Loris* 16(2): 73-79. (Includes checklist of endemic woody plants at Peradeniya.)

Werner, W.L. (1981). From far and near... a plea for the conservation of three unique forests in Sri Lanka. *Loris* 15(6): 330-331. (Notes on threatened swamp-forest in south-west, montane forest in Dimbula and a dry forest near Madugoda.)

Werner, W.L. (1984). *Die Höhen- und Nebelwälder auf der Insel Ceylon (Sri Lanka).* Tropische und subtropische Pflanzenwelt 46. Akademie der Wissenschaften und der Literatur, Mainz, and Steiner, Wiesbaden. 200 pp. (Analysis of vegetation of Central Highlands with land use map; in German and English.)

Sudan

Area 2,505,815 sq. km

Population 20,945,000

Floristics c. 3200 species (I. Friis, 1984, pers. comm.); Andrews (1950-1956) gives 3137 species. Endemism estimated at 50 (Brenan, 1978, cited in Appendix 1). Especially important are the Imatong Mountains, which hold about half of Sudan's flora (Friis, pers. comm.), and Jebel Marra.

Northern half of area: flora with Saharan and Sahelian affinities; southern half: Sudanian region, with small area of transition between that and Guinea-Congolian region in south-west corner, and small area of Somalia-Masai region in south-east corner.

Vegetation North of 17°N (excluding Red Sea Hills): desert with little or no perennial vegetation. Between 14 and 17°N (including Red Sea Hills): semi-desert with grasses and herbs with or without woody vegetation (species of *Acacia*). These two zones cover approximately half of Sudan. Further south is wooded grassland covering something over one third of Sudan. A flood zone with different types according to frequency of flooding occupies about 11% of the country along the White Nile valley. Montane vegetation best represented on the Imatong mountains where there are substantial areas of montane forest dominated by *Podocarpus*, *Olea* and *Syzygium*. Estimated rate of deforestation for closed broadleaved forest 40 sq. km/annum out of 6400 sq. km (FAO/UNEP, 1981).

For vegetation map see White (1983), cited in Appendix 1.

Checklists and Floras Sudan north of c. 16°N is included in the computerized *Atlas der Pflanzenwelt des Nordafrikanischen Trockenraumes* (Frankenberg and Klaus, 1980), cited in Appendix 1. See also:

Andrews, F.W. (1950-1956). *The Flowering Plants of the (Anglo-Egyptian) Sudan*, 3 vols. Published for the Sudan Govt by Buncle, Arbroath, Scotland. (Keys to families and genera; descriptions, distributions, line drawings.)

Broun, A.F. and Massey, R.E. (1929). *Flora of the Sudan*. Sudan Govt Office, London. 502 pp. (Key to families; descriptions, distributions.)

Macleay, K.N.G. (1953). The ferns and fern allies of the Sudan. *Sudan Notes and Records* 34(2): 286-298.

Sahni, K.C. (1968). *Important Trees of the Northern Sudan*. Forestry Research and Education Centre, Khartoum; United Nations/FAO. 138 pp. (Descriptions, distributions, line drawings.)

Wickens, G.E. (1969). Some additions and corrections to F.W. Andrews' Flowering Plants of the Sudan. *Sudan Forest Department Bulletin* 14 (New series). 49 pp.

Information on Threatened Plants

Hedberg, I. (1979), cited in Appendix 1. (List for Sudan, on pp. 85-88, by G. Wickens, contains 258 species and infraspecific taxa "of restricted distribution which are endangered or likely to be so because of present or future agricultural activities".)

IUCN has records of 38 species and infraspecific taxa believed to be endemic; no categories assigned.

Three species which occur in the Sudan are included in *The IUCN Plant Red Data Book* (1978).

Useful Addresses

CITES Management and Scientific Authorities: Wildlife Conservation and National Parks Forces, Central Administration, P.O. Box 336, Khartoum.

Additional References

Andrews, F.W. (1948). The vegetation of the Sudan. In Tothill, J.D. (Ed.), *Agriculture in the Sudan*. Oxford Univ. Press, London. Pp. 32-61. (With small-scale vegetation map and 18 black and white photographs.)

Bari, E.A. (1968). Sudan. In Hedberg, I. and O. (1968), cited in Appendix 1. Pp. 59-63.

Jackson, J.K. (1956). The vegetation of the Imatong Mountains, Sudan. *J. Ecol.* 44: 341-374.

Jenkin, R.N., Howard, W.J., Thomas, P., Abell, T.M.B. and Deane, G.C. (1977). *Forestry Development Prospects in the Imatong Central Forest Reserve, Southern Sudan*, 2 vols. Land Resource Study 28, Land Resources Division, Ministry of Overseas Development, Tolworth Tower, U.K. (Vol. 1 – maps, with vegetation map 1:50,000; vol. 2 – main report, 217 pp.)

Quézel, P. and Bourreil, P. (1968). Botanique. In Sarre, H., Reyre, Y., Quézel, P. and Bourreil, P., Missions au Dar-Fur (1967). *Dossiers de la Rech. Coop. Prog.* 45(2). Centre National de la Recherche Scientifique, Paris. 23 pp.

Wickens, G. (1977). *The Flora of Jebel Marra (Sudan Republic) and its Geographical Affinities*. Kew Bull. Addit. Ser. 5. 368 pp. (Includes annotated checklist of 982 species and infraspecific taxa; distribution maps.)

Suriname

Area 163,820 sq. km

Population 352,000

Floristics Estimated at 4500 species (J.C. Lindeman, 1984, pers. comm.); endemics few, most of the 293 recorded by Pulle (1906) having become synonyms or found in neighbouring countries.

Vegetation Towards the coast a belt of wet savanna of the Sabon-Pasi type; on the coast mangrove and swamp vegetation; the interior is forested, with 148,300 sq. km of rain forest, mostly lowland, except for montane savanna on small isolated granite domes and on the Tafelberg, a table mountain of Roraima sandstone; along the centre of the southern border, a large hilly savanna (the Sipaliwini), with affinities to the Brazilian cerrado.

Estimated rate of deforestation for closed broadleaved forest 25 sq. km/annum out of 148,300 sq. km (FAO/UNEP, 1981).

Checklists and Floras Covered by the family and generic monographs of *Flora Neotropica*, described in Appendix 1. Country Floras and checklists are:

Kramer, K.U. (1978). *The Pteridophyta of Suriname: an enumeration with keys of the ferns and fern-allies*. Natuurwetenschappelijke Studiekring voor Suriname en de Nederlandse Antillen. No. 93. Utrecht.

Pulle, A.A. (1906). *An Enumeration of the Vascular Plants Known From Surinam*. E.J. Brill, Leiden. 555 pp. (List of 2101 recorded species.)

Pulle, A.A. (1932-1953) (Ed.) *Flora of Suriname*. Now published by E.J. Brill, Leiden. By 1943 first parts of Vols 1-4 and some fascicles of 2(2), 3(2) and 4(2) published. Since Pulle's death edited by J. Lanjouw to 1968, by J. Lanjouw and A.L. Stoffers to 1976 and since by A.L. Stoffers and J.C. Lindeman. 1(2), 2(2) and 5(1) completed, additions and corrections to 3(1) and 3(2) in prep. (137 out of 173 recorded families covered so far.)

Werkhoven, M.C.M. (in press). *Orchids of Suriname*. (Illustrated, bilingual, semi-popular work.)

A 30-year project to prepare the *Flora of the Guianas* is being coordinated by the Institute of Systematic Botany, University of Utrecht, The Netherlands, and the Smithsonian Institution, Washington, D.C., in collaboration with Office de la Recherche Scientifique et Technique Outre-Mer, Cayenne, French Guiana, and other leading botanical institutions. Part 1 (Cannaceae, Musaceae and Zingiberaceae by P.J.M. Maas) is in press.

Field-guides

Lindeman, J.C. and Mennega, A.M.W. (1963). *Bomenboek voor Suriname*. Uitgave Dinst's Lands Bosbeheer, Paramaribo. Also as *Mededel. Bot. Mus. Herb. Utrecht* 200. 312 pp. (Includes 96 drawings by W.H.A. Hekking; keys; about 400 tree species covered.)

Roosmalen, M.G.M. van (1977). *Surinaams Vruchtenboek*. (1 – descriptions of fruits of woody plants; 2 – drawings; an English version, covering all 3 Guianas, is in print and will be available from Utrecht, address below.)

Wessels Boer, J.G., Hekking, W.H.A. and Schulz, J.P. (1976). *Fa Joe Kan Tak' Mi No Moi: Surinaamse Wandelflora* 1 (2 parts). Natuurgids Serie B No. 4, STINASU, Paramaribo. 293 pp. (The most common plants, line drawings.)

Information on Threatened Plants None.

Laws Protecting Plants According to the Timber Law of 1947 (Official Journal No. 103, 1952), a permit is required to collect forest byproducts, defined as "vegetable products that are collected in the forest without cutting, scoring or otherwise damaging wood-like vegetation", in addition to *Quassia amara* and *Hibiscus tiliaceus*. A permit is also required to collect wood, seeds and other products from Letterwood (Moraceae spp.), *Hymenaea courbaril*, *Dipteryx odorata*, *Carapa* sp., *Lonchocarpus* sp., *Bombax* sp. and forest lianes. The following timbers may not be cut without a special permit: *Manilkaria bidentata*, *Dipteryx odorata*, *Copaifera guianensis* and *Aniba rosaeodora* (M. Werkhoven, 1984, pers. comm.).

Voluntary Organizations

Foundation for Conservation of Nature in Suriname (STINASU), P.O. Box 436,
    Paramaribo.
The Suriname Orchid Society, P.O. Box 9189, Paramaribo.

Botanic Gardens There is a garden with many economic plants at the Agricultural Experimental Station, Paramaribo.

Useful Addresses

Institute of Sytematic Botany, University of Utrecht, Heidelberglaan 1, P.O. Box
    80102, 3508 TC Utrecht, Netherlands.
National Herbarium of Suriname, University of Suriname, Leysweg, P.O. Box 9212,
    Paramaribo.
CITES Management Authority: Head of Suriname Forest Service, c/o Head Nature
    Conservation Division, P.O. Box 436, 10 Cornelis Jongbawstraat, Paramaribo.
CITES Scientific Authority: Nature Conservation Commission, c/o Suriname Forest
    Service, P.O. Box 436, Paramaribo.

Additional References

Schulz, J.P. (1968). Nature Preservation in Suriname – a review of the present status.
    Suriname Forest Service, P.O. Box 436, Paramaribo. (Principally covers the
    protected areas.)
Teunissen, P.A. (1978). *Reconnaissance Map of Surinam Lowland Ecosystems (Coastal
    Plain and Savanna Belt)*. STINASU, Paramaribo. (1:200,000.)
Teunissen, P.A. and Werkhoven, M.C.M. (1979). *Biological Bibliography of Suriname*.
    I.O.L., Paramaribo.
The vegetation of Suriname. A set of 11 papers by various authors, 1953-1971, some
    reprinted from elsewhere. Van Eedenfonds, Amsterdam. Also distributed as
    numbered parts in *Mededel. Bot. Mus. & Herb. Utrecht* (between 113 and 350). 1(1)
    and 4 contain annotated lists of species.

Svalbard

An archipelago in the Arctic Ocean belonging to Norway, comprising 4 main islands (Vestspitsbergen, Nordaustlandet, Edgeøya and Barentsoya) and numerous islets.

Area 62,000 sq. km.

Population 3737 (1980) (*Times Atlas*, 1984)

Floristics Based on *Flora Europaea*, D.A. Webb (1978, cited in Appendix 1) estimates 150-175 native vascular species. Most of these plants (133 species) grow on the island of Spitsbergen (Resvoll-Holmsen, 1927). In general, species diversity is greatest on lower mountain slopes especially surrounding fjords, e.g. Isfjorden on Spitsbergen, one of the richest parts of the archipelago. Elements: Arctic/alpine.

Vegetation Most of land surface covered by ice sheet; tracts of bare rock common; remainder is a treeless landscape, dominated by arctic tundra of heathland species, mostly Dwarf Birch (*Betula nana*), dwarf willows (*Salix reticulata, S. herbacea*), ericaceous species and other arctic/alpines such as *Dryas octopetala*. Cryptogams abundant.

Checklists and Floras Covered by *Flora Europaea* (Tutin *et al.*, 1964-1980), cited in Appendix 1.

Resvoll-Holmsen, H. (1927). *The Flora of Svalbard (Spitsbergen and Bjørnøy)*. J.W. Cappelens, Oslo. 64 pp. (Translated from the Norwegian into English by M. Rathbone (unpublished), 53 pp.; a descriptive account of the vegetation, with keys.)
Rønning, O.I. (1964). *Svalbards Flora*. Norsk Polarinstituut Polarhandbok Nr. 1, Norsk Polarinstitutt, Oslo. 123 pp. (In Norwegian; illus.)

See also:

Lid, J. (1967). Synedria of twenty vascular plants from Svalbard. *Bot. Jb.* 86(1-4): 481-493.
Neilson, A. (1970). *Vascular plants of Edgeøya, Svalbard*. Norsk Polarinstituut, Skrifter 150. Norsk Polarinstitutt, Oslo. 71 pp. (Describes history of botanical studies; checklist of vascular plants and vegetation types on the island of Edgeøya.)
Rønning, O.I. (1961). *Some New Contributions to the Flora of Svalbard*. Norsk Polarinstitutt, Skrifter 124. Norsk Polarinstitutt, Oslo. 21 pp. (Illus.)

Information on Threatened Plants The flora of Svalbard was screened for the publication '*List of Rare, Threatened and Endemic Plants in Europe*' (Threatened Plants Unit, 1983, cited in Appendix 1) but no threatened taxa were identified.

Additional References
Eurola, S. (1968). Über die Fjeldheidevegetation in den Gebieten von Isfjorden und Hornsund in Westspitzbergen. *Aquilo, Ser. Botanico* 7. 56 pp. (Includes species lists, maps, black and white photographs.)
Ministry of Environment (1981). *Environmental Regulations for Svalbard*. Ministry of Environment. (Revised edition.)

Swaziland

Area 17,366 sq. km

Population 630,000

Floristics 2715 indigenous vascular species, 110 naturalized exotics; 4 endemic species (Kemp, 1983a).

Flora in eastern half with South African coastal affinities; western half in Kalahari-Highveld region.

Vegetation Mostly scrub woodland without characteristic dominants; also area of Afromontane vegetation in north-west. West of country occupied by mosaic of Afromontane scrub forest, scrub woodland and secondary grassland.

For vegetation map see White (1983), cited in Appendix 1.

Checklists and Floras Swaziland is included in the incomplete *Flora of Southern Africa*, and in *The Genera of Southern African Flowering Plants* (Dyer, 1975, 1976), both cited in Appendix 1.

Compton, R.H. (1966). *An Annotated Check List of the Flora of Swaziland.* J. S. Afr. Bot. Suppl. 6. Mbabane. 191 pp. (List of 2350 species, annotated with growth form, habitat and broad distribution; extensive descriptive notes on the flora.)

Compton, R.H. (1976). *The Flora of Swaziland.* J. S. Afr. Bot. Suppl. 11. Kirstenbosch. 684 pp. (Keys, descriptions, distributions, specimens.)

Dlamini, B. (1981). *Swaziland Flora: their Local Names and Uses.* Ministry of Agriculture and Co-operatives, Mbabane. 72 pp. (Edited by D.K. Rycroft.)

Kemp, E.S. (1981). *Additions and Name Changes for the Flora of Swaziland.* Swaziland National Trust Commission, Lobamba. 74 pp.

Kemp, E.S. (1983a). *A Flora Checklist for Swaziland.* Occasional Paper No. 2, Swaziland National Trust Commission, Lobamba. 101 pp.

Kemp, E.S. (1983b). *Trees of Swaziland.* Occasional Paper No. 3, Swaziland National Trust Commission, Lobamba. 59 pp. (Checklist with broad geographical distributions.)

Information on Threatened Plants

Hall, A.V. *et al.* (1980), cited in Appendix 1. (List for Swaziland on pp. 82-83 contains 5 endemic: V:1, I:1, K:3; and 18 non-endemic: R:10 (regional category), I:5, K:3 species and infraspecific taxa.)

Hedberg, I. (1979), cited in Appendix 1. (List for Swaziland on pp. 101-103, by E.S. Kemp, contains 155 species and infraspecific taxa: E:2, V:16, R:137.)

Two species which occur in Swaziland are included in *The IUCN Plant Red Data Book* (1978).

Laws Protecting Plants The Flora Protection Act No. 45 of 1952 (Laws of Swaziland, Chapter 144) gives a limited amount of protection to certain listed rare or attractive plants (mostly bulbs, succulents, cycads and some ferns); now being amended to include the category "Specially Protected Flora", with a list which so far includes only cycads.

The Forest Legislation (1979) prohibits cutting down, removal, sale or purchase of indigenous timber without permission from the Minister of Agriculture.

Voluntary Organizations

Natural History Society, Mbabane.

Additional References

Compton, R.H. (1968). Swaziland. In Hedberg, I. and O. (1968), cited in Appendix 1. Pp. 256-257.

Werger, M.J.A. (1978), cited in Appendix 1. Citation includes list of relevant chapters.

342

Sweden

Area 449,790 sq. km

Population 8,284,000

Floristics 1600-1800 native vascular species estimated by D.A. Webb (1978, cited in Appendix 1), from *Flora Europaea*. 3 endemic species; 1 endemic subspecies (IUCN figures). Elements: Atlantic, Boreal, and Arctic/alpine. Areas of diversity: dry meadows and sand steppe of south-east Skane (e.g. Sandhammeren and Brösarp hills); Hallands Vadero (virgin deciduous forest); islands of Öland and Gotland.

Vegetation Forest, mostly plantations, occupies c. 56% of land area, with Scots Pine and spruce predominant in the northern and central Boreal zone, with pockets of deciduous forest. In southern Sweden, broadleaved deciduous forest is the natural climax, dominated by beech (in the far south) and oak, but today largely replaced by agriculture. Birch forest in mountains of north and west, but south of the tundra of the Arctic/alpine zone. Only 4% of all forests are not or are only slightly affected by forestry activities, and less than 1% is protected by law. In the west, extensive alpine heath with willow, *Cassiope* and lichens. Marshland, bogs and mires total 5 m ha, mostly in northern Sweden.

For a vegetation map see Andersson *et al.* (1981).

Checklists and Floras Included in the completed *Flora Europaea* (Tutin *et al.*, 1964-1980), cited in Appendix 1. Below, in order, are 3 Floras: 1 regional, 1 national and 1 of a region of southern Sweden, Skane, where species diversity is especially high. Also relevant is Lindman (1964) and the plant atlas by Hultén (1971), both cited in Appendix 1.

Lid, J. (1974). *Norsk og Svensk Flora*, 4th Ed. Norske Samlaget, Oslo. 808 pp. (Covers Norway and Sweden; line drawings.)

Lindman, C.A.M. (1926). *Svensk fanerogamflora*, 2nd Ed. Norstedt and Söner, Stockholm. 644 pp. (Illus.)

Weimarck, H. (1963). *Skanes Flora*. Corona, Lund. 720 pp. (Flora of the region Skane; summarizes distribution of each species throughout Sweden and Denmark; colour plates.)

Bibliographies:

Hylander, N. (1963). Report on the floristic and taxonomic studies on the Swedish flora during the period 1945-1960. *Webbia* 18: 473-495.

Jonsell, B. (1975). Report on floristic and taxonomic studies on the Swedish vascular flora during the period 1961-1972. *Mem. Soc. Brot.* 24(2): 799-834.

Relevant journals: *Nordic Journal of Botany*, Copenhagen; *Sveriges Natur*, Svenska Naturskyddsföreningen (address below); *Fauna och flora*, Naturhistoriska Riksmuseet (address below).

Field-guides

Krok, Th. and Almqvist, S. (1983). *Svensk Flora*, 25th Ed. by E. Almqvist. Esselte Studium, Uppsala. 405 pp. (Keys; colour plates; 1st Ed. 1883.)

Ursing, B. (1947). *Svenska Växter*. Nordisk Rotogravyer, Stockholm. 398 pp. (Keys; illus.)

Information on Threatened Plants In 1972 a long-term project entitled 'Project Linnaeus' was launched, grant-aided by WWF-Sweden, to investigate the conservation status of Sweden's native flora and promote its protection. 170 taxa were identified as threatened nationally and published in:

Nilsson, Ö. and Gustafsson, L.-A. (1976-1982). Projekt Linné rapporterar. *Svensk Bot. Tidskr.* 70(2) onwards. (170 separate species accounts; each includes distribution throughout Scandinavia, known localities in Sweden since 1850, ecological data, recommendations.)

Other publications about Project Linnaeus include:

Gustafsson, L.-A. (1976). Projekt Linné samlar landets botanister till kraftinsats (Project Linnaeus unites the botanists of Sweden in a powerful effort). *Fauna och flora* 71(5): 189-201.
Nilsson, Ö. (1981). Project Linnaeus: Assessing Swedish plants threatened with extinction. In Synge (1981), cited in Appendix 1. Pp. 105-112. (Describes the project.)

The first in a plant Red Data Book series has recently been published:

Ingelög, T., Thor, G. and Gustafsson, L. (Eds) (1984). *Floravard i Skogsbruket. Del 2 - Artdel.* (Plant Conservation and Forestry. Part 2 - Species Data.). Författarna och Skogsstyrelsen, Jönköping. (Data sheets and maps for 270 threatened species covering vascular plants, mosses, lichens and fungi; in Swedish.)

A computerized threatened plant data-base is now being developed at the Swedish Agricultural University, Uppsala (address below), with support from SNEPB (see below), WWF-Sweden and the University. The aim is to keep up to date the official lists of threatened vascular plants, bryophytes, lichens and fungi. Using this data, committees of experts will be planning appropriate conservation measures. For more details see:

Ingelög, T. (1981). *Floravard i Skogsbruket* (Plant Conservation and Forestry). Skogsstyrelsen, Jönköping. 153 pp. (Describes state of plant conservation in Sweden with reference to integrating conservation with forestry; lists 278 native and threatened forest species of flowering plants, ferns, mosses, fungi and lichens; photographs.)
Ingelög, T., Gustafsson, L. and Thor, G. (1984). *Skyddsvärda Skogsväxter i Sverige, Floravard i skogsbruket, No. 3 - Fotoflora i färg.* Skogsstyrelsen. 64 pp. (Briefly describes 73 vascular plants threatened by forestry; short descriptions; colour photographs for each species; maps.)

Sweden is included in the Nordic Council of Ministers' threatened plant list (Ovesen *et al.*, 1978, 1982) and in the European list (Threatened Plants Unit, 1983), both cited in Appendix 1); latest IUCN statistics, based upon this latter work: endemic taxa - nt:4; non-endemics rare or threatened worldwide - V:6, R:6, I:1 (world categories).

Laws Protecting Plants The Nature Conservancy Act 1964, amended in 1973, prohibits the removal of, or damage to, any plant if the survival of the species concerned is believed threatened. It lists 134 taxa for protection in the different provinces. For details see Koester (1980), cited in Appendix 1, and:

Ryberg, M. (1979). *Fridlysta växter* (Protected Plants), 4th Ed. Svenska Naturskyddsföreningen, Stockholm. 52 pp. (Lists taxa protected in the provinces.)

Voluntary Organizations
Svenska Botaniska Föreningen (Swedish Botanical Society), Naturhistoriska Riksmuseet, Box 50007, 104 05 Stockholm.

Svenska Naturskyddsföreningen (SNF) (Swedish Society for the Conservation of
Nature), Box 6400, 113 82 Stockholm.

Världsnaturfonden (WWF-Sweden), Ulriksdals Slott, 171 71 Solna.

Botanic Gardens

Bergius Botanic Garden, Box 50017, 104 05 Stockholm.

Botanic Garden, Helsingborgs Museum, S. Storgatan 31, 252 23 Helsingborg.

Gothenburg Botanical Garden, Carl Skottsbergs Gata 22, 413 19 Göteborg.

University of Lund Botanical Garden, Ö Vallgatan 18, 223 61 Lund.

Uppsala Botanic Garden, Villavägen 8, 752 36 Uppsala.

Useful Addresses

Data-base for Threatened Plants, The Swedish Agricultural University, 750 07 Uppsala.

Naturhistoriska Riksmuseet, Box 50007, 104 05 Stockholm.

Statens Naturvardsverk (Swedish National Environment Protection Board, SNEPB),
Box 1302, 171 25 Solna.

Swedish NGO Secretariat on Acid Rain, c/o SNF, see above.

CITES Management Authority: Lantbruksstyrelsen (National Board of Agriculture),
Vallgatan 8, 551 83 Jönköping.

CITES Scientific Authority: SNEPB, see above.

Additional References

Andersson, L., Rafstedt, T. and Sydow, U. von. (1981). Vegetationskartor över de
Svenska fjällen (Vegetation maps of the Swedish mountains). *Svensk Bot. Tidskr.*
75(3): 147-155.

Selander, S. (1957). *Det Levande Landskapet i Sverige* (The living landscape of
Sweden), 2nd Ed. Stockholm. 492 pp. (Describes main vegetation types and effects
of human interference.)

Sjörs, H. (Ed.) (1965). The Plant Cover of Sweden. *Acta Phytogeogr. Suec.* (Uppsala)
50. 314 pp.

Switzerland

Area 41,287 sq. km

Population 6,309,000

Floristics 2600-2750 native vascular species, estimated by D.A. Webb (1978, cited
in Appendix 1) from *Flora Europaea*; 1 endemic species (IUCN figure). Elements: Central
European, alpine. For a detailed floristic account see Becherer (1972b).

Vegetation In lowland Switzerland, semi-natural vegetation very scarce,
consisting mainly of beech forest and mixed deciduous forest, the former also widespread
at lower altitudes in the Alps (except central Alps) and Jura. Fir and spruce at higher
levels, the spruce dominant in the subalpine zone. In central Alps, beech replaced by large
stands of Scots Pine, with spruce in subalpine zone giving way to Arolla Pine (*Pinus
cembra*) and larch in the upper subalpine zone. Above treeline, rich alpine meadows with
snowbed communities. Drier parts of central Alps and rocky slopes of the Jura support
dry meadows with affinities to the steppe grasslands of east and south Europe. Numerous
scattered wetland communities throughout country, many protected.

For a vegetation map see Schmid (1944-1950).

Checklists and Floras Included in the completed *Flora Europaea* (Tutin *et al.*, 1964-1980) and the *Illustrierte Flora von Mitteleuropa* (Hegi, 1935-1979), cited in Appendix 1. National Floras are:

Binz, A. and Thommen, E. (1953). *Flore de la Suisse*, 4th Ed. Griffon, Neuchâtel. (German edition also available.)

Hess, H.E., Landolt, E. and Hirzel, R. (1967-). *Flora der Schweiz und angrenzender Gebiete*, 3 vols to date. Birkhäuser, Basel. (Covers all Switzerland and parts of Austria, France, Federal Republic of Germany and Italy; 1 – pteridophytes and dicotyledons; 2 and 3 – dicotyledons and monocotyledons; line drawings, and detailed historical and ecological introduction.)

Hess, H.E., Landolt, E. and Hirzel, R. (1976). *Bestimmungsschlüssel zur Flora der Schweiz*. Birkhäuser, Basel. 657 pp. (Illustrated key, based on the Flora cited above.)

See also:

Landolt, E. (1975). Floristiche und Zytotaxonomische Arbeiten an der Flora der Schweiz zwischen 1960 und 1972. *Mem. Soc. Brot.* 24(2): 777-798.

Zoller, H. (1964). Flora des schweizerischen Nationalparks und seiner Umgebung. *Ergebnisse der Wissenschaftlichen Untersuchungen im Schweizerischen Nationalpark* 9(51): 1-408.

For a bibliography see Hamann and Wagenitz (1977), cited in Appendix 1, and for plant atlases see:

Thommen, E. (1973). *Taschenatlas der Schweizer Flora*, 5th Ed. by A. Becherer. Birkhäuser, Basel. 303 pp. (Small pocket atlas; line drawings with brief text on distributions; French edition also available.)

Welten, M. and Sutter, R. (1982). *Verbreitungsatlas der Farn- und Blütenpflanzen der Schweiz*, 2 vols. Birkhäuser, Stuttgart. 716 and 698 pp. (Not seen.)

Field-guides

Binz, A. (1980). *Schul- und Exkursionsflora für die Schweiz*, 17th Ed. by A. Becherer and C. Heitz. Schwabe, Basel. 422 pp. (Keys; illus.; French edition, 1976.)

Hegi, G., Merxmüller, H. and Reisigl, H. (1977). *Alpenflora: Die wichtigeren Alpenpflanzen Bayerns, Osterreichs und der Schweiz*. Parey, Berlin. 194 pp. (Covers Bavaria, Austria and Switzerland; introduction includes ecological descriptions of plant communities; lists protected plants; illus.; maps.)

Landolt, E. (1984). *Unsere Alpenflora*, 5th Ed. Verlag Schweizer Alpen-Club, Zürich. 318 pp. (Detailed introduction including plant geography of alpine areas; keys; colour illus.; maps.)

Also see Oberdorfer (1983) and Grey-Wilson (1979), both cited in Appendix 1.

Information on Threatened Plants National threatened plant list:

Landolt, E. *et al.* (1982). Bericht über die gefährdeten und seltenen Gefässpflanzenarten der Schweiz ('Rote Liste') (Report on threatened and rare vascular plants of Switzerland – Red List). *Ber. Geobot. Inst. ETH.* 49: 195-218. (Lists 773 species, 28% of Swiss flora.)

Also relevant:

Becherer, A. (1972a). Erloschene Arten der Schweizer Flora. *Ber. Schweiz. Bot. Ges.* 82: 300-301. (Disappearing species in the Swiss flora.)

Ritter, M. and Waldis, R. (1983). Vue d'ensemble des périls menaçent la flore ségétale et rudérale: Avec la liste rouge de la flore ségétale et rudérale. *Contributions à la protection de la nature en suisse*, No. 5. Ligue suisse pour la protection de la nature (LSPN).

The Institut Fédéral de Recherches Forestières (address below) is developing a computerized landscape information system to maintain a floristic inventory of vascular plant species and data on their conservation. Field records and data collection throughout the country is now being centralized (Wildi, 1981). For further details see Anon (1985), cited in Appendix 1.

Switzerland is included in the European threatened plant list (Threatened Plants Unit, 1983, cited in Appendix 1); latest IUCN statistics, based upon this work: endemic taxa – R:1; non-endemics rare or threatened worldwide – Ex:1, E:1, V:10, R:13, I:3 (world categories).

In 1982, the LSPN (address below) launched a campaign "Aktion Kornblume" to promote awareness about plants and their conservation. The publication below presents the arguments behind the campaign:

Rohrer, N. (1982). *Coquelicots et bluets* (German translation entitled *'Un-Kraut' in Feld und Acker*). Ligue suisse pour la protection de la nature (LSPN), Numéro spécial 1. 25 pp. (Describes threatened plant communities; lists threatened species; colour photographs.)

Laws Protecting Plants The inventory of federal and cantonal law relating to the protection of wild plants was first drawn up in November 1974 and revised in October 1975. The Federal Law of 1966 on the Protection of Nature and Landscape (Loi fédérale sur la protection de la nature et du paysage du 1 juillet 1966) provides protection for 60 taxa. Partial protection is also provided for a further 300 taxa in one or more of the Cantons. For details see:

Fehlmann, S.R. (1979). *Geschützte Pflanzen*. Silva, Zürich. 124 pp. (Describes protected species by habitat; lavishly illustrated with colour photographs.)

Landolt, E. (1982). *Geschützte Pflanzen in der Schweiz* (Protected plants in Switzerland), 3rd Ed. Schweizerischer Bund für Naturschutz, Basel. 215 pp. (Describes over 150 protected taxa with colour illus.; detailed introduction with vegetation descriptions and legislation details in each Canton; French edition also available.)

Ligue Suisse pour la Protection de la Nature (1975). *Inventaire des plantes protégées par la loi fédérale et les lois cantonales, dressé en novembre 1974, révisé octobre 1975*. (Mimeo; lists protected taxa in Latin, German and French, and those districts in which protection is enforced.)

Voluntary Organizations

Ligue suisse pour la protection da la nature (LSPN), (= Schweizerischer Bund für Naturschutz – SBN), Postfach 73, 4020 Basel.

WWF-Switzerland, Förrlibickstrasse 66, 8005 Zürich (Postfach, 8037 Zürich).

Botanic Gardens Numerous; listed in Henderson (1983), cited in Appendix 1; only subscribers to the Botanic Gardens Conservation Co-ordinating Body are given below:

Botanischer Garten der Universität Berne, Altenbergrain 21, 3013 Berne.
Isole di Brissago Botanic Garden, "Al Poggio", 6622 Ronca 5/a.
Jardin Botanique de Lausanne, 14 bis, Avenue de Cour, 1007 Lausanne.
Jardin Botanique de Neuchâtel, Rue E-Argand 11, 2000 Neuchâtel 7.
Städtische Sukkulentensammlung Zürich, Mythenquai 88, 8002 Zürich.
University of Zürich Botanic Gardens, Zollikerstrasse 107, 8008 Zürich.

Useful Addresses

Division de la protection de la nature et du paysage, Office fédéral des forêts,
    Laupenstrasse 20, 3001 Berne.
Institut Fédéral de Recherches Forestières, 8903 Birmensdorf, Zürich.
IUCN and WWF-International, World Conservation Centre, Avenue du Mont Blanc,
    1196 Gland.
CITES Management Authority (for flora): Office Fédéral de l'Agriculture, Section de
    la protection des végétaux, Mattenhofstrasse 5, 3003 Berne.
CITES Secretariat: 6 rue de Maupas, Case postale 78, 1000 Lausanne 9.

Additional References

Becherer, A. (1972b). *Führer durch die Flora der Schweiz*. Schwabe, Basel. 206 pp.
    (Floristic account of main plant communities.)
Ellenberg, H. and Klötzli, F. (1972). Waldgesellschaften und Waldstandorte der
    Schweiz. *Mitt. Schweiz. Anst. Forstl. Vers. Wesen* 48(4): 589-930.
Schmid, E. (1944-1950). *Vegetationskarte der Schweiz* 1: 200,000. Hans Huber, Berne.
Wildi, O. (1981). Grundzüge eines Landschaftsdatensystems. *Eidg. Anst. Forstl.
    Versuchswes.* 233. 56 pp.

Syria

Area 185,170 sq. km

Population 10,189,000

Floristics No figure for Syria, but Lebanon and Syria together have about 3000 species; 11% of the flora of Lebanon and Syria is endemic (Zohary, 1973, cited in Appendix 1). Many endemics confined to the isolated mountain chains in the Syrian Desert. Most of Syria's flora comprises Mediterranean and Irano-Turanian elements, the latter mainly confined to the Syrian Desert.

Vegetation Steppes and deserts cover most of Syria, with large tracts of the Syrian Desert having no vegetation at all, or else with rather diffuse *Artemisia* scrub and scattered *Pistacia atlantica* and *Amygdalus orientalis* trees. In the west, the narrow Mediterranean coastal plain has evergreen maquis scrub. Further inland are the Jebel el Ansariyah mountains rising to 1385 m, with *Sarcopoterium* maquis scrub and *Astragalus* thorn scrub on the lower slopes, and remnants of *Cedrus libani* (Cedar of Lebanon) and *Abies cilicica* forest at higher elevations. In the south, the Anti-Lebanon Mountains, rising to 2814 m at Mt Hermon, support *Pistacia/Amygdalus* scrub and remnants of steppe-coniferous forests, with *Abies cilicica*, *Pinus brutia* and *Quercus*, now much depleted.

Subalpine and alpine communities are found above 2000 m. For detailed description of vegetation and plant communities see Zohary (1973), cited in Appendix 1.

Checklists and Floras Syria is included in the *Flora of Syria, Palestine and Sinai* (Post, 1932), cited in Appendix 1, and will be covered by the *Med-Checklist*, cited in Appendix 1. See also:

Bouloumoy, L. (1930). *Flore du Liban et de la Syrie*, 2 vols. Vigot Freres, Paris. (1 – keys; 2 – plates.)

Mouterde, P. (1966-). *Nouvelle Flore du Liban et de la Syrie*, 3 vols so far. Dar El-Machreq, Beirut. (Vols 1-2 – pteridophytes, gymnosperms, monocotyledons and dicotyledons to Umbelliferae and Cornaceae; 3 – so far 3 fascicles, including Ericaceae, Labiatae, Scrophulariaceae. In addition there are 2 supplementary volumes with line drawings.)

Mouterde, P. (1973). Novitates florae libano-syriacae. *Saussurea* 4: 17-25. (17 new species and 2 varieties described from Syria and Lebanon.)

Thiébaut, J. (1936-1953). *Flore Libano-Syrienne*, 3 vols. Centre National de la Recherche Scientifique, Paris.

Information on Threatened Plants None. The section on Syria in the draft list for North Africa and the Middle East produced by IUCN Threatened Plants Committee Secretariat (1980), cited in Appendix 1, contains only 93 endemic species without categories. The list was taken from Mouterde (1966-), cited above.

Additional References

Charpin, A. and Greuter, W. (1975). Données disponibles concernant la flore de la Syrie et du Liban. In CNRS (1975) cited in Appendix 1. Pp. 115-117.

Gómez-Campo, C. (Ed.) (1985), cited in full in Appendix 1.

Taiwan

Area 35,988 sq. km

Population 19,000,000 (1984 estimate)

Floristics 3577 vascular plant species comprising 565 ferns, 20 gymnosperms and 2992 angiosperms (T.-C. Huang, quoted in Li *et al.*, 1979). 25% species endemism (Li, *et al.*, 1979). Floristic affinities mainly with southern China and Japan. There are also elements from north and central China, the Indian plains and the Malay Peninsula.

Vegetation Subtropical evergreen forests in lowlands; broadleaved evergreen forests with abundant Lauraceae, Fagaceae and Theaceae at medium altitudes; mixed forest above 1800 m; coniferous forests with *Abies* spp., *Chamaecyparis formosensis*, *Juniperus* spp., *Pinus* spp., and *Tsuga chinensis*, between 1800-3600 m; extensive grasslands above 2500 m.

Checklists and Floras

Kanehira, R. (1936). *Formosan Trees Indigenous to the Island*. Dept of Forestry, Taihoku. 574 pp. (Revised edition with English and Chinese descriptions; notes on distribution.)

Li, H.-L. (1963). *Woody Flora of Taiwan*. Livingstone, Pennsylvania. 974 pp. (c. 1030 species are woody.)

Li, H.-L., Liu, T.-S., Huang, T.-C., Koyama, T. and DeVol, C.E. (1975-1979). *Flora of Taiwan*, 6 vols. Epoch, Taipei. (1 – Introduction, keys, descriptions of ferns, gymnosperms; 2-5 – angiosperms; 6 – checklist of vascular species, statistics on flora, bibliography.)

Su, H.-J. (1975). *Native Orchids of Taiwan*. Harvest Farm Magazine, Taipei. 204 pp. (90 ornamental wild orchids described and illustrated.)

Yang, T.-I. (1982). *A List of Plants in Taiwan*. Natural Publ. Co., Taipei. 1632 pp. (5998 species listed; statistics on native and introduced taxa.)

Field-guides

Cheng, C. (1979). *Formosan Orchids*. Chow Cheng Orchids, Taichung. 132 pp. (Taiwan has 360 orchid species; 115 briefly described, many colour plates.)

Ho, F.-C. (1977-). *Tropical Plants of Taiwan in Colour*, 5 vols planned, 3 published so far. Heng-Chun Tropical Botanical Garden, Taiwan. (In Chinese.)

Hsu, C.-C. (1975). *Illustrations of Common Plants of Taiwan, 1. Weeds*, 2nd Ed. Taiwan Provincial Education Assoc., Taipei. 558 pp.

Hsu, C.-C. (1975). *Illustrations of Common Plants of Taiwan, 7. Grasses*. Taiwan Provincial Education Assoc., Taipei. 884 pp. (In Chinese and English.)

Hsu, K.-S. (Ed.) (1982). *Pteridophytes of Taiwan*. Taiwan Provincial Dept of Education. 138 pp. (119 colour plates, descriptions in Chinese.)

Hsu, K.-S. *et al.* (1984). *The Wild Woody Plants of Taiwan*. Taiwan Provincial Dept of Education. 169 pp. (150 colour plates, descriptions in Chinese.)

Liu, T.-S. (1960, 1962). *Illustrations of Native and Introduced Ligneous Plants of Taiwan*, 2 vols. National Taiwan Univ., Taipei. (Line drawings; short descriptions in Chinese.)

Ying, S.-S. (1975). *Alpine Plants of Taiwan in Color*, 2 vols. (223 alpines described, in Chinese and English.)

Information on Threatened Plants

Hsu, K.-S. (Ed.) (1982). *The Rare and Threatened Plants of Taiwan*. Taiwan Provincial Junior High School, Keelung. 100 pp. (Lists 37 ferns, 8 gymnosperms, 293 angiosperms, with codes for habitat loss, overcollecting and restricted geographical ranges; includes non-endemic threatened plants; many colour plates, text in Chinese.)

Liu, T. and Hsu, K.-S. (1971). The rare and threatened plants and animals of Taiwan. *Quart. J. Chinese Forestry* 4(4): 89-96. (In Chinese, not seen.)

Su, H.-J. (1980). Studies on the rare and threatened forest plants of Taiwan. *Bull. Exp. Forest National Taiwan Univ.* 125: 165-205. (Lists 336 taxa, including non-endemics; in Chinese with English summary.)

Laws Protecting Plants The Law of Cultural Heritage Preservation (1982) covers the preservation of the natural heritage, including nature reserves, endangered plants and animals. The type of protection and taxa covered are under review (H.-J. Su, 1984, *in litt.*).

Voluntary Organizations

Wildlife and Nature Conservation Society, 4f 23 Alley 36, Lane 305, Min Chyuan E. Road, Taipei.

Botanic Gardens

Botanical Garden, National Taiwan University, Taipei.

Botanical Garden, Taiwan Forestry Research Institute, Po-A Road, Taipei 107.

Heng-Chun Tropical Botanical Garden, Taiwan Forestry Research Institute, Heng-Chun, Pingtung, Taiwan 924.

Additional References The Taiwan Nature Conservation Strategy was prepared in 1984 (and translated into English in 1985) by the Ministry of Interior.

Tanzania

Area 939,762 sq. km

Population 21,710,000

Floristics c. 10,000 species of vascular plants (Polhill, 1968). Brenan (1978, cited in Appendix 1) estimates 1122 endemic species from a sample of the *Flora of Tropical East Africa*.

Flora affinities predominantly Zambezian with broad coastal strip of Zanzibar-Inhambane regional mosaic; Lake Victoria regional mosaic in north-west corner; Somalia-Masai region in north-central area; Afromontane region on the mountains and small area of Afroalpine region on the highest peaks. Endemism is concentrated in the coastal forests, and the eastern mountain forests, which have Guinea-Congolian affinities.

Vegetation Wide range of vegetation types. Predominantly deciduous woodland with *Brachystegia* (Miombo) becoming *Acacia/Commiphora* thicket in drier areas, and forest in the high rainfall of the mountains and coast. Grassland occurs on the volcanic soils in the north and on the high altitude plateaux with ericaceous heaths notable for giant *Lobelia* and *Senecio* and best developed on Kilimanjaro. Patches of fresh water and halophytic swamp vegetation are found inland. Mangrove is well developed in some coastal areas, notably the Rufigi delta. The moist forests deserve special mention due to their high degree of endemism; they are very varied, ranging from wet lowland to dry montane.

Forests cover only 1-2% of the country; estimated rate of deforestation for closed broadleaved forest 100 sq. km/annum out of 14,400 sq. km (FAO/UNEP, 1981). According to Myers (1980, cited in Appendix 1), remaining moist forest amounts to 9400 sq. km.

For vegetation map see White (1983), cited in Appendix 1.

Checklists and Floras Tanzania is included in the incomplete *Flora of Tropical East Africa*. Tanzania's plants of high altitudes are listed in *Afroalpine Vascular Plants* (Hedberg, 1957). Both of these are cited in Appendix 1. See also:

Bjornstad, A. (1976). *The Vegetation of Ruaha National Park, Tanzania. 1. Annotated Check-List of the Plant Species*. Publication No. 215 of the Serengeti Research Institute.

Brenan, J.P.M. and Greenway, P.J. (1949). *Check-Lists of the Forest Trees and Shrubs of the British Empire. No. 5: Tanganyika Territory*. Part 2. Imperial Forestry Institute, Oxford. 653 pp. (Short descriptions, citation of specimens.)

Engler, A. *et al.* (1895). *Die Pflanzenwelt Ost-Afrikas und der Nachbargebiete*. 3 volumes: vol. A – Characteristic features of plant distribution (154 pp., 8 black and white photographs), vol. B – Useful plants (536 pp., illus.), vol. C – Annotated checklist (433 pp. + 40 page index, 45 illustrations). Berlin. (Also covers Rwanda, Burundi, northern Mozambique, southern Uganda and southern Kenya.)

Gilbert, V.C. (1970). *Plants of Mt Kilimanjaro*. U.S. National Park Service, Washington, D.C. 117 pp.

Greenway, P.J. and Vesey-Fitzgerald, D.F. (1972). Annotated checklist of plants occuring in Lake Manyara National Park. *J. East Afr. Nat. Hist. Soc. and Nat. Mus.* 28: 1-29.

Kerfoot, O. (1964). A first check-list of the vascular plants of Mbeya Range, Southern Highland region of Tanzania. *Tanganyika Notes and Records* 62: 1-17.

Rodgers, W.A., Hall, J.B., Mwasumbi, L. and Vollesen, K. (1984). The conservation values and status of the Kimbosa Forest Reserve, Tanzania. University of Dar es Salaam. 84 pp. Mimeo. (Includes checklist.)

Steele, R.C. (1966). A check-list of the trees and shrubs of the South Kilimanjaro forests. Part 1. *Tanzania Notes and Records* 65: 97-102.

Vollesen, K. (1980). Annotated check-list of the vascular plants of the Selous Game Reserve, Tanzania. *Opera Botanica* 59: 1-117.

Field-guides A very useful key to families is included in Lind and Tallantire (in press), cited under Uganda.

Cribb, P.J. and Leedal, G.P. (1982). *The Mountain Flowers of Southern Tanzania*. Balkema, Rotterdam. 244 pp. (Descriptive notes, 19 colour photographs, 72 line drawings.)

Information on Threatened Plants

Hedberg, I. (1979), cited in Appendix 1. (List for Tanzania, pp. 95-99, by R.C. Wingfield, contains about 390 endemic species and infraspecific taxa generally not included in Polhill's list of 1968. All are 'seemingly rare', but relatively few are given IUCN threatened categories.)

Polhill, R.M. (1968). Tanzania. In Hedberg, I. and O. (1968), cited in Appendix 1. Pp. 166-178. (Lists about 200 trees and shrubs endemic to the Usambaras, the Ulugurus or Lindi District.)

White, F. (1979). Some interesting and endangered plant species in the West Usambara Mountains, Tanzania. *Notes from the Forest Herbarium, University of Oxford* 1. 5 pp. (Mimeo.)

Data sheets of 2 species occuring in Kenya and Tanzania are published in *The IUCN Plant Red Data Book* (1978).

IUCN has records of c. 1200 species and infraspecific taxa believed to be endemic. Only a few of these known to be rare or threatened, information lacking for the rest: E:9, V:26, R:72, I:2.

Laws Protecting Plants Ten forest trees are specifically protected from felling on any unreserved land under the Forests (Reserved Trees) Order of 1972. Much of the natural forest is quite well protected under Forest Reserve status.

Botanic Gardens A long-established arboretum at the Silviculturist's Head-quarters, Forest Division, Lushoto, Usambaras (and another in the East Usambaras at Amani) has important collections of African and other tropical woody plants.

Useful Addresses

Director of Forests, Ministry of Natural Resources, Box 932, Dar es Salaam.

The Herbarium, Dept of Botany, University of Dar es Salaam, Box 35091, Dar es Salaam.

CITES Management Authority: The Director of Wildlife, Ministry of Natural Resources and Tourism, P.O. Box 1994, Dar es Salaam.

Additional References

Gillman, C. (1949). A vegetation-types map of Tanganyika Territory. *Geog. Rev.* 39(1): 7-37. (With coloured map 1:2,000,000.)

Mshigeni, K.E. (1979). Exploitation and conservation of biological resources in Tanzania. In Hedberg, I. (Ed.) (1979), cited in Appendix 1. Pp. 140-143. (Illus.)

Rodgers, W.A. and Homewood, K.M. (1982a). Species richness and endemism in the Usambara mountain forests, Tanzania. *Biol. J. Linn. Soc.* 18: 197-242. (Includes a list of endemic forest trees.)

Rodgers, W.A. and Homewood, K.M. (1982b). Biological values and conservation prospects for the forests and primate populations of the Uzungwa Mountains, Tanzania. *Biol. Conserv.* 24: 285-304.

Werth, E. (1901). *Die Vegetation der Insel Sansibar.* 97 pp. (With coloured vegetation map 1:300,000.)

Williams, R.O. (1949). *The Useful and Ornamental Plants in Zanzibar and Pemba.* Govt Printer, Zanzibar. 497 pp. (Illus.)

Thailand

Area 514,000 sq. km

Population 50,584,000

Floristics c. 12,000 vascular plant species, including 25 gymnosperms (Larsen, 1979) and c. 600 fern species (Parris, 1985, cited in Appendix 1).

Vegetation Tropical lowland evergreen rain forest originally covered much of central and southern Thailand, now reduced to fragments in south Peninsular Thailand, covering c. 30,000 sq. km (Myers, 1980, cited in Appendix 1); semi-evergreen forest dominated by Dipterocarpaceae in centre and north below 1000 m; hill evergreen forest with *Quercus* and *Castanopsis* above 700 m, much disturbed by shifting cultivation; dry dipterocarp forest with bamboo, and mixed deciduous forests with Teak (*Tectona grandis*), are the most extensive forest types, covering 39,000 sq. km; teak forests extensively logged or burnt; few patches of pine forest in north and east; mangroves extensive on west coast and Chao Phya delta; freshwater swamp-forest in centre and south (FAO/UNEP, 1981). Much of the central plain densely populated and cleared for rice cultivation.

Estimated rate of deforestation of closed broadleaved forest 2440 sq. km/annum out of a total of 81,350 sq. km (FAO/UNEP, 1981). However, between 1973 and 1978, Myers (1980) estimates an average of c. 13,800 sq. km/annum were "grossly disrupted or cleared" as a result of commercial logging, to which he adds a further 4000 sq. km/annum converted by shifting cultivators. For vegetation maps see:

Department of Forestry (1952, 1961). *Types of Forests of Thailand.* Bangkok. (Two forestry maps at 1:2,500,000 and 1:1,000,000.)

Checklists and Floras The Flora is:

Smitinand, T. and Larsen, K. (Eds) (1970-). *Flora of Thailand.* Applied Science Research Corporation and Thailand Institute of Scientific and Technological Research, Bangkok. (1 – Introduction, plant geography; in prep.; 2 – covers mainly smaller families so far; 3 – ferns; 4(1) – Leguminosae-Caesalpiniodeae, 1984; 4(2) – Leguminosae-Mimosoideae, in press; 5(1) – various families, in press.)

Eastern Thailand is included in the rather dated, and now incomplete *Flore Générale de L'Indo-Chine* (1907-1951), cited in Appendix 1. See also:

Craib, W.G. *et al.* (1911-1940). Contributions to the Flora of Siam. *Kew Bulletin* (many parts).

Craib, W.G. and Kerr, A.F.G. (Eds) (1925-1962). *Florae Siamensis Enumeratio: A List of Plants Known from Siam with Records of their Occurrence*, 3 vols. Siam Soc., Bangkok. (Introduction includes notes on the flora; not completed, covers Ranunculaceae to Gesneriaceae.)

Seidenfaden, G. and Smitinand, T. (1959-1965). *The Orchids of Thailand, A Preliminary List*. Parts 1-4(2). Siam Society, Bangkok. (Keys, descriptions of about 750 species, notes on distribution.)

Suvatti, C. (1978). *Flora of Thailand*, 2 vols. Royal Institute, Bangkok. (1 – Enumeration of 222 pteridophyte taxa, 47 gymnosperms, 108 monocotyledons, 1121 dicotyledons; 2 – 2487 dicotyledons.)

Various authors have contributed papers on the flora of Thailand in the series 'Studies in the Flora of Thailand' in the journal *Dansk Botanisk Arkiv*, Copenhagen (1961-1969) and in the *Thai Forest Bulletin (Botany)* from 1972.

Precursor treatments of various orchid genera by G. Seidenfaden have been published in the series 'Contributions to the orchid flora of Thailand' (10 parts) in *Botanisk Tidsskrift* and *Nordic Journal of Botany* (1969-1982). See also the series 'Orchid genera in Thailand' (10 parts, by the same author) in *Dansk Botanisk Arkiv* and *Opera Botanica* (1968-1982).

Field-guides
Smitinand, T. (Ed.) (1975). *Wild Flowers of Thailand*. Bangkok. 228 pp. (142 coloured photographs of indigenous plants; notes in Thai and English.)

Information on Threatened Plants No national Red Data Book, but see:

Bain, J.R. and Humphrey, S.R. (1982). *A Profile of the Endangered Species of Thailand*, 2 vols. Office of Ecological Services, Gainesville, USA. (Lists 53 plant taxa rare or threatened in Thailand, with 3 data sheets and 14 dot maps.)

Ng, F.S.P. and Low, C.M. (1982). *Check List of Endemic Trees of the Malay Peninsula*. Forest Research Institute, Kepong. 94 pp. (Lists 654 trees endemic to the Malay Peninsula, including Peninsular Thailand; rarity is based on numbers of herbarium specimens.)

Laws Protecting Plants *Tectona grandis* and *Dipterocarpus* spp. are protected
under the Forest Act. Some 250 tree species and several orchids are protected under the Royal Decree (T. Smitinand, 1984, *in litt.*).

Voluntary Organizations
Wildlife Fund Thailand, 8 Sukhumvit 12, Bangkok 10110.

Botanic Gardens
Khao Chong Botanic Garden, Trang, Peninsular Thailand.
Mae Sa Botanic Garden, Chiang Mai, Northern Region.
Phu Khae Botanic Gardens, Saraburi.

Useful Addresses

Royal Thai Forest Department, Bangkok 10900.

CITES Management Authority: Agricultural Regulation Division, Dept of Agriculture, Paholyothin Road, Bagkhen, Bangkok 10900.

Additional References

Hansen, B. (1973). Bibliography of Thai botany. *Nat. Hist. Bull. Siam Soc.* 24: 319-408.

Larsen, K. (1979). Exploration of the flora of Thailand. In Larsen, K. and Holm-Nielsen, L.B. (Eds), *Tropical Botany*. Academic Press, London. Pp. 125-133.

Ogawa, H., Yoda, K. and Kira, T. (1961). A preliminary survey on the vegetation of Thailand. In Kira, T. and Umesao, T. (Eds), *Nature and Life in Southeast Asia*, 1. Pp. 21-157. Fauna and Flora Research Society, Kyoto.

Vidal, J.E. (1979). Outline of ecology and vegetation of the Indochinese Peninsula. In Larsen, K. and Holm-Nielsen, L.B. (Eds), *Tropical Botany*. Academic Press, London. Pp. 109-123.

Togo

Area 56,785 sq. km

Population 2,838,000

Floristics 2000 species (Scholz and Scholz, 1983); 2302 species excluding ferns (Brunel *et al.*, 1985); 20 endemic species (Brenan, 1978, cited in Appendix 1), but flora of adjoining countries too poorly known to be sure.

Flora in northern half of country with Sudanian affinities; flora in south with Guinea-Congolian/Sudanian affinities.

Vegetation Sudanian woodland with *Isoberlinia* in north; lowland rain forest and forest interspersed with secondary grassland and cultivation in south. Substantial mangrove and coastal mosaic along coast. Estimated rate of deforestation for closed broadleaved forest 21 sq. km/annum out of 3040 sq. km (FAO/UNEP, 1981).

For vegetation map see White (1983), cited in Appendix 1.

Checklists and Floras Togo is included in the *Flora of West Tropical Africa*, cited in Appendix 1.

Brunel, J.F., Hiepko, P. and Scholz, H. (Eds) (1985). *Flore Analytique du Togo. Phanérogames*. Deutsche Gesellschaft für Technische Zusammenarbeit, Eschborn, F.R.G. 751 pp. (2302 species.)

Scholz, H. and U. (1983). *Flore Descriptive des Cypéracées et Graminées du Togo*. Phanerogamarum Monographiae 15. Cramer, FL-9490, Vaduz, Liechtenstein. 360 pp.

Information on Threatened Plants No published lists of rare or threatened plants; IUCN has records of 25 species and infraspecific taxa believed to be endemic; no categories assigned.

Useful Addresses

CITES Management Authority: Direction des Forêts et Chasses, B.P. 355, Lomé.

Additional References

Aubréville, A. (1937). Les forêts du Dahomey et du Togo. *Bull. Com. Etud. Hist. Scient. Afr. Occid. Fr.* 20. 112 pp. (With 18 black and white photographs.)

Ern, H. (1979). Die Vegetation Togos. Gliederung, Gefährdung, Erhaltung. *Willdenowia* 9: 295-312. (With 9 colour photographs.)

Tokelau

A territory of New Zealand in the south-west Pacific Ocean, 3900 km south-west of Hawaii, between longitudes 160-170°W and latitudes 5-12°S, comprising 3 inhabited atolls. Area 10 sq. km; population 1620 (1980 estimate, *Times Atlas*, 1983). The atolls are only 5 m above sea level. Nukunona (5.5 sq. km) consists of 24 islets around a large lagoon and contains small areas of *Cordia*, *Pisonia* and *Guettarda* woodland. Fakaofo (2.5 sq. km) consists of 61 islets with beach scrub (*Scaevola*, *Tournefortia*) and coconut groves. Accounts on the extent and composition of the vegetation are given in Hinckley (1969) and Parham (1971).

40 vascular plant species (including introductions); most are widespread throughout the Pacific. No information on threatened plants.

References

Hinckley, A.D. (1969). Ecology of terrestrial arthropods on the Tokelau atolls. *Atoll Res. Bull.* 124. 18 pp.

Parham, B.E.V. (1971). The vegetation of the Tokelau Islands with special reference to plants of Nukunonu Atoll. *N.Z. J. Bot.* 9: 576-609. (Includes checklist.)

Whistler, W.A. (1981). A naturalist in the South Pacific north to Tokelau. *Bull. Pacific Tropical Bot. Garden* 11(2): 29-37.

Tonga

A scattered group of 171 limestone and volcanic islands in the central South Pacific Ocean, between latitudes 18°-23°S, and longitudes 173°-177°W. 3 groups: Vava'u in the north, Ha'apai and Tongatapu in the south. 36 of the islands are permanently inhabited; about half the resident population is on Tongatapu, the largest island. The limestone islands are mostly below 100 m, flat topped and fertile. The volcanic islands are more rugged; the highest point is 1030 m on the island of Kao, in the Ha'apai Group. Some islands are still volcanically active.

Area 699 sq. km

Population 107,000

Floristics c. 770 vascular plant taxa (including introductions). According to Yuncker (1959), there are c. 70 species of ferns (of which *Dryopteris macroptera* is the only endemic), 3 species of gymnosperms (of which *Podocarpus pallidus* is the only endemic), and 698 angiosperm taxa (including 9 endemics).

Vegetation Lowland rain forests, dominated by *Calophyllum*, on limestone islands, but much has been cleared on the larger islands for settlements and the cultivation of yams, bananas and coconuts (Yuncker, 1959); 'Eua, in the Tongatapu Group, has the best examples of volcanic island rain forest in Tonga (Dahl, 1980, cited in Appendix 1); moss forests on the summit of Kao, and on Tafahi, to the north of the Vava'u Group; coastal scrub, with *Barringtonia* and *Scaevola*, on most islands; *Casuarina* woodlands on recent lava flows; large areas of secondary vegetation, including *Lantana* and *Psidium* scrub, and *Sorghum* and *Panicum* grasslands, especially on the islands of Tongatapu, 'Eua, and Uta Vava'u in the Vava'u Group; mangroves and *Cyperus* swamps, the latter especially around the crater lake on Niuafo'ou, an isolated volcanic island to the north-west of the Vava'u Group. The crater zone of most volcanic islands has a distinct but sparse herbaceous flora.

Checklists and Floras

Hemsley, W.B. (1894). The Flora of the Tonga or Friendly Islands. *J. Linn. Soc. Bot.* 30: 158-217.

Hürlimann, H. (1967). Bemerkenswerte Farne und Blütenpflanzen van den Tonga-Inseln. *Bauhinia* 3: 189-202. (Describes 6 fern species, one gymnosperm and 39 angiosperm taxa.)

St John, H. (1977). The flora of Niuatoputapo Island, Tonga. Pacific Plant Studies 32. *Phytologia* 36(4): 374-390. (Annotated checklist; 211 vascular plant taxa, of them 92 indigenous, including one endemic.)

Yuncker, T.G. (1959). Plants of Tonga. *Bull. Bernice P. Bishop Mus.* 220. 83 pp. (Annotated checklist.)

Information on Threatened Plants None.

Laws Protecting Plants The Forest Act (providing for forest reserves) has provisions to protect the diminishing stocks of trees used in making handicrafts, cosmetic oils, folk medicines and traditional costumes (Pulea, 1984).

Additional References

Pulea, M. (1984). Environmental legislation in the Pacific region. *Ambio* 13(5-6): 369-371.

Trinidad and Tobago

Trinidad, the most southerly of the West Indian Islands, is separated from the north coast of South America by a 11.3 km wide strait. It is mountainous, 80.5 km long by 59.5 km wide.

Tobago lies 30.4 km north-east of Trinidad and is 41.84 km long by 12 km wide.

Area Trinidad: 4829 sq. km; Tobago: 300.5 sq. km

Population 1,105,000

Floristics 2281 species of flowering plants, of which 215 are endemic (Adams and Baksh, 1981). Main affinities are with South America, rather than the Antilles.

Vegetation On Trinidad, montane forests in the north – mainly lower montane rain forest but also montane rain forest and an area of elfin woodland; in the south-east evergreen and semi-evergreen seasonal forest; in the north-east and south-west almost pure stands of *Mora excelsa* forest; marsh grassland of Aripo savanna type, rich in sedges and terrestrial orchids, at the foot of the Northern Range; in patches on the coast mangrove woodland and swamp. On Tobago rain forest in sheltered valleys on interior mountains; lower montane forest on red clay soils on slopes and "xerophytic" rain forest (dry evergreen formation associated with shallow rocky soils, drying winds and moderately high rainfall); some elfin woodland. (Beard, 1946a,b, 1949, the latter cited in Appendix 1.) 44.1% forested according to FAO (1974, cited in Appendix 1).

Checklists and Floras Covered by the family and generic monographs of *Flora Neotropica* (cited in Appendix 1). The Flora is:

Williams, R.O. and Cheesman, E.E. *et al.* (1928-). *Flora of Trinidad and Tobago.* Ministry of Agriculture, Lands and Fisheries, Trinidad and Tobago. Vols. I and II (polypetalous and gamopetalous dicotyledons) complete. Vol. III (monocotyledons) two thirds complete; sedges, grasses, palms and ferns in preparation, the rest in press.

See also:

De Freitas, K. and Pacquet, J. (1980). *Manual of Dendrology*, vol. 4. Inventory of the indigenous forests of Trinidad and Tobago. Wildlife Conservation Committee, Ministry of Agriculture, Trinidad and Tobago. 181 pp. (Descriptions and photographs of the bole, bark and foliage of mostly indigenous trees of potential commercial value.)

Field-guides
Marshall, R.C. (1934). *Trees of Trinidad and Tobago.* Govt Printer, Port of Spain. 101 pp.

Information on Threatened Plants
Adams, C.D. and Baksh, Y.S. (1981). What is an endangered plant? *Living World* 1981-2: 9-14. Journal of the Trinidad and Tobago Field Naturalists Club. (Includes distribution tables of 648 threatened non-endemic species and of 215 endemic species, with criteria for ranking the selected species.)

The IUCN Plant Red Data Book has two data sheets for Trinidad and Tobago.

Voluntary Organizations
Trinidad and Tobago Field Naturalists' Club, 1 Palm Avenue, Petit Valley, Diego Martin, Trinidad.

Botanic Gardens
Royal Botanic Gardens, Port of Spain.

Useful Addresses
The Conservator of Forests, Forestry Division, Ministry of Agriculture, Lands and Fisheries, Long Circular Road, Port of Spain.
CITES Management Authority: Ministry of Agriculture, Lands and Food Production, Forestry and Veterinary Services Division, St Clair Circle, Port of Spain.

Additional References
Bacon, P.R. and ffrench, R.P. (Eds) (1972). *The Wildlife Sanctuaries of Trinidad and Tobago.* Wildlife Conservation Committee, Ministry of Agriculture, Lands and Fisheries, Trinidad and Tobago. 80 pp, 29 figs, map.

Beard, J.S. (1944). The natural vegetation of Tobago, B.W.I. *Ecol. Monogr.* 14: 135-163.

Beard, J.S. (1946a). The Mora forests of Trinidad. *J. Ecol.* 33: 173-193.

Beard, J.S. (1946b). The natural vegetation of Trinidad. *Oxford Forestry Memoirs* No. 20. Clarendon Press, Oxford. 152 pp.

Marshall, R.C. (1934). Physiography and vegetation of Trinidad and Tobago. *Oxford Forestry Memoirs* No. 17. Clarendon Press, Oxford. 56 pp.

Richardson, W.D. (1963). Observations on the vegetation and ecology of the Aripo savannas, Trinidad. *J. Ecol.* 51: 295-313. (A vegetation unique in the Caribbean, comparable with similar formations on the S. American mainland.)

Thelen, K.D. and Faizool, S. (Eds) (1980). *Policy for the Establishment and Management of a National Park System in Trinidad and Tobago.* 26 pp. Forestry Division, Ministry of Agriculture, Lands and Fisheries, Port of Spain.

Tristan da Cunha

A group of four islands in the south Atlantic c. 2800 km from South Africa and c. 3200 km from the nearest point of South America. Tristan da Cunha (37°05'S 12°20'W), Inaccessible and Nightingale lie close together, Gough about 350 km SSE of the others (40°20'S 10°00'W). The group is a dependency of St Helena, itself a British Dependent Territory.

Area 159 sq. km total; Tristan: 86 sq. km; Inaccessible: 12 sq. km; Nightingale: 4 sq. km; Gough: 57 sq. km

Population 299 (*Whitaker's Almanack*, 1984)

Floristics Total flora c. 157 species; 33 native species of ferns and 41 of flowering plants (Wace and Dickson, 1965); 11 species and 5 varieties of ferns endemic, 29 species and one subspecies of flowering plants endemic (IUCN figures, based on Groves, 1981).

Vegetation Tussock grassland on the coast, with fern-bush vegetation on the lower parts; over much of the island and above the fern-bush there is wet-heath; over much of the uplands on the larger islands there are extensive open communities of cushion, mat-forming and crevice plants. All the native vegetation types except the last-mentioned are, to some extent, peat-forming.

Checklists and Floras

Groves, E.W. (1981). Vascular plant collections from the Tristan da Cunha group of islands. *Bull. Brit. Mus. (Nat. Hist.), Bot.* 8(4): 333-420. (With several maps, line drawings and black and white photographs.)

Rudmose Brown, R.N. (1905). The Botany of Gough Island – I. Phanerogams and ferns. *Bot. J. Linn. Soc.* 37: 238-250. (With 2 black and white photographs.)

Wace, N.M. and Dickson, J.H. (1965). The terrestrial botany of the Tristan da Cunha Islands. *Phil. Trans. Roy. Soc.* 249B: 273-360. (With 15 black and white photographs.)

Information on Threatened Plants Few if any species threatened. For details of the islands' natural history see:

Wace, N.M. and Holdgate, M.W. (1976). *Man and Nature in the Tristan da Cunha Islands*. Monograph No. 6, IUCN, Switzerland.

Latest IUCN statistics: 46 endemic taxa, including R:6, I:11. Others not threatened or insufficiently known.

Additional References

Wace, N.M. (1961). The vegetation of Gough Island. *Ecol. Monog.* 31: 337-367. (With several black and white photographs.)

Trobriand Islands

Low coral islands, 145 km north of eastern New Guinea in the Solomon Sea, south-west Pacific Ocean. The main islands are Trobriand or Kiriwina (a raised atoll to 30 m), Kaileuna, Vakuta and Kitava. Area 440 sq. km; population 12,700 (1971, *Encyclopedia Britannica*, 1974). The islands are politically part of Papua New Guinea.

The Trobriand Islands are covered by the incomplete *Flora Malesiana* (1948-), cited in Appendix 1. No separate Flora or checklist has been published. No figure for size of flora. No information on threatened plants.

The Trobriand Islands have limestone rain forest, coastal brackish and mangrove forests according to the vegetation map of Malesia (Whitmore, 1984, cited in Appendix 1). See also the *Vegetation Map of Malaysia* (van Steenis, 1958), covering the *Flora Malesiana* region at scale 1:5,000,000 and cited in Appendix 1.

Tromelin

A small (c. 1 sq. km) coral islet 480 km NNW of Mauritius and 390 km west of Antongil Bay, Madagascar, in the Indian Ocean, 15°51'S 54°25'E. 6 species of pantropical or Indo-Pacific species of plant have been recorded (Staub in Stoddart, 1970; Renvoize, 1979). The vegetation consists of two main communities: a shrub layer 0.5-1 m high and occasionally to 2.5 m; a herb mat. There is a feral rabbit population which may account for the paucity of plant species.

References

Renvoize, S.A. (1979). The origins of Indian Ocean island floras. In Bramwell, D. (Ed.), *Plants and Islands*. Academic Press, London. Pp. 107-129.

Stoddart, D.R. (Ed.) (1970). Coral Islands of the western Indian Ocean. *Atoll Res. Bull.* 136. 224 pp. (Includes species inventories of several islands by F.R. Fosberg and S.A. Renvoize; see especially paper by F. Staub, pp. 197-209, on the geography and ecology of Tromelin, with 10 black and white photographs.)

Tuamotu Archipelago

76 islands situated in the Pacific Ocean between longitudes 130-155°W and latitudes 14-23°S. Most of the islands are low coral atolls not more than 7 m above sea level, many enclosing central lagoons. Makatéa, a raised atoll, reaches 111 m. Area 842 sq. km; population c. 4250 (Douglas, 1969, cited in Appendix 1). The Tuamotus, together with the Gambier Islands, form an administrative division of French Polynesia.

4 native fern taxa; 52 native angiosperm taxa; 36% species endemism (*Flora of Southeastern Polynesia*, 1931-1935). Floristic affinities are with the floras of neighbouring coral islands in Polynesia, and also with the flora of Malesia. No information on threatened plants.

Most islands have *Pandanus*, *Pisonia* and *Cordia* scrub; some mangroves on northern islands (Dahl, 1980, cited in Appendix 1). Coconut plantations on many islands. Little native vegetation remaining on Makatéa mainly due to phosphate mining; some islands such as Mururoa greatly damaged by testing of nuclear weapons (Douglas, 1969, cited in Appendix 1).

References The Tuamotus are included in the *Flora of Southeastern Polynesia* (Brown and Brown, 1931-1935), cited in Appendix 1. See also:

Copeland, E.B. (1932). Pteridophytes of the Society Islands. *Bull. Bernice P. Bishop Mus.* 93. 86 pp. (Descriptions and keys; notes on distribution.)
Doty, M.S. *et al*. (1954). Floristics and plant ecology of Raroia atoll, Tuamotus. *Atoll Res. Bull.* 33. 58 pp. (Includes annotated list of 51 native and introduced vascular plants; keys to common species.)
Sachet, M.-H. (1983). Takapoto Atoll, Tuamotu Archipelago: terrestrial vegetation and flora. *Atoll Res. Bull.* 277. 41 pp. (Includes annotated checklist of 4 native fern species and 136 angiosperm taxa, including 100 introductions.)
Stoddart, D.R. and Sachet, M.-H. (1969). Reconnaissance geomorphology of Rangiroa Atoll, Tuamotu Archipelago, with list of vascular flora of Rangiroa. *Atoll Res. Bull.* 125. 44 pp. (Lists 121 taxa of which 41 indigenous; notes on habitats, occurrence.)
Wilder, G.P. (1934). The Flora of Makatea. *Bull. Bernice P. Bishop Mus.* 120. 49 pp. (Incomplete; descriptions and notes on about 200 taxa, including introduced plants; notes on physical features; vegetation.)

Tubuai

Tubuai, or the Austral Islands, comprise 6 volcanic islands and a low coral atoll (Maria Island). The islands are c. 3000 km east of New Caledonia in the South Pacific Ocean, between latitudes 20°S and 30°S, and longitudes 140-150°W. Area 142.5 sq. km; population c. 3830 (Douglas, 1969, cited in Appendix 1). The highest point is 633 m, on Rapa Iti. Tubuai is an administrative division of French Polynesia.

c. 700 flowering plant species; less than 20% of the vascular flora is endemic (J. Jérémie, 1984, *in litt*.). 2 endemic genera – *Metatrophis* and *Oparanthus*. Rapa Iti has 76 ferns and 86 angiosperms; 62% species endemism (Dahl, 1980, cited in Appendix 1). The flora of Tubuai is mainly related to that of the Society Islands and Marquesas (*Flora of*

Southeastern Polynesia, 1931-1935, cited in Appendix 1), but the flora of Rapa Iti is more related to those of New Zealand, Hawaii and Easter Island. No information on threatened plants.

Most of the high islands have lowland and montane forests. The lower slopes support *Casuarina* and *Cocos* whereas the upper slopes are mainly grassland and scrub. Much of the lowlands have been cleared for coffee, citrus and copra production. Rurutu and Rimatara have elevated reef limestone (makatea) with limestone woodland. Maria Island has dense *Cocos* and *Pandanus* forest.

References The Austral Islands are included in the *Flora of Southeastern Polynesia* (Brown and Brown, 1931-1935), cited in Appendix 1. See also:

Copeland, E.B. (1932). Pteridophytes of the Society Islands. *Bull. Bernice P. Bishop Mus*. 93. 86 pp. (Descriptions and keys; notes on distribution.)

Fosberg, F.R. and St John, H. (1952). Végétation et flore de l'atoll Maria, îles Australes. *Rev. Sci. Bourbonn*. 1951: 1-7. (Not seen, citation from Frodin.)

Hallé, N. (1980). Les orchidées de Tubuai (archipel des Australes, Sud Polynésie): suivies d'un catalogue des plantes à fleurs et fougères des îles Australes. *Cah. Indo-Pacifique* 2(3-4): 69-130. (Includes checklist of 627 vascular plant species of Austral Islands and Rapa Iti.)

Tunisia

Area 164,148 sq. km

Population 7,042,000

Floristics 2200 species (Lebrun, 1976, cited in Appendix 1); 2120 species (Le Houérou, 1975).

Northern part of country with Mediterranean flora; Saharan flora in south; transition between the two in centre of country.

Vegetation Desert with little or no perennial vegetation in the south; semi-desert grassland and shrubland in centre; some Mediterranean sclerophyllous forest, e.g. of cork oak, survives along the north coast.

For vegetation map see White (1983), cited in Appendix 1.

Checklists and Floras Tunisia is included in the incomplete *Flore de l'Afrique du Nord*, the computerized *Atlas der Pflanzenwelt des Nordafrikanischen Trockenraumes* (Frankenberg and Klaus, 1980), *Flore du Sahara* (Ozenda, 1977), and is being covered in *Med-Checklist*. These are all cited in Appendix 1.

Battandier, J.A. and Trabut, L. (1902/1904). *Flore Analytique et Synoptique de l'Algérie et de la Tunisie*. Alger. 460 pp.

Bonnet, E. and Barratte, G. (1896). *Catalogue Raisonné des Plantes Vasculaires de la Tunisie*. Exploration Scientifique de la Tunisie publiée sous les auspices du Ministère de l'Instruction Publique, Botanique. Paris. 519 pp.

Cuénod, A., Pottier-Alapetite, G. and Labbe, A. (1954). *Flore Analytique et Synoptique de la Tunisie*. Imprimerie SEFAN, Tunis. 287 pp. (One volume, covering Gymnosperms and Monocotyledons only.)

Pottier-Alapetite, G. (1979). *Flore de la Tunisie: Angiospermes – Dicotylédones*. Imprimerie Officielle, Tunis. 651 pp. (The continuation of Cuénod *et al.*, 1954; gamopetalous dicotyledons still to follow. Cited by Frodin.)

Information on Threatened Plants Tunisia is included in the draft list for North Africa and the Middle East produced by IUCN Threatened Plants Committee Secretariat (1980), cited in Appendix 1.

Mathez, J., Quézel, P. and Raynard, C. (1985). The Maghrib countries. In Gómez-Campo, C. (Ed.), *Plant Conservation in the Mediterranean Area*.

Pottier-Alapetite, G. (1959). Espèces végétales rares ou menacées de Tunisie. In *Animaux et Végétaux Rares de la Région Méditerranéenne*. Proceedings of the IUCN 7th Technical Meeting, 11-19 September 1958, Athens, vol. 5. IUCN, Brussels. Pp. 135-139.

Latest IUCN statistics: endemic taxa – E:1, R:1, nt:1; non-endemics rare or threatened worldwide: V:1, R:7, I:9 (world categories).

Botanic Gardens
Service Botanique et Agronomique de Tunisië, Service Botanique, Ariana.

Useful Addresses
CITES Management Authority: Direction des forêts, Ministère de l'Agriculture, 36 rue Alain Savary, Tunis.
CITES Scientific Authority: Institut National de Recherches Forestières, Tunis.

Additional References
Gaussen, H. and Vernet, A. (1958). *Carte Internationale du Tapis Végétal. Feuille de Tunis-Sfax* (1:1,000,000, in colour). Tunisian Govt.
Le Houérou, H.-N. (1959). Recherches écologiques et floristiques sur la végétation de la Tunisie méridionale. *Mém. Inst. Rech. Sahar.* 8. Part 1: Les milieux naturels, la végétation, 281 pp. (Contains description of the vegetation.) Part 2: La flore, 229 pp. (Contains checklist of vascular plants for central and south Tunisia. An additional part contains maps, including a bioclimatic map 1:2,000,000, and tables.)
Le Houérou, H.-N. (1967). *Carte Phytoécologique de la Tunisie Méridionale*. 2 feuilles au 1:500,000 en couleurs. Institut National de Recherche Agronomique de Tunisie, Tunis.
Le Houérou, H.-N. (1975). Etude préliminaire sur la compatibilité des flores nord-africaine et palestinienne. In CNRS (1975), cited in Appendix 1. Pp. 345-350.
Quézel, P. and Bounaga, D. (1975). Aperçu sur la connaissance actuelle de la flore d'Algérie et du Tunisie. In CNRS (1975), cited in Appendix 1. Pp. 125-130.

Turkey

Area 779,452 sq. km

Population 45,358,000

Floristics c. 8000 native vascular species; 2000-2400 endemic species (Davis, 1975). Meeting-point of 3 phytogeographical regions: Euro-Siberian, Mediterranean and Irano-Turanian; also Balkan and alpine elements.

Endemics scattered throughout country, except Turkey-in-Europe where almost absent. Endemism highest in Irano-Turanian region, especially near Erzincan, Erzurum, mountains south of Lake Van and on gypsacaceous chalk near Cankiri and Sivas; also the Lycian and Cilician Taurus in the Mediterranean region. Boreal and Tertiary relicts abundant east of Melet River in north-east.

Anatolia renowned as an important centre of floristic diversity, where many European cultivated plants have their origin; e.g. a vital gene pool of wild crop relatives (e.g. *Lathyrus chrysanthus*, *Vicia assyriaca*) present in the southern Irano-Turanian region. Also many fodder plants of economic value grow in the arid areas (Davis, 1965-1985).

Vegetation Great range of topography, climate and soils has resulted in many diverse, but 3 major, vegetation types:

Irano-Turanian (largest region): comprises 2 main phytogeographical areas – a wide peripheral and disjunct montane area of degraded sub-Mediterranean scrub and forest and an inner treeless steppe, centred on central Anatolia. West of the 'Anatolian Diagonal', a floristic demarcation line almost halving the region, *Artemisia fragrans* steppe is dominant, giving way to *Pinus nigra* ssp. *pallasiana* forest and *Cistus laurifolius* scrub in the east.

Euxine (Euro-Siberian): a northern belt of broadleaved deciduous forest from shore of Black Sea to 1200 m, dominated by Oriental Beech (*Fagus orientalis*) with several oak species (*Quercus frainetto*, *Q. hartwissiana*, *Q. robur*, *Q. petraea*), hornbeam, chestnut and lime. Between 1200 m and 1500 m, a transition zone with conifers (*Abies bornmulleriana*, *A. nordmanniana* and *Pinus sylvestris*). Above 1500 m, montane coniferous forest dominant, showing affinities with equivalent zones in Caucasus and mountains of central Europe.

Mediterranean: occurs in west and south Anatolia, extending as enclaves along Black Sea coast; in Turkey-in-Europe confined to Gallipoli peninsula. Between 1000-1200 m maquis once covered large areas, but degradation now widespread, becoming replaced by phrygana. Above 1000 m conifers dominant, giving way to a spiny cushion community of *Astragalus* and *Acantholimon* species at higher altitudes, especially in the Taurus.

Checklists and Floras Turkey-in-Europe is covered in the completed *Flora Europaea* (Tutin *et al.*, 1964-1980); the whole country will be covered by the *Med-Checklist* (both cited in Appendix 1). The authoritative national Flora, however, is:

Davis, P.H. (Ed.) (1965-1985). *Flora of Turkey and the East Aegean Islands*, 9 vols. Edinburgh Univ. Press, Edinburgh. (1 – introduction covers topography, climate, excellent phytosociological description; pteridophytes; gymnosperms; angiosperms (dicotyledons: Ranunculaceae to Polygalaceae); 2 – Portulacaceae to Celastraceae; 3 – Leguminosae; 4 – Rosaceae to Dipsacaceae; 5 – Compositae; 6 – Lobeliaceae to Scrophulariaceae; 7 – Orobanchaceae to Rubiaceae; 8 – monocotyledons (Butomaceae to Typhaceae); 9 – Juncaceae to Gramineae; a supplement is planned; maps; illus.)

Also relevant:

Webb, D.A. (1966). The Flora of European Turkey. *Proc. Roy. Irish Acad.* 65 Sec. B(1) 100 pp. (Checklist of 2006 vascular species, with localities for rare and local species; 2 maps.)

Atlases:

Anon. (1962). *Distribution of Forest Trees and Shrubs in Turkey*. Harita Genel
  Müdürlügü, Istanbul. (Maps.)
Anon. (1982-1983). *Türkiye Florasi Atlasi* (Atlas Florae Turcicae), 4 fasc. Istanbul.
Yakar, N. (1964-1966). *Renkli Türkiye Bitkileri Atlasi* (Coloured Atlas of Turkish
  Plants), 3 fasc. Istanbul.

Bibliographies:

Aytug, N. and Cakman, A. (1972). *Türkiye Flora Bibliografyasi (A Bibliography of
  Turkish Flora) 1843-1968*. Türdok Bibliografya Ser. 5., Türdok, Ankara.
Davis, P.H. and Edmondson, J.R. (1979). Flora of Turkey: a floristic bibliography.
  Notes Roy. Bot. Gard. Edinburgh 37(2): 273-283. (The literature base for the Flora
  of Turkey project.)
Demiriz, H. (in prep.). Bibliographia Phytotaxonomica et Geobotanica Turcica. (An
  annotated bibliography of c. 4000 entries.)
Zeybek, N. (1972). *A Bibliography of the Papers on the Taxonomy and Ecology of
  Turkish Flora (1841-1971)*. University of Ege, Izmir.

A list of endemic species, taken from Vols 1-6 of Davis' *Flora of Turkey* (1965-) is
included in the draft list of rare, threatened and endemic plants for North Africa and the
Middle East produced by the IUCN Threatened Plants Committee Secretariat (1980), cited
in Appendix 1.

Relevant journals: *Istanbul Eczacilik Fakültesi Mecmuasi* (Journal of the Faculty of
Pharmacy), Istanbul University; *Istanbul Üniversitesi Fen Fakültesi Mecmuasi* (Review of
the Faculty of Science), Istanbul Üniversitesi; *Ege Ü. Fen Fakültesi Dergisi* (Journal of the
Faculty of Sciences), Izmir; *L.U. Orman Fakültesi Dergisi* (Journal of the Faculty of
Forestry).

Field-guides
Baytop, T. and Mathew, B. (1984). *The Bulbous Plants of Turkey*. Batsford, London.
  132 pp. (Colour photographs.)
Taylor, S. (1984). *A Traveller's Guide to the Woody Plants of Turkey*. Redhouse
  Press, Istanbul. (Keys; line drawings.)
Yaltirik, F. (1981). *Dendroloji – I. Orman ve Parklarimidaki Bazi Yaprakli Agaç ve
  Calilarin Kisin Taninmasi* (The Identification of Native and Exotic Deciduous Trees
  and Shrubs in Winter). Bozak Matbasi, Istanbul. 205 pp. (Keys in English and
  Turkish; black and white photographs; line drawings.)
Yaltirik, F. (1984). *Türkiye Meseleri Teshis Kilavuzu* (Turkish Oaks). Yenilik Basimevi,
  Istanbul. (Keys; line drawings and colour figures.)

Information on Threatened Plants A nationally threatened plant list is currently
being prepared by many Turkish botanists under the co-ordination of T. Ekim of Ankara
University and with the support of the Turkish Association of Nature and Natural
Resources (address below). The conservation categories will follow those of IUCN. This is
the first list of its kind for Turkey and is due for publication in 1986. The following papers,
all but the last very short, are relevant:

Baytop, T. (1959). Les plantes rares de l'Anatolie et les précautions prises en vue et
  leur protection. In *Animaux et Végétaux Rares de la Région Méditérranéenne*.
  Proceedings of the IUCN 7th Technical Meeting, 11-19 September, 1958, Athens,
  vol. 5. IUCN, Brussels. Pp. 197-198.

Baytop, A. and Demiriz, H. (1980). Rare plants and endemics in Turkey-in-Europe. *Istanbul Univ. Fen Fak. Mec. Seri B* 45: 109-111. (Brief floristic account; lists 29 rare and threatened taxa.)

Birand, H. (1959). La végétation Anatolienne et la nécessité de sa protection. In *Animaux et Végétaux Rares de la Région Méditérranéenne*. Proceedings of the IUCN 7th Technical Meeting, 11-19 September, 1958, Athens, vol. 5. IUCN, Brussels. Pp. 192-196.

Demiriz, H. (1981). Nature Conservation in Turkey. *Threatened Plants Committee – Newsletter* 7: 21-22.

Demiriz, H. and Baytop, T. (1985). The Anatolian Peninsula. In Gómez-Campo, C. (Ed.) (1985), cited in Appendix 1. (Discusses endemism, threats to the flora, species case studies, local uses of plants.)

Laws Protecting Plants Two laws are relevant: the Forestry Law No. 6831 (1956) and the Law Protecting Cultural Heritage and Natural Resources No. 2863 (1983).

Voluntary Organizations

Cevre Koruma ve Yesillendirme Dernegi (Environment and Woodlands Protection Society), Mühürdarbagi Sokak 6, Kadiköy, Istanbul.

Dogal Hayati Koruma Dernegi (Society for the Protection of Wildlife), P.K. 18, Bebek, Istanbul.

Türkiye Cevre Sorunlari Vakfi (Environmental Problems Foundation of Turkey), Kennedy Caddesi 33-7, Kavaklidere, Ankara.

Türkiye Tabiatini Koruma Dernegi (Turkish Association for Conservation of Nature and Natural Resources), Menekse Sokak 29-4 Kizilay, Ankara.

Botanic Gardens

Atatürk Arboretum, Istanbul Üniversitesi, Orman Fakültesi Büyükdere, Istanbul.

Cukurova Üniversitesi Botanik Bahçesi, Ziraat Fakültesi, Peyzaj, Mimarisi Bölümü, Adana.

Ege Universitesi, Fen Fakültesi, Botanik Bahçesi, Bornova-Izmir.

Istanbul Üniversitesi Botanik Bahçesi, Süleymaniye, Istanbul.

Useful Addresses

Department of Wildlife and Hunting, The General Directorate of Forests, Ministry of Agriculture, Forest and Rural Development, Gazi Ciftliei, Ankara.

Additional References

Davis, P.H. (1971). Distribution patterns in Anatolia with particular reference to endemism. In Davis, P.H. *et al.* (Eds), *Plant Life of South-West Asia*. Botanical Society of Edinburgh, Edinburgh. Pp. 15-27. (Detailed floristic account; maps.)

Davis, P.H. (1975). Turkey: present state of floristic knowledge. In CNRS, 1975, cited in Appendix 1. Pp. 93-113. (Includes extensive bibliography; in English; French summary.)

Davis, P.H. (1979). Towards a supplement for the Flora of Turkey. *Webbia* 34(1): 135-141.

Kayacik, H. and Yaltirik, F. (1971). General aspects of Turkish forestry. In P.H. Davis *et al.* (Eds), *Plant Life of South-West Asia*. Botanical Society of Edinburgh, Edinburgh. Pp. 283-291. (Describes 8 major forest types and their conservation.)

Türköz, N. (1968). Conservation of Natural Resources: the example of forest management in Turkey. In *Proc. Tech. Meeting on Wetland Conservation*, 9-16 October, 1967. Ankara-Bursa-Istanbul. IUCN Publications, New Series No. 12. Morges, Switzerland. Pp. 23-24.

Wagenitz, G. (1963). Zur Kenntnis der Flora und Vegetation Anatoliens (Ergebnisse einer Reise im Herbst 1957). *Willdenowia* 3(2): 221-288.

Walter, H. (1956). Vegetationsgliederung Anatoliens. *Flora* 143(2): 295-326.

Turks and Caicos Islands

A group of islands south-east of the Bahamas and north of Haiti, covering 430 sq. km, with a population of 8000. A British Dependent Territory. Of the 8 Turks islands, two, Grand Turk and Salt Cay, are inhabited. The Caicos group has 6 principal islands, 35 km north-west of the Turks.

For botanical information, see under the Bahamas, with which the Turks and Caicos Islands are usually treated, e.g. in the Flora of Correll and Correll, 1982. No information on threatened plants. A conservation reference:

Ray, C. and Sprunt IV, A. (1971). *Parks and conservation in the Turks and Caicos Islands: A report on the ecology of the Turks and Caicos with particular emphasis upon the impact of development upon the natural environment.* Mimeo. 45 pp, 3 maps. Address of first author: John Hopkins University, 615 N. Wolfe St., Baltimore, Maryland 21205. (Mainly marine and littoral issues.)

Tuvalu

Tuvalu (formerly the Ellice Islands) consists of 9 inhabited atolls, of area 24.6 sq. km and population 8000, in the south-west central Pacific Ocean, between latitudes 5°-12°S and longitudes 175°-180°E. The largest island, Vaitupu, is only 6 sq. km. Most of the natural vegetation of Tuvalu has been cleared for cultivation, especially for copra production (Douglas, 1969, cited in Appendix 1). Funafuti (an atoll with 30 islets) contains a central mangrove swamp; *Scaevola* scrub on some islands (Douglas, 1969, cited in Appendix 1; Dahl, 1980, cited in Appendix 1).

No published Flora. No information on threatened plants.

References
Maiden, J.H. (1904). The botany of Funafuti, Ellice Group. *Proc. Linn. Soc. N.S.W.* 29: 539-556. (Includes annotated list of 55 vascular plant taxa.)

Uganda

Area 236,578 sq. km

Population 15,150,000

Floristics c. 5000 species (J.B. Gillett, 1984, pers. comm.). Endemism low compared with Kenya or Tanzania; Brenan (1978, cited in Appendix 1) estimates 30 endemic species from a sample of the *Flora of Tropical East Africa*. Species richness highest in Guinea-Congolian forests in south-west.

Floristic affinities predominantly with Lake Victoria region, but flora in c. 1/3 of area in north with Sudanian affinities. Extreme north-east: Somalia-Masai region. Forests 1000-1800 m in south-west have strong West African (Guinea-Congolian) affinities. Afroalpine and Afromontane elements on three separate mountain groups over 4000 m in the south-west and on Mt Elgon in the west.

Vegetation Most of the country is fire-climax secondary grassland and cultivated land, but much of this would develop into some sort of forest or denser woodland type in the absence of fire. The high mountains exhibit clear zonation of vegetation types from forest at lower altitudes, through bamboo thicket to heath thicket, moorland and finally to exposed rock with foliaceous lichens.

Estimated rate of deforestation for closed broadleaved forest 100 sq. km/annum out of 7500 sq. km (FAO/UNEP, 1981). Myers (1980, cited in Appendix 1) gives the same figure for the remaining amount of moist forest.

For vegetation map see White (1983), cited in Appendix 1.

Checklists and Floras Uganda is included in the incomplete *Flora of Tropical East Africa*. Uganda's plants of high altitudes are listed in *Afroalpine Vascular Plants* (Hedberg, 1957). These are both cited in Appendix 1.

Eggeling, W.J. and Dale, I.R. (1952). *The Indigenous Trees of the Uganda Protectorate*. (1st Ed. 1940, by W.J. Eggeling; 2nd Ed. revised and enlarged by I.R. Dale.) Govt Printer, Entebbe. 491 pp. (Keys; short descriptions, citation of specimens; local names; illus.)

Field-guides
Hamilton, A.C. (1981). *A Field Guide to Uganda Forest Trees*. Published privately, and available from the author, University, Coleraine, Londonderry, N. Ireland. 279 pp. (Keys, descriptions, local names, line drawings.)

Lind, E.M. and Tallantire, A.C. (in press). *Some Common Flowering Plants of Uganda*, 4th Ed. (first published 1962). Oxford Univ. Press, Nairobi. (Keys, line drawings, short descriptions.)

Information on Threatened Plants No published lists of rare or threatened plants; IUCN has records of 75 species and infraspecific taxa believed to be endemic, of which nine are known to be rare or threatened, information lacking for all the rest.

Katende, A.B. (1976). The problem of plant conservation and the endangered plant species in Uganda. In Miège, J. and Stork, A,L. (1975, 1976), cited in Appendix 1. Pp. 451-456.

Botanic Gardens
Entebbe Botanic Gardens, P.O. Box 40, Entebbe.
Makerere University Botanical Gardens, Botany Dept, Makerere University, P.O. Box 7026, Kampala.

Useful Addresses
The National Research Council, P.O. Box 6884, Kampala.
The Uganda Academy of Science, P.O. Box 16606, Kampala.
The Uganda Institute of Ecology, P.O. Box 22, Lake Katwe.

Additional References

Langdale-Brown, I., Osmaston, H.A. and Wilson, J.G. (1964). *The Vegetation of Uganda and its Bearing on Land-Use*. Govt Printer, Entebbe. 159 pp. (With coloured vegetation map 1:500,500.)

Osmaston, H.A. (1968). Uganda. In Hedberg, I. and O. (1968), cited in Appendix 1. Pp. 148-151.

Union of Soviet Socialist Republics

Area 22,400,000 sq. km

Population 275,761,000

Floristics According to a recent checklist (Czerepanov, 1981), the U.S.S.R. flora contains 21,119 species of higher plants, from 1865 genera in 200 families. No endemic families, but 92 endemic genera (5.5% of the flora), dominated by the Umbelliferae (26), Compositae (19), Cruciferae (7) and Campanulaceae (6) (Komarov and Shishkin, 1933-1964).

Species diversity and degree of endemism greatest in the south, especially the mountains of Middle Asia (Soviet Central Asia) and the Caucasus, which support 62 and 12 endemic genera respectively. This compares with 7 endemic genera in Kazakhstan, and only 4 in the Soviet Far East. Species diversity falls dramatically in the vast expanses of the north, which includes much of Siberia, where permanently frozen sub-soils or permafrost, occupy about 50% of the total land area. In European U.S.S.R. endemics are concentrated in the south-east; many are relict species.

19 floristic provinces throughout the country have been defined, mostly containing circum-Boreal floristic elements, but also sub-tropical ones, especially in the southern part of the Soviet Far East, together with Mediterranean and Irano-Turanian.

Main centres of species diversity: Carpathian Mts; southern shore of the Crimea; west and east Transcaucasus Region; western Kopet-Dag; Tian-Shan; Pamir mountain ranges; and Primorskiy Region. Many nationally rare species occur in these localities.

Vegetation 4 distinct vegetation zones (and subzones): tundra, forest, steppe and desert, with corresponding altitudinal belts. Unevenly distributed forests, both coniferous ('taiga') and mixed broadleaf and coniferous, occupy c. 33% of the country. Cold steppe and prairie vegetation dominate much of south central U.S.S.R., although prairie declines abruptly in East Siberia and becomes almost absent in the Soviet Far East. Desert and semi-desert vegetation covers the Caspian-Turanian Lowland (in the south-west of Asiatic U.S.S.R.) and large parts of Middle Asia; alpine vegetation confined to mountainous regions of the south and east, e.g Carpathians, Caucasus, Kopet-Dag, Pamirs including

Mt Communism (7495 m.), Dzungarskii Alatau, Tarbagataj, Tien Shan and the Altay and Sayan ranges. For a vegetation map see Herbich *et al.* (1970).

Checklists and Floras The national Flora, and associated papers are:

Czerepanov, S.K. (1973). *Additamenta et corrigenda ad "Floram URSS"*. Nauka, Leningrad. 668 pp. (Lists all taxonomic changes made throughout the U.S.S.R. since the publication of Komarov's *Flora URSS*, below.)

Czerepanov, S.K. (1981). *Plantae Vasculares URSS*. Nauka, Leningrad. 509 pp. (A checklist.)

Komarov, V.L. and Shishkin, B.K. (1933-1964). *Flora URSS*, 30 vols. Academy of Sciences of the U.S.S.R., Moscow and Leningrad. (A comprehensive and complete account covering the entire country; vols 1-13 reprinted in 1963 by Cramer, Weinham; vols 1-21 and 24 translated into English, 1963-1979, by N. Landau, Israel Program for Scientific Translations, Jerusalem.)

Flora Europaea (Tutin *et al.*, 1964-1980), cited in Appendix 1, covers European U.S.S.R., eastwards to the Ural Mountains, southwards to the Ural River and the Caspian Sea.

More than 30 Republican and other regional Floras have been published. For details see:

Frodin, D.G. (1984), cited in Appendix 1. Pp. 362-386.

Lebedev, D.V. (1956). *Introduction to the Botanical Literature of the USSR: manual for geobotanists*. Academy of Sciences of the U.S.S.R., Moscow and Leningrad. 382 pp.

Lipschitz, S.Y. (1975). *Florae URSS Fontes*. Nauka, Leningrad. 231 pp.

Sapiraite, S. (1971). *A Bibliography of Botanical Literature 1800-1965*. Lietuvos tsr Mokslu Akad. Centriue Biblioteka. 528 pp.

Other useful publications:

Federov, A. (1974-). *Flora Partis Europaeae URSS*, 5 vols to date. Nauka, Leningrad. (Incomplete; 1 – pteridophytes, gymnosperms, Gramineae; 2 – Orchidaceae to Commelinaceae; 3 – Caprifoliaceae to Lobeliaceae; 4 – Capparaceae to Typhaceae; 5 – Salicaceae to Plantaginaceae.)

Herbich, A.A. *et al.* (1970). Karta rastitel'nosti SSR m 1: 2,500,000 (principy, metody, sostojanie raboty po erropejskoj casti strany). (The vegetation map of the U.S.S.R., scale 1: 2,500,000 (the principles and methods of mapping; the status of the work relating to the European part of the U.S.S.R.).) *Bot. Zurn.* 55(11): 1634-1643. (In Russian.)

Sokolov, S.Y. *et al.* (1977-). *The Areas of Trees and Shrubs of the USSR*, 2nd Ed., 3 vols. Nauka, Leningrad. (Incomplete; 1 – (1977), 164 pp., 91 maps; 2 – (1980), 172 pp., maps; not seen.)

Field-guides

Alekseev, Y.E. *et al.* (1971). *Herbaceous Plants of the USSR*, 2 vols. Mysl, Moscow. (In Russian; illus.; map.)

Borodina, N.A. *et al.* (1966). *Trees and Shrubs of the USSR*. Mysl, Moscow. 637 pp. (In Russian; illus.)

Information on Threatened Plants 2 nationwide Red Data Books have been published:

Borodin, A.M. *et al.* (Eds) (1984). *Red Data Book of the USSR. Rare and Endangered Species of Animals and Plants*, 2nd Ed. Lesnaya Promyshlennost, Moscow. 478 pp. (In Russian; describes c. 600 plant taxa; maps; illus.)

Takhtajan, A. (Ed.) (1981). *Rare and Vanishing Plants of the USSR to be Protected*, 2nd Ed. Nauka, Leningrad. 261 pp. (In Russian; assesses more than 700 taxa rare or threatened throughout the U.S.S.R., covers distribution and threats; maps; illus.)

Also relevant:

Beloussova, L. and Denisova, L. (1981). The USSR Red Data Book and its compilation. In Synge (1981), cited in Appendix 1. Pp. 93-99. (Refers to the 1st edition of Borodin (1984).)

Denisova, L.V. and Beloussova, L.S. (1974). *Rare and Disappearing Plants of the USSR*. Moscow. 149 pp. (Illus.)

All 15 Republics have published threatened plant lists, and most Republics have official state plant Red Data Books:

Akademiya Nauk Arm. SSR. Botanicheskii Institut (1979). *List of Rare and Disappearing Species of the Flora of Armenia*. Erevan. 27 pp. (In Armenian.)

Andreev, G.N. *et al.* (1979). *The Animals and Plants of the Murmansk Region which are Rare and require Protection*. Murmansk. 160 pp. (In Russian; not seen.)

Bassiev, T.U. *et al.* (1981). *Red Data Book of the Northern Osetia*. Ordjonikidze. 88 pp. (In Russian; not seen.)

Beloussova, L.S. and Denisova, L.V. (1974). *Okhrana redkikh vidov rastenii v SSSR*. U.S.S.R. Ministry of Agriculture, Moscow. 70 pp.

Bijaschev, G.S. *et al.* (Eds) (1981). *Red Data Book of Kazakh SSR. Rare and Endangered Species of Animals and Plants. Part 2. Plants*. Nauka, Alma-Ata. 260 pp. (Maps; line drawings.)

Charkevitch, S.S. and Katchura, N.N. (1981). *Rare Plant Species of the Soviet Far East*. Nauka, Moscow. 230 pp. (Species accounts with maps; illus.; in Russian.)

Chopik, V.I. (1978). *Rare and Threatened Species of the Ukraine*. Kiev. 211 pp. (Describes over 150 species; maps; illus.; in Russian.)

Geideman, T.S. *et al.* (1982). *Rare Species of the Flora of Moldavia*. Kishinev. 80 pp. (Includes biological, ecological and geographical data; in Russian.)

Gorchakovsky, P.I. and Shurova, E.A. (1982). *Rare and Disappearing Plants of the Urals and Pre-Urals*. Nauka, Moscow. 207 pp. (Describes about 100 vascular plants.)

Jankevičius *et al.* (1981). *Red Data Book of the Lithuanian SSR. Rare and Endangered Species of Animals and Plants*. Mosklas, Vilnius. 84 pp. (Plant section prepared by J. Balevičiene and K. Balevičius; in Russian and Lithuanian.)

Katcharava, V.Ja. (Ed.) (1982). *Red Data Book of the Georgian SSR. Rare and Endangered Species of Animals and Plants*. Sabchota Sakartvelo, Tbilisi. 256 pp. (Editor of the plant section: N.N. Ketzkhoveli.)

Kononov, V.N. and Shabanova, G.A. (1978). *The Rare and Endangered Plants of Moldavia*. Moldavskoe Obshchestvo Okhrany Prirody, Kishinev. 27 pp. (Colour drawings; in Russian.)

Kozlov, V.L. *et al.* (Eds) (1981). *Red Data Book of the Byelorussian SSR. Rare and Endangered Species of Animals and Plants*. Byelorussian Soviet Encyclopedia, Minsk. 288 pp. (In Byelorussian and Russian; editors of the plant secton: V.I. Parfenov and N.V. Koslovsckaya.)

Kumari, E. (1982). *Red Data Book of the Estonian SSR*. Valgus, Tallinn. 248 pp. (In Estonian; Russian and English summaries.)

Litvinskaya, S.A., Tilka, A.P. and Filimonova, R.G. (1983). *Rare and Threatened Plants of Kuban*. Krasnodarskoye knizhnoe izdatelstvo, Krasnodar. 159 pp.

Lukss, Y.A., Privalova, L.A. and Kryukova, I.V. (Eds) (1975). *Catalogue of Rare, Vanishing and Extinct Plants of the Crimean Flora Recommended for Preservation.* Bulletin of the State Nikita Botanical Gardens, Yalta. 20 pp. and appendix of 174 pp. (In Russian, English summary; not seen.)

Malyshev, L.I. and Peshkova, G.A. (1979). *They Stand in Need of Conservation: Rare and Endangered Plants of Central Siberia.* Nauka, Siberian Branch, Novosibirsk. 174 pp. (In Russian.)

Malyshev, L.I. and Sobolenskaya, K.A. (Eds) (1980). *Rare and Endangered Plant Species of Siberia.* Nauka Novosibirsk. 223 pp. (Describes over 300 species; line drawings; maps; in Russian.)

Nabiev, M.M. (Ed.) (1984). *Red Data Book of the Uzbek SSR. Rare and Endangered Species of Animals and Plants. Vol. 2 – Plants.* FAN, Tashkent. 151 pp. (In Russian, with Uzbek and English preface.)

Nikitin, V.V. and Klyushkin, E.A. (1975). Plant species of the Turkmen SSR to be entered in the Red Book. *Izv. Akad. Nauk Turkmen SSR, Ser. Biol. Nauk.* 2: 73-76. (In Russian.)

Roshchevsky, M.P. *et al.* (Eds) (1982). *Rare Animals and Plants of the Komi ASSR in Need of Protection.* Syktyvkar. 152 pp. (In Russian; not seen.)

Seredin, N.M. (1980, 1981). Materials for the Red Data Book of the North Caucasus, Pre-Caucasus, Dagestan. *Izvestiya Severo-Kavkazskogo nauchnogo centra Vysshei Shkoly.* 2 parts, pp. 78-85, pp. 90-94. (In Russian; not seen.)

Shvedchikova, N.K. (1983). On the new and rare species in the Crimean flora. *Byull. Mosk. Obschch. Ispyt. Prir. Biol.* 88(22): 122-128.

Sytnik, K.M. *et al.* (Eds) (1979). *Red Data Book of the Ukrainian SSR.* Academy of Sciences of the Ukrainian SSR. Naukova Dumka, Kiev. 497 pp. (Covers 151 species of higher vascular plants; maps and line drawings.)

Tkachenko, V.I., Assorina, I.A. (1978). *Rare and Endangered Plant Species of the Kirgizia Wild Flora.* Frunze. 128 pp. (In Russian; not seen.)

Wintergoller, B.A. (1976). *Rare Plants of Kazakstan.* Nauka. 197 pp. (In Russian.)

Yunushov, S. and Kamelin, R.V. (1980). *Materials for the Red Data Book of the Tadzhik SSR. Rare and Endangered Animals and Plants.* Dushanbe. 32 pp. (In Russian; not seen.)

A plant Red Data Book for the Armenian SSR is in press (E. Gabrielian, 1984, *in litt.*).

In addition to those above, lists of plants in need of protection have been published for many individual administrative regions of the Russian Soviet Federal Socialist Republic (RSFSR), namely Bryansk, Murman, Ryazan, Yaroslavl, Ivanovo, Vladimir, Saratov, Leningrad, Tomsk, Gorky, Kuibishev, the Stravopol territory and the Karelian Autonomous Soviet Socialist Republic (Tikhomirov, 1981). These registers are essentially scientific recommendations to encourage local authorities to pass special ordinances to regulate the use of wild plants, especially to prohibit their sale, give protection to certain species and to establish a system of protected areas.

Eventually it is intended that all Republics will have state Red Data Books in addition to the official national one produced by the Ministry of Agriculture (Borodin, *et al.*, 1984). The smaller Autonomous Socialist Republics also have the authority to publish their own official Red Data Books. To date, only the North Ossetian and Daghestan Autonomous Socialist Republics have done so.

Other general works:

Beloussova, L. (1977). Endangered plants of the USSR. *Biol. Conserv.* 12: 1-11.

Beloussova, L.S., Denisova, L.V. and Nikitina, S.V. (1979). *Rare Plants of the USSR*. Lesnaya Promyshlennost, Moscow. c. 150 pp. (In Russian.)

Chopik, V.I. (1970). Scientific grounds for the protection of rare species in the Ukrainian flora. *Ukr. Bot. Zh.* 27: 693-704.

Elias. T.S. (1983). Rare and endangered species of plants – the Soviet side. *Science* 219: 19-23.

Golubev, V.N. and Molchanov, E.F. (1978). *Systematic Instructions on the Population Quantity and Eco-biological Studies of Rare, Disappearing and Endemic Plants of the Crimea*. Nikita Botanical Garden, Yalta. (In Russian.)

Lapin, P.I. (1975). Our endangered environment – the Russian view: rare and endangered plant species in the USSR. *Gard. J.* 25(6): 171-175.

Shelyag-Sosonko, Y.R., Sytnik, K.M. *et al.* (1980). *Ukraine, White Russia, Moldavia. The Protection of Important Botanical Regions*. Akademiya Nauk Ukrainskoi SSR. Naukova Dumka, Kiev. (In Russian; pp. 332-370 include lists of threatened plants by region.)

Tikhomirov, V.N. (1981). Regional rare plant conservation schemes in the USSR. In Synge, H. (1981), cited in Appendix 1. Pp. 101-104.

Laws Protecting Plants In 1977, the Supreme Soviet adopted the new Constitution of the U.S.S.R. in which nature conservation was elevated to the highest state level. Articles 18 and 64 are particularly relevant. In 1983, an official nature conservation statement entitled 'About the USSR Red Data Book' (no. 313, 12th April 1983) was published in 'A Collection of Decisions of the Government of the USSR' (Sobranie postanovleniy Pravitelstva Soyusa Sovetskich Sotsialistitcheskick Respublik) N.12, Clause 56, pp. 220-221.

Each Republic has passed its own plant protection laws. Details of some of these are as follows: in the Byelorussian SSR 85 taxa receive protection; Estonia SSR: c. 100; Kirghiz SSR: 45; Latvia SSR: 109; Lithuania SSR: 177; Moldavia SSR: 60; Uzbek SSR: 80.

Under RSFSR Environmental Law: article 6 provides for special consideration to be given to the conservation and protection of existing or potential wild genetic stocks of food plants.

Relevant publications:

Birkmane, K. (1974). *Okhraniaemye vidy rastenii Latviiskoi SSR* (Protected plants in Latvia). Riga, Zinatne. 58 pp. (In Russian.)

Kryukova, I.V., Lukss, Y.A. and Privalova, L.A. (1980). *Preserved Plants of the Crimea: A Handbook*. Tavriya, Simferopol. 96 pp. (In Russian; not seen.)

Zaveruha, B.V., Andrienko, T.L. and Protopopova, V.V. (1983). *Protected Plants of the Ukraine*. Naukova dumka, Kiev. 175 pp.

Voluntary Organizations
All-Russia Society for the Protection of Nature, Proezd Kujbysheva, 103012 Moscow.
All-Union Botanical Society, 2 Prof. Popov Street, 197022 Leningrad.

Botanic Gardens Numerous (see Henderson, 1983, cited in Appendix 1). 116 U.S.S.R. botanic gardens collaborate with the IUCN Botanic Gardens Conservation Coordinating Body through the Moscow Main Botanic garden:

Moscow Main Botanic Garden of the Academy of Sciences, Herbarium, Moscow U-274.

A major work, the first of its kind, contains contributions from many of these gardens:

Lapin, P.I. (Ed.) (1983). *Rare and Endangered Species of the USSR Flora Cultivated in Botanical Gardens and other Introduction Centres of the USSR*. Nauka, Moscow. 301 pp. (Lists threatened plant holdings in 94 Soviet botanic gardens; English summary.)

See also:

Antonyuk, N.E., Borodina, R.N., Sobko, V.G. and Skvortsova, L.S. (1982). *Rare Plants of the Flora of the Ukraine in Cultivation*. Central Republic Botanic Gardens, Kiev. 212 pp. (Based on information received from 94 botanic gardens and contains data on 117 native threatened plant species cultivated in the U.S.S.R.)
Gogina, E.E. (1979). USSR: The policies of botanic gardens and their activities in the conservation of threatened plants. In Synge and Townsend (1979), cited in Appendix 1. Pp. 141-148.
Winterholler, B.A. (1979). USSR: Rare and threatened plants and their conservation in the botanic gardens of Kazakhstan. In Synge and Townsend (1979), cited in Appendix 1. Pp. 149-152.

Useful Addresses

All-Union Research Institute for Nature Conservation and Nature Reserves Studies, Ministry of Agriculture of the U.S.S.R., Sadki-Znamenskoye, Leninsky District, P/O Vilar, Moscow Region 142790.
Komarov Botanical Institute, Academy of Sciences of the U.S.S.R., 2 Prof. Popov Street, Leningrad 197022.
CITES Management Authority: Main Administration for Nature Conservation, Nature Reserves, Forestry and Game Management (GLAVPRIRODA), Ministry of Agriculture of the U.S.S.R., Orlikov per 1/11, Moscow 107139.
CITES Scientific Authority: All-Union Research Institute (as above).

Additional References

Kirpicznikov, M.E. (1969). The Flora of the USSR. *Taxon* 18(6): 685-708.
Pryde, P.R. (1972). *Conservation in the Soviet Union*. Cambridge Univ. Press, London. 301 pp.

United Arab Emirates

Includes Abu Dhabi, Ajman, Dubai, Fujairah, Ras al Khaimah, Sharjah and Umm al Qaiwain; formerly the Trucial States.

Area 75,150 sq. km

Population 1,255,000

Floristics Flora likely to be small, with few, if any, endemics; similar to that of Iran.

Vegetation Mostly desert with little or no vegetation; some wooded steppes between the desert foreland and the Hajar Mts in the east (bordering Oman); the hillsides of the Hajar Mts are mostly barren, but there is some dry scrub with occasional trees in wadis and damp hollows (Satchell, 1978; Vesey-Fitzgerald, 1957).

Checklists and Floras None specifically on U.A.E. Works relating to the Arabian peninsula as a whole are outlined under Saudi Arabia.

Field-guides Occasional field-guides to specific groups are published in the *Bulletin of the Emirates Natural History Group.*

Information on Threatened Plants None.

Voluntary Organizations
Emirates Natural History Group (Abu Dhabi), P.O. Box 2687, Abu Dhabi. (Publishes a bulletin three times a year.)

Useful Addresses
CITES Management Authority: Ministry of Agriculture and Fisheries, P.O. Box 213, Abu Dhabi; or P.O. Box 1509, Dubai.

Additional References
Kahn, M.I.R. (1982). Mangrove forest in the U.A.E. *Pakistan J. Forest.* 32(2): 36-39.
Satchell, J.E. (1978). Ecology and environment in the United Arab Emirates. *J. Arid Environ.* 1: 201-226. (Includes physical description, provisional vegetation map and extensive bibliography, mostly on fauna.)
Vesey-Fitzgerald, D.F. (1957). The vegetation of central and eastern Arabia. *J. Ecol.* 45: 779-798. (With 4 black and white photographs and small-scale vegetation map.)

United Kingdom

(Great Britain and Northern Ireland)

Area 244,754 sq. km

Population 55,624,000

Floristics 1700-1850 native vascular species, estimated by D.A. Webb (1978, cited in Appendix 1), from *Flora Europaea*; 16 endemic species (IUCN figures). Elements: Atlantic, with small alpine component in mountains of north and west.

Vegetation Principally an agricultural landscape, mainly of cultivated land in the lowlands and moorland in the uplands. Originally, two-thirds covered by natural forest, principally of oak (*Quercus robur* and *Q. petraea*), lime, birch and ash giving way to beech in the south, and Scots Pine in the north. Much forest cleared in neolithic times; today only small areas remain ('Ancient Woodland'), most of it modified by centuries of coppicing and browsing. Of the forest cover of 8.5% (Anon, 1983), most is recent plantation forestry, poor in native species. Plant-rich communities: calcareous grasslands, coastal sand-dunes, salt-marshes, shingle beaches, cliff communities, lowland heathlands, peat bogs and fenland.

Checklists and Floras Included in the completed *Flora Europaea* (Tutin *et al.*, 1964-1980, cited in Appendix 1), but Northern Ireland is treated as part of Ireland rather than the British Isles. Numerous national Floras, the present standard work being:

Clapham, A.R., Tutin, T.G. and Warburg, E.F. (1962). *Flora of the British Isles*, 2nd Ed. (3rd Ed. in prep.) Cambridge Univ. Press, London. 1269 pp. (For a more recent and abbreviated paperback edition see *Excursion Flora of the British Isles*, 3rd Ed., 1981, 499 pp.)

'County Floras', in effect detailed checklists with localities, are too numerous to list. Earliest is Ray's *Flora of Cambridgeshire* (1660), although most 'County Floras' were compiled during the 19th century. Today, many continue to be revised by amateur botanists.

Since 1941 more than 100 detailed ecological species case-studies have been published in the *Journal of Ecology* in a series 'The Biological Flora of the British Isles'. (Includes many accounts of rare and threatened taxa.)

For Wales see:

Ellis, R.G. (1983). *Flowering Plants of Wales*. National Museum of Wales, Cardiff. 338 pp. (An annotated checklist of vascular plants; habitat details; distribution dot maps; a valuable historical and floristical introduction.)

For Northern Ireland see:

Scannell, M.J.P. and Synnott, D.M. (1972). *Census Catalogue of the Flora of Ireland*. Stationery Office, Dublin. 127 pp. (Checklist for both the Republic and Northern Ireland; natives and aliens; new edition in prep.)
Stewart, S.A. and Corry, T.H. (1938). *A Flora of the North-East of Ireland*, 2nd Ed. (By R.L. Praeger and W.R. Megaw). Quota Press, Belfast. 472 pp.
Webb, D.A. (1977). *An Irish Flora*, 6th Ed. Dundalgan Press, Dundalk. 277 pp.

Bibliography:

Kent, D.H. (1967). *Index to Botanical Monographs. A Guide to Monographs and Taxonomic Papers Relating to Phanerogams and Vascular Cryptogams found Growing Wild in the British Isles*. Academic Press, London. 163 pp.

Atlases:

Fitter, A. (1978). *An Atlas of the Wild Flowers of Britain and Northern Europe*. Collins, London. 272 pp.
Jermy, A.C., Arnold, H.R., Farrell, L. and Perring, F.H. (1978). *Atlas of Ferns of the British Isles*. Botanical Society of the British Isles and the British Pteridological Society, London. 101 pp. (Dot maps and distribution notes.)
Perring, F.H. and Walters, S.M. (Eds) (1982). *Atlas of the British Flora*, 3rd Ed. Ebury Press, Wakefield. 432 pp.

National botanical journal: *Watsonia* (Journal of the Botanical Society of the British Isles), covers taxonomy, floristics, conservation and history of flora.

Field-guides Innumerable field-guides published in recent years. A selection are listed below:

Fitter, R., Fitter, A. and Blamey, M. (1974). *The Wild Flowers of Britain and Northern Europe*. Collins, London. 336 pp.
Keble Martin, W. (1982). *The New Concise British Flora*. Ebury Press and Michael Joseph, London. 247 pp.
McClintock, D. and Fitter, R.S.R. (1965). *The Pocket Guide to Wild Flowers*. Collins, London. 340 pp.
Mitchell, A. (1974). *A Field Guide to the Trees of Britain and Northern Europe*. Collins, London. 415 pp.

Page, C.N. (1982). *The Ferns of Britain and Northern Ireland*. Cambridge Univ. Press, Cambridge. 447 pp. (Maps; illus.)

Phillips, R. (1977). *Wild Flowers of Britain*. Ward Lock, London. 191 pp.

Rose, F. (1981). *The Wild Flower Key*. Warne, London. 480 pp. (Covers British Isles and northwest Europe.)

Schauer, T. (1982). *A Field Guide to the Wild Flowers of Britain and Northern Europe*. Collins, London. 464 pp.

Information on Threatened Plants The Biological Records Centre (BRC), at Monks Wood Experimental Station, Huntingdon, co-ordinates a computerized botanical monitoring scheme for all species throughout the country. Its foundation was a mapping scheme organized by the Botanical Society of the British Isles (BSBI), the results of which were published in 1962 as the first edition of the *Atlas of the British Flora*. The scheme used a 10 km square matrix system, used today by the Nature Conservancy Council to monitor the status of threatened plants. In 1977 BRC produced Britain's first national plant Red Data Book, recently revised:

Perring, F.H. and Farrell, L. (1983). *British Red Data Books: 1. Vascular Plants*, 2nd Ed. RSNC, Lincoln (address below). 99 pp. (Identifies more than 300 rare and threatened taxa, 17.6% of the native flora, and summarizes status of and threats to each species.)

A vast quantity of literature is available, so only a selection of the major works are included here:

Crompton, G. (1981). Surveying rare plants in Eastern England. In Synge (1981), cited in Appendix 1. Pp. 117-124.

Ellis, E.A., Perring, F. and Randall, R.E. (1977). *Britain's Rarest Plants*. Jarrold, Norwich. 41 pp. (Ecological species descriptions; photographs.)

Milne-Redhead, E. (Ed.) (1963). *The Conservation of the British Flora, BSBI Conference Report No. 8*. BSBI, London (address below). 90 pp.

Perring, F. (Ed.) (1970). *The Flora of a Changing Britain, BSBI Conference Report No. 11*. Classey, Hampton. 157 pp. (Examines factors responsible for change in the British flora since 1945 and predicts further changes.)

Perring, F.H. (1971). Rare plant recording and conservation in Great Britain. *Boissiera* 19: 73-79.

Perring, F. and Randall, R.E. (1981). *Britain's Endangered Plants*. Jarrold, Norwich. 42 pp. (Describes 42 taxa; illus.)

Perring, F.H. and Walters, S.M. (1971). Conserving rare plants in Britain. *Nature* 229(5284): 375-377.

Ratcliffe, D.A. (1979). The role of the Nature Conservancy Council in the conservation of rare and threatened plants in Britain. In Synge and Townsend (1979), cited in Appendix 1. Pp. 31-35.

Wells, D.A. (1981). The protection of British rare plants in nature reserves. In Synge (1981), cited in Appendix 1. Pp. 475-480.

Included in the European threatened plant list (Threatened Plants Unit, 1983, cited in Appendix 1); latest IUCN statistics, based upon this work: endemic taxa – Ex:1, E:1, V:2, R:7, I:1, nt:4; non-endemics rare or threatened worldwide – E:1, V:4, R:3 (world categories).

In 1982 IUCN, under contract to the EEC through the U.K. Nature Conservancy Council, prepared a report (unpublished), *Threatened Plants, Amphibians and Reptiles, and*

Mammals (excluding Marine Species and Bats) of the European Economic Community, which includes data sheets on 2 plants Endangered in Britain.

Laws Protecting Plants The 1981 Wildlife and Countryside Act repeals and re-enacts the 1975 Conservation of Wild Creatures and Wild Plants Act. It lists 62 plant species for special protection and prohibits intentional picking, uprooting, destruction or sale of these plants, or any part of them including their seed. Under the Act, uprooting any native British plant is also prohibited, unless permission has been granted from the land-owner or occupier. The introduction of 4 non-native species is also prohibited: *Heracleum mantegazzianum*, *Polygonum cuspidatum*, *Macrocystis pyrifera* and *Sargassum muticum*. For further details see:

Council for Environmental Conservation (CoEnCo) (1982). *Wildlife and the Law No. 1: Wild Plants*. 4 pp. (Summarizes Act and lists protected species; illus.)
Donald, D. (1982). Conservation. The BSBI's policy towards plant conservation today. *BSBI News* 31: 8-9. (A code of conduct and list of rare species is given on pp. ii-vii.)
Nature Conservancy Council (1982). *Wildlife. The Law and You*. Interpretative Branch, NCC (address below). 15 pp. (A popular guide to wildlife legislation; lists plant species protected; illus.)
Sands, T.S. (1981). Wildlife and Countryside Act 1981. *BSBI News* 29: 16-17. (Summarizes plant sections of Act; lists protected species.)

In Northern Ireland there is no legal protection for plants, but a list is in preparation.

Voluntary Organizations Numerous. Each county in England and Wales has its own trust for nature conservation, co-ordinated by the Royal Society for Nature Conservation (address below). Other voluntary bodies include:

Botanical Society of the British Isles (BSBI), c/o Department of Botany, British Museum (Natural History), Cromwell Road, London SW7 5BD.
Botanical Society of Edinburgh (National Botanical Society for Scotland), c/o Royal Botanic Garden, Inverleith Row, Edinburgh EH3 5LR.
British Ecological Society, Burlington House, Piccadilly, London WIV 0LQ.
British Pteridological Society, 42 Lewisham Road, Smethwick, Worly, West Midlands B66 2BS.
Fauna and Flora Preservation Society (FFPS), c/o The Zoological Society of London, Regent's Park, London NW1 4RY.
National Council for the Conservation of Plants and Gardens (NCCPG), Royal Horticultural Society's Garden, Wisley, Woking, Surrey GU23 6QB.
National Trust, 42 Queen Anne's Gate, London SW1H 9AS.
National Trust for Scotland, 5 Charlotte Street, Edinburgh EH2 4DU.
Royal Society for Nature Conservation (RSNC), The Green, Nettleham, Lincoln, Lincolnshire LN2 2NR.
Scottish Wildlife Trust, 25 Johnston Terrace, Edinburgh EH1 2NH.
Woodland Trust, Westgate, Grantham, Lincolnshire NG31 6LL.
WWF-United Kingdom, Panda House, 11-13 Ockford Road, Godalming, Surrey GU7 IQU.

Botanic Gardens Numerous; listed in Henderson (1983), cited in Appendix 1; only subscribers to the Botanic Gardens Conservation Coordinating Body are given below:

Bath Botanical Gardens, Bath City Council, Pump Room, Bath BA1 1LZ.
Cambridge University Botanic Garden, 1 Brookside, Cambridge CB2 1JF.

Chelsea Physic Garden, 66 Royal Hospital Road, London SW3 4HS.

City of Liverpool Botanic Garden, The Mansion House, Calderstones Park, Liverpool L18 3JD.

Glasgow Botanic Gardens, Glasgow G12 OUE.

Harlow Car Gardens, The Northern Horticultural Society, Harrogate, North Yorkshire HG3 1QB.

New University of Ulster Botanic Garden, Coleraine, Co. Londonderry NT52 1SA, Northern Ireland.

Paignton Botanical Gardens, Totnes Road, Paignton, Devon.

Plant Science Botanic Garden, University of Reading, Whiteknights, Reading RG6 2AS.

Royal Botanic Garden, Inverleith Row, Edinburgh EH3 5LR.

Royal Botanic Gardens, Kew, Richmond, Surrey TW9 3AE.

Royal Horticultural Society's Garden (see NCPPG, under Voluntary Organizations).

Tresco Abbey Gardens, Tresco, Isles of Scilly, Cornwall.

University of Birmingham Botanic Garden, Department of Plant Biology, The University, P.O. Box 363, Birmingham B15 2TT.

University of Bristol Botanic Garden, Department of Botany, Woodland Road, Bristol BS8 1UG.

University of Durham Botanic Garden, Surveyor's Department, Hollow Drift, Green Lane, Durham DH1 3LA.

University of Hull Botanic and Experimental Garden, Department of Plant Biology, University of Hull, Hull HU6 7RX.

University of Liverpool Botanic Gardens, Ness, South Wirral L64 4AY.

Westonbirt Arboretum (Forestry Commission Research), Tetbury, Glos. GL8 8QS.

Useful Addresses

Biological Records Centre (BRC), Institute of Terrestrial Ecology, Monks Wood Experimental Station, Abbots Ripton, Huntingdon, Cambridgeshire PE17 2LS.

Countryside Commission for England and Wales, John Dower House, Crescent Place, Cheltenham, Gloucestershire Gl50 3RA.

Countryside Commission for Scotland, Battleby, Redgorton, Perth PH1 3EW.

Eastern England Rare Plant Project, 1 Brookside, Botanic Gardens, Cambridge, Cambridgeshire CB2 1JF.

Nature Conservancy Council (NCC), Northminster House (National Headquarters and Headquarters for England), Peterborough PE1 1UA; Ffordd Penrhos, Bangor, Gwynedd LL57 2LQ (Headquarters for Wales); 12 Hope Terrace, Edinburgh EH9 2AS (Headquarters for Scotland).

CITES Management Authority: Department of the Environment, Tollgate House, Houlton Street, Bristol BS2 9DJ.

CITES Scientific Authority: The Herbarium, Royal Botanic Gardens, Kew, Richmond, Surrey TW9 3AE.

Additional References

Anon (1983). A conservation/development programme for UK forestry. In Anon, *The Conservation and Development Programme for the UK: a response to the World Conservation Strategy*. Kogan Page, London. Pp. 215-224.

Harvey, J.L. and Lewis, D.H. (1985). *The Flora and Vegetation of Britain: Origins and Changes – the Facts and the Interpretation*. Proceedings of a symposium to honour A.R. Clapham, University of Sheffield, May 1984. Academic Press, London. 129 pp. (Includes floristic and ecological aspects of contemporary British vegetation, including conservation topics; reprinted from *New Phytologist* 1 (1984).)

Hawksworth, D.L. (Ed.) (1974). *The Changing Flora and Fauna of Britain*. Academic Press, London. 461 pp.

Kent, D.H. (1974). Progress in the study of the British flora, 1961-1971. *Mem. Soc. Brot.* 24(1): 353-375.

Mabey, R. (1980). *The Common Ground*. Hutchinson, London. 280 pp. (Also in paperback by Arrow Books, 1981, 266 pp.)

Pennington, W. (1974). *The History of British Vegetation*, 2nd Ed. English Universities Press, London. 152 pp.

Peterken, G.F. (1981). *Woodland Conservation and Management*. London. 328 pp. (Illus; maps.)

Pigott, C.D. (1984). The Flora and Vegetation of Britain: Ecology and Conservation. In Proceedings of a Symposium "The Flora and Vegetation of Britain: Origins and Changes – the facts and their interpretation", 19 May 1984. *New Phytologist* 98(1): 119-128.

Polunin, O. and Walters, M. (1985), cited in Appendix 1.

Praeger, R.L. (1934). *The Botanist in Ireland*. Hodges, Figgis and Co., Dublin. 587 pp. (Detailed habitat descriptions; maps; black and white photographs.)

Rackham, O. (1976). *Trees and Woods in the British Landscape*. Dent, London. 204 pp. (Describes woodland evolution, management and conservation; black and white photographs; maps.)

Ratcliffe, D. (Ed.) (1977). *A Nature Conservation Review. The Selection of Biological Sites of National Importance to Nature Conservation in Britain*, 2 vols. Cambridge Univ. Press, Cambridge. (1 – vegetation communities; 2 – site accounts, with evaluations of national conservation importance.)

Tansley, A.G. (1968). *Britain's Green Mantle*, 2nd Ed. revised by M.C.F. Proctor. Allen and Unwin, London. 326 pp. (Detailed descriptions of all major British habitats; black and white photographs; line drawings.)

United Kingdom: Channel Islands

Five inhabited islands – Jersey, Guernsey, Alderney, Sark and Herm – and numerous uninhabited islets, 16-48 km off the north-west coast of France.

Area 194 sq. km

Population 137,000

Floristics About 1800 vascular taxa (Bichard and McClintock, 1975); 1340 taxa in Guernsey (McClintock, 1975); over 1500 species in Jersey (Le Sueur, 1985). The islands support a large alien flora. Close affinities with the floras of Normandy and Brittany.

Vegetation Considerable cover of semi-natural vegetation still present, including sand dune communities, cliff-top heathlands, grassland, wetlands (including water meadows) and woodlands, although the latter are rather scarce. Sand dunes and coastal grasslands increasingly threatened by tourist developments; the few remaining wetlands now threatened by drainage schemes.

Checklists and Floras See under United Kingdom and:

Le Sueur, F. (1985). *Flora of Jersey*. Société Jersiaise. 243 pp. (Over 1500 species described, including ferns; introduction describes botanical history of the island and influence of man on the flora; text covers localities, habitats and species abundance; 600 distribution maps; colour paintings; black and white photographs.)

McClintock, D. (1975). *The Wild Flowers of Guernsey with Notes of the Frequencies of all Species Recorded for the Channel Islands*. Collins, London. 288 pp. (Pteridophytes, gymnosperms, angiosperms; line drawings.)

Field-guides

Bichard, J.D. and McClintock, D. (1975). *Wild Flowers of the Channel Isles*. Chatto and Windus, London. 33 pp. (Colour photographs and descriptions of 100 selected species.)

Ellis, E.A. (1977). *Wild Flowers of the Channel Islands*. Jarrolds, Norfolk. (2 parts.)

Page, J. and Ryan, P. (1982). *Wild Flowers of Guernsey*. (Not seen.)

Information on Threatened Plants See the British Plant Red Data Book (Perring and Farrell, 1983), under United Kingdom.

Laws Protecting Plants Not covered by U.K. legislation. No protected plant list, although on the island of Herm picking of wild flowers is prohibited.

Voluntary Organizations

Jersey Wildlife Preservation Trust, Les Augrès Manor, Jersey.

National Trust of Guernsey, Les Mouilpiedes, St Martins, Guernsey.

National Trust of Jersey, The Elms, St Mary, Jersey.

Société Guernesiaise, La Couteur House, Collings Road, St Peter Port, Guernsey.

Société Jersiaise, 9 Pier Road, St Helier, Jersey.

Useful Addresses

Conservation Officer, Planning Department, South Hill, St Helier, Jersey.

Additional References

Le Sueur, F. (1973). Changes in the Flora of Jersey, 1873-1973. *Bull. Soc. Jersiaise* 21(1): 33-40. (Descriptive account of species and habitat loss; distribution dot maps for 12 species.)

Le Sueur, F. (1976). *A Natural History of Jersey*. Phillimore, London and Chichester. 221 pp.

United States

By Patrick R. Gregerson and Robert A. DeFilipps

Area 9,363,132 sq. km

Population 235,681,000

Floristics Approximately 20,000 species of flowering plants native to continental U.S.A. Major centres of narrow endemism are in: Florida (James, 1961, Long, 1974), the southern Appalachian Mountains (Holt, 1971), Trans-Pecos Texas, California (Stebbins, 1978) and Hawaii (covered separately); also many endemics in the southwestern semi-arid region (Nevada-Arizona-Utah), and in the Pacific Northwest (Oregon and Washington);

limited endemism in the northeastern states, with some in the Ohio River Valley and also very localized in the unglaciated parts of the Central States region (Hartley, 1966); endemism lowest in the extensive Great Plains region.

Vegetation There are three standard references to the overall vegetation of the United States: Küchler (1964), Bailey (1976, 1978), and Gleason and Cronquist (1964). See also regional accounts such as Waggoner (1975) and Braun (1964), and basic texts such as Oosting (1956) and Benson (1979). Below is a simplified summary of the major vegetation types outlined in these works:

In Alaska, tundra of low scrub dominated by willows and birches, plant cover decreasing northwards. Also in Alaska large areas of coniferous forest, mainly of pines, spruces and firs, in a belt stretching over much of Canada and down the Pacific coast to central California; montane coniferous forest occurs on the Rockies, the Appalachians and on other mountain ranges; in eastern North America, original cover was of deciduous forest, of many different associations that variously include bald cypress, hemlock, hickory, maple and oak, much now cleared or recently secondary; in the centre, from the Rockies east to Indiana, and from Canada south to Mexico, originally a massive belt of grassland (prairie) of which only small relicts have survived conversion to agriculture; in the west, from central Washington south to Mexico, deserts, including the Great Basin, Mojave and Sonoran Deserts, many rich in endemic succulents; on coastal California, the chaparral, a 1-3 m high dense scrub similar in structure (though not in flora) to that in other regions of Mediterranean climate; in the southern tip of Florida, subtropical vegetation including mangrove.

Anderson (1977) gives the area covered by each vegetation type (following the classification in Oosting, 1956). Klopatek *et al.* (1979) present a map of the U.S. showing loss in natural vegetation types in each county, based on Küchler's 1964 map of potential natural vegetation. Twenty-three of Küchler's 106 predominant potential vegetation types have lost more than 50% of their potential area, including the Florida Everglades, California steppe, southern floodplain forest, bluestem prairie and beech-maple forest types. Additionally, the *Forest Atlas of the Midwest* by Merz (1978) depicts the remaining forested land in the counties of 9 states, but this includes commercial forests.

Approximately one-third (740 million acres) of the United States is owned and administered (therefore, under nominal protection) by federal government agencies (National Geographic Society, 1982), much of it in the western third of the nation, for example including nearly all Nevada and much of Idaho, Utah and Alaska.

Checklists and Floras There is no modern national Flora. The *Flora North America* Project was suspended in 1972; a new effort to write such a Flora through collaboration of many botanical institutions is being organized from the Missouri Botanical Garden. There are, however, three recent checklists, in order of appearance:

Kartesz, J.T. and Kartesz, R. (1980). *A Synonymized Checklist of the Vascular Flora of the United States, Canada and Greenland.* Univ. of North Carolina Press, Chapel Hill, North Carolina. 544 pp. (Currently being updated for a new edition.)

Shetler, S.G. and Skog, L.E. (Eds) (1978). *A Provisional Checklist of Species for Flora North America (Revised).* Missouri Botanical Garden, St Louis, Missouri. 199 pp. Flora North America Report 84. (Computer printout of 16,274 species of native and naturalized vascular plants.)

U.S.D.A. Soil Conservation Service (1982). *National List of Scientific Plant Names,* 2 vols. U.S. Soil Conservation Service, Washington, D.C. SCS-TP-159. 416 pp, 438 pp. (1 – List of Plant Names; 2 – Synonymy.)

See also:

Little, E.L. (1971-1981). *Atlas of United States Trees*, 6 vols. U.S. Forest Service, Washington, D.C. (Base maps, dot maps.)

The principal state and regional Floras, and a few of a more specialized nature, are given below. Others are listed in U.S.D.A. Soil Conservation Service (1982), vol. 1, pp. 2-3, cited above.

Barkley, T.M. (1978). *A Manual of the Flowering Plants of Kansas*, 2nd Ed. State Univ. Endowment Assoc., Manhattan.

Benson, L. (1982). *The Cacti of the United States and Canada*. Stanford Univ. Press, Palo Alto, California. 1044 pp.

Correll, D.S. and Johnston, M.C. (1970). *Manual of the Vascular Plants of Texas*. Texas Research Foundation, Renner, Texas. 1881 pp.

Cronquist, A. *et al.* (1972-). *Intermountain Flora: Vascular Plants of the Intermountain West, U.S.A.* Hafner Publishing and Columbia Univ., New York.

Davis, R. (1952). *Flora of Idaho*. Brown, Dubuque.

Dorn, R.D. (1977). *Manual of the Vascular Plants of Wyoming*. Garland Publishing.

Fernald, M.L. (1970). *Gray's Manual of Botany*, 8th Ed. Revised by R.C. Rollins. Van Nostrand Reinhold, New York. (Corrected printing of 1950 edition.)

Gleason, H.A. (1952, 1963). *The New Britton and Brown Illustrated Flora of the Northeastern United States and Adjacent Canada*. Hafner, New York.

Gleason, H.A. and Cronquist, A. (1963). *Manual of Vascular Plants of the Northeastern United States and Adjacent Canada*. Van Nostrand, Princeton, New Jersey. 810 pp.

Hitchcock, C.L. and Cronquist, A. (1973). *Flora of the Pacific Northwest*. Univ. of Washington Press, Seattle. 730 pp.

Holmgren, A.H. and Reveal, J.L. (1966). *Checklist of the Vascular Plants of the Intermountain Region*. Research Paper INT-32. U.S. Forest Service. (Dicotyledons only.)

Hultén, E. (1973). *Flora of Alaska and Neighboring Territories*. Stanford Univ. Press, Palo Alto, California.

Kearney, T. and Peebles, R.H. (1960). *Arizona Flora*, 2nd Ed. Univ. California Press, Berkeley.

Little, E.L. (1979). *Checklist of United States Trees*. Agriculture Handbook 541. U.S. Department of Agriculture, Washington, D.C. 375 pp.

Long, R.W. and Lakela, O. (1971). *A Flora of Tropical Florida*. Univ. of Miami Press, Florida. 962 pp.

McGregor, R.L. *et al.* (1977). *Atlas of the Flora of the Great Plains*. Iowa State Univ., Ames, Iowa.

Martin, W.C. and Hutchins, C.R. (1980, 1981). *A Flora of New Mexico*, 2 vols. Cramer, FL-9490, Vaduz, Liechtenstein. 2591 pp.

Munz, P.A. (1974). *A Flora of Southern California*. Univ. of California Press, Berkeley, Los Angeles. 1086 pp.

Munz, P.A. and Keck, D.D. (1959). *A California Flora*. Univ. of California Press, Berkeley, California. 1681 pp.

Peck, M.E. (1961). *A Manual of the Higher Plants of Oregon*, 2nd Ed. Binfords and Mort, Portland.

Phillips, W.L. and Stuckey, R.L. (1976). *Index to Plant Distribution Maps in North American Periodicals Through 1972*. Hale, Boston.

Radford, A.E. *et al.* (1968). *Manual of the Vascular Flora of the Carolinas*. Univ. of North Carolina Press, Chapel Hill. 1183 pp.

Radford, A.E. *et al.* (Eds) (1980). *Vascular Flora of the Southeastern United States*. Univ. N. Carolina Press, Chapel Hill. 5 vols planned; Vol 1, Asteraceae, by A. Cronquist.

Seymour, F.C. (1969). *The Flora of Vermont*, 4th Ed. Vermont Agric. Exp. Sta. Bull. 660.

Small, J.K. (1933). *Manual of the Southeastern Flora*. Published by the author, New York. 1554 pp.

Steyermark, J.A. (1963). *Flora of Missouri*. Iowa State Univ. Press, Ames, Iowa. 1725 pp.

Strausbaugh, P.D. and Core, E.L. (1978). *Flora of West Virginia*, 2nd Ed. Seneca Books, Grantsville. 1079 pp.

Van Bruggen, T. (1976). *The Vascular Plants of South Dakota*. Iowa State Univ. Press, Ames.

Weber, W.A. (1976). *Rocky Mountain Flora*, 5th Ed. Colorado Assoc. Univ. Press, Boulder, Colorado. 479 pp.

Weishaupt, C.G. (1971). *Vascular Plants of Ohio*, 3rd Ed. Kendall/Hunt Publishing, Dubuque.

Field-guides

Brockman, C.F. (1979). *Trees of North America: A Guide to Field Identification*. Golden Press, New York. 280 pp.

Craighead, J.J., Craighead, F.C. and Davis, R.J. (1963). *A Field Guide to Rocky Mountain Wildflowers*. Houghton Mifflin, Boston. 277 pp.

Elias, T.S. (1980). *The Complete Trees of North America: Field Guide and Natural History*. Van Nostrand Reinhold, New York. 948 pp.

Little, E.L. (1980). *The Audubon Society Field Guide to North American Trees*, 2 vols (Western region, Eastern Region). Knopf, New York. 639 pp, 714 pp.

Newcomb, L. (1977). *Newcomb's Wildflower Guide*. Little, Brown & Co., Boston-Toronto. 490 pp. (1375 species illustrated.)

Niehaus, T.F. and Ripper, C.L. (1976). *A Field Guide to Pacific States Wildflowers*. Houghton Mifflin, Boston. 432 pp.

Peterson, R.T. and McKenny, M. (1968). *A Field Guide to Wildflowers of Northeastern and Northcentral North America*. Houghton Mifflin, Boston. 420 pp.

Spellenberg, R. (1979). *The Audubon Society Field Guide to North American Wildflowers – Western Region*. Knopf, New York.

Information on Threatened Plants Information can be found in many reports and papers developed by individuals, specific states or by the national government. The recommended latest version of the threatened plant list is U.S. Fish and Wildlife Service (1980), with supplement (1983). National and regional lists are:

Ayensu, E.S. and DeFilipps, R.A. (1978). *Endangered and Threatened Plants of the United States*. Smithsonian Institution and World Wildlife Fund-U.S., Washington, D.C. 403 pp. (Lists 90 'Extinct', 839 'Endangered' and 1211 'Threatened' taxa for the continental U.S., including Alaska.)

Kral, R. (1983). *A Report on Some Rare, Threatened, or Endangered Forest-Related Vascular Plants of the South*, 2 vols. U.S.D.A. Forest Service, Atlanta. 1305 pp. (Descriptions of 322 taxa, dot maps and management practices; outlined by J. Lamlein in *Threatened Plants Newsletter* 13: 18, 1984.)

Nelson, B.B. and Arndt, R.E. (1980). *Eastern States Endangered Plants*. U.S. Bureau of Land Management, Alexandria, Virginia. 109 pp. (State listings, dot maps, annotated descriptions and drawings.)

U.S. Fish and Wildlife Service (1980). Endangered and Threatened Wildlife and Plants: Review of Plant Taxa for Listing as Endangered or Threatened Species. *Federal Register* 45(242): 82480-82569. 15 December. (Lengthy list, principally of species to be considered for listing under the Endangered Species Act.)

U.S. Fish and Wildlife Service (1983). Endangered and Threatened Wildlife and Plants; Supplement to Review of Plant Taxa for Listing; Proposed Rule. *Federal Register* 48(229): 53640-53670. 28 November. (Changes to the 1980 list.)

U.S. Fish and Wildlife Service (1984). *Endangered and Threatened Wildlife and Plants*. U.S. Fish and Wildlife Service, Washington, D.C. 50 CFR 17.12. 20 July. (Republication of taxa officially listed as 'Endangered' and 'Threatened' under the Endangered Species Act, published annually.)

See also:

Ayensu, E.S. (1981). Assessment of threatened plant species in the United States. In Synge, H. (Ed.), *The Biological Aspects of Rare Plant Conservation*. Wiley, Chichester. Pp. 19-58. (Contains list of state lists, updated below.)

Crow, G.E. (1982). *New England's Rare, Threatened and Endangered Plants*. U.S. Fish and Wildlife Service, Washington, D.C. 130 pp. (101 species listed, ink drawings, colour photos, dot maps and descriptions.)

Endangered Species Technical Bulletin. Issued monthly by the Endangered Species Program of the U.S. Fish and Wildlife Service, Washington, D.C. 9 vols. to date. (Reviews the status of species proposed for listing under the Endangered Species Act.)

Mohlenbrock, R.H. (1983). *Where Have All the Wildflowers Gone?* Macmillan, New York, and Collier Macmillan, London. 239 pp.

Morse, L.E. and Henefin, M.S. (Eds) (1981). *Rare Plant Conservation: Geographical Data Organization*. New York Botanical Garden, New York. 377 pp.

Prance, G.T. and Elias, T.S. (Eds) (1977), cited in Appendix 1.

The Proceedings of the Symposium "Rare and Endangered Plant Species in New England" (1979). Various authors. Held at Harvard University Science Center. *Rhodora* 82(829): 1-237.

For bibliographies see:

Miasek, M.A. and Long, C.R. (1978). *Endangered Plant Species of the World and Their Endangered Habitats*. Library, New York Botanical Garden, Bronx, New York. 46 pp.

Wood, D.A. (1977). *Endangered Species: A Bibliography*. Environmental Series, No. 3, Oklahoma State University, Stillwater. 85 pp.

Wood, D.A. (1981). *Endangered Species Concepts, Principles, and Programs: A Bibliography*. Florida Game and Fresh Water Fish Commission. 228 pp.

State Lists of Threatened Plants (updated from Ayensu, 1981)

Alabama

Freeman, J.D., Causey, A.S., Short, J.W. and Haynes, R.R. (1979). *Endangered, Threatened and Special Concern Plants of Alabama*. Departmental Series No. 3, Auburn University, Auburn. 25 pp. (Colour photos, over 300 species listed, annotated descriptions.)

Alaska

Murray, D.F. (1980). *Threatened and Endangered Plants of Alaska.* U.S. Department of Agriculture, Forest Service and U.S. Department of the Interior, Bureau of Land Management. 59 pp. (42 species, dot maps, black ink drawings.)

Arizona

Anon. (1973). *Rare and Endangered Plants of Arizona.* 33 pp. (Characteristics and location.)

Arkansas

Tucker, G.E. (1974). Threatened Native Plants of Arkansas. In *Arkansas Natural Areas Plan* (Arkansas Department of Planning). Little Rock. Pp. 39-65. (Descriptions, characteristics, dot maps and photos.)

California

California Department of Fish and Game (198-?). *Natural Diversity Data Base: Special Plants.* California Dept of Fish and Game, Sacramento. 29 pp.

California Department of Fish and Game (1984). *Potential Candidates for Listing.* California Department of Fish and Game Endangered Plant Program. 4 pp. (Listing of 103 species.)

California Native Plant Society (1984). *Top 'Candidates' for State Listing.* California Native Plant Society Rare Plant Program. 5 pp. (183 species listed and entered into computer data base system.)

Natural Diversity Data Base (1984). *Elements with Occurrence.* California Department of Fish and Game. Pp. 13-31. (Listing of 391 special plants, categories of endangerment and entered into computer data base system.)

Smith, J.P., Jr. and York, R. (1984). *Inventory of Rare and Endangered Vascular Plants of California*, 3rd Ed. California Native Plant Society Special Publication No. 1, CNPS, Berkeley. 174 pp. (List of 34 taxa 'presumed extinct in CA', 604 'rare or endangered in CA and elsewhere', 198 'R/E in CA, more common elsewhere', 144 'needing more information', and 499 'of limited distribution'; outlined by T. Messick in *Threatened Plants Newsletter* No. 14, 1985.)

Colorado

Peterson, J.S. (1982). *Threatened and Endangered Plants of Colorado.* U.S. Department of the Interior, Fish and Wildlife Service, Denver, Colorado. 25 pp. (43 species listed, photos, ink drawings and descriptions.)

Connecticut

Dowhan, J.J. and Craig, R.J. (1976). *Rare and Endangered Species of Connecticut and Their Habitats.* Report of Investigations, No. 6. State Geological and Natural History Survey of Connecticut, Hartford. 137 pp. (Descriptions.)

Mehrhoff, L.J. (1978). *Rare and Endangered Vascular Plant Species in Connecticut.* U.S. Fish and Wildlife Service, Newton Corner, Mass. 41 pp. (List of species and habitats.)

Delaware

Tucker, A.O., Dill, N.H., Broome, C.R, Phillips, C.E. and Maciarello, M.J. (1979). *Rare and Endangered Vascular Plant Species in Delaware.* U.S. Fish and Wildlife Service, Newton Corner, Mass. 89 pp. (List of 449 rare plants with annotations.)

Florida

Ward, D.B. (Ed.) (1979). Volume 5: Plants. In Pritchard, P.C.H. (Ed.), *Rare and Endangered Biota of Florida*. Univ. Presses of Florida, Gainesville. (Described in *Threatened Plants Committee – Newsletter* No. 7: 23-24, 1981.)

Georgia

McCollum, J.L. and Ettman, D.R. (1977). *Georgia's Protected Plants*. Georgia Department of Natural Resources, Atlanta. 64 pp. (58 species described, dot maps and ink drawings.)

Idaho

Henderson, D.M., Johnson, F.D., Packard, P. and Steele, R. (1977). *Endangered and Threatened Plants of Idaho – A Summary of Current Knowledge*. College of Forestry, Wildlife and Range Sciences Bulletin 21, University of Idaho Forest, Wildlife and Range Experiment Station, Moscow, Idaho. 161 pp. (Updated in 1983; dot maps and annotations.)

Steele, R., Brunsfeld, S., Henderson, D., Holte, K., Johnson, F. and Packard, P. (1981). *Rare and Endangered Vascular Plant Species of Concern by the Plant Technical Committee of the Idaho Natural Areas Council*, Supplement: *1983 Status Changes and Addition to Bulletin 34*. College of Forestry, Wildlife and Range Sciences Bulletin 34, Univ. of Idaho Forest, Wildlife and Range Experiment Station, Moscow, Idaho.

Illinois

Paulson, G.A. and Schwegman, J. (1976). Endangered, Vulnerable, Rare and Extirpated Vascular Plants in Illinois – Interim List of Species. Illinois Nature Preserves Commission and Department of Conservation, Rockford and Springfield (unpublished manuscript). 189 pp and appendices. (Descriptions and dot maps.)

Indiana

Bacone, J.A. and Hedge, C.L. (1980). A preliminary list of endangered and threatened vascular plants in Indiana. In Moulton, B. (Ed.), *Proc. Indiana Acad. Sci.* 89: 359-371.

Barnes, W.B. (1975). Rare and Endangered Plants in Indiana. Indiana Department of Natural Resources, Indianapolis (unpublished manuscript).

Iowa

Roosa, D.M. and Eilers, L.J. (1978). *Endangered and Threatened Iowa Vascular Plants*. State Preserves Advisory Board Special Report 5, State Conservation Commission, Des Moines. 93 pp. (Descriptions and Iowa Endangered Species Act.)

Kansas

McGregor, R.L. (1977). *Rare Native Vascular Plants of Kansas*. Technical Publications of the State Biological Survey of Kansas 5, Lawrence, Kansas. 44 pp. (114 species described.)

Kentucky

Branson, B.A. *et al.* (1981). Endangered, threatened and rare animals and plants of Kentucky. *Trans. Kentucky Acad. Sci.* 42: 77-89.

Louisiana

Curry, M.G. (1976). Rare Vascular Plants of Louisiana. VTN Louisiana Inc., Metairie, Louisiana (unpublished manuscript).

Maine

Gawler, S. (1981). *An Annotated List of Maine's Vascular Plants*. Critical Areas Program, Augusta, Maine.

Maryland

Broome, C.R., Tucker, A.O., Reveal, J.L. and Dill, N.H. (1979). *Rare and Endangered Vascular Plant Species in Maryland*. U.S. Fish and Wildlife Service, Newton Corner, Mass. 64 pp. (List of 237 species and annotations.)

Maryland Natural Heritage Program (1984). *Threatened and Endangered Plants and Animals of Maryland*. Maryland Dept of Natural Resources, Annapolis, Maryland. 476 pp. (Proceedings of symposium held 3-4 September 1981, including 7 papers on endangered plants and habitats.)

Massachusetts

Coddington, J. and Field, K.G. (1978). *Rare and Endangered Vascular Plant Species in Massachusetts*. U.S. Fish and Wildlife Service, Newton Corner, Mass. 62 pp. (Annotated list 243 species; county distribution, threats and habitats.)

Sorrie, B.A. (1983). *Native Plants for Special Consideration in Massachusetts*. Massachusetts Natural Heritage Program, Div. of Fisheries and Wildlife, Boston. 14 pp. (List of 250 rare plants species.)

Michigan

Beaman, J.H. (1977). Commentary on Endangered and Threatened Plants in Michigan. *Michigan Botanist* 16: 110-122.

Wagner, W.H., Voss, E.G., Beaman, J.H., Bourdo, E.A., Case, F.W., Churchill, J.A. and Thompson, P.W. (1977). Endangered, Threatened, and Rare Vascular Plants in Michigan. *Michigan Botanist* 16: 99-110.

Minnesota

Morley, T. (1972). Rare or Endangered Plants of Minnesota. Department of Botany, University of Minnesota, Minneapolis (unpublished manuscript).

Mississippi

Pullen, T.M. (1975). Rare and Endangered Plant Species in Mississippi. Department of Biology, University of Mississippi (unpublished manuscript).

Missouri

Morgan, S. (1984). *Select Rare and Endangered Plants of Missouri*. Missouri Dept of Conservation, Jefferson City, Missouri. 29 pp. (Not seen.)

Montana

Lesica, P., Moore, G., Peterson, K.M. and Rumely, J.H. (1984). *Vascular Plants of Limited Distributions*. Monograph No. 2, *Montana Acad. Sci.* 43: 1-61.

Watson, T.J. (1976). *An Evaluation of Putatively Threatened or Endangered Species from the Montana Flora*. Univ. of Montana. 31 pp.

Nebraska

U.S.D.A. Soil Conservation Service (1975). Threatened and Endangered Species of Vascular Plants in Nebraska. U.S.D.A. Soil Conservation Service, Lincoln (unpublished manuscript).

Nevada

Beatley, J.C. (1977). *Endangered Plant Species of the Nevada Test Site, Ash Meadows, and South Central Nevada; Threatened Species; Addendum*. Department of

Biological Sciences, University of Cincinnati. 150 pp. (Dot map, photos and descriptions.)

Mooney, M. and Pinzl A. (compilers) (1984). *Sensitive Plant List for Nevada*. March 1984, Threatened and Endangered Plant Workshop, Reno, Nevada. 8 pp.

Mozingo, H. and Williams, M. (1980). *Threatened and Endangered Plants of Nevada; an Illustrated Manual*. U.S. Fish and Wildlife Service, Portland, Oregon.

Rhoads, W.A., Cochrane, S.A. and Williams, M.P. (1978). *Status of Endangered and Threatened Plant Species on Nevada Test Site – A Survey, Part 2: Threatened Species*. Santa Barbara Operations, EG & G Inc., Goleta, California. 148 pp. (Descriptions and dot maps.)

Rhoads, W.A. and Williams, M.P. (1977). *Status of Endangered and Threatened Plant Species on Nevada Test Site – A Survey, Part 1: Endangered Species*. Santa Barbara Operations, EG & G. Inc., Goleta, California. 102 pp. (Descriptions, dot maps and photos.)

New Hampshire

Storks, I.M. and Crow, G.E. (1978). *Rare and Endangered Vascular Plant Species in New Hampshire*. U.S. Fish and Wildlife Service, Newton Corner, Mass. 69 pp. (List of species and annotations.)

New Jersey

Snyder, D.B. and Vivian, V.E. (1981). *Rare and Endangered Vascular Plant Species in New Jersey*. U.S. Fish and Wildlife Service, Newton Corner, Mass.

New Mexico

Fletcher, R., Isaacs, B., Knight, P., Martin, W., Sabo, D., Spellenberg, R. and Todsen, T. (1984). *A Handbook of Rare and Endemic Plants of New Mexico*. Univ. of New Mexico Press, Albuquerque. 291 pp. (List of over 130 species, ink drawings, descriptions and dot maps.)

New York

Mitchell, R.S. and Sheviak, C. (1984). *Rare Plants of New York State*. New York State Museum Bull. 445, Albany, New York. 96 pp. (Reprint of 1981 issue.)

North Carolina

Massey, J., Otte, D., Atkinson, T. and Whetstone, R. (1983). *An Atlas and Illustrated Guide to the Threatened and Endangered Vascular Plants of the Mountains of North Carolina and Virginia*. U.S.D.A. Forest Service, Southeastern Forest Experiment Station, Asheville, North Carolina. 218 pp. (45 species listed with descriptions, habitat details, ink drawings, dot maps.)

Sutter, R.D., Mansberg, L. and Moore, J. (1983). Endangered, threatened and rare plant species of North Carolina: A revised list. *ASB Bulletin* 30(4): 153-163. (List of over 300 species.)

North Dakota

U.S.D.A. Soil Conservation Service (1972). Rare and Endangered Plant Species in North Dakota. U.S.D.A. Soil Conservation Service, Bismarck (unpublished manuscript).

Ohio

Andreas, B., Burns, J., Cusick, A., Emmitt, D., Marshall, J. and Spooner, D. (1984). *Ohio Endangered and Threatened Vascular Plants: Abstracts of State-listed Taxa*. Ohio Dept of Natural Resources, Columbus, Ohio. 635 pp. (367 species described, including habitat, threats, dot maps.)

Cooperrider, T.S. (1982). Endangered and threatened plants of Ohio. *Ohio Biological Survey Biological Notes No. 16.* 92 pp. (821 species listed with descriptions.)

Stuckey, R.L. and Roberts, M.L. (1977). Rare and Endangered aquatic vascular plants of Ohio: An annotated list of the imperiled species. *Sida* 7(1): 24-41.

Oklahoma

Zanoni, T.A., Gentry, J.L., Tyrl, R.J. and Risser, P.G. (1979). *Endangered and Threatened Plants of Oklahoma.* Department of Botany and Microbiology, University of Oklahoma, Norman. 64 pp. (26 species listed with ink drawings, descriptions and dot maps.)

Oregon

Meinke, R.J. (1982). *Threatened and Endangered Vascular Plants of Oregon: An Illustrated Guide.* U.S. Fish and Wildlife Service, Portland, Oregon.

Siddall, J.L., Chambers, K.L. and Wagner, D.H. (1979). *Rare, Threatened and Endangered Vascular Plants in Oregon – An Interim Report.* Natural Area Preserves Advisory Committee, Salem. 109 pp. (Over 600 species listed with descriptions.)

Soper, C., Kagan, J. and Yamamoto, S. (1983). *Rare and Endangered Plants and Animals of Oregon.* Oregon Chapter, The Nature Conservancy, Portland, Oregon.

Pennsylvania

Genoways, H. and Brenner, F. (Eds) (1984). *Species of Special Concern in Pennsylvania.* Carnegie Museum of Natural History, Pittsburgh, Pennsylvania. 430 pp. (21 plants listed with descriptions, habitat details, recommendations for action, maps, colour photos.)

Wiegman, P.G. (1979). *Rare and Endangered Vascular Plant Species in Pennsylvania.* U.S. Fish and Wildlife Service, Newton Corner, Mass. 94 pp. (List of species and bibliography.)

Rhode Island

Church, G.L. and Champlin, R.L. (1978). *Rare and Endangered Vascular Plant Species in Rhode Island.* U.S. Fish and Wildlife Service, Newton Corner, Mass. 17 pp. (120 species listed and annotated.)

South Carolina

Forsythe, D.M. and Ezell, W.B., Jr. (Eds) (1979). *Proceedings of the First South Carolina Endangered Species Symposium.* Nongame-Endangered Species Section, South Carolina Wildlife and Marine Resources Department and The Citadel, Charleston. 201 pp. (147 species listed.)

South Dakota

Schumacher, C.M. (1979). *Status of Endangered and Threatened Plants in South Dakota.* Technical Notes, Environment, 9, U.S.D.A. Soil Conservation Service, Huron.

Tennessee

Committee for Tennessee Rare Plants (1978). The rare vascular plants of Tennessee. *J. Tennessee Acad. Sci.* 53(4): 128-133.

Texas

Beaty, H.E. *et al.* (1983). *Endangered, Threatened & Watch Lists of Plants of Texas.* Texas Organization for Endangered Species, Austin. 7 pp. (Lists habitats and reason for status.)

Johnston, M.C. (1974). *Rare and Endangered Plants Native to Texas.* Rare Plant Study Center, University of Texas, Austin. 12 pp.

Utah

Welsh, S.L. (1978). Endangered and Threatened plants of Utah: A reevaluation. *The Great Basin Naturalist* 38(1): 1-18.

Welsh, S.L. (1979). *Illustrated Manual of Proposed Endangered and Threatened Plants of Utah.* Denver Federal Center, U.S. Fish and Wildlife Service, Denver, Colorado. 318 pp. (148 species listed, with ink drawings, descriptions and dot maps.)

Welsh, S.L., Atwood, N.E. and Reveal, J.L. (1975). Endangered, Threatened, Extinct, Endemic, and Rare or Restricted Utah vascular plants. *The Great Basin Naturalist* 35(4): 327-376.

Vermont

Countryman, W.D. (1978). *Rare and Endangered Vascular Plant Species in Vermont.* U.S. Fish and Wildlife Service, Newton Corner, Mass. 68 pp. (Species listed and annotated.)

Virginia

Massey, J., Otte, D., Atkinson, T. and Whetstone, R. (1983). *An Atlas and Illustrated Guide to the Threatened and Endangered Vascular Plants of the Mountains of North Carolina and Virginia.* U.S.D.A. Forest Service, Southeastern Forest Experiment Station, Asheville, North Carolina. 218 pp. (45 species listed with descriptions, habitat details, ink drawings, dot maps.)

Porter, D.M. (1979). *Rare and Endangered Vascular Plant Species in Virginia.* U.S. Fish and Wildlife Service, Newton Corner, Mass. 52 pp. (350 taxa listed.)

Porter, D.M. (1979). Vascular plants. In Linzey, D.W. (Ed.), *Endangered and Threatened Plants and Animals of Virginia.* Virginia Polytechnic Institute and State University, Blacksburg. Pp. 31-122. (333 species listed, dot maps and descriptions.)

Washington

Sheehan, M. and Schuller, R. *et al.* (1984). *Endangered, Threatened and Sensitive Vascular Plants of Washington.* Washington Natural Heritage Program, Olympia. 29 pp.

West Virginia

Clarkson, R.B., Evans, D.K., Fortney, R.H., Grafton, B. and Rader, L. (1981). *Rare and Endangered Vascular Plant Species in West Virginia.* U.S. Fish and Wildlife Service, Newton Corner, Mass.

Wisconsin

Read, R.H. (1976). *Endangered and Threatened Vascular Plants in Wisconsin.* Scientific Areas Prevention Council Technical Bulletin 92, Department of Natural Resources, Madison. 58 pp. (268 taxa included with photos and descriptions.)

Wyoming

Dorn, R.D. (1977). Rare and endangered species. In *Manual of the Vascular Plants of Wyoming*, Vol. 2. Harland, New York. Pp. 1394-1400.

Dorn, R. (1980). *Illustrated Guide to Special Interest Vascular Plants of Wyoming.* U.S. Fish and Wildlife Service and Bureau of Land Management. 67 pp. (Lists 13 plants considered of special interest, with dot maps and illus.)

Laws Protecting Plants The Endangered Species Act of 1973 (Public Law 93-205, 987 Stat. 884), as amended in 1983, directs that no federally funded activity shall jeopardize the existence of species once officially determined as 'Endangered' or 'Threatened', as defined by the Act. By January 1985, 67 native U.S. plants, 5 plants from the U.S. and elsewhere, and 1 plant not occurring in the U.S. had been officially listed as 'Endangered' under the Act; the figures for 'Threatened' plants are 9, 2 and 2 respectively.

Many states have their own laws protecting plants. Nilsson (1983) includes a useful table indicating the form and type of legal protection in each state; according to this, the following plant species are protected under state laws: California (174), Georgia (58), Michigan (208), Missouri (365), Ohio (417), South Carolina (1), Texas (10), Wisconsin (56).

Useful references:

Berger, T.J. and Neuner, A.M. (1979). *Directory of State Protected Species: A Reference to Species Controlled by Non-Game Regulations.* Association of Systematics Collections, Lawrence, Kansas.

Council on Environmental Quality (1981). *A Summary of the Legal Authorities for Conserving Wild Plants.* Council on Environmental Quality, Washington, D.C. 156 pp.

Kartesz, J.T. and Kartesz, R. (1977). *The Biota of North America, Part 1: Vascular Plants, Volume I. Rare Plants.* Biota of North America Committee, Pittsburgh, PA. (State Laws.) 361 pp.

McMahan, L. (1980). Legal protection for rare plants. *The American University Law Review* 29(3): 515-569.

Nilsson, G. (1983). *The Endangered Species Handbook.* Animal Welfare Institute, P.O. Box 3650, Washington, D.C. 20007. See in particular M. Bean on questions and answers about the Endangered Species Act (pp. 114-116) and the section on state endangered species programs with useful table on numbers of protected plants in each state (pp. 119-120).

Voluntary Organizations

American Orchid Society, Executive Director-Editor, 84 Sherman Street, Cambridge, Massachusetts 02140.

Botanical Society of America, Dept of Biological Sciences, Florida International University, Miami, Florida 33199.

Cactus and Succulent Society of America, Inc., Conservation Committee, Huntington Botanical Gardens, address below.

California Native Plant Society, 2280 Ellsworth, Suite D, Berkeley, California 94704.

Environmental Defense Fund, 1525 18th St. NW, Washington, D.C. 20036.

Garden Club of America, Endangered Species Advisor, 369 Atherton Ave., Atherton, California 94025.

Natural Resources Defense Council, 1330 New York Ave. NW, Suite 300, Washington, D.C. 20005.

The Nature Conservancy (TNC), 1800 N. Kent St., Suite 800, Arlington, Virginia 22209. (Nearly all states have Natural Heritage Programs; TNC has a computer database on vascular plants considered rare, threatened and endangered in the states of U.S.A.)

World Wildlife Fund-U.S., Plant Conservation Program, 1601 Connecticut Ave. NW, Washington, D.C. 20009.

For additional information see also:

National Wildlife Federation Conservation Directory, 1412 16th St. NW, Washington, D.C. 20036. (Lists government and non-governmental organizations interested in conservation.)

Botanic Gardens There are several botanic gardens in each of the major floristic regions of the country. For information on them see Henderson, 1983, cited in Appendix 1, or refer to:

American Association of Botanical Gardens and Arboreta, Inc. (AABGA), National Office, P.O. Box 206, Swarthmore, Pennsylvania 19081.

Some of the principal gardens that are subscribers to the Botanic Gardens Conservation Co-ordinating Body are listed below:

Berry Botanic Garden, 11505 S.W. Summerville Avenue, Portland, Oregon 97219.
Chicago Botanic Garden, P.O. Box 400, Glencoe, Illinois 60022.
Desert Botanical Garden, 1201 N. Galvin Parkway, Phoenix, Arizona 85010.
Fairchild Tropical Garden, 10901 Old Cutler Road, Miami, Florida 33156.
Fullerton Arboretum, California State University, Fullerton, California 92634.
Huntington Botanical Gardens, 1151 Oxford Road, San Marino, California 91108.
Los Angeles State and County Arboretum, 301 North Baldwin Avenue, Arcadia, California 91006.
Missouri Botanical Garden, P.O. Box 299, St Louis, Missouri 63166.
Mitchell Park Conservatory, 524 South Layton Boulevard, Milwaukee, Wisconsin 53215.
Rancho Santa Ana Botanic Garden, 1500 N. College Avenue, Claremont, California 91711.
Strybing Arboretum, 9th Avenue at Lincoln Way, San Francisco, California 91006.
The Arnold Arboretum of Harvard University, The Arborway, Jamaica Plain, Massachusetts 02130.
The New York Botanical Garden, Bronx, New York 10458.
U.C.I. Arboretum, School of Biological Sciences, University of California, Irvine, California 92717.
University of Washington Arboretum, Seattle, Washington 98195.

The Center for Plant Conservation, at the Arnold Arboretum (address above), was created in 1984 to: establish a national network of programmes at botanic gardens and arboreta on protection, cultivation and study of U.S. endangered plants; create an endangered species data bank on the biology and horticulture of these species; and develop live plant collections at regional gardens.

Useful Addresses
Enforcement of U.S. laws: Division of Law Enforcement, U.S. Fish and Wildlife Service, P.O. Box 28006, Washington, D.C. 20005.
Office of Endangered Species, U.S. Fish and Wildlife Service, Department of the Interior, Washington, D.C. 20240. (This office can provide the addresses of the regional and field offices.)
Smithsonian Institution, Department of Botany, Plant Conservation Unit, National Museum of Natural History, Washington, D.C. 20560.
TRAFFIC-USA, World Wildlife Fund-U.S., address above.
CITES Management Authority: Chief, Federal Wildlife Permit Office, 1000 North Glebe Road, Room 611, Arlington, Virginia 2201.
CITES Scientific Authority: Office of the Scientific Authority, U.S. Fish and Wildlife Service, Department of the Interior, Washington, D.C. 20240.

Additional References
Anderson, J.R. (1977). Land use and land cover changes – A framework for monitoring. *J. Res. U.S. Geol. Survey* 5(2): 143-153.

Bailey, R.G. (1976). *Ecoregions of the United States* (Map). U.S. Forest Service, Ogden. Utah. (1:7,500,000.)

Bailey, R.G. (1978). *Description of the Ecoregions of the United States* (Manual). U.S. Forest Service, Ogden, Utah. 77 pp.

Benson, L. (1979). *Plant Classification*, 2nd Ed. Heath, Boston. 901 pp.

Braun, E.L. (1964). *Deciduous Forests of Eastern North America*. Hafner, New York. 596 pp.

Gleason, H.A. and Cronquist, A. (1964). *The Natural Geography of Plants*. Columbia Univ. Press, New York and London. 420 pp.

Hartley, T.G. (1966). The Flora of the "Driftless Area". *Univ. of Iowa Studies in Natural History* 21(1).

Holt, P.E. (Ed.) (1971). *The Distributional History of the Biota of the Southern Appalachians, Part II. Flora*. Research Division Monograph 2. Virginia Polytechnic Institute and State Univ., Blacksburg, Virginia. 414 pp.

James, C.W. (1961). Endemism in Florida. *Brittonia* 13: 225-244.

Klopatek, J.M., Olson, R.J., Emerson, C.J. and Joness, J.L. (1979). Land-use conflicts with natural vegetation in the United States. *Envir. Conserv.* 6(3): 191-199.

Küchler, A.W. (1964). Potential Natural Vegetation of the Conterminous United States (Map and Manual). *American Geographical Society Special Publication* 36. 116 pp. (1:3,168,000.)

Long, R.W. (1974). The vegetation of southern Florida. *Florida Scientist* 37(1): 33-45.

Merz, R.W. (1978). *Forest Atlas of the Midwest*. U.S. Forest Service, St Paul, Minnesota. 48 pp.

National Geographic Society (September 1982). *America's Federal Lands* (Map). National Geographic Society, Washington, D.C. (1:5,889,000.)

Oosting, H.J. (1956). *The Study of Plant Communities*, 2nd Ed. Freeman, San Francisco. 440 pp.

Stebbins, G.L. (1978). Why are there so many rare plants in California? I. Environmental factors. *Fremontia* 5(4): 6-10; II. Youth and age of species, *Fremontia* 6(1): 17-20.

Waggoner, G.S. (1975). *Eastern Deciduous Forest, Vol. I: Southeastern Evergreen and Oak-Pine Region*. U.S. National Park Service, Washington, D.C. 206 pp.

United States: Miscellaneous Islands

Uninhabited coral atolls and reefs lying between longitudes 150-180°W and latitudes 6°N and 1°S in the Pacific Ocean. Baker Island reaches 6 m above sea-level; all the rest are below 3 m. The islands have an arid tropical climate. The flora consists of widespread pan-Pacific species.

Baker Island (1.5 sq. km) – atoll with fringing reef; 15 vascular plants, all herbaceous.

Howland Island (1.6 sq. km) – atoll with 6 vascular species, including scattered *Cordia* trees.

Jarvis Island (4.1 sq. km) – atoll with fringing reef; scanty vegetation with no trees. 8 vascular species.

Kingman Reef (0.03 sq. km) – triangular reef with deep lagoon and small coral islet.

References
Christophersen, E. (1927). Vegetation of Pacific Equatorial Islands. *Bull. Bernice P. Bishop Mus.* 44. 79 pp. (Includes annotated checklists for each island.)

United States Virgin Islands

A territory of the United States comprising three mountainous islands – St Croix (218 sq. km), St Thomas (72.5 sq. km), St John (52 sq. km) – and 50 islets and cays.

Area 345 sq. km

Population 103,000

Floristics 890 native seed plant species; 27 endemic, 4 to St Croix (Fosberg, 1974).

Vegetation Severely modified by man to mostly dry scrub woodland, with cacti and agaves predominant; on high ground a reduced type of evergreen forest – 'xerophytic rain forest'; along the coast small patches of stunted mangrove and some beach forest; St Croix greatly altered by sugar cane cultivation; 75% of St John is National Park where forest of mixed native and introduced species has regenerated; overall 5.9% forested (FAO, 1974, cited in Appendix 1).

Checklists and Floras
Britton, N. and Wilson, P. (1923-1930). Botany of Porto Rico and the Virgin Islands. *Scientific survey of Porto Rico and the Virgin Islands*, 5 (626 pp.) and 6 (663 pp.). New York Academy of Sciences, New York. (Keys, descriptions, general ranges and distributions by island.)

See also:

Fosberg, F.R. (1974). Sketch of the St Croix flora. In Multer, H.G. and Gerhard, L.C. (Eds), *Guidebook to the geology and ecology of some marine and terrestrial environments, St Croix, U.S. Virgin Islands*. Spec. publ. No 5, Farleigh Dickinson University, St Croix. Pp. 239-244.

Fosberg, F.R. (1976). Revisions in the flora of St Croix, U.S. Virgin Islands. *Rhodora* 78: 79-119.

Kartesz, J.T. and R. (1980). A synonymized check list of the vascular flora of the United States, Canada and Greenland. *The Biota of North America*, vol. 2. Univ. of N. Carolina Press. 498 pp. (Includes Puerto Rico and the U.S. Virgin Islands.)

Liogier, A.H. (1965). Nomenclatural changes and additions to Britton and Wilson's "Flora of Porto Rico and the Virgin Islands". *Rhodora* 67(772): 315-361.

Liogier, A.H. (1967). Further changes and additions to the flora of Porto Rico and the Virgin Islands. *Rhodora* 69(779): 372-376.

Little, E.L., Jr. and Wadsworth, F.H. (1964). *Common trees of Puerto Rico and the Virgin Islands*. Agriculture Handbook No. 249, U.S.D.A. Forest Service, Washington, D.C. 548 pp. (Keys, mainly to families; descriptions, illus., distributions.) Spanish edition by authors and J. Marrero, Editorial UPR, Puerto Rico, 1967.

Little, E.L., Jr. and Woodbury, R.O. (1976). *Flora of Buck Island Reef National Monument (U.S. Virgin Islands)*. Forest Service Research Paper 19. Institute of Tropical Forestry, Río Piedras, Puerto Rico. 27 pp. (Illus., map.)

Little, E.L., Jr. *et al.* (1974). *Trees of Puerto Rico and the Virgin Islands, Second volume*. Agriculture Handbook No. 449, U.S.D.A. Forest Service, Washington, D.C. 1024 pp. (2nd vol. to Little and Wadsworth, 1964, above; includes endemic, rare and endangered tree species.)

Information on Threatened Plants

Ayensu, E.S. and DeFilipps, R.A. (1978). *Endangered and Threatened Plants of the United States*. Smithsonian Institution and WWF-U.S., Washington, D.C. Pp. 225-232. (Lists 102 'Endangered' and 'Threatened' taxa from Puerto Rico and the Virgin Islands, both U.S. and British, with bibliography; 16 of them are from the U.S. Virgin Is.)

Little, E.L., Jr. and Woodbury, R.O. (1980). *Rare and Endemic Trees of Puerto Rico and the Virgin Islands*. Conservation Research Report No. 27, U.S.D.A. Forest Service, Washington, D.C. 26 pp.

Laws Protecting Plants Covered under the U.S. Endangered Species Act, see the United States.

Voluntary Organizations

The Virgin Islands Conservation Society (address not known).

Additional References

Multer, H.G. and Gerhand, L.C. (Eds) (1974). *Guidebook to the geology and ecology of some marine and terrestrial environments, St Croix, U.S. Virgin Islands*. Spec. Publ. No. 5, Farleigh Dickinson University, St Croix. 303 pp. (Vegetation section on pp. 201-244.) (Not seen.)

Uruguay

Area 186,925 sq. km

Population 2,990,000

Floristics No information; the northernmost limit for several Argentine species and the southernmost for many Brazilian and Paraguayan ones.

Vegetation In the north, occupying the greater part of the country, medium-tall grassland with few trees; in the south various croplands and grazing pastures, with only a few islands of original savanna remaining. Along the banks of the Río Uruguay, Río Negro and on coastal areas palm grasslands predominate. The few forested areas are on banks of these two rivers, also on the Río Parano, their chief tributaries and on the eastern side of the Laguna Mirim.

Checklists and Floras

Flora of Uruguay (various authors) (1958-1972). Museo Nacional de Historia Natural, Montevideo. (Incomplete; 4 small fascicles so far covering ferns and 7 angiosperm families.)

Herter, G. (1939-1957). *Flora Ilustrada del Uruguay*. 13 fascicles, 600 pp. in all. (Line drawings, 4 per page.)

Herter, G. (1949-1956). *Flora del Uruguay*. Montevideo. 10 fascicles.

Lombardo, A. (1964). *Flora Arbórea y Arborescente del Uruguay*. Consejo Departamental de Montevideo. 151 pp. (Illus.)

Lombardo, A. (1982-1984). *Flora Montevidensis*, 3 vols. Intendencia Municipal de Montevideo.

Information on Threatened Plants There is no national Red Data Book. Threatened plants are mentioned in:

Ravenna, P. (1977). Neotropical species threatened and endangered by human activity in Iridaceae, Amaryllidaceae and allied bulbous families. In Prance, G.T. and Elias, T.S. (Eds) (1977), cited in Appendix 1. Pp. 257-266.

Voluntary Organizations
CIPFE, Canelones 1164, Montevideo.

Botanic Gardens
Jardín Botánico, Laboratorio Botánica, Facultad de Agronomía, Casilla de Correo 1238, Montevideo.

Useful Addresses
Museo Nacional de Historia Natural, Buenos Aires 652, Casilla 399, Montevideo.

CITES Management and Scientific Authority: Instituto Nacional para la Preservación del Medio Ambiente (INPMA), c/o Ministerio de Educación y Cultura, Ituzaingo 1255, Montevideo.

Additional References
Rosengurtt, B. (1945). La vegetación del Uruguay. In Verdoorn, F. (Ed.) (1945), cited in Appendix 1. Pp. 142-143.

Vanuatu

Vanuatu, formerly the New Hebrides, consists of 80 islands forming a double chain, 724 km long, in the south-west Pacific Ocean, between latitudes 12°-21°S and longitudes 166°-170°E. Most of the islands are volcanic, some actively so. The highest point is 1889 m on Espíritu Santo, the major island of the group.

Area 14,763 sq. km

Population 136,000

Floristics c. 1000 vascular plant species (Schmid, 1978, cited in Appendix 1), of which c. 150 are endemic (Chew, 1975). Floristic affinities with Fiji, Solomons and New Caledonia (van Balgooy, 1971, cited in Appendix 1).

Vegetation Tropical lowland evergreen rain forest, with *Castanospermum*, *Euodia* and *Hernandia*; small areas of broadleaved deciduous forest; closed conifer forest, dominated by *Agathis*, restricted to western parts of Espíritu Santo, Eromanga and Aneityum; montane rain forest between 1000-1500 m; cloud forest with *Metrosideros*, above 1500 m; extensive coastal forest with *Casuarina*, *Hibiscus* and *Pandanus*; swamp forest with *Barringtonia* and *Pandanus*, on Efate; scattered mangrove forests. Extensive areas have been cleared for plantations and pastures particularly on the plateaux of Efate and Espíritu Santo, but Eromanga still has about 180 sq. km of closed forest (Schmid, 1978).

Checklists and Floras No comprehensive Flora, but see:

Gowers, S. (1976). *Some Common Trees of the New Hebrides and Their Vernacular Names*. Dept of Agriculture, Port Vila. 189 pp. (Keys, descriptions and line drawings of c. 60 trees.)

Guillaumin, A. (1948). Compendium de la Flora phanérogamique des Nouvelles-Hébrides. *Ann. Mus. Colon. Marseille* 6: 5-56.

Information on Threatened Plants None.

Additional References

Chew, W.L. (1975). The phanerogamic flora of the New Hebrides and its relationships. *Phil. Trans. Roy. Soc. London B* 272: 315-328.

Schmid, M. (1978) The Melanesian forest ecosystems (New Caledonia, New Hebrides, Fiji Islands and Solomon Islands). In Unesco/UNEP/FAO (1978), cited in Appendix 1. Pp. 654-683.

Venezuela

Area 912,047 sq. km

Population 17,819,000

Floristics Presently calculated at 15,000 to 20,000 species of vascular plants, estimated to reach possibly 25,000 (Steyermark, 1979b). Although better explored botanically than most other South American countries, Huber (1984, pers. comm.) states that only about 10% of the country has been covered so far.

The principal centres for endemic and relict taxa are in the Andes, the Coastal Cordillera, the Serranía del Interior (Interior Coastal Range), the Guayana Highland and the Amazonian lowlands in the Upper Orinoco-Upper Río Negro region (Huber, pers. comm.). The Guayana Highland, sparsely populated territory in the south covering over half of Venezuela and reaching into Guyana, is especially rich: over 4000 endemic species (and 100 endemic genera) described; possibly over 6000 endemics present, out of a predicted flora of around 8000 species; affinities to Africa, Malaysia and Australia (Maguire, 1970).

Vegetation On the western littoral (Anzoótequi Province) deciduous thorn woodland and subdesert deciduous scrub. The central littoral is predominantly under cultivation. On the east coast and on the Orinoco Delta, swamp forest and flooded tall grasslands. In the Guayana Highland, between the coastal Andean Cordilleras and the Guyanese Shield, tall grasslands with broadleaved trees (Unesco, 1981, cited in Appendix 1). In central Venezuela, south of the Río Orinoco, lowland tropical evergreen seasonal forest. Scattered throughout northern and central Venezuela are areas of deciduous thorn forest. In the south, predominantly tropical submontane rain forest and lowland tropical rain forest. Estimated rate of deforestation for closed broadleaved forest 1250 sq. km/annum out of 318,700 sq. km (FAO/UNEP, 1981).

Checklists and Floras Venezuela is covered by the family and generic monographs of *Flora Neotropica* (cited in Appendix 1). The country Floras are:

Aristeguieta, L. (1973). *Familias y géneros de los arboles de Venezuela*. Instituto Botánico, Caracas. 845 pp.

Badillo, V.M. and Schnee, L. (1965). *Clave de las Familias de Plantas Superiores de Venezuela*. Rev. Fac. Agron. Venez. No. 6, Maracay. 255 pp. (Keys to all families of higher plants in Venezuela.)

Lasser, T. (Ed.) (1964-). *Flora de Venezuela*. Instituto Botánico, Caracas. 13 vols. so far, covering 25 families, including ferns, Orchidaceae, Rubiaceae, Compositae.

Maguire, B. *et al.* (1953-). The botany of the Guayana Highland. *Mem. New York Bot. Gard.* 12 parts, between vols. 8 and 38. Various family treatments resulting from field activities begun in 1944. Parts 13 and 14 (in prep.) will conclude the systematic treatment of the flora of the Roraima Formation in Guyana; other reports will be issued as separate papers.

Pittier, H. *et al.* (1945-1947). *Catálogo de la Flora Venezolana*, 2 vols. Vargas, Caracas.

Steyermark, J. (1951-57). Contributions to the Flora of Venezuela. *Fieldiana Bot.* 28: 1-1225. (List of new species resulting from explorations made from 1943-1945.)

See also:

Ramia, M. (1974). *Flora of Sabanas Llaneras*. Monte Avila Editores, Caracas. 287 pp. (Species list, line drawings.)

Steyermark, J. (1968). Contribuciónes a la flora de la Sierra de Imataca, Altiplanicie de Nuria y region adyacente del Territorio Federal Amacuro al sur del río Orinoco. *Acta Bot. Ven.* 3: 49-175.

Steyermark, J. (1979a). Flora of the Guayana Highland: endemicity of the generic flora of the summits of the Venezuela tepuis. *Taxon* 28: 45-54. (Evaluates the endemicity of the generic flora on the summits of various tepuis of Venezuela.)

Steyermark, J. *et al.* (1972). The Flora of the Meseta del Cerro Jaua. *Mem. New York Bot. Gard.* 23: 833-892. (List of algae, mosses, hepaticas, ferns and 35 angiosperm families.)

Steyermark, J. and Huber, O. (1978). *Flora del Avila*. Sociedad Venezolana de Ciencias Naturales, Caracas. 971 pp. (Description of vegetation types and of 1892 species.)

Steyermark, J. *et al.* (in press). Flora del Parque Nacional Mochima. *Acta Bot. Venez.*

Steyermark, J. *et al.* (in press). Flora del Parque Nacional Morrocoy. *Acta Bot. Venez.*

Vareschi, V. (1970). *Flora de los Páramos de Venezuela*. Universidad de los Andes, Mérida. 429 pp.

The Missouri Botanical Garden, in cooperation with the Instituto Botánico and the National Herbarium of Venezuela, in 1983 completed a project to collect plants in botanically poorly known areas under threat. Funding for a second phase is being sought.

Field-guides

Dunsterville, G. and Garay, L. (1979). *Orchids of Venezuela, An Illustrated Field Guide*. 34 vols. Botanical Museum of Harvard University, Allston, Mass. 1055 pp. (Line drawings from the previous 6 vols of *Venezuelan Orchids Illustrated* with 50 additions and many old ones redrawn and improved.)

Hoyos, J. (1979). *Los Arboles de Caracas*, 2nd Ed. Sociedad de Ciencias Naturales, La Salle, Caracas. 381 pp. (Species descriptions, plant uses, colour illus.)

Ramia, M. (1974). *Plantas de las Sabanas Llaneras*. Monte Avila, Caracas. (Describes 555 species from the plains, line drawings.)

Schnee, L. (1973). *Plantas comunes de Venezuela*, 2nd Ed. Universidad Central de Venezuela, Instituto de Botánica Agricola, Maracay (Aragua). 822 pp.

Information on Threatened Plants There is no national Red Data Book. 20 species are included in the annex to the Convention on Nature Protection and Wildlife Preservation in the Western Hemisphere (1940). Papers on threatened plants include:

Steyermark, J. (1973). Preservamos las cumbres de la Península de Paria. Defensa de la Naturaleza. *Año* 2(6): 33-35.

Steyermark, J. (1977). Future outlook for threatened and endangered species in Venezuela. In Prance, G.T. and Elias, T.S. (Eds) (1977), cited in Appendix 1. Pp. 128-135.

Steyermark, J. (1979b). Plant refuge and dispersal centres in Venezuela: their relict and endemic element. In Larsen, K. and Holm-Nielsen, L.B. (Eds), *Tropical Botany*. Academic Press, London. Pp. 185-221. (Includes an appendix of endemic and relict species in the Coastal Cordillera and Serranía del Interior.)

Steyermark, J. (1982). Relationships of some Venezuelan forest refuges with lowland tropical floras. In Prance, G.T. (Ed.) (1982), cited in Appendix 1. Pp. 182-220. (List of endemics in each of the 5 principal forest refuges.)

The IUCN Plant Red Data Book has 2 data sheets for Venezuelan plants. Threatened plants are also mentioned in several papers in:

Prance, G.T. and Elias, T.S. (Eds) (1977), cited in Appendix 1. See in particular J.T. Mickel on rare and endangered ferns (pp. 323-328).

Voluntary Organizations

Fundación para la Defensa de la Naturaleza (FUDENA), Apdo 70376, Caracas 1070-A.

Sociedad Conservacionista Audubon de Venezuela, Apto 80450, Caracas 1080-A.

Sociedad Venezolana de Ciencias Naturales, Apdo 1521, Calle Arichuana con Cumaco, El Marqués, Caracas 1010-A.

Botanic Gardens

Jardín Botánico de Barinas, UNELLEZ Campus (Universidad Nacional Experimental de los Llanos Ezequiel Zamora), Barinas.

Jardín Botánico de Maracaibo, Autopista al Acropuerto "La Chinita", Apto Postal 10.123, Maracaibo.

Jardín Botánico del Instituto Nacional de Parques, Caracas.

Jardín Botánico, Instituto Botánico, Apto 2156, Caracas.

Jardín Botánico San Juan de Lagunillas, Facultad de Ciencias Forestales, Departamento de Botánica, Universidad de los Andes, Mérida 5101.

Jardín Botánico Xerofítico de Coro, Coro, Falcoñ.

Useful Addresses

Depto. de Conservación, Universidad de los Andes, Facultad de Ciencias Forestales, Mérida.

Herbário, Facultad de Agronomía, Instituto de Botánica Agricola, Universidad Central de Venezuela, Maracay, Aragua.

Herbário Universitário, Escuela de Recursos Naturales, Universidad Nacional Experimental de los Llanos Occidentales "Ezequiel Zamora", Buanare, Portuguesa.

Instituto Botánico, Apto 2156, Caracas.

Instituto de Botánico, Facultad de Agronomía, Universidad Central de Venezuela, El
   Limón, Maraxay, Aracuay 2101.
CITES Management and Scientific (for flora) Authorities: Ministerio del Ambiente y de
   los Recursos Naturales Renovables, Torre Sur, Piso 19, Centro Simón Bolivar-El
   Silencio, Caracas 1010.

Additional References

Maguire, B. (1970). On the Flora of the Guayana Highland. *Biotropica* 2(2): 85-100.
Steyermark, J. (1974). Situación actual de las exploraciones botánicas en Venezuela.
   Acta Bot. Venez. 9: 241-243.
Veillon, J.P. (1976). Deforestations in the western llanos of Venezuela since 1950 to
   1975. In Hamilton, L.S. (Ed.), *Tropical Rain Forest Use and Preservation: A study
   of problems and practices in Venezuela.* Sierra Club, Int. Ser. No. 4, San Francisco.
   115 pp.

Venezuela: Islands

Venezuela has a number of small islands and islets in the Caribbean Sea. Mostly coral, flat
and dry, with few inhabitants.

Isla de Margarita is the largest and most important, with a more varied flora than the
others. Area 1072 sq. km; population c. 160,000; 67 km long by 32 km wide, composed of
two conical mountains separated by a stretch of lowland.

Floristics About 858 vascular plants (Johnston, 1909). Less than five endemics,
confined to upper slopes, not endangered at present.

Vegetation In the lowlands mangrove, dry cactus scrub and thorn woodland,
mostly very disturbed. On the mountains dry deciduous woodlands, evergreen cloud
forests and thickets, not very disturbed.

Checklists and Floras See under Venezuela, above. See also:

Johnston, J.R. (1909). Flora of the Islands of Margarita and Coche. *Proc. Boston Soc.
   Nat. Hist.* 34(7): 163-312. Also as *Contr. Gray Herb. Harvard Univ.*, No. 37.

The Sociedad de Ciencias Naturales La Salle, Caracas, is preparing an updated Flora of
Margarita. The coordinator is Dr Jesus Hoyos (A. Sugden, 1984, pers. comm.).

Information on Threatened Plants None.

For other information, see under Venezuela, above.

Viet Nam

By Professor Dao Van Tien and Dr Phan Ke Loc

Area 329,566 sq. km

Population 58,307,000

Floristics Over 8000 indigenous vascular plant species known; at least 10% species endemism. It is estimated that the total number of indigenous vascular plants amounts to c. 12,000. Viet Nam, Kampuchea and Laos together have c. 600 fern species (Parris, 1985, cited in Appendix 1).

Vegetation Tropical rain forests in submontane and montane regions; few lowland rain forests below 500 m remain undisturbed. Tropical evergreen closed forests widely distributed in north and along the Truong Son (Annamite Chain); semi-deciduous closed forests in north-west and south; dry deciduous dipterocarp forests in south. Pine forests mainly restricted to Tay Nguyen highlands; bamboo forests, scrub and grassland throughout the whole of Viet Nam; mangroves mostly in south.

78,169 sq. km of forests and woodlands (data from interpretation of LANDSAT images, 1973-1982, and aerial photographs, 1970-1981, together with field checks carried out by the Forestry Inventory and Planning Institute in 1982). Estimated rate of deforestation of closed broadleaved forests 600 sq. km/annum out of a total of 74,000 sq. km (FAO/UNEP, 1981).

For vegetation map see the *National Atlas of Viet Nam* (in press). See also:

Rollet, B. (1969). *Vegetation Map, Republic of Vietnam*. Scale 1:1,000,000. Nat. Geographic Service, Dalut, Viet Nam.

Checklists and Floras No national Flora. Viet Nam is included in *Flore du Cambodge, du Laos, et du Vietnam* (1960-) and *Flore Générale de L'Indo-Chine* (1907-1951), cited in Appendix 1. See also:

Pham Hoang Ho (1970, 1972). *An Illustrated Flora of South Viet Nam*, 2 vols. Saigon. (In Vietnamese.)

Field-guides

Forestry Inventory and Planning Institute (1972-1982). *Forest Trees of Viet Nam*, 5 vols. Ha Noi.
Le Kha Ke *et al.* (1969-1976). *Common Plants of Viet Nam*, 6 vols. Ha Noi.
Pham Hoang Ho (1972). *Common Plants of South Viet Nam*. Lua Thieng, Saigon.

Information on Threatened Plants

Phan Ke Loc (1983a). Preliminary results of the discovery of some rare and endangered plant species to be protected in Viet Nam. Report in the symposium of the national workshop on rational use of resources and environmental protection, 19-23 November 1983, Ha Noi.
Phan Ke Loc (1983b). Map showing the distribution of rare and endangered plant species to be protected in Viet Nam (with list of 568 species). Report in the Symposium on the National Atlas of Viet Nam, 26-27 November, 1983, Ha Noi.
Westing, A.H. and C.E. (1981). Endangered species and habitats of Viet Nam. *Envir. Conserv.* 8(1): 59-62. (Summary of extent of forest destruction in southern Viet Nam; mentions 2 rare gymnosperms.)

IUCN has a preliminary list, prepared by Phan Ke Loc and Ngu Ven Tien Ban, of 14 Rare endemic taxa (including 2 cycads, 4 orchids and 8 palms).

Voluntary Organizations

Society of Environmental Conservation, Ha Noi.

Botanic Gardens

Zoo-Botanical Garden, Ha Noi.
Zoo-Botanical Garden, Ho Chi Minh City.

Useful Addresses

Department of Botany, University of Ha Noi.

Department of Survey of National Resources, State Committee for Science and Technology, Ha Noi.

Department of Wildlife, Forest Inventory and Planning Institution, Ha Noi.

Additional References

Legris, P. (1974). Vegetation and floristic composition of humid tropical continental Asia. In Unesco, *Natural Resources of Humid Tropical Asia*. Natural Resources Research 12. Paris. Pp. 217-238.

Pfeiffer, E.W. (1984). The conservation of nature in Viet Nam. *Envir. Conserv.* 11(3): 217-221.

Schmid, M. (1974). *Végétation du Viet-Nam: Le Massif Sud-Annamitique et Les Régions Limitrophes*. ORSTOM, Paris. 243 pp.

Vidal, J.E. (1979). Outline of ecology and vegetation of the Indochinese Peninsula. In Larsen, K. and Holm-Nielsen, L.B. (Eds), *Tropical Botany*. Academic Press, London. Pp. 109-123.

Wake Island

Three low coral atolls (Wake, Wilkes and Peale) surrounding a lagoon and enclosed by a reef at its north-western end, situated 550 km north of Pokak (Marshall Islands) at latitude 19°18'N, longitude 166°35'E. Area 7.4 sq. km; population 302 (1980). Wake is an unincorporated territory of the United States.

The atolls are covered by an open scrub forest (mainly *Tournefortia*); *Sesuvium* flats and *Pemphis* scrub along the lagoon margins; little, if anything, remains of the original *Pisonia* forest (Fosberg and Sachet, 1969; Fosberg, 1973, cited in Appendix 1). 123 vascular plant taxa recorded on the island of which only 21 are indigenous.

References Wake is included in the regional checklists of Fosberg, Sachet and Oliver (1979, 1982), and the *Flora of Micronesia* (1975-), cited in Appendix 1. See also:

Christophersen, E. (1931). Vascular plants of Johnston and Wake Islands. *Occ. Papers Bernice P. Bishop Mus.* 9(13). 20 pp.

Fosberg, F.R. (1959). Vegetation and flora of Wake Island. *Atoll Res. Bull.* 67. 20 pp. (Includes annotated checklist.)

Fosberg, F.R. and Sachet, M.-H. (1969). Wake Island vegetation and flora, 1961-1963. *Atoll Res. Bull.* 123. 15 pp. (Includes annotated checklist.)

Wallis and Futuna

Wallis (13°16'S, 176°15'W) and Futuna or Horne (14°25'S, 178°20'W) are a French overseas territory in the south-west Pacific Ocean. Area 255 sq. km; population 10,000. The islands are volcanic. Wallis attains 146 m; Futuna 500 m.

More than 400 vascular plant taxa of which c. 250 indigenous (Morat, Veillon and Hoff, 1983). 5 endemic flowering plant species; 3 restricted to Futuna (St John, 1977). Floristic affinities to Samoa and Fiji (St John and Smith, 1971). No information on threatened plants.

Evergreen closed forest covers about 15% of Wallis; secondary forest and scrub ('toafa') with *Pandanus*, grasses and ferns (Morat, Veillon and Hoff, 1983). Much of the original vegetation has been cleared for coconuts, breadfruit and bananas (Douglas, 1969, cited in Appendix 1). About 30% of Futuna still has dense forest, particularly in the montane regions above 400 m; scrub ('toafa') and grassland in deforested areas.

References Wallis is included in the *Flora of Southeastern Polynesia* (Brown and Brown, 1931-1935), cited in Appendix 1. See also:

Morat, Ph., Veillon, J.M. and Hoff, M. (1983). *Introduction à La Végétation et à La Flore du Territoire de Wallis et Futuna*. ORSTOM, Nouméa. 53 pp. (Account of 3 botanical missions; descriptions and map of vegetation; annotated checklist of species with geographical and ecological data.)
St John, H. and Smith, A.C. (1971). The vascular flora of the Horne and Wallis Islands. *Pacific Science* 25(3): 313-348. (Annotated checklist with geographic and ecological data, indigenous and endemic species indicated.)
St John, H. (1977). Additions to the flora of Futuna Island, Horne Islands. *Phytologia* 36(4): 367-373.

Western Sahara

Area 266,000 sq. km

Population 151,000

Floristics 330 species (quoted in Lebrun, 1976, cited in Appendix 1); Saharan Mauritania and Western Sahara together have a flora of roughly 600 species (Quézel, 1978, cited in Appendix 1).

Flora essentially Saharan, but with a Mediterranean element in the extreme north-west corner.

Vegetation Mostly desert with little or no perennial vegetation. Along the coast is a narrow band of so-called Atlantic coastal desert; vegetation cover is relatively dense and relatively rich in species, consisting mostly of succulent shrubs and dense growth of epiphytic and terrestrial lichens. In the extreme north, the coastal area is occupied by sub-Mediterranean grassland and succulent shrubland, composed especially of three species of shrubby *Euphorbia*.

For vegetation map see White (1983), cited in Appendix 1.

Checklists and Floras Western Sahara is included in the computerized *Atlas der Pflanzenwelt des Nordafrikanischen Trockenraumes* (Frankenberg and Klaus, 1980), and *Flore du Sahara* (Ozenda, 1977), both cited in Appendix 1.

Guinea, E. (1948). Catálogo razonado de las plantas del Sáhara español. *An. Jard. Bot. Madrid* 8: 357-442.

Guinea López, E. (1949). Geobotánico. In Hernández-Pacheco, E. *et al.* (Eds), *El Sáhara Español: Estudo Geológico, Geográfico y Botánico*, pp. 631-806. Consejo Superior de Investigaciones Científicas, Instituto Estudios Africanos, Madrid.

Sauvage, C. (1946). Notes botaniques sur le Zemmour oriental (Mauritanie septentrionale). *Mém. Off. Nat. Anti-acrid.* 2. 46 pp. (With four black and white photographs.)

Sauvage, C. (1949). *Nouvelles Notes Botaniques sur le Zemmour Oriental (Mauritanie Septentrionale).* In Travaux Botaniques dédiés à René Maire, Mém. Soc. Hist. Nat. Afr. N., hors sér. 2: 279-290. (With 2 black and white photographs.)

Information on Threatened Plants None.

Additional References

Guinea, E. (1945a). *Aspecto Forestal del Desierto. La Vegetacion Leñosa y los Pastos del Sáhara Español.* Instituto Forestal de Investigaciones y Experiencias, Madrid. 152 pp. (With vegetation map in colour; several black and white photographs.)

Guinea López, E. (1945b). *España y el Desierto. Impresiones Saharianas de un Botánico Español.* Colección España ante el Mundo, Instituto de Estudios Políticos, Madrid. 279 pp. (With numerous black and white photographs.)

Murat, M. (1939). La végétation du Sahara occidental en zone espagnole. *Compt. Rend. Somm. Séanc. Soc. Biogéog.* (It is not clear if this was ever published.)

Western Samoa

The Samoan Archipelago is a chain of tropical, volcanic islands extending in a west-northwesterly direction in the South Pacific Ocean, 4200 km south-west of Hawaii and 1000 km north-east of Fiji. The archipelago is divided politically into Western Samoa and American (or Eastern) Samoa. Western Samoa (13°50'S, 172°W) comprises Savai'i and Upolu, and 7 smaller islands, 5 of which are uninhabited. Of these, Nu'utele, Nu'ulua, Namu'a and Fanuatapu are relatively undisturbed tuff cone islands forming the Aleipata Islands (total area 1.75 sq. km), situated to the east of Upolu (Whistler, 1983). The highest point is 1857 m, on Savai'i. American Samoa is covered separately.

Area 2841 sq. km

Population 163,000

Floristics No figure for number of flowering plants, but c. 200 fern species (Parris, 1985, cited in Appendix 1). Overall endemism for the Samoan archipelago (including American Samoa) is estimated at c. 25% (Whistler, 1980). Much of the flora is related to that of Fiji and Tonga. The flora of the Aleipata Islands is distinct from that of Upolu and Savai'i.

Vegetation On Savai'i and Upolu most of the original lowland tropical forest has been cleared or highly modified. Montane forests are less damaged and still contain a rich endemic flora. Cloud forests, montane lava flow scrub and montane meadows are found in the upland regions. The Aleipata Islands include a number of littoral communities, as well as *Diospyros* coastal forests and *Dysoxylum* lowland forests, which are otherwise rare in Western Samoa (Whistler, 1983).

Checklists and Floras

Christensen, C. (1943). A revision of the Pteridophyta of Samoa. *Bishop Mus. Bull.* 177. 138 pp. (Includes American Samoa.)

Christophersen, E. (1935, 1938). Flowering plants of Samoa. *Bull. Bernice P. Bishop Mus.* 128. 221 pp.; 154. 77 pp. (Includes American Samoa.)

Parham, B.E.V. (1972). *Plants of Samoa*. DSIR Information Series no. 85. Govt Printer, Wellington, N.Z. 162 pp. (Short descriptions of many plants used locally, listed alphabetically by local names; includes American Samoa.)

Whistler, W.A. (1978). Vegetation of the montane region of Savai'i, Western Samoa. *Pacific Science* 32(1): 79-94. (Includes annotated checklist of 86 taxa, the majority of the species on Savai'i.)

Whistler, W.A. (1983). Vegetation and flora of the Aleipata Islands, Western Samoa. *Pacific Science* 37(3): 227-249. (Includes checklist of 260 species of which 178 native or naturalized.)

Information on Threatened Plants None.

Botanic Gardens

Arboretum of the South Pacific Regional College of Tropical Agriculture, Alafua, P.O. Box 890, Apia, Western Samoa.

Additional References

Whistler, W.A. (1980). The vegetation of Eastern Samoa. *Allertonia* 2(2): 46-190.

Yemen, Democratic

(SOUTH YEMEN)

Area 287,680 sq. km

Population 2,066,000

Floristics c. 1000 species (J.R.I. Wood, 1984, pers. comm.); probably at least 30 endemics (A.G. Miller, 1984, *in litt.*), particularly in succulent genera; most endemics on isolated mountains along the coast. Likely to be the richest part of the Arabian peninsula for succulent species (Wood, pers. comm.).

Vegetation Mostly poor *Acacia* bushland with dissected plateaux in the east (Hadramaut), and some fertile cultivated valley bottoms.

Checklists and Floras Floristically one of the least known countries in the world.

Blatter, E. (1914-1916). Flora of Aden. *Rec. Bot. Surv. India* 7. 418 pp., + 19 page index. (Keys, descriptions, notes on distributions.)

Wood, J.R.I. (in press). *Handbook of the Flora of Yemen*. To be published in 1985 by Routledge and Kegan Paul, London. (Does not actually cover South Yemen, but includes many of the same species.)

South Yemen is also included in the following:

Schwartz, O. (1939). *Flora des Tropischen Arabien*. Mitteilungen aus dem Institut für allgemeine Botanik in Hamburg 10. 393 pp. (Annotated checklist with distributions for tropical Arabia.)

Vierhapper, F. (1907). *Beiträge zur Kenntnis der Flora Südarabiens und der Inseln Sokótra, Sémha und 'Abd el Kûri*. Denkschr. K. Akad. Wiss. Wien, Math.-Naturwiss. 71: 321-490. (Descriptions; no keys.)

Other works relating to the Arabian peninsula as a whole are outlined under Saudi Arabia.

Information on Threatened Plants One species from South Yemen (*Wissmannia carinensis*) is included in *The IUCN Plant Red Data Book* (1978).

Additional References
Scott, H. (1944). *In the High Yemen*. London.

Yemen Arab Republic

(NORTH YEMEN)

Area 189,850 sq. km

Population 6,386,000

Floristics About 1700 species of vascular plants (Wood, in press). Endemism about 5-10% (J.R.I. Wood, 1984, pers. comm.) with a high proportion in succulent genera and the high montane flora. Many endemic species extend into isolated localities in Saudi Arabia or South Yemen.

Floristic affinities: the lowland flora with tropical Africa, especially Ethiopia and Somalia; the montane flora with the East African highlands; the eastern desert flora with the Sahara-Sindian region.

Vegetation On the coast: mangrove, abundant in places. On the coastal plain: desert vegetation, mainly of dune grasses and occasional shrubs. Further inland: extensive cultivation, with trees appearing in places. On the western escarpment of the main mountain range: thickets, often of succulents, with open woodland in places at lower altitudes; quite dense woodland is found in the foothills of the escarpment mountains where cultivation is sparse. At higher altitudes there has been much clearing as a result of cultivation; now scattered perennials of rocky places, and scrub, often heavily grazed. On the montane plains at around 2400 m: steppe with much cereal cultivation and few trees. In the mountains and desert east and north of Sana'a, the vegetation is more sparse and less well known (Hepper, 1977; Hepper and Wood, 1979). An important element of the vegetation is the succulent scrub (up to 2-3 m) which is being rapidly cleared in many areas.

Checklists and Floras Works relating to the Arabian peninsula as a whole are outlined under Saudi Arabia. See also:

Schwartz, O. (1939). *Flora des Tropischen Arabien*. Mitteilungen aus dem Institut für allgemeine Botanik in Hamburg 10. 393 pp. (Annotated checklist with distributions for tropical Arabia.)
Wood, J.R.I. (in press). *Handbook of the Flora of Yemen*. To be published in 1985 by Routledge and Kegan Paul, London.

Information on Threatened Plants None.

Useful Addresses

Gesellschaft für Technische Zusammenarbeit Forestry Project (HARAZ), Eschborn 4; or c/o Ministry of Agriculture, Sana'a.

Ministry of Agriculture (Forestry Division), Sana'a.

Additional References

Hepper, F.N. (1977). Outline of the vegetation of the Yemen Arab Republic. *Publ. Cairo Univ. Herb.* 7/8: 307-322. (Includes account of botanical exploration, and six black and white photographs.)

Hepper, F.N. and Wood, J.R.I. (1979). Were there forests in the Yemen? *Proc. Seminar for Arabian Studies* 9: 65-69, with ten black and white photographs. (c/o Institute of Archaeology, 31-34 Gordon Square, London WC1H OPY.)

Scott, H. (1944). *In the High Yemen.* London.

Wood, J.R.I. (1983). Vegetation of the Yemen Arab Republic. In *Soil Survey of North Yemen.* Cornell University. c. 50 pp.

Yugoslavia

Area 255,803 sq. km

Population 22,420,000 (1981)

Floristics 4750-4900 native vascular species, estimated by D.A. Webb (1978, cited in Appendix 1), from *Flora Europaea*; 137 endemics (IUCN figures). Elements: Mediterranean; sub-Mediterranean; Central European in the Alps of the north-west and on the large Pannonian plain in the north-east.

Floristically diverse areas: Macedonian Mts, supporting a mixture of alpine species and Balkan endemics; mountains of north-west and central Yugoslavia, including the high limestone plateau of the Dinaric range; and Dalmatia.

Vegetation Mixed deciduous oakwoods, dominated by *Quercus frainetto* and *Q. cerris*, cover the karstlands of the far north and south between 200 and 700 m. These are replaced by beech at higher altitudes. In the central mountains, coniferous forest is dominant, with riverine forests of alder, willow and ash at lower levels. Coastal maquis still widespread with scattered patches of Aleppo Pine (*Pinus halepensis*), Holm Oak (*Q. ilex*), Manna Ash (*Fraxinus ornus*) and evergreen shrubs. Some White Oak (*Q. pubescens*) along the Croatian coast. On the northern Pannonian plain, mostly an agricultural landscape, but some relicts of steppe flora still present on saline soils.

For a vegetation map see Matvejev (1961).

Checklists and Floras Covered by the completed *Flora Europaea* (Tutin *et al.*, 1964-1980) and the partially completed *Med-Checklist*, both cited in Appendix 1.

National Floras:

Horvatić, S. and Trinajstić, I. (Eds) (1967-). *Analitička flora Jugoslavije*, 4 vols. Institut za Botaniku Sveučilišta, Zagreb. (Incomplete; 1 – pteridophytes, gymnosperms; 2 – Lauraceae to Brassicaceae; 3 – Ulmaceae to Juglandaceae; 4 – Juglandaceae to Caryophyllaceae; introduction covers geography and vegetation; habitats; maps.)

Mayer, E. *et al*. (1964-). *Catalogus Florae Jugoslaviae*. Acad. Sci. RP Socialista Foedradivae Jugoslav., Ljubljana. (Covers vascular plants and mosses; no keys; in Serbo-Croat; 2 vols to date: 1 – pteridophytes, gymnosperms; 2(1) – mosses.)

Regional Floras:

Beck, G. and Malý, K. (1950-1983). *Flora Bosnae et Hercegovinae*, 4 vols. Sarajevo and Beograd. (In Serbo-Croat.)

Domac, R. (1950). *Flora: za odredjivanje i upoznavanje bilja* (Flora of the Danube Basin and adjacent uplands). Institut za Botaniku Sveučilišta, Zagreb. 552 pp. (Illus.)

Domac, R. (1973). *Mala Flora Hrvatske i Susjednik Područja*. 'Skolska Knijga', Zagreb. 543 pp. (Covers Croatia only.)

Josifović, M. (Ed.) (1970-1977). *Flora SR Srbije (Flore de la République Socialiste de Serbie)*, 9 vols. Srpska Akademija Nauka i Umetnosti, Beograd. (In Serbo-Croat; native, naturalized and commonly cultivated plants of Serbia, Novi Pazar and Kosovo; details on distribution, habitat and ecology.)

Mayer, E. (1952). *Seznam Praprotnic in cvetnic slovenskega ozemlja* (List of ferns and flowering plants in the district of Slovenia). Slovenska Akad. Znanosti, Ljubljana. 427 pp.

A bibliography:

Pulević, V. (1980). Bibliografija o flori i vegetaciji Crne gore (Bibliography of the flora and vegetation of Montenegro). *Crnog. akad. nauk. umjetn., Odj. prir. nauk.*, Titograd. Bibliografije 1. 235 pp.

Field-guides

Domac, R. (1979). *Mala flora Hrvatske i susjednih područja* (The small Flora of Croatia and adjacent regions). Zagreb. 543 pp.

Lakušić, R. (1982). *Planinske biljke* (Mountain plants). Svjetlost, Sarajevo. 204 pp.

Martinčić, A. and Sušnik, F. (Ed.) (1984). *Mala flora Slovenije*, 2nd Ed. Ljubljana. (Keys and descriptions; taxa names in Latin and Slovene.)

Polunin (1980), cited in Appendix 1.

Šilić, Č. (1983a). *Atlas Drveća i Grmlja* (Atlas of trees and shrubs), 2nd Ed. Svjetlost, Sarajevo. 218 pp.

Šilić, Č. (1983b). *Šumske Zeljaste Biljke* (Herbaceous plants of the forests), 2nd Ed. Svjetlost, Sarajevo. 272 pp.

Information on Threatened Plants The Institute of Botany in the Republic of Croatia is currently compiling a Yugoslav Red Data Book. Another is in preparation for the Republic of Slovenia. Other relevant publications:

Anon (1984). *Naučen sobir Florata i vegetacijata na Jugoslavija i problemot na nivnata zaštita* (Symposium on the flora and vegetation in Yugoslavia and their protection). Skopje, 2-4 June 1982. *Maked. akad. nauk. umetn., Oddel. biol. med. nauk*. 162 pp.

Blečić, V. (1957). Endemične i retke biljke u Srbiji (Endemic and rare plants in Serbia). *Zašt. Prir.* 9: 1-6.

Broz, V. (1963). Rad na zaštiti retke i ugrožene flore (The work on the protection of rare and threatened flora). *Zašt. Prir.* 26: 125-130.

Fukarek, P. (1959). Arbres et arbustes rares et menacés de la flore de Yougoslavie. In *Animaux et Végétaux Rares de la Région Méditerranéenne*. Proceedings of the IUCN 7th Technical Meeting, 11-19 September 1958, Athens, vol. 5. IUCN, Brussels. Pp. 159-165.

Godicl, L. (1981). The protection of rare plants in nature reserves and national parks in Yugoslavia. In Synge (1981), cited in Appendix 1. Pp. 491-502.

Pevalek, I. (1959). Sur les plantes rares et menacées de la région Méditerranéenne de la Yougoslavie. In *Animaux et Végétaux Rares de la Région Méditerranéenne*. Proceedings of the IUCN 7th Technical Meeting, 11-19 September 1958, Athens, vol. 5. IUCN, Brussels. Pp. 166-167.

Pulević, V. (1984). Zaštićene biljne vrste u SR Crnoj Gori (Protected plant species in SR Montenegro). *Glasnik* (Bulletin of the Republic Institution for the Protection of Nature and the Museum of Natural History in Titograd) 16: 33-54. (Discusses the status of 57 threatened species; English abstract.)

Wraber, T. (1965). Nekaj misli o varstvu narave, posebej še rastlinstva (Some ideas about nature protection, especially of plants). *Varstvo Narave* (Ljubljana) 2-3: 75-88. (Discusses 56 species protected in Slovenia; English summary.)

Included in the European threatened plant list (Threatened Plants Unit, 1983, cited in Appendix 1); latest IUCN statistics, based upon this work: endemic taxa – Ex:1, E:1, V:6, R:85, I:3, K:21, nt:20; doubtful endemics – K:2; non-endemics rare or threatened worldwide – E:1, V:27, R:76, I:9 (world categories).

Laws Protecting Plants Legislation has been enacted in each of the Republics; the following figures, where available, indicate number of taxa protected, followed in brackets by date of legislation:

Slovenia 28 (1976); Croatia 32 (1972), see Plavšić-Gojković (1976); Bosnia and Hercegovina 7 (?); Serbia 17 (1955 and 1973/74); Montenegro; Macedonia. Legislation has also been passed in the Autonomous Provinces of Kosovo and Vojvodina.

Plavšić-Gojković, N. (1972). Zaštićene biljne vrste u SR Hrvatskoj (Protected plant species in the Socialist Republic of Croatia). *Mala hortikulturna biblioteka* 2. 68 pp. (Distribution maps and line drawings for each species.)

Plavšić-Gojković, N. (1976). Seltene Geschutzte Pflanzen der Sozialistischen Republic Kroatien (Jugoslawien), (Rare protected plants of the Croatia, Yugoslavia.) *Poljopr. znan. smotra* 36(46): 61-71. (In German; summary and abstract in Serbo-Croat.)

Skoberne, P. (1984). *Zavarovane rastline* (Protected plants). Škofja Loka. 32 pp. (Colour photographs with brief distribution and conservation notes for 28 species protected in Slovenia.)

Botanic Gardens

Alpine Botanic Garden Juliana, Prirodoslovni muzej Slovenije, Prešernova 20, 61000 Ljubljana.

Arboretum Volčji potok, 61335 Radomlje.

Biological Institute, Botanical Garden "Lokrum", P.O. Box 39, 50001 Dubrovnik.

Botanical Garden, Marulićev trg 9a, 41007 Zagreb.

Botanic Garden of the Botanic Institute, University of Skopje, P.O. Box 439, 91000 Skopje-Gazibaba, Makedonija.

Botanic Garden, Universitatis Labacensis, Ižanska 15, 61000 Ljubljana.

Botanical Institute and Garden, Takovska 43, 11000 Beograd.

Botanički vrt Zemaljskog muzeja, BiH Sarajevo.

Insular Karst Garten and Flora Exsicata Adriatica, Buličeva 8-v, 41000 Zagreb.

Velebit Botanic Garden, Kod Doma na Zavižanu, 51284 Jurjevo kod Senja.

Useful Addresses Republican Institutes for Nature Conservation:

Pokrajinski zavod za zaštitu prirode, Petrovaradinska tvrdjava, 21000 Novi Sad.
Republički zavod za zaštitu prirode, Ilica 44, 41000 Zagreb.
Republički zavod za zaštitu prirode, Ilije Milačića 22, 81001 Titograd.
Republički zavod za zaštitu prirode, Treći bulevar 106, 11000 Beograd.
Zavod SR Slovenije za varstvo naravne in kulturne dediščine, Plečnikov trg 2, 61001 Ljubljana.
Zavod za zaštitu spomenika kulture BiH, Obala 27. jula 11a, 71000 Sarajevo.

Additional References

Bertović, S., Kamenarović, M. and Kevo, R. (1961). *The Protection of Nature in Croatia.* Zagreb. (Not seen.)

Matvejev, S. (1961). *Biogeografija Jugoslavije.* Biološki Institut N.R. Srbije. Posebna izdanja (monograph) 9. 232 pp. (Account of biogeography throughout country, describing floristic regions; English summary; black and white photographs; maps.)

Šilić, Č. (1984). *Endemične Biljke.* Svjetlost, Sarajevo. 227 pp. (Describes endemic taxa; 164 colour illus.)

Zaïre

Area 2,345,410 sq. km

Population 32,084,000

Floristics 11,000 species (quoted in Lebrun, 1976, cited in Appendix 1); 10,000 species (Robyns, 1947); c. 3200 endemic species (Brenan, 1978, cited in Appendix 1, estimated from a sample of the *Flore du Congo Belge et du Ruanda-Urundi*); this is the highest percentage endemism in tropical Africa. Levels of diversity and endemism highest around edges of Congo basin, principally to the west (Forestier Central) and to the east (Haut Katanga).

Floristic affinities mostly Guinea-Congolian but flora in southern c. 1/3 of area with Zambezian affinities or transition between the two. Considerable areas of high ground along eastern border with Afromontane flora.

Vegetation Central and western parts: lowland rain forest and secondary grassland. Southern and south-western parts: higher rainfall *Brachystegia-Julbernardia* (Miombo) woodland. Also extensive areas of Afromontane communities and transitional rain forest in the east. Edaphic grassland in a fragmentary band between Miombo and lowland rain forest.

Estimated rate of deforestation for closed broadleaved forest 1800 sq. km/annum out of 1,056,500 sq. km (FAO/UNEP, 1981). Myers (1980, cited in Appendix 1) quotes a range of figures for covering of moist forests: from 750,000 sq. km to 1,000,000 sq. km.

For vegetation map see White (1983), cited in Appendix 1.

Checklists and Floras Zaïre is included in the incomplete *Flore du Congo Belge et du Ruanda-Urundi*, continued since 1972 as *Flore d'Afrique Centrale (Zaïre – Rwanda – Burundi)*. Zaïre's plants of high altitudes are listed in *Afroalpine Vascular Plants* (Hedberg, 1957). Both are cited in Appendix 1.

Pieters, A. (1977). *Essences Forestières du Zaïre.* Publ. Univ. Gent. 349 pp.

Robyns, W. (1947-1955). *Flore des Spermatophytes du Parc National Albert*, 3 vols. Institut des Parcs Nationaux du Congo Belge, Bruxelles. 745, 627 and 571 pp. (Descriptive keys, distributions, specimens.)

Troupin, G. (1956). *Flore des Spermatophytes du Parc National de la Garamba*, 1 vol. Institut des Parcs Nationaux du Congo Belge, Bruxelles. 349 pp. (Incomplete, covering gymnosperms and monocotyledons only; descriptions, distributions, specimens, line drawings and black and white photographs.)

Field-guides

Lebrun, J. (1935). *Les Essences Forestières des Régions Montagneuses du Congo Oriental.* Publ. INEAC, Sér. Scient. No. 1, Bruxelles. 263 pp. (Keys, descriptions, distributions, specimens; some line drawings.)

Information on Threatened Plants

Hedberg, I. (1979), cited in Appendix 1. (Only a few examples of threatened plants given, p. 92, but also indication of threatened vegetation types.)

IUCN has records of c. 1500 species and infraspecific taxa believed to be endemic; no categories available.

Botanic Gardens

Jardin Botanique de Eala, B.P. 278, Mbandaka.

Jardin Botanique de Kisantu, B.P. 65, Kisantu-Inkisi.

There is also a botanic garden at Yangambi (INERA), but address not known.

Useful Addresses

CITES Management and Scientific Authority: Department de l'environnement, conservation de la nature et tourisme, Bureau du Commissaire d'Etat, B.P. 12348 Kin 1.

Additional References

Devred, R. (1958). La végétation forestière du Congo belge et du Ruanda-Urundi. *Bull. Soc. R. For. Belg.* 65: 409-468. (With vegetation map.)

Germain, R. (1968). Congo-Kinshasa. In Hedberg, I. and O. (1968), cited in Appendix 1. Pp. 121-125.

Lebrun, J. (1960). *Etude sur la Flore et la Végétation des Champs de Lave au Nord du Lac Kivu (Congo Belge).* Inst. Parcs Nat. Congo Belge, Bruxelles. 352 pp.

Lebrun, J. and Gilbert, G. (1954). *Une Classification Ecologique des Forêts du Congo.* Publ. INEAC, Sér. Scient. No. 63, Bruxelles. 89 pp.

Léonard, J. (1950). Botanique du Congo Belge. 1. Les groupements végétaux. In *Encyclopédie du Congo Belge* 1: 345-389. Bieleveld, Bruxelles.

Robyns, W. (1950a). La flore et la végétation. In *Encyclopédie du Congo Belge* 1: 390-409. Bieleveld, Bruxelles.

Robyns, W. (1950b). Les territoires phytogéographiques. In *Encyclopédie du Congo Belge* 1: 409-424. Bieleveld, Bruxelles.

Schmitz, A. (1963). Aperçu sur les groupements végétaux du Katanga. *Bull. Soc. R. Bot. Belg.* 96: 233-447.

Schmitz, A. (1971). *La Végétation de la Plaine de Lubumbashi (Haut-Katanga).* Publ. INEAC, Sér. Scient. No. 113, Bruxelles. 388 pp. (With 32 black and white photographs.)

Schmitz, A. (1977). *Atlas des Formations Végétales du Shaba (Zaïre).* Fondat. Univ. Luxembourg, Sér. Documents 4. 95 pp.

There is a series of vegetation and soil maps covering Zaïre, Rwanda and Burundi in c. 25 parts, published between 1954 and c. 1970 by the Institut National pour l'Etude Agronomique du Congo (INEAC); each is accompanied by a descriptive memoir, and several of the maps are to different scales. The series is called *Carte des Sols et de la Végétation du Congo Belge et du Ruanda-Urundi*, or, more recently, ... *du Congo, du Rwanda et du Burundi*.

Zambia

Area 752,617 sq. km

Population 6,445,000

Floristics 4600 species (quoted in Lebrun, 1960, cited in Appendix 1). Brenan (1978, cited in Appendix 1) estimates 211 endemic species from a sample of *Flora Zambesiaca*. Diversity highest in Solwezi-Mwinilunga area and Mbala District near Tanzania.

Flora predominantly Zambezian; also small patches of Afromontane flora in the northeast.

Vegetation Predominantly higher rainfall *Brachystegia-Julbernardia* (Miombo) woodland, but drier types in the Zambezi and Luangwa valleys, where there is also extensive Mopane woodland (dominated by *Colophospermum mopane*). Also extensive areas of dry evergreen forest and edaphic grassland mosaic with semi-aquatic vegetation in west. Montane communities are confined to the Nyika Plateau and the Mafinga Mountains on the border with Malawi. Both areas consist mostly of grassland, but with some montane forest, thicket and shrubland; montane forest is largely restricted to remnant patches at the heads of river valleys. Estimated rate of deforestation for closed broadleaved forest 400 sq. km/annum out of 30,100 sq. km (FAO/UNEP, 1981).

For vegetation maps see Wild and Barbosa (1968), and White (1983), both cited in Appendix 1.

Checklists and Floras Zambia is included in the incomplete *Flora Zambesiaca*, and also in *Trees of Central Africa* (Coates Palgrave *et al.*, 1957), both cited in Appendix 1.

Fanshawe, D.B. (1973). *Check List of the Woody Plants of Zambia Showing their Distribution*. Forest Research Bulletin No. 22, Ministry of Lands and Natural Resources, Lusaka. 48 pp. (Checklist with distributions.)
Nath Nair, D.M. (1967). A numbered dichotomous key to the selected families of Zambian flowering plants. Cyclostyled. 62 pp.
White, F. (1962). *Forest Flora of Northern Rhodesia*. Oxford Univ. Press, London. 455 pp. (Keys, descriptions, distributions, specimens, line drawings.)

Information on Threatened Plants No published lists of rare or threatened plants; IUCN has records of c. 270 species and infraspecific taxa believed to be endemic; no categories assigned.

413

Useful Addresses

CITES Management and Scientific Authority: The Conservation Division of the Dept of National Parks and Wildlife Service, Private Bag 1, Chilanga.

Additional References

Edmonds, A.C.R. (1976). *The Republic of Zambia Vegetation Map 1:500,000*. Institut für Angewandte Geodäsie, Frankfurt, F.R.G./Govt, Republic of Zambia.

Fanshawe, D.B. (1961). Evergreen forest relics in Northern Rhodesia. *Kirkia* 1: 20-24.

Fanshawe, D.B. (1969). *The Vegetation of Zambia*. Forest Research Bulletin No. 7, Ministry of Rural Development, Kitwe. 67 pp. (With 19 black and white photographs.)

Kornas, J. (1979). *Distribution and Ecology of the Pteridophytes in Zambia*. Panstwowe Wydawnictwo Naukowe, Warsaw. 207 pp. (Gives 146 species; distribution maps, 11 black and white photographs.)

Trapnell, C.G., Martin, J.D. and Allan, W. (1950). *Vegetation-Soil Map of Northern Rhodesia*. Govt Printer, Lusaka. 20 pp. (With coloured map 1:1,000,000.)

Werger, M.J.A. (1978), cited in Appendix 1. Citation includes list of relevant chapters.

White, F. (1968). Zambia. In Hedberg, I. and O. (1968), cited in Appendix 1. Pp. 208-215.

Zimbabwe

Area 390,310 sq. km

Population 8,461,000

Floristics 5428 species and infraspecific taxa of vascular plants (Gibbs Russell, 1975); c. 4200 species (quoted in Lebrun, 1960, cited in Appendix 1); largely woodland and grassland species. Brenan (1978, cited in Appendix 1) estimates 95 endemic species, from a sample of *Flora Zambesiaca*. Wild (1964) lists 42 species believed to be endemic to the Chimanimani Mountains and (1965a) 20 endemic to the serpentine soils of the Great Dyke.

Flora predominantly Zambezian with some Afromontane elements in the eastern highlands.

Vegetation Predominantly dry *Brachystegia-Julbernardia* (Miombo) woodland, but with large areas of *Colophospermum mopane* (Mopane) woodland along the Zambezi and Limpopo valleys. Woodland without characteristic dominants occurs over a large area in the south-west; dry semi-evergreen forests dominated by *Baikiaea plurijuga* occur on Kalahari sand in the north-west and dry forest dominated by species of *Combretum* and *Commiphora* are found on deep sandy soils in the Zambezi valley. An unusual edaphic community is the grassland confined to the mineral-rich soils of the Great Dyke. Extensive hydromorphic grasslands occur along the central watershed. Open grasslands interspersed with ericoid thickets cover the higher parts of the mountains on the eastern border, with montane rain forest occuring mainly on the eastern and south-eastern slopes between altitudes of 2100 and 2400 m; the vegetation of the quartzites of the Chimanimani Mountains is of particular interest.

For vegetation maps see Wild and Barbosa (1968), and White (1983), both cited in Appendix 1.

414

Checklists and Floras Zimbabwe is included in the incomplete *Flora Zambesiaca*, and also in *Trees of Central Africa* (Coates Palgrave *et al.*, 1957), both cited in Appendix 1. See also:

Boughey, A.S. (1964). A check list of the trees of Southern Rhodesia. *J. S. Afr. Bot.* 30: 151-176.

Drummond, R.B. (1975). A list of trees, shrubs and woody climbers indigenous or naturalised in Rhodesia. *Kirkia* 10(1): 229-286. (Gives 1172 species.)

Drummond, R.B. (1981). *Common Trees of the Central Watershed (Including National Tree List)*. The Department of Natural Resources, Zimbabwe.

Eyles, F. (1916). A record of plants collected in Southern Rhodesia. *Trans. R. Soc. S. Afr.* 5(4): 273-564. (Lists 2372 vascular species.)

Pardy, A.A. (1951-56). *Notes on Indigenous Trees and Shrubs of Southern Rhodesia*. Govt Printer, Salisbury. Reprinted from several volumes (48/3-53/6) of the *Rhodesia Agricultural Journal*. (With numerous black and white photographs.)

Field-guides

Biegel, H.M. (1979). *Rhodesian Wild Flowers*. Thomas Meikle Series No. 4, National Museums and Monuments, Salisbury. 77 pp. (Numerous colour illustrations by Margaret Tredgold.)

Guy, G.L. and Elkington, B.D. (1972). *The Bundu Book of Trees, Flowers and Grasses*, 2nd Ed., revised by R.B. Drummond (first published 1965). Longman Rhodesia, Salisbury. 136 pp. (Photographs and drawings throughout.)

Linley, K. and Baker, B. (1972). *Flowers of the Veld*. Bundu Series, Longman Rhodesia, Salisbury. 120 pp. (Colour photographs throughout.)

Plowes, D.C.H. and Drummond, R.B. (1976). *Wild Flowers of Rhodesia*. Longman Rhodesia, Salisbury. (Colour photographs throughout.)

Wild, H. (1972). *A Rhodesian Botanical Dictionary of African and English Plant Names*, revised and enlarged by Biegel, H.M. and Mavi, S. Govt Printers, Salisbury. 281 pp.

Information on Threatened Plants

Hedberg, I. (1979), cited in Appendix 1. (List for Zimbabwe, pp. 99-100, by H. Wild and T. Müller, contains 84 species and infraspecific taxa: E:18 (country category) including 3 endemics, V:26 inc. 4 endemics, R:40 inc. 1 endemic.)

Kimberley, M.J. (1980). Specially protected plants in Zimbabwe. *Excelsa* 9: 53-54.

IUCN has records of c. 170 species and infraspecific taxa believed to be endemic, 8 of which are known to be rare or threatened.

Laws Protecting Plants The Parks and Wildlife Act of 1976 provides specific protection for 23 listed taxa (list amended 1979), including all aloes, cycads and epiphytic orchids. See also Kimberley (1980), cited above.

Botanic Gardens

Ewanrigg Botanic Gardens, P.O. Box 8119, Causeway, Harare.
National Botanic Garden, P.O. Box 8100, Causeway, Harare.
University of Zimbabwe Botanic Garden, P.O. Box MP167, Harare.
Vumba Botanical Garden, P/Bag V7472, Umtali.

Useful Addresses

CITES Management and Scientific Authority: Dept of National Parks and Wildlife Management, P.O. Box 8365, Causeway, Harare.

Additional References

Gibbs Russell, G.E. (1975). Comparison of the size of various African floras. *Kirkia* 10: 123-130.

Rattray, J.M. (1961). Vegetation types of Southern Rhodesia. *Kirkia* 2: 68-93.

Werger, M.J.A. (1978), cited in Appendix 1. Citation includes list of relevant chapters.

Wild, H. (1964). The endemic species of the Chimanimani Mountains and their significance. *Kirkia* 4: 125-157.

Wild, H. (1965a). The flora of the Great Dyke of S. Rhodesia with special reference to the serpentine soils. *Kirkia* 5(1): 49-86. (With 8 black and white photographs.)

Wild, H. (1965b). Vegetation map of Rhodesia. In Collins, M.O. (Ed.), *Rhodesia, its Natural Resources and Economic Development*. Salisbury.

Wild, H. (1968). Rhodesia. In Hedberg, I. and O. (1968), cited in Appendix 1. Pp. 202-207.

Appendix 1: General and Regional References

This appendix contains references, with appropriate cross-references, to all relevant works which do not appear in full elsewhere. As well as citing works referred to only briefly in the country sheets, it contains references to a number of works of general importance for a given subject or area. Refer to Appendix 2 for an index to this appendix, based on geographical relevance and subject matter.

Adam, J.-G. (1971-1984). See *Flore Descriptive des Monts Nimba.*

Adumbratio Florae Aethiopicae (1953-), by Chiarugi, A. *et al.* (currently under the direction of G. Moggi.) Introduction + 32 families (mostly small ones) so far; published in *Webbia*, starting with vol. 9 (1953), and also published separately. Most recent part is in *Webbia* 33(1), 1978. Covers Ethiopia, Somalia, Djibouti and Socotra; also includes ferns.

AETFAT. For proceedings of recent meetings, see Hedberg and Hedberg (1968); Merxmüller (1971); Miège and Stork (1975, 1976); Kunkel (1979); Killick (1983).

Allen, R. (1980). *How To Save The World*. Kogan Page, London. 150 pp. (Popular version of the *World Conservation Strategy* (IUCN, UNEP and WWF, 1980), cited in full below.)

Ambio 12(6), 1983. Special issue on the Indian Ocean and adjoining East Africa. (See in particular articles by P. O'Keefe, pp. 302-305, and C.G. Wenner, pp. 305-307, both on soil erosion in Kenya; P. Randrianarijaona, pp. 308-311, on soil erosion in Madagascar.)

Ambio 13(5-6), 1984. Special issue on the South Pacific. (See in particular, D.A. Ballendorf on Micronesia, pp. 294-295; A.L. Dahl on environmental problems, pp. 296-301; A.B. Viner on environmental protection in Papua New Guinea, pp. 342-344; G. Seddon on the effects of logging in the Gogol Valley, Papua New Guinea, pp. 351-354; M. Pulea on environmental legislation in the Pacific region , pp. 369-371.

Anon. (1984). *Conservation de la Nature en Europe. Vingt Années d'Activités*. Division de l'environnement et des ressources naturelles. Council of Europe, Strasbourg. 111 pp. (Includes description of vegetation, and a conservation data sheet for each country.)

Anon. (1985). *Catalogue of Data Banks in the Field of Nature Conservation*. 2nd Colloquy on Computer Applications in the Field of Nature Conservation, Strasbourg, 26 and 27 February 1985, Council of Europe, Strasbourg. 109 pp.

Atlas Florae Europaeae. See Jalas and Suominen (1972-).

Atlas der Pflanzenwelt des Nordafrikanischen Trockenraumes. See Frankenberg and Klaus (1980).

Aubréville, A. (1950). *Flore Forestière Soudano-Guinéenne*. Societé d'Editions Géographiques Maritimes et Coloniales, Paris. (Reprinted 1975 by Centre Technique de Forestier Tropical, Nogent-sur-Marne, according to Frodin.) 523 pp. (Illustrated. 115 plates, + distribution maps. Covers Mali, Senegal, Gambia, Niger, Burkina Faso, Mauritania.)

Aubréville, A. and Leroy, J.-F. (Eds) (1960-). See *Flore du Cambodge, du Laos, et du Vietnam.*

Ayensu, E.S. (1984). Keynote Address: the Afrotropical Realm. In McNeely, J.A. and Miller, K.R. (Eds), *National Parks, Conservation and Development. The Role of*

Protected Areas in Sustaining Society. Proceedings of the World Congress on National Parks, Bali, Indonesia, 11-22 October, 1982. Smithsonian Institution Press, Washington. Pp. 80-86. (Covers all tropical Africa.)

Ayensu, E.S., Heywood, V.H., Lucas, G.L. and De Filipps, R.A. (1984). *Our Green and Living World – The Wisdom To Save It*. Cambridge University Press, Cambridge. 255 pp. (Illustrated account of plant diversity, emphasizing economic values of plants; the need for and goals of conservation.)

Baker, J.G. (1877). *Flora of Mauritius and the Seychelles: A Description of the Flowering Plants and Ferns of those Islands*. Reeve, London; reprinted 1970 by Cramer, Lehre, according to Frodin. 557 pp. (Primitive keys, descriptions, short notes on distributions.)

Balgooy, M.M.J. van (1971). *Plant-geography of the Pacific as Based on a Census of Phanerogam Genera*. Blumea Suppl. 6. Rijksherbarium, Leiden. (Contains list of 1666 Pacific genera with their distributions.)

Baumann, H. and Künkele, S. (1982). *Die Wildwachsenden Orchideen Europas*. Kosmos, Stuttgart. 432 pp. (Geographical coverage as *Flora Europaea*; maps, illus.)

Beard, J.S. (1949). The natural vegetation of the Windward and Leeward Islands. *Oxford Forestry Memoirs No. 21*. Clarendon Press, Oxford. 192 pp.

Beddome, R.H. (1892). *Handbook to the Ferns of British India, Ceylon and the Malay Peninsula*, Calcutta. 610 pp. (Reprinted by Today and Tomorrow's Printers, 1969; supplement in 1972 by B.K. Nayar and S. Kaur. Illustrated; keys.)

Belousova, L.S. and Denisova, L.V. (1983). *Rare Plants of the World*. Timber Industry, Moscow. 340 pp. (Data sheets on 2000 selected rare and threatened plants from around the world with emphasis on Europe and Russia; over 250 colour illustrations; in Russian.)

Boardman, R. (1981). *International Organization and the Conservation of Nature*. Macmillan Press, London. 215 pp. (History of international conservation with emphasis on the founding of IUCN; case studies include conservation in the Arctic, Antarctic and East Africa regions.)

Bramwell, D. (Ed.) (1979). *Plants and Islands*. Academic Press, London. (See in particular S.A. Renvoize on the origin of Indian Ocean Island floras, pp. 107-129; V.H. Heywood on the future of island floras, pp. 431-441; R. Melville on endangered island floras, pp. 361-377.)

Brenan, J.P.M. (1978). Some aspects of the phytogeography of tropical Africa. *Ann. Missouri Bot. Gard.* 65(2): 437-478.

Brosse, J. (1977). *Atlas des Arbres de France et d'Europe Occidentale*. Bordas, Paris. 239 pp. (Illus.)

Brown, F.B.H. and E.D.W. *See Flora of Southeastern Polynesia*.

Burkill, H.M. (1985-). *The Useful Plants of West Tropical Africa*, 1 vol., 2nd Ed. (1st Ed. 1937, by J. Hutchinson and J.M. Dalziel). Royal Botanic Gardens, Kew. 960 pp. (Families A-D only; the first of 4 vols. Based on the *Flora of West Tropical Africa*, this extensive 'Flora' enumerates the vernacular names, with some translations, and uses of each species covered.)

Cabanis, Y., Cabanis, L. and Chabouis, F. (1969-1970). *Végétaux et Groupements Végétaux de Madagascar et des Mascareignes*, 4 vols. Tananarive. 260 plates. (Cited by Frodin.)

Cabrera, A. and Willink, A. (1980). *Biogeografía de América Latina*, 2nd Ed. Serie Biología, Monografía 13. Secretaria General, Organización de los Estados Americanos. Washington, D.C. 122 pp.

Carlquist, S. (1965). *Island Life: A Natural History of the Islands of the World*. Natural History Press, New York. 451 pp. (Origin, evolution and adaptations of island flora and fauna; Galápagos and Hawaiian Islands well covered.)

Carlquist, S. (1974). *Island Biology*. Columbia University Press, New York. 660 pp. (Dispersal, evolution and adaptive radiation of island flora and fauna; separate chapters on the flora of the Galápagos, Hawaiian Islands, Macaronesia, New Caledonia, New Zealand, Southwestern Australia, and continental islands (e.g. Madagascar); insular woodiness and equatorial highland biota.)

Castroviejo, S. (1979). Synthèse des progrès dans le domaine de la recherche floristique et littérature sur la flore de la region méditerranéenne. *Webbia* 34(1): 117-131. (Includes European and non-European Mediterranean.)

Clark, M.R. and Dingwall, P.R. (1985). *Conservation of Islands in the Southern Ocean: A Review of the Protected Areas of Insulantarctica*. IUCN, Switzerland. 193 pp. (Directory of information on physical and biological features of the southern islands; conservation status, problems and priorities; details of administration and management of protected areas.)

CNRS (1975). *La Flore du Bassin Méditerranéen. Essai de Systématique Synthétique*. Colloques Internationaux du Centre National de la Recherche Scientifique, Editions du CNRS, 15 quai Anatole-France, 75700 Paris. 576 pp.

Coates Palgrave, K. (1977). *Trees of Southern Africa*. Struik, Cape Town. 959 pp. (With keys, distribution maps, line drawings, colour illustrations, colour and black and white photographs throughout. Covers South Africa and its states, Lesotho, Swaziland, Namibia, Botswana, Zimbabwe, and Mozambique south of the Zambezi River.)

Coates Palgrave, O.H. *et al.* (1957). *Trees of Central Africa*. Text by K. Coates Palgrave, illustrations by O.H. Coates Palgrave, photographs by D. and P. Coates Palgrave. National Publications Trust, Rhodesia and Nyasaland. 466 pp. (110 colour illus. of indigenous trees of Central Africa with short descriptions and black and white photographs. Covers Zambia, Zimbabwe and Malawi.)

Consolidated Index to Flora Europaea. See Halliday and Beadle (1983).

Costin, A.B. and Groves, R.H. (Eds) (1973). *Nature Conservation in the Pacific*. IUCN, Switzerland, and Australian National Univ. Press, Canberra. 337 pp. (Evaluation of land for conservation, problems facing nature conservation.)

Cufodontis, G. (1953-1972). See *Enumeratio Plantarum Aethiopiae Spermatophyta*.

Dahl, A.L. (1980). *Regional Ecosystems Survey of the South Pacific Area*. Technical Paper No. 179. South Pacific Commission, Nouméa, New Caledonia. 99 pp.

Derrick, L.N. and Scheepen, J. van (1984). The European taxonomic, floristic and biosystematic documentation system. *Webbia* 38: 681-685.

Distr.Pl.Afr. See *Distributiones Plantarum Africanarum*.

Distributiones Plantarum Africanarum, (1969-). Unbound distribution maps grouped in 'fascicles'; 23 fascicles, 820 species or infraspecific taxa so far. Jardin Botanique National de Belgique, Bruxelles.

Douglas, G. (1969). Draft check list of Pacific Oceanic Islands. *Micronesica* 5(2): 327-463. (Includes brief notes on physical character, vegetation, past and present land use, references to species lists; 39 islands proposed for international scientific supervision.)

Drake del Castillo, E. (1893). See *Flore de la Polynésie Française*.

Dyer, R.A. (1975, 1976). *The Genera of Southern African Flowering Plants*, 2 vols. 3rd Ed. Flora of Southern Africa, Botanical Research Institute, Private Bag x144, Pretoria. 1040 pp. (Covers South Africa and its states, Lesotho, Swaziland and Namibia.)

Eckholm, E. (1976). *Losing Ground – Environmental Stress and World Food Prospects*. Pergamon Press, Oxford. 223 pp. (Chapters on deforestation, the spread of deserts, fuelwood crisis, soil erosion.)

Eckholm E. (1978). *Disappearing Species: the Social Challenge*. Worldwatch Paper 22. Worldwatch Institute, Washington, D.C. 38 pp. (Social costs of loss of species.)

Ehrlich, P. and A. (1981). *Extinction: the Causes and Consequences of the Disappearance of Species*. Random House, New York. 305 pp.

Eig, A., Zohary, M. and Feinbrun-Dothan, N. (1931). *The Plants of Palestine: An Analytical Key*. Univ. Press, Jerusalem. 426 pp. (In Hebrew.)

Ellenberg, H. (1978). *Vegetation Mitteleuropas mit den Alpen in Ökologischer Sicht*. Ulmer, Stuttgart. 981 pp. (Black and white photographs and line drawings.)

Engler, A. (1892). *Über die Hochgebirgsflora des Tropischen Afrika*. Abh. Königl. Preuss. Akad. Wiss. Berlin vom Jahre 1891. 461 pp. (Reprinted in 1975 by Koeltz, Königstein/Ts., according to Frodin. Covers mountains throughout tropical Africa.)

Enumeratio Plantarum Aethiopiae Spermatophyta, (1953-1972), by G. Cufodontis. (Appeared first in various volumes of *Bull. Jard. Bot. Etat Bruxelles* starting with vol. 23(3/4). In 1967 the journal changed name to *Bull. Jard. Bot. Nat. Belg.* and publication of *Enumeratio* finished in 1972 in vol. 42(3). In 1974 the whole work was reprinted in 2 vols; 1675 pp. Checklist with distributions. Covers Ethiopia, Somalia and Djibouti.)

FAO (1974). See Lugo *et al.* (1981).

FAO/UNEP (1981). *Tropical Forest Resources Assessment Project (in the Framework of the Global Environment Monitoring System – GEMS)*. UN 32/6.1301-78-04. Technical Reports nos. 1-3, Food and Agriculture Organization of the United Nations, Rome. (Comprises 3 separate reports: *Los Recursos Forestales de la América Tropical*. 343 pp. (Forest Resources of Tropical America; in Spanish); 2 – *Forest Resources of Tropical Africa*. 108, 586 pp. (In English and French); 3 – *Forest Resources of Tropical Asia*. 475 pp. (In English and French). Each report comprises 2 parts. Part 1: Regional Synthesis – includes classification of forest types; methodologies used to assess current and future trends; extent and distribution of each forest type; rates of deforestation. Tables provide detailed summaries of the extent and rates of loss for each country and forest type. Part 2: Country Briefs – detailed accounts for each country, including introduction to physical geography; description of vegetation types; present areas of natural and plantation forests; ownership, legal status and management of forests; present trends of deforestation, degradation and forest utilization.)

Favarger, C. (1956, 1958). *Flore et Végétation des Alpes*, 2 vols. Delachaux and Niestlé, Neuchâtel.

Fitter, R., Fitter, A. and Blamey, M. (1974). *The Wild Flowers of Britain and Northern Europe*. Collins, London. 336 pp. (Colour illus.)

Fl.Afr.Cent. See *Flore d'Afrique Centrale*.

Fl.Afr.N. See *Flore de l'Afrique du Nord*.

Fl.Brit.Ind. See *Flora of British India*.

Fl.Camb.Lao.Viet. See *Flore du Cambodge, du Laos, et du Vietnam*.

Fl.Descr.Mt.Nimba. See *Flore Descriptive des Monts Nimba*.

Fl.E.Him. See *Flora of Eastern Himalaya*.

Fl.Eur. See *Flora Europaea*.

Fl.Gén.Ind.-Chin. See *Flore Générale de L'Indo-Chine*.

Fl.Iran. See *Flora Iranica*.

Fl.Less.Ant. See *Flora of the Lesser Antilles, Leeward and Windward Islands*.

Fl.Mac.Check. See *Flora of Macaronesia: Checklist of Vascular Plants*.

Fl.Males. See *Flora Malesiana*.

Fl.Males.Bull. See *Flora Malesiana Bulletin*.

Fl.Masc. See *Flore des Mascareignes: La Réunion, Maurice, Rodrigues*.

Fl.Mesoam. See *Flora Mesoamericana*.

Fl.Micronesia. See *Flora of Micronesia*.

Fl.Micronesica. See *Flora Micronesica*.

Fl.Neotrop. See *Flora Neotropica*.

Fl.Palaes. See *Flora Palaestina*.

Fl.Pol.Fr. See *Flore de la Polynésie Française*.

Fl.S.Afr. See *Flora of Southern Africa*.

Fl.Sahara. See *Flore du Sahara*.

Fl.SE.Pol. See *Flora of Southeastern Polynesia*.

Fl.Trop.Afr. See *Flora of Tropical Africa*.

Fl.Trop.E.Afr. See *Flora of Tropical East Africa*.

Fl.W.Trop.Afr. See *Flora of West Tropical Africa*.

Fl.Zamb. See *Flora Zambesiaca*.

Flora Europaea (1964-1980). Edited by T.G. Tutin, V.H. Heywood, N.A. Burges, D.H. Valentine, S.M. Walters, D.A. Webb., P.W. Ball and D.M. Moore. Cambridge Univ. Press, Cambridge. (Vol. 1 – gymnosperms, pteridophytes, angiosperms (Salicaceae to Platanaceae); Vol. 2 – Rosaceae to Umbelliferae; Vol. 3 – Diapensiaceae to Myoporaceae; Vol. 4 – Plantaginaceae to Compositae; Vol. 5 – monocotyledons; maps.) (Volume 1 is under revision, timescale of 4 years. A provisional check-list shows 151 new species, 96 new subspecies, 14 new records for Europe, 7 newly naturalized records and 17 taxa to move from in obs. to numbered species. This represents a 10% increase in taxa. Includes European Russia.)

Flora Iranica (1963-). Edited by K.H. Rechinger. Graz, Austria. (About 150 parts so far out of a projected 170, covering the Iranian highlands, and surrounding mountain ranges of Afghanistan, parts of West Pakistan, northern Iraq, Azerbaijan and Turkmenistan. About 2500 pages of text with line drawings and photographs of selected species.)

Flora Malesiana (1948-). Edited by C.G.G.J. van Steenis. Flora Malesiana Foundation, Leiden. (20 volumes planned, of which 8 published, including history of botanical exploration, cyclopedia of collectors and annotated bibliography for the botanical region of Malesia, i.e. Brunei, Indonesia, Malaysia, Papua New Guinea, Philippines; Pacific, and Australasia. 125 families revised so far, including Cyatheaceae, Anacardiaceae, Dipterocarpaceae, Ericaceae, Fagaceae and many other smaller families.)

Flora Malesiana Bulletin. Flora Malesiana Foundation and the Rijksherbarium, Leiden. (Invaluable source of information on current research activities; includes annotated bibliography of publications and papers on botany and plant conservation in S.E. Asia and adjacent regions. Issued annually.)

Flora Mesoamericana. A project, begun in 1980, to provide a concise Flora for the region from the Isthmus of Tehuantepec region in Mexico through the Central American Republics to the border of Panama with Colombia. Planned to comprise 7 volumes (225 plant families) to be published over 16 years. Coordinated by the Missouri Botanical Garden, Universidad Nacional Autonoma de México (UNAM) and the British Museum (Natural History). Will be in Spanish and published by UNAM. First volume (6) due for publication in 1984-5.

Flora Micronesica (1933). Compiled by R. Kanehira. South Seas Bureau, Tokyo. 505 pp. (In Japanese; introduction covers vegetation, floristics.)

Flora Neotropica (1968-). A 30-year project to produce a series of monographs of the entire flora of the Neotropics between the Tropics of Cancer and Capricorn. So far some 39 monographs covering individual genera, parts of families or whole families have been published by the New York Botanical Garden, Bronx, New York. They include, for example, Bromeliaceae by L.B. Smith and R.J. Downs (Parts 14(1-3)), and Flacourtiaceae by H.O. Sleumer (Part 22). Over 4000 species of flowering plants covered. Secretary: Missouri Botanical Garden, St Louis, Missouri.

Flora of British India (1872-1897), by J.D. Hooker. 7 vols. (Covers India, Bangladesh, western Tibet, and parts of southern Burma and the Malay Peninsula; nomenclature dated.)

Flora of Eastern Himalaya. Vols 1 (1966) and 2 (1971) edited by H. Hara, Vol. 3 (1975) edited by H. Ohashi. Univ. of Tokyo Press, Tokyo. (Enumeration of plants collected in Bhutan, eastern Nepal, Sikkim and W. Bengal during botanical expeditions to the Eastern Himalaya.)

Flora of Macaronesia: Checklist of Vascular Plants, 2nd Ed. (1979). By A. Hansen and P. Sunding. 2 parts. Botanical Garden and Museum, University of Oslo, Oslo. 93, 55 pp. (Covers Azores, Madeira, Salvage Islands, Canary Islands and Cape Verde Islands.)

Flora of Mauritius and the Seychelles. See Baker (1877).

Flora of Micronesia (1975-). By F.R. Fosberg and M.-H. Sachet. Smithsonian Contrib. Bot. 20: 15 pp. (Gymnosperms); 24: 28 pp. (Casuarinaceae, Piperaceae, Myricaceae); 36: 34 pp. (Convolvulaceae); 46: 71 pp. (Caprifoliaceae to Compositae). (Keys, descriptions, synonymy of Micronesian plants.) See also *ibid*. 45: 40 pp. (1980). Systematic studies of Micronesian plants, by the same authors. (Covers Caroline, Gilbert, Mariana, and Marshall Islands, Minimi-Tori-Shima (Marcus), Nauru and Wake.)

Flora of Southeastern Polynesia (1931-1935). Bull. Bernice P. Bishop Mus. (Complete for Marquesas, with additional records for the Gambier Islands, Pitcairn group, Tuamotus, and Tubuai, formerly the Austral Islands. *Ibid*. 84: 194 pp. (monocotyledons, by F.B.H. Brown); *ibid*. 89: 123 pp. (pteridophytes, by F.B.H. and E.D.W. Brown); *ibid*. 130: 386 pp. (dicotyledons, by F.B.H. Brown).

Flora of Southern Africa (1963-). Various authors. Editors: R.A. Dyer, L.E. Codd, H.B. Rycroft, B. de Winter, J.H. Ross and O.A. Leistner. Govt Printer for Botanical Research Institute, Private Bag x144, Pretoria. 51 families plus parts of 4 others published, 7 in press, 142 plus parts of 4 others remaining. (Covers South Africa and its states, Lesotho, Swaziland, Namibia and Botswana.)

Flora of Syria, Palestine and Sinai. See Post (1932).

Flora of the Lesser Antilles, Leeward and Windward Islands, edited by R.A. Howard. Vol. 1, Orchidaceae, by L.A. Garay and H.R. Sweet (1974), Vol. 2, Pteridophyta by G.R. Proctor (1977), Vol. 3, Monocotyledoneae by R.A. Howard *et al.* (1979). Arnold Arboretum, Jamaica Plain, Massachusetts, U.S.A. (Covers the Leeward Islands from Anguilla and St Martin to Guadeloupe and Dominica, and the Windward Islands – Martinique, St Lucia, St Vincent, Barbados, The Grenadines and Grenada.)

Flora of the Mongolian Steppe and Desert Areas. See Norlindh (1949).

Flora of Tropical Africa (1868-1937). By Oliver, D. *et al.* up to 1877, vols 1-3; by Thistleton-Dyer, W.T. *et al.* (Eds) from 1897, vols 4-9, 10(1). London.

Flora of Tropical East Africa (1952-). Edited by W.B. Turrill and E. Milne-Redhead (1952-57), E. Milne-Redhead and R.M. Polhill (1966-1972) and R.M. Polhill (1973-). Published by Crown Agents, London (1952-1975) and Balkema, Rotterdam (1982-). 113 families published, 24 in press, 68 remaining; 5500 species revised out of an estimated 10,000 (perhaps a little high); residue mainly Malvales, Gamopetalae and Liliales. (Covers Kenya, Tanzania and Uganda.)

Flora of West Tropical Africa (1927-36), by J. Hutchinson and J.M. Dalziel. 2nd Ed., 2 vols: 1 revised by R.W.J. Keay (1954); 2 and 3 (parts 1 and 2) revised by F.N. Hepper (1963, 1968 and 1973); pteridophytes by A.H.G. Alston (1959). Crown Agents, London. (Covers West Africa from Nigeria, and north to 18°N, so including the southern parts of Mauritania, Mali and Niger.)

Flora Palaestina (1966-). Vols 1 (1966) and 2 (1972) edited by M. Zohary; vol. 3 (1977) edited by N. Feinbrun-Dothan. Vol. 4 in press. Israel Academy of Sciences, Jerusalem. (Includes Israel and Jordan.)

Flora Zambesiaca (1960-), vols 1, 2, 3(1), 4, 7(1), 10(1), Supplement (Pteridophytes). Edited by A.W. Exell, H. Wild, A. Fernandes, J.P.M. Brenan and E. Launert. Vol. 1 - Gymnosperms, Angiosperms (Ranunculaceae to Sterculiaceae); Vol. 2 - Tiliaceae to Connaraceae; Vol. 3, Part 1 - Leguminosae (Mimosoideae); Vol. 4 - Rosaceae to Cornaceae; Vol. 7, Part 1 - Escalloniaceae to Salvadoraceae, Vol. 10, Part 1 - Gramineae (tribes Bambuseae-Paphoreae); Pteridophyta. Crown Agents, London (Distributed by HMSO). (Covers Zambia, Zimbabwe, Botswana, Malawi, Mozambique and the Caprivi Strip. Estimated to be 65% complete, with several families due out next year.)

Flore d'Afrique Centrale (Zaire - Rwanda - Burundi). Began as *Flore du Congo Belge et du Ruanda-Urundi*, 10 vols. (1948-63). Continued (1967-1971) as *Flore du Congo du Rwanda et du Burundi*, and continues (1972-) as *Flore d'Afrique Centrale (Zaire - Rwanda - Burundi).* By Robyns, W. *et al.* (Eds) (1948-1963), and by Jardin Botanique National de Belgique (1967-). Published by L'Institute National Pour L'Etude Agronomique du Congo Belge, Bruxelles. (Out of a total of 202 families of seed plants, 52 still to be published, including Compositae, Gramineae, Orchidaceae and Rubiaceae. Out of 39 pteridophyte families, 10 have been published.)

Flore de l'Afrique du Nord (1952-) by R. Maire. 15 vols published so far out of 22 expected; incomplete. Paris. (Pteridophytes, Gymnosperms, Monocotyledons and Dicotyledons to Rosaceae. Covers North Africa bounded by c. 21°N, 25°E, so including parts of Mauritania, Mali, Niger and Chad, but not Egypt.)

Flore de la Polynésie Française: Description des Plantes Vasculaires aux Iles de la Société (1893). By E. Drake del Castillo. Paris. 352 pp. (Covers Society Islands, Tubuai, Wallis.)

Flore des Mascareignes: La Réunion, Maurice, Rodrigues (1976-) edited by J. Bosser, Th. Cadet, H.R. Julien and W. Marais. The Sugar Industry Research Institute, Mauritius; ORSTOM, Paris; Royal Botanic Gardens, Kew, U.K. (Keys, descriptions, distributions, etc.; line drawings. Eight volumes covering 73 families published so far, 130 families remaining. Palms in press as 9th volume. Covers Mauritius, Reunion and Rodrigues.)

Flore Descriptive des Monts Nimba (Côte d'Ivoire, Guinée, Libéria), 6 vols (1971-1984), by J.-G. Adam. Editions CNRS, 15 quai Anatole France, 75700 Paris. 2181 pp. Vols 1-4 published in *Mém. Mus. Nat. Hist. Nat. Paris* 20 (1971); 22 (1971); 24 (1975); 25 (1975). (Includes 1096 plates of line drawings by the author. Covers all of Mt Nimba on the border of Ivory Coast, Liberia and Guinea.)

Flore du Cambodge, du Laos, et du Vietnam (1960-). Edited by A. Aubréville and J.-F. Leroy. Muséum National d'Histoire Naturelle, Paris. (20 fascicles to date including Annonaceae, Dipterocarpaceae, Guttiferae, Ranunculaceae and many other smaller families.)

Flore du Congo Belge et du Ruanda-Urundi. See *Flore d'Afrique Centrale.*
Flore du Congo du Rwanda et du Burundi. See *Flore d'Afrique Centrale.*

Flore du Sahara (1977) by P. Ozenda, 2nd Ed. Editions du CNRS, 15 quai Anatole-France, 75700 Paris. 622 pp. (Line drawings and black and white photographs throughout. Covers North Africa from central Libya and northern Chad west as far as Mauritania and the southern end of the Atlas Mountains.)

Flore et Plantes des Antilles. See Fournet (1976).

Flore et Végétation des Alpes. See Favarger (1956, 1958).

Flore Forestière Soudano-Guinéenne. See Aubréville (1950).

Flore Générale de L'Indo-Chine (1907-1951). Edited by M.H. Lecomte. 8 fascicles and Supplement, Paris. (Now very incomplete and taxonomy dated. Covers Kampuchea, Laos, Thailand and Viet Nam.)

Flowers of Europe: a Field Guide. See Polunin (1969).

Flowers of Greece and the Balkans: a Field Guide. See Polunin (1980).

Flowers of South-west Europe: a Field Guide. See Polunin and Smythies (1973).

Flowers of the Mediterranean. See Polunin and Huxley (1978).

Fosberg, F.R. (1973). On present condition and conservation of forests in Micronesia. In Pacific Science Association, *Planned Utilization of the Lowland Tropical Forests.* Pp. 165-171. (Proceedings of the Pacific Science Standing Committee Symposium, August 1971, Bogor, Indonesia; covers Caroline Islands, Gilbert Islands, Guam, Minami-Tori-Shima, Marianas, Marshalls, Nauru and Wake.)

Fosberg, F.R. and Sachet, M.-H. (1975-). See *Flora of Micronesia.*

Fosberg, F.R. and Sachet, M.-H. (1972). Status of floras of Western Indian Ocean Islands. In IUCN, *Comptes Rendus de la Conférence Internationale sur la Conservation de la Nature et de ses Ressources à Madagascar, 1970.* Publications UICN Nouvelle Série 36. Pp. 152-155. Covers the western Indian Ocean.

Fosberg, F.R., Sachet, M.-H. and Oliver, R. (1979). A geographical checklist of the Micronesian Dicotyledonae. *Micronesica* 15(1-2): 41-295. (Lists 1342 taxa, of which 622 native to Micronesia; distribution indicated.)

Fosberg, F.R., Sachet, M.-H. and Oliver, R. (1982). Geographical checklist of the Micronesian Pteridophyta and Gymnospermae. *Micronesica* 18(1): 23-82. (Enumeration of 207 pteridophyte taxa, of which 198 indigenous, and 19 gymnosperms, including one endemic to Micronesia.)

Fournet, J. (1976). *Fleurs et Plantes des Antilles.* Les éditions du Pacifique. 142 pp. (In French, with descriptions of vegetation types and a list of 144 species of flowering plants, each with a colour photograph and short description. Republished in English in 1977.)

Frankel, O.H. and Soulé, M.E. (1981). *Conservation and Evolution.* Cambridge University Press, Cambridge. 327 pp. (Genetic resources and their conservation.)

Frankenberg, P. and Klaus, D. (1980). *Atlas der Pflanzenwelt des Nordafrikanischen Trockenraumes: Computerkarten wesentlicher Pflanzenarten und Pflanzenfamilien.* Arb. Geog. Inst. Univ. Bonn. Unpaginated. (Distribution maps of 473 species. Covers all of Africa north of c. 16°N.)

Frodin, D.G. (1984). *Guide to Standard Floras of the World.* Cambridge University Press, Cambridge. 619 pp. (An annotated, geographically arranged systematic bibliography of the principal Floras, enumerations, checklists and chorological atlases of the world.)

Fuller, K.S. and Swift, B. (1984). Latin American Wildlife Trade Laws (Leyes del Comercio de Vida Sylvestre en América Latina). WWF-US, Washington, D.C. Mimeograph. (Accounts for each country in South and Middle America on wildlife laws, in English and Spanish, with species lists; mostly faunal but flora to be covered in more detail in a revised version in preparation.)

Gómez-Campo, C. (1979). The role of seed banks in the conservation of Mediterranean flora. *Webbia* 34(1): 101-107.

Gómez-Campo, C. (Ed.) (1985). *Plant Conservation in the Mediterranean Area.*

Geerling, C. (1982). *Guide de Terrain des Ligneux Sahéliens et Soudano-Guinéens.* Veenman and Zonen, Wageningen. 340 pp. (Key; line drawings throughout. Covers Africa west of and including Cameroon, and south of 18°N.)

Gentry, A. (1978). Floristic knowledge and needs in Pacific Tropical America. *Brittonia* 30: 134-153. (Covers Middle America, Colombia, Ecuador, Peru; includes composite vegetation map.)

Given, D.R. (Ed.) (1983). *Conservation of Plant Species and Habitats.* Nature Conservation Council, Wellington, N.Z. 128 pp. (Proceedings of symposium held at 15th Pacific Science Congress, Dunedin, February 1983; papers deal with conservation problems in New Zealand, South Pacific islands and South East Asia; monitoring and strategies for conserving threatened plants.)

Goodwillie, R. (1980). *European Peatlands.* Nature and Environment Series No. 19, Council of Europe, Strasbourg. 75 pp.

Götz, E. (1975). *Die Gehölze der Mittelmeerländer: ein Bestimmungsbuch nach Blattmerkmalen.* Ulmer, Stuttgart. 114 pp. (Illustrated key to woody Mediterranean (European) plants based on vegetative features; map.)

Graham, A. (Ed.) (1973). *Vegetation and Vegetational History of Northern Latin America.* Papers presented as part of a symposium, 'Vegetation and Vegetational History in Northern Latin America', at the American Institute of Biological Sciences meetings, Bloomington, Ind. (U.S.A.), 1970. Elsevier, Amsterdam. 393 pp. (Maps; illus.)

Grey-Wilson, C. (1979). *The Alpine Flowers of Britain and Europe.* Collins, London. 384 pp. (Covers north-west, south-west and central Europe as far as central Italy, north-west Yugoslavia, south-west Austria and the Pyrenees; colour and line drawings.)

Grubov, V.I. (Ed.) (1963-). *Rasteniya Tsentral'noi Azii. (Plantae Asiae Centralis).* AN SSSR Press, Moscow. (15 fascicles planned, according to Frodin; detailed account of flora with keys. 1 – ferns and regional bibliography covering the territory between Soviet Asia, 'China Proper' and the Indian subcontinent, with the exception of the Mongolian People's Republic; 2-7 – treatments of Cyperaceae, Juncaceae, Liliaceae and Orchidaceae.)

Hall, A.V., de Winter, M. and B., van Oosterhout, S.A.M. (1980). *Threatened Plants of Southern Africa.* South African National Scientific Programmes Report No. 45, Pretoria. 244 pp. (Covers South Africa and its states, Lesotho, Swaziland, Namibia and Botswana.)

Halliday, G. and Beadle, M. (1983). *Consolidated Index to Flora Europaea.* Cambridge Univ. Press, Cambridge. 210 pp. (Compiled from the separate indices of vols 1 to 5 of *Flora Europaea.* Includes European Russia.)

Hamann, U. and Wagenitz, G. (1977). *Bibliographie zur Flora von Mitteleuropa,* 2nd Ed. Parey, Berlin. 374 pp. (Includes Austria, Federal Republic of Germany, German Democratic Republic, northern Italy, Luxembourg and north-west Yugoslavia.)

Hamilton, A.C. (1982). *Environmental History of East Africa. A Study of the Quaternary.* Academic Press, London. 328 pp. (Covers Kenya, Tanzania, Uganda.)

Hansen, A. and Sunding, P. (1979). See *Flora of Macaronesia: Checklist of Vascular Plants.*

Hartmann, F.-K. and Jahn, G. (1967-). *Waldgesellschaften des Mitteleuropäischen Gebirgsraumes Nördlich der Alpen. Tabellen, Grundlagen und Erläuterungen*

(Woodland Associations of the Middle European Mountain Areas north of the Alps), 2 vols. Fischer, Stuttgart. (1 vol. planned.)

Hayek, A. von (1924-1933). See *Prodromus Florae Peninsulae Balcanicae.*

Hedberg, I. (Ed.) (1979). *Systematic Botany, Plant Utilization and Biosphere Conservation.* Almqvist & Wiksell International, Stockholm. 157 pp. (See especially paper by I. Hedberg, pp. 83-104, on the possibilities and needs for conservation of plant species and vegetation in Africa.)

Hedberg, I. and O. (Eds) (1968). *Conservation of Vegetation in Africa South of the Sahara.* Acta Phytogeogr. Suec. 54. 320 pp. (Proceedings of a symposium held at the 6th plenary meeting of AETFAT in Uppsala, 1966.) (Covers all of Africa south of and including Mauritania, Mali, Chad, Sudan, but excluding Burkina Faso, Togo, Guinea Bissau, Gambia and Niger; also includes several oceanic and offshore islands.)

Hedberg, O. (1951). Vegetation belts of the East African mountains. *Svensk Bot. Tidskr.* 45(1): 140-202. (With 12 black and white photographs. Covers the following mountains: Virunga, Ruwenzori, Aberdare, Elgon, Kenya, Kilimanjaro and Meru.)

Hedberg, O. (1957). Afroalpine vascular plants. *Symb. Bot. Upsal.* 15(1). 411 pp. (With 24 black and white photographs. Covers the high mountains of eastern Africa.)

Hegi, G. (1935-1979). See *Illustrierte Flora von Mitteleuropa.*

Hegi, G., Merxmüller, H. and Reisigl, H. (1977). *Alpenflora. Die Wichtigeren Alpenpflanzen Bayerns, Österreichs und der Schweiz.* Parey, Berlin. 194 pp. (Lists protected plants; colour illus.; maps.)

Henderson, D.M. (1983). *International Directory of Botanical Gardens IV*, 4th Ed., (first published 1963 as *Regnum Vegetabile* vol. 28). Koeltz Scientific Books, D-6240 Koenigstein, W. Germany. 288 pp.

Heywood, V.H. (1971). Preservation of the European flora. The taxonomist's role. *Bull. Jard. Bot. Nat. Belg.* 41(1): 153-166.

Holdridge, L. (1967). *Life Zone Ecology*, Rev. Ed. Tropical Science Center, San José, Costa Rica.

Holloway, C.W. (1976). Conservation of threatened vertebrates and plant communities in the Middle East and South West Asia. In IUCN, *Ecological Guidelines for the Use of Natural Resources in the Middle East and South West Asia.* IUCN, Switzerland. Pp. 179-188.

Holzner, W., Werger, M.J.A., and Ikusima, I. (Eds) (1983). *Man's Impact on Vegetation*, Geobotany 5. Junk Publishers, The Hague. 370 pp. (Human impact on major vegetation types around the world; separate chapters deal with the Central High Andes, Central Europe, Middle East, Himalayas, China and Japan.)

Hooker, J.D. (1872-1897). See *Flora of British India.*

Horvat, I., Glavacáv, V. and Ellenberg, H. (1974). *Vegetation Südosteuropas.* Fischer, Stuttgart. 768 pp. (Covers Albania, Bulgaria, Greece, Yugoslavia and European Turkey; detailed vegetation map; black and white photographs and line drawings.)

Howard, R.A. (Ed.) (1974-). See *Flora of the Lesser Antilles, Leeward and Windward Islands.*

Hultén, E. (1971). *Atlas över Växternas Utbredning i Norden* (Atlas of the distribution of vascular plants in northwestern Europe), 2nd Ed. Generalstabens Litografiska Anstalt, Stockholm. 531 pp. (Maps for Denmark, Finland, Norway, Sweden and Soviet Baltic republics; habitat data; ferns and flowering plants; colour plates for each species.)

Hutchinson, J. and Dalziel, J.M. (1927-1936). See *Flora of West Tropical Africa.*

Huxley, A. (1985). *Green Inheritance: The World Wildlife Fund Book of Plants.* Anchor Press/Doubleday, Garden City, New York. 193 pp. (Illustrated account of plant diversity with special emphasis on plants of economic importance and the need for conservation. Complements the IUCN/WWF Plants Programme and Campaign 1984-86.)

Ikonographie der Flora des Südöstlichen Mitteleuropa. See Jávorka and Csapody (1979).

Ill.Fl.Mitteleur. See *Illustrierte Flora von Mitteleuropa.*

Illustrierte Flora von Mitteleuropa: mit Besonderer Berücksichtigung von Deutschland, Österreich und der Schweiz, 2nd Ed. (1935-1979), 6 vols. By G. Hegi. (Revised and edited by K. Suessenguth *et al.*; various publishers. 3rd Ed. (1980-), edited by H.J. Conert, U. Hamann, W. Schultze-Motel, and G. Wagenitz – incomplete; 5 vols planned; covers Austria, FRG, Switzerland and some adjacent regions; maps; line drawings; black and white photographs. Standard Flora for central Europe.)

Island Bibliographies. See Sachet and Fosberg (1955, 1971).

IUCN Conservation Monitoring Centre and WWF (1984). *The IUCN/WWF Plants Conservation Programme 1984-85.* IUCN, Switzerland. 29 pp.

IUCN Plant Red Data Book, The. See Lucas and Synge (1978).

IUCN Threatened Plants Committee Secretariat (1980). First Preliminary Draft of the List of Rare, Threatened and Endemic Plants for the Countries of North Africa and the Middle East. Mimeo, IUCN, Kew. 170 pp. (Covers Algeria, the Azores, the Canaries, Cyprus, Egypt, Israel, Jordan, Lebanon, Libya, Madeira, Morocco, Salvage Islands, Syria and Tunisia. Includes regional and country lists of threatened plants; select bibliography of Floras, conservation references.)

IUCN, UNEP and WWF (1980). *World Conservation Strategy: Living Resource Conservation for Sustainable Development.* IUCN, Switzerland. (Sections deal with the objectives of conservation, including maintenance of essential ecological processes and life-support systems, preservation of genetic diversity, and sustainable utilization of species and ecosystems; priority requirements at national and international levels. For popular account see Allen, 1980.)

Jacobs, M. (1982). Assessment of the deforestation problem in Malesia. Rijksherbarium, Leiden. 7 pp. (Significance of tropical rain forests, species richness, causes of deforestation.)

Jalas, J. and Suominen, J. (Eds) (1972-). *Atlas Florae Europaeae: Distribution of Vascular Plants in Europe*, 6 vols. The Committee for Mapping the Flora of Europe and Societas Biologica Fennica Vanamo, Helsinki. (Dot maps; incomplete; 1 – pteridophytes; 2 – gymnosperms; 3-6 – angiosperms (Salicaceae to Caryophyllaceae.) Includes European Russia.)

Jalas, J. and Suominen, J. (Eds) (1984). Proceedings of the VII meeting of the Committee for the Mapping of Flora Europaea, August 23-24, 1983, Helsinki. *Norrlinia* 2. 119 pp. (Progress report of mapping programme and papers about mapping at national level.)

Jávorka, S. and Csapody, V. (1979). *Ikonographie der Flora des Südöstlichen Mitteleuropa.* Fischer, Stuttgart. 703 pp. (Covers Austria, Czechoslovakia, Hungary, Poland, Romania, Yugoslavia and European USSR; 4090 illus.)

Kanehira, R. (1933). See *Flora Micronesica.*

Kent, D.H. and Brummitt, R.K. (1966-1971). *Index to European and Taxonomic Literature for 1965-1970*, 6 vols. Bentham-Moxon Trustees, Royal Botanic Gardens, Kew. (Geographical coverage as *Flora Europaea*; since 1971 subsumed in *Kew Record.*)

Killick, D.J.B. (Ed.) (1983). The Proceedings of the Xth AETFAT congress held at the CSIR Conference Centre, Pretoria, Republic of South Africa, from 19-23 January 1982. *Bothalia* 14(3/4). 1023 pp. (Several relevant papers are indicated under individual countries, but see also reports on the progress of various Floras, pp. 1015-1023, by various authors.)

Knapp, R. (1973). *Die Vegetation von Afrika*. Fischer, Stuttgart. 626 pp. (With 823 illustrations in the text.)

Koester, V. (1980). *Nordic Countries' Legislation on the Environment, with Special Emphasis on Conservation: A Survey*. IUCN Environmental Policy and Law Paper, IUCN, Gland, Switzerland. (Summarizes plant legislation.)

Kornas, J. (1976). Decline of the European flora – facts, comments and forecasts. *Phytocenosis* 5: 173-185.

Krüssmann, G. (1979). *Die Bäume Europas. Ein Taschenbuch für Naturfreunde*, 2nd Ed. Parey, Berlin. 172 pp. (Native, naturalized and commonly cultivated European trees; habitat details; maps; colour and black and white illus.)

Kunkel, G. (Ed.) (1979). *Taxonomic Aspects of African Economic Botany. Proceedings of the IX Plenary Meeting of AETFAT, Las Palmas de Gran Canaria, 18-23 March, 1978*. Excmo. Ayuntamiento de Las Palmas de Gran Canaria, Islas Canarias; obtainable from: The Secretary, Bentham-Moxon Trust, Royal Botanic Gardens, Kew, Richmond, England TW9 3AE. 250 pp. (See especially the reports on the progress of various Floras, pp. 157-195, by various authors, including F.N. Hepper's map showing the extent of floristic exploration in Africa south of the Sahara.)

La Flore du Bassin Méditerranéen. See CNRS (1975).

Lack, H.W. (Ed.) (1984). *Current Projects on the Mediterranean Flora*. OPTIMA Secretariat, Botanical Garden and Botanical Museum Berlin-Dahlem and the OPTIMA Publications Commission, Berlin. 152 pp. (A project register.)

Lall, J.S. and Moddie, A.D. (Eds) (1981). *The Himalaya: Aspects of Change*. India International Centre, Oxford Univ. Press, Delhi. (Geology, vegetation, flora and fauna, impact of man. See in particular K.C. Sahni on Eastern Himalayan flora, pp. 32-49; M.A. Rau on Western Himalayan flora, pp. 50-63.)

Landolt, E. (1984). *Unsere Alpenflora*, 5th Ed. Verlag Schweizer Alpen-Club, Zürich. 318 pp. (Detailed introduction including plant geography of alpine areas; keys; colour illus.; maps.)

Larsen, K. and Holm-Nielsen, L.B. (Eds) (1979). *Tropical Botany*. Academic Press, London. 453 pp.

Lebrun, J. (1960). Sur la richesse de la flore de divers territoires africains. *Bull. Séances Acad. Roy. Sci. d'Outre-Mer* 6(2): 669-690.

Lebrun, J.-P. (1977-1979). *Eléments pour un Atlas des Plantes Vasculaires de l'Afrique Sèche*, 2 vols. Etudes Botaniques Nos 4, 6, Institute d'Elevage et de Médecine Vétérinaire des Pays Tropicaux, 10 rue Pierre Curie, 94 704 Maisons Alfort. 265, 255 pp.; 50, 40 maps with overlays.

Lebrun, J.-P. and Stork, A.L. (1978). *Index Général des 'Contributions à l'Etude de la Flore de l'Afrique du Nord' du Dr René Maire*. Etude Botanique No. 5, Institute d'Elevage et de Médecine Vétérinaire des Pays Tropicaux, 10 rue Pierre Curie, 94 704 Maisons Alfort. 365 pp. (An index to species given in R. Maire's 1918-1949 work. Stands as a checklist of plants of the region. Covers all North Africa except Egypt.)

Lebrun, J.P. (1976). Richesses spécifiques de la flore vasculaire des divers pays ou régions d'Afrique. *Candollea* 31: 11-15.

Lecomte, M.H. (Ed.) (1907-1951). See *Flore Générale de L'Indo-Chine.*

Léonard, J. (1980). Noms de plantes et de groupements végétaux cités dans Pierre Quézel: La végétation du Sahara. Du Tchad à la Mauritanie. Jardin Botanique National de Belgique, Meise. 45 pp. Unpublished mimeo. (Index to page numbers in Quézel, 1965.)

Letouzey, R. (1969-1972). *Manuel de Botanique Forestière: Afrique Tropicale,* 3 vols. Centre Technique Forestier Tropical, 45 bis avenue de la Belle Gabrielle, 94 Nogent-sur-Marne. 461 pp. (Good introduction to the botany and forestry of tropical Africa, especially Cameroon and Gabon. An English version is in preparation. Line drawings throughout.)

Lind, E.M. and Morrison, M.E.S. (1974). *East African Vegetation.* Longman, London. 257 pp. (Covers Kenya, Tanzania and Uganda.)

Lindman, C.A.M. (1964). *Nordens Flora,* 3 vols. Wahlström and Widstrand, Stockholm. (Covers Denmark, Finland, Norway and Sweden; colour plates; maps.)

Lucas, G. and Synge, H. (1978). *The IUCN Plant Red Data Book.* IUCN, Switzerland. 540 pp. (Data sheets on 250 vascular plant taxa selected to illustrate the various types of threats, habitats, distributions, plant groups and protective measures, with emphasis on species of particular interest or importance.)

Lugo, A.E., Schmidt, R. and Brown, S. (1981). Tropical forests in the Caribbean. *Ambio* 10(6): 318-324. (References to *FAO (1974)* refer to figures for deforestation on individual Caribbean islands in a table in the above paper, attributed to this source.)

Maire, R. (1952-). See *Flore de l'Afrique du Nord.*

Med-Checklist: a Critical Inventory of Vascular Plants of the Circum-mediterranean Countries. (1981-). 2nd Edition of vol. 1 published in 1984. Edited by W. Greuter, H.M. Burdet and G. Long. Conservatoire et Jardin botaniques, Ville de Genève, and the Organization for the Phyto-Taxonomic Investigation of the Mediterranean Area (OPTIMA), Berlin. (Incomplete; 1 – pteridophytes.)

Melville, R. (1970). Endangered plants and conservation in the islands of the Indian Ocean. *Papers and Proceedings of the IUCN 11th Technical Meeting,* New Delhi, India, 25-28 November 1969. IUCN, Switzerland. Pp. 103-107.

Merrill, E.D. (1924). Bibliography of Polynesian botany. *Bull. Bernice P. Bishop Mus.* 13. 68 pp.

Merrill, E.D. (1945). *Plant Life of the Pacific World.* Macmillan, New York. 295 pp. (Introduction to natural history of Malesia and Pacific; notes on some island groups; botanical history and selected bibliography.)

Merrill, E.D. (1947). *A Botanical Bibliography of the Islands of the Pacific.* Smithsonian Institution, Washington, D.C. 404 pp.

Merrill, E.D. and Walker, E.H. (1938). *A Bibliography of Eastern Asiatic Botany.* Arnold Arboretum, Harvard Univ., Mass. 719 pp. (Covers China, Japan, Korea, central and east Siberia and the Soviet Far East. Supplement 1, 1960, by the American Institute of Biological Sciences. 552 pp.)

Merxmüller, H. (Ed.) (1971). Proceedings of the 7th plenary meeting of the AETFAT, München, 7th-12th September, 1970. *Mitt. Bot. Staatssamml. München* 10. 638 pp. (See especially the reports on the progress of various Floras and maps, pp. 13-164.)

Meusel, H. (Ed.) (1965-). *Vergleichende Chorologie der Zentral-Europäischen Flora.* 4 vols forming 2 parts; 3 parts planned. Fischer, Jena. (Systematic atlas of vascular plants native to central Europe; includes world range and ecological details; vols 1 and 2 – pteridophytes, gymnosperms, monocotyledons, dicotyledons to Plantaginaceae; maps; illus.)

Miège, J. and Stork, A.L. (Eds) (1975, 1976). Comptes Rendus de la VIIIe Réunion de l'AETFAT, 2 vols. Proceedings of the 8th plenary meeting of AETFAT in Geneva, 16-21 September, 1974. *Boissiera* 24a and 24b. 692 pp. (Several relevant papers are indicated under individual countries, but see also paper by I. Hedberg, pp. 437-441, with a follow-up of the 1966 AETFAT meeting 'Conservation of vegetation in Africa south of the Sahara', and also the progress reports on current Floras and vegetation maps, pp. 519-666.)

Moore, D.M. (Ed.) (1982). *Green Planet: The Story of Plant Life on Earth*. Cambridge University Press, Cambridge. 288 pp. (Illustrated encyclopedia of plant ecology and geography; major vegetation types; human impact on vegetation.)

Morse, L.E. and Henifin, M.S. (Eds) (1981). *Rare Plant Conservation: Geographical Data Organization*. New York Botanical Garden, New York. 377 pp. (27 papers based on symposium at the New York Botanical Garden, November 1977. Covers information sources; data storage and handling techniques; evaluation of rarity and threats; guidelines for the preparation of status reports on rare and endangered plants with particular reference to the U.S. Endangered Species Act.)

Myers, N. (1979). *The Sinking Ark – A New Look at the Problem of Disappearing Species*. Pergamon Press, Oxford. 307 pp. (Human impact on extinction rates; reasons for species conservation; rates and consequences of deforestation of tropical forests, with profiles on Brazil, Costa Rica, Indonesia and Kenya; regional reviews on Amazonia, South East Asia and Tropical Africa; *in situ* and *ex situ* conservation.)

Myers, N. (1980). *Conversion of Tropical Moist Forests*. (A report prepared for the Committee on Research Priorities in Tropical Biology of the National Research Council.) National Academy of Sciences, Washington, D.C. 205 pp. (Analysis of data on deforestation, with useful reviews for each country with tropical moist forests.)

Myers, N. (1983). *A Wealth of Wild Species – Storehouse for Human Welfare*. Westview Press, Boulder, Colorado. 272 pp. (Economic benefits of wildlife to agriculture, medicine and industry; final chapter on genetic engineering.)

Myers, N. (1984). *The Primary Source: Tropical Forests and Our Future*. Norton, New York. 399 pp. (Reflects the goals and objectives of the WWF/IUCN Tropical Forest Campaign 1982-84. Covers importance of tropical forests as the richest ecosystems on Earth; economic values; human impact and rates of deforestation; consequences of forest destruction; recent conservation initiatives.)

Nayar, B.K. and Kaur, S. (1972). *Companion to R.H. Beddome's 'Handbook to the Ferns of British India, Ceylon and the Malay Peninsula'*. Pama Primlane, New Delhi. 196 pp.

Nordens Flora. See Lindman (1964).

Norlindh, T. (1949). *Flora of the Mongolian Steppe and Desert Areas*. Report of the Scientific Expedition to N.W. Provinces, China, 31. Stockholm. 155 pp. (Includes enumeration of plants collected in north-west China and Mongolia; history of botanical exploration.)

Oberdorfer, E. (1983). *Pflanzensoziologische Exkursionsflora für Süddeutschland und die Angrenzenden Gebiete*, 5th Ed. Ulmer, Stuttgart. 1051 pp. (Covers Austria, GDR, FRG and Switzerland; keys; line drawings.)

OPTIMA-Projekt "Kartierung der Mediterranean Orchideen" (1979-). 3 vols. 1 – Index der Verbreitungskarten für die Orchideen Europas und der Mittelmeerländer, by E. and B. Willing; 2 – Orchideenforschung und Naturschutz im Mittelmeergebiet Internationales Artenschutzprogramm, by H. Baumann *et al.*; 3 – Die

Orchideenflora von Euböa (Griechenland), by S. Künkele and K. Paysan. *Beih. Veröff. Naturschutz Landschaft. Bad.-Württ.* 14: 163 pp.; 19: 189 pp.; 23: 140 pp.

Org.Est.Amer. (1967). See Organización de los Estados Americanos.

Organización de los Estados Americanos (1967). La Convención para la protección de la flora, de la fauna y de la bellezas escénicas naturales de los Estados Americanos: Lista de especies de fauna y flora en vías de extinción en los Estados Miembros. OEA, Washington, D.C. Mimeograph. (Short flora lists for 8 countries; covers all of the Americas except for the Caribbean.)

Ovesen, C.H. *et al.* (1978). *Hotade Djur och Växter i Norden. Uhanalaiset Eläimet ja Kasvit Pohjoismaissa.* (Threatened animals and plants in the Nordic countries). *NU A* 9: 1-73.

Ovesen, C.H. *et al.* (1982). *Hotade Djur och Växter i Norden. Pohjolan Uhanalaiset Eläimet ja Kasvit. NU* 4: 1-194.

Ozenda, P. (1977). See *Flore du Sahara.*

Ozenda, P. *et al.* (1979). *Vegetation Map (scale 1:3,000,000) of the Council of Europe Member States.* Nature and Environment Series No. 16. Council of Europe, Strasbourg. 99 pp. (3 colour maps.)

Pac.Pl.Areas. See *Pacific Plant Areas.*

Pacific Plant Areas. Vol. 1 (1963) edited by C.G.G.J. van Steenis, includes bibliography of Pacific and Malesian plant maps; vol. 2 (1966) by C.G.G.J. van Steenis and M.M.J. van Balgooy; vols 3-4 (1975, 1984) by M.M.J. van Balgooy, contain distribution maps of Pacific genera, notes on ecology, taxonomy and supplement to bibliography. Vol. 1 published by National Institute of Science and Technology, Manila; vols 2-4 – Rijksherbarium, Leiden.

Palmer, E. and Pitman, N. (1972). *Trees of Southern Africa*, 3 vols. Balkema, Cape Town. 2235 pp. (Keys, descriptions, distributions; line drawings, black and white, colour photographs. Covers South Africa and its states, Botswana, Lesotho, Namibia and Swaziland.)

Parris, B.S. (1985). Ecological aspects of distribution and speciation in Old World tropical ferns. *Proc. R. Soc. Edin.* 86: 341-346.

Pételot, A. (1955). *Bibliographie botanique de L'Indo-Chine.* 102 pp. Arch. Rech. Agron. Cambodge, Laos, Viet Nam, 24. Saigon.

Peters, A.J. and Lionnet, J.F.G. (1973). Central western Indian Ocean bibliography. *Atoll Res. Bull.* 165. 322 pp. (With 3 maps. Covers 2-11°S × 45-75°E.)

Pichi-Sermolli, R.E.G. (1979). A survey of pteridological flora of the Mediterranean region. *Webbia* 34(1): 175-242.

Polunin, O. (1969). *Flowers of Europe: a Field Guide.* Oxford Univ. Press, London. 662 pp. (Keys; colour photographs and line drawings. Includes European Russia.)

Polunin, O. (1976). *Trees and Bushes of Europe.* Oxford Univ. Press, London. 280 pp. (Colour photographs and line drawings.)

Polunin, O. (1980). *Flowers of Greece and the Balkans: a Field Guide.* Oxford Univ. Press, Oxford. 592 pp. (Covers Albania, Bulgaria, Greece and Yugoslavia; concise vegetation descriptions of national parks; colour photographs and line drawings.)

Polunin, O. and Huxley, A. (1978). *Flowers of the Mediterranean.* Chatto and Windus, London. 260 pp. (Keys; colour photographs and line drawings.)

Polunin, O. and Smythies, B.E. (1973). *Flowers of South-west Europe: a Field Guide.* Oxford Univ. Press, London. 480 pp. (French and Spanish translations available.)

Polunin, O. and Stainton, A. (1984). *Flowers of the Himalaya.* Oxford University Press, Oxford. 580 pp. (Describes 1500 species; 690 colour photographs; 315 line drawings. Covers eastern Himalaya from Nepal to Sikkim.)

Polunin, O. and Walters, M. (1985) *A Guide to the Vegetation of Britain and Europe*. Oxford Univ. Press, Oxford. (Soils, climate, vegetation history, plant communities, national parks and nature reserves; illus.)

Poore, D. and Gryn-Ambroes, P. (1980). *Nature Conservation in Northern and Western Europe*. UNEP, IUCN and WWF, Gland, Switzerland. 408 pp.

Post, G.E. (1932). *Flora of Syria, Palestine and Sinai*, 2 vols, 2nd Ed. American Press, Beirut. (Revised by J.E. Dinsmore.)

Prance, G.T. (Ed.) (1982). *Biological Diversification in the Tropics*. Proceedings of the 5th International Symposium of the Association for Tropical Biology held at Macuto Beach, Caracas, Venezuela, Feb. 8-13, 1979. Columbia Univ. Press, New York. 714 pp. (Maps; illus.)

Prance, G.T. and Elias, T.S. (Eds) (1977). *Extinction is Forever*. New York Botanical Garden, Bronx, New York 10458, U.S.A. 437 pp. (Proceedings of a symposium entitled Threatened and Endangered Species of Plants in the Americas and their Significance in Ecosystems Today and in the Future; essays on state of knowledge on this topic for most parts of the Americas.)

Prescott-Allen, R. and C. (1982). *What's Wildlife Worth?* International Institute for Environment and Development, London. 92 pp. (Economic contributions of wild plants and animals to developing countries, covering food crops, fuelwood, fibres, herbs and medicinal plants, among other topics.)

Prescott-Allen, R. and C. (1983). *Genes From The Wild*. International Institute for Environment and Development, London. 101 pp. (Covers use of wild genetic resources for improving food, forage, and timber crops; threats to wild genetic resources; *in situ* gene banks.)

Priszter, S. (Ed.) (1983). *Arbores Fruticesque Europae Vocabularium Octo Linguis Redactum* (Trees and shrubs of Europe: dictionary in eight languages: Latin, English, French, German, Hungarian, Italian, Spanish, Russian.) Akadémiai Kiadó, Budapest. 300 pp.

Prod.Fl.Pen.Balc. See *Prodromus Florae Peninsulae Balcanicae*.

Prodromus Florae Peninsulae Balcanicae (1924-1933). By A.von Hayek. *Feddes Repert.*: 30(1-3). (Covers Albania, Bulgaria, Greece and Yugoslavia; 1 – pteridophytes, gymnosperms, dicotyledons; 2 – dicotyledons; 3 – monocotyledons; reprinted 1968.)

Quézel, P. (1965). *La Végétation du Sahara. Du Tchad à la Mauritanie*. Fischer, Stuttgart. 333 pp. (With 15 maps, 72 black and white and 18 colour photographs. Covers from Chad to Mauritania.)

Quézel, P. (1978). Analysis of the flora of Mediterranean and Saharan Africa. *Ann. Missouri Bot. Gard.* 65(2): 479-534.

Radovsky, F.J., Raven, P.H. and Sohmer, S.H. (1984). *Biogeography of the Tropical Pacific*. Bernice P. Bishop Mus. Special Publ. no. 72. Honolulu, Hawaii. 221 pp. (See in particular T.F. Stuessy, R.W. Sanders and M. Silva on the phytogeography and evolution of the flora of Juan Fernández, pp. 55-69; and P. Morat, J.-M. Veillon and H.S. Mackee on the floristic relationships of New Caledonian rain forest phanerogams, pp. 71-128.)

Ranjitsinh, M.K. (1979). Forest destruction in Asia and the South Pacific. *Ambio* 8(5): 192-201.

Rechinger, K.H. (Ed.) (1963-). See *Flora Iranica*.

Reed, C.F. (1969). *Bibliography to Floras of Southeast Asia*. Harrod, Maryland. 191 pp. (Covers Burma, Kampuchea, Laos, Malay Peninsula, Singapore, Thailand, Viet Nam.)

Sachet, M.-H. and Fosberg, F.R. (1955, 1971). *Island Bibliographies*. Nat. Academy Sciences, Washington, D.C. 577 pp. (Micronesian botany, ecology, vegetation of tropical Pacific islands; includes Banaba Island, the Carolines, Gilberts and Marianas Islands, Minami-Tori-Shima, Nauru and Wake. *Supplement*, 1971, 427 pp.)

Schacht, W. (1976). *Blumen Europas. Ein Naturführer für Blumenfreunde*. Parey, Berlin. 203 pp. (Covers all Europe including European USSR; distribution maps and colour photographs of each taxon.)

Schnell, R. (1976, 1977). *La Flore et la Végétation de l'Afrique Tropicale*, 2 vols. Vols 3 and 4 of Introduction à la Phytogéographie des Pays Tropicaux. Gauthier-Villars, Paris. 459, 378 pp.

Simmons, J.B., Beyer, R.I., Brandham, P.E., Lucas, G.Ll. and Parry, V.T.H. (Eds) (1976). *Conservation of Threatened Plants*, Nato Conference Series 1: Ecology, vol. 1. 336 pp. (Proceedings of the Conference on the Functions of Living Plant Collections in Conservation and Conservation-Orientated Research and Public Education, held at the Royal Botanic Gardens, Kew, 2-6 September 1975. Papers cover the rôle of botanic gardens in conservation and public education; techniques of collecting and cultivating selected threatened plant groups; seed banks; documentation of living plant collections; recent developments in international co-operation and legislation, including CITES.)

Smithsonian Institution (1969). *A Bibliography of the Botany of South East Asia*. Washington, D.C. 161 pp.

Soulé, M.E. and Wilcox, B.A. (Eds) (1980). *Conservation Biology: An Evolutionary-Ecological Perspective*. Sinauer Associates, Sunderland, Mass. 395 pp. (Ecological principles of plant and animal conservation, with emphasis on the tropics; island biogeography, including evolutionary changes in small populations and causes of extinctions; *ex situ* conservation, mainly dealing with animals; tropical rain forest conservation.)

Steenis, C.G.G.J. van (1934-1936). On the origin of the Malaysian mountain flora, 1-3. *Bull. Jard. Bot. Buitenzorg* 13: 135-262; 13: 287-417; 14: 56-72. (Includes checklist of plant genera occurring above 1000 m; notes on species giving details of distribution, localities and altitudinal range.)

Steenis, C.G.G.J. van (1958). *Vegetation Map of Malaysia (Scale 1:5,000,000)*. Unesco. (See also the accompanying 8-page booklet *Commentary on the Vegetation Map of Malaysia*.)

Steenis, C.G.G.J. van (Ed.) (1948-). See *Flora Malesiana*.

Stehlé, H. (1945). Forest types of the Caribbean islands. *Caribbean Forester* 6 (suppl.): 273-408.

Stork, A.L. and Lebrun, J.-P. (1981). *Index des Cartes de Répartition des Plantes Vasculaires d'Afrique: Complément 1935-1976, Supplément 1977-1981 (avec Addendum A-Z)*. Etude Botanique No. 8, Institute d'Elevage et de Médecine Vétérinaire des Pays Tropicaux, 10 rue Pierre-Curie, 94 704 Maisons Alfort. 98 pp.

Sutlive, V.H., Altshuler, N. and Zamora, M.D. (1981). *Where Have All The Flowers Gone? Deforestation in the Third World*. Studies in Third World Societies no. 13. College of William and Mary, Williamsburg, Virginia. 278 pp. (14 papers covering rates of deforestation in the tropics, separate chapters on Brazil, Gabon, Ghana, Indonesia and Malaysia.)

Synge, H. (1980). Endangered Monocotyledons in Europe and south west Asia. In Brickell, C.D., Cutler, D.F. and Gregory, M. (Eds), *Petaloid Monocotyledons: Horticultural and Botanical Research*. Academic Press, London. Pp. 199-206.

Synge, H. (Ed.) (1981). *The Biological Aspects of Rare Plant Conservation*. Wiley, London. 588 pp. (Proceedings of an international conference held at King's College, Cambridge, 14-19 July 1980. Sections deal with the survey and assessment of rare and threatened species; tropical forests, including conservation needs and opportunities, and techniques for identification and conservation of threatened species; the meaning of 'rarity' and monitoring rare plant populations; ecological studies on rare plants; introductions and re-introductions; protected areas. A selected bibliography of Red Data Books and threatened plant lists is included as an Appendix.)

Synge, H. and Townsend, H. (Eds) (1979). *Survival or Extinction: The Practical Role of Botanic Gardens in the Conservation of Rare and Threatened Plants*. Bentham-Moxon Trust, Royal Botanic Gardens, Kew. 250 pp. (Proceedings of a conference held at Kew, 11-17 September 1978. Papers deal with the rôle of botanic gardens and conservation organizations in collecting, propagating, distributing and re-introducing threatened plants; national policies and activities; education and the role of the media in conservation; seed banks; and the conservation of special groups, e.g. orchids.)

The Genera of Southern African Flowering Plants. See Dyer (1975, 1976).

Thirgood, J.V. (1981). *Man and the Mediterranean Forest. A History of Resource Depletion*. Academic Press, London. 194 pp.

Thonner, F. (1980). *Exkursionsflora von Europa. Anleitung zum Bestimmen der Gattungen der Europäischen Blütenpflanzen*. Rijksherbarium, Leiden. 461 pp. (Geographical coverage as *Flora Europaea*.)

Threatened Plants Unit, IUCN Conservation Monitoring Centre (1983). *List of Rare, Threatened and Endemic Plants in Europe (1982 edition)*, 2nd Ed. Nature and Environment Series No. 27, Council of Europe, Strasbourg. 357 pp. (Lists over 2000 threatened vascular taxa, with IUCN conservation categories for each, at national and European level; introduction includes analysis of data and provides details of plant laws in each country; country coverage as in *Flora Europaea*, plus Canary Islands, Azores and Madeira.)

Toledo, V.M. (1985). A critical evaluation of the floristic knowledge in Latin America and the Caribbean. Report presented to The Nature Conservancy International Program. Washington, D.C. 95 pp.

Trees and Bushes of Europe. See Polunin (1976).

Trees of Central Africa. See Coates Palgrave *et al.* (1957).

Trees of Southern Africa. See Coates Palgrave (1977).

Trees of Southern Africa. See Palmer and Pitman (1972).

Über die Hochgebirgsflora des Tropischen Afrika. See Engler (1892).

Unesco (1974). *Natural Resources of Humid Tropical Asia*. Natural Resources Research 12. Paris. 456 pp. (See in particular, R.A. de Rosayro on vegetation of humid tropical Asia, pp. 179-196; P. Legris on vegetation and floristic composition of humid tropical continental Asia, pp. 217-238; M. Jacobs on vegetation and botany of Malesia, pp. 263-294.)

Unesco (1981). *Vegetation map of South America: Explanatory notes*. Unesco, Paris.

Unesco/FAO (1970). *Ecological Study of the Mediterranean Zone. Vegetation Map of the Mediterranean Zone*. Unesco, France. 90 pp. (In French and English; explanatory notes and a 1:5,000,000 vegetation map of the Mediterranean zone.)

Unesco/UNEP/FAO (1978). *Tropical Forest Ecosystems: a State-of-Knowledge Report*. Natural Resources Research 14, Unesco, Paris. 683 pp.

US Council of Environmental Quality and Department of State (1980). *The Global 2000 Report to the President: Entering the Twenty-first Century*, 3 vols. U.S.

Government Printing Office, Washington, D.C. (1 – Interpretive Report; 2 – Technical Report; 3 – Documentation on the Governments' Global Sectoral Models: The Government's "Global Model". Analysis of global trends in population, resources and the environment, including food and agriculture, forestry and loss of genetic resources.)

Verdoorn, F. (Ed.) (1945). *Plants and Plant Science in Latin America*. Chronica Botanica Co., Waltham, Mass., U.S.A. (*Chron. Bot.* 16). 381 pp.

Vidal, J.E. (1972). *Bibliographie Botanique Indochinoise*. Bull. de la Société des Etudes Indochinoises 47(4): 748 pp. (Additions and amendments to Pételot, 1955.)

Walters, S.M. (1971). Index to the rare, endemic vascular plants of Europe. *Boissiera* 19: 87-89.

Walters, S.M. (1976). The conservation of threatened vascular plants in Europe. *Biol. Conserv.* 10(1): 31-41.

Walters, S.M. (1979). The role of Mediterranean botanic gardens in plant conservation. *Webbia* 34(1): 109-116.

Walters, S.M. (Ed.) (1975). *European Floristic and Taxonomic Studies*. BSBI Conference Report No. 15. A conference held in Cambridge, 19 June to 2 July 1974. Classey, Faringdon. 144 pp.

Webb, D.A. (1978). Flora Europaea – a retrospect. *Taxon* 27(1): 3-14.

Werger, M.J.A. (Ed.) (1978). *Biogeography and Ecology of Southern Africa*, 2 vols. Junk, The Hague. 1439 pp. (Several chapters relevant to southern Africa as a whole: see especially chapters by D.I. Axelrod and P.H. Raven, pp. 77-130, on the Late Cretaceous and Tertiary vegetation history of Africa; B.J. Huntley, pp. 1333-1384, on the conservation of ecosystems in southern Africa; D.J.B. Killick, pp. 515-560, on the Afro-alpine region; E.J. Moll and F. White, pp. 561-598, on the Indian Ocean coastal belt; H.C. Taylor, pp. 171-229, on the Cape region; M.J.A. Werger, pp. 145-170, on the biogeographical division of southern Africa, and pp. 231-299 on the Karoo-Namib region; M.J.A. Werger and B.J. Coetzee, pp. 301-462, on the Sudano-Zambezian region; F. White, pp. 463-513, on the Afro-montane region; H. Wild, pp. 1301-1332, on the vegetation of toxic soils; and E.M. van Zinderen Bakker, pp. 131-143, on Quaternary vegetation changes in southern Africa.)

White, F. (1981). The history of the Afromontane archipelago and the scientific need for its conservation. *Afr. J. Ecol.* 19: 33-54.

White, F. (1983). *The Vegetation of Africa. A Descriptive Memoir to Accompany the Unesco/AETFAT/UNSO Vegetation Map of Africa*. Natural Resources Research 20, Unesco, Paris. 356 pp. (The accompanying map 1:5,000,000 comprises three sheets in colour. The descriptive memoir has a very extensive bibliography, including a geographical bibliography. Covers all of Africa and the surrounding islands.)

Whitmore, T.C. (1975a). *Conservation Review of Tropical Rain Forests, General Considerations and Asia*. IUCN, Switzerland. 116 pp. (Introduction to forest types and their distribution, protected area coverage and conservation priorities for rain forest countries from India east to Polynesia.)

Whitmore, T.C. (1975b). *Tropical Rain Forests of the Far East*. Clarendon Press, Oxford. 282 pp. (Ecology, classification and distribution of forest types, impact of man.)

Whitmore, T.C. (1984). A vegetation map of Malesia at scale 1:5 million. *J. Biogeography* 11: 461-471. (10 forest types depicted; deforested areas and conservation areas over 200 sq. km shown; explanatory text.)

Wild, H. and Barbosa, L.A. Grandvaux (1967, 1968). *Vegetation Map of the Flora Zambesiaca Area*. Flora Zambesiaca Supplement, Collins, Salisbury. (Descriptive memoir of 71 pp. and colour map 1:2,500,000.)

Wolkinger, F. and Plank, S. (1981). *Dry Grasslands of Europe*. Nature and Environment Series No. 21, Council of Europe, Strasbourg. 56 pp. (11 maps.)

World Conservation Strategy: Living Resource Conservation for Sustainable Development. See IUCN, UNEP and WWF (1980).

Yon, D. and Tendron, G. (1981). *Alluvial Forests of Europe*. Nature and Environment Series No. 22, Council of Europe, Strasbourg. 65 pp. (Maps.)

Zohary, M. (1973). *Geobotanical Foundations of the Middle East*, 2 vols. Fischer, Stuttgart; Swets and Zeitlinger, Amsterdam. (Includes geobotanical map of Middle East; covers the Arabian Peninsula, north-east Egypt, Iran, Iraq, Israel, Jordan, Lebanon, Syria and Turkey.)

Appendix 2: Index to Bibliography

This appendix contains a geographical index to the references given in Appendix 1. The regions are subdivided where appropriate. Each reference appears as many times as is necessary to indicate its geographical coverage; a work which covers the whole of the Americas, for example, appears under the heading 'Americas'; one which covers South and Central America occurs once under each of those headings. The Mediterranean region is made up of parts of Europe, Africa and the Middle East, and so a work relating to it must, by definition, appear under one of the other regions as well; for example, a work covering the north coast of Africa would be found under both North Africa and Mediterranean.

The letters in square brackets [ ] after each reference are a crude indication of subject matter:

C: Conservation; reference contains a list of threatened plants or a descriptive account of some aspect of conservation.

F: Flora; reference contains an enumeration of species (usually fairly comprehensive); with or without descriptions, maps, illustrations.

V: Vegetation; reference contains a descriptive or cartographic account of vegetation.

O: Other; e.g. bibliography, description of phytogeography, etc.

Where more than one letter is given, the first indicates the prime subject matter. Any other letters indicate additional subject matter and are in alphabetic order and do not imply order of importance.

The regions and subregions are mostly self-explanatory, but a few are defined as follows:

Africa: The line between North and Tropical Africa is taken as the cut-off point of the Flora of West Tropical Africa, which is approximately 18°N; Southern Africa comprises Namibia (except the Caprivi Strip), South Africa and its states, Lesotho and Swaziland.

Americas: Middle America stretches from Mexico to Panama; the Caribbean includes the islands of the Caribbean sea but not the surrounding mainland coasts.

Australasia comprises Australia, New Zealand and associated islands. Continental Asia excludes Peninsular Malaysia, which is included in Malesia.

Europe: Scandinavia comprises Denmark, Norway, Sweden and Finland. The line between Eastern and Western Europe (taken in a purely geographical sense, i.e. with no political implications) is the eastern border of F.R.G., Austria and Italy. European Turkey is included in Europe.

Malesia includes Malaysia, Indonesia, Philippines and the island of New Guinea.

Middle East stretches from Turkey to the Arabian Peninsula and Iran.

General References:
Ayensu et al. (1984) [C]; Belousova and Denisova (1983) [C]; Boardman (1981) [OC]; Carlquist (1965) [O]; Carlquist (1974) [O]; Derrick et al. (1984) [O]; Eckholm (1976) [C]; Eckholm (1979) [C]; Ehrlich and Ehrlich (1981) [C]; FAO/UNEP

(1981) [VC]; Frankel and Soulé (1981) [C]; Frodin (1984) [O]; Hedberg (1979) [OC]; Henderson (1983) [O]; Holdridge (1967) [V]; Holzner *et al.* (1983) [VC]; Huxley (1985) [C]; IUCN-CMC and WWF (1984) [C]; IUCN, UNEP and WWF (1980) [COV]; Larsen and Holm-Nielsen (1979) [OV]; Lucas and Synge (1978) [C]; Moore (1982) [VCO]; Morse and Henifin (1981) [CO]; Myers (1979) [COV]; Myers (1980) [CV]; Myers (1983) [CO]; Myers (1984) [CV]; Prance (1982) [OCV]; Prescott-Allen (1982) [CO]; Prescott-Allen (1983) [CO]; Simmons *et al.* (1976) [C]; Soulé and Wilcox (1980) [C]; Sutlive *et al.* (1981) [VCO]; Synge (1981) [COV]; Synge and Townsend (1979) [C]; Unesco/UNEP/FAO (1978) [OCV]; US C.E.Q. and Dept of State (1980) [COV].

Africa:
Ayensu (1984) [CO]; Knapp (1973) [V]; Kunkel (1979) [OC]; Lebrun (1960) [O]; Lebrun (1976) [O]; Merxmüller (1971) [OFV]; Miège and Stork (1975, 1976) [OCFV]; Stork and Lebrun (1981) [O]; White (1983) [V].

North Africa:
Castroviejo (1979) [O]; CNRS (1975) [OFV]; Fl.Afr.N. [F]; Fl.Sahara [FOV]; Frankenberg and Klaus (1980) [F]; Gómez-Campo (1979) [C]; Gómez-Campo (1985) [CV]; IUCN (1980) [CF]; Lack (1979) [O]; Lebrun (1977-1979) [F]; Lebrun and Stork (1978) [F]; Léonard (1980) [V]; Med-Checklist [F]; Pichi-Sermolli (1979) [O]; Polunin and Huxley (1978) [F]; Quézel (1965) [V]; Quézel (1978) [O]; Synge (1980) [C]; Unesco/FAO (1970) [V].

Southern Africa:
Ambio (1983) [CO]; Coates Palgrave (1977) [F]; Dyer (1975, 1976) [F]; Fl.S.Afr. [F]; Hall *et al.* (1980) [C]; Hedberg and Hedberg (1968) [CV]; Killick (1983) [OCFV]; Palmer and Pitman (1972) [F]; Werger (1978) [OCV].

Tropical Africa:
Adumbratio [F]; Ambio (1983) [CO]; Aubréville (1950) [F]; Brenan (1978) [O]; Burkill (1985-) [FO]; Coates Palgrave (1977) [F]; Coates Palgrave *et al.* (1957) [F]; Distr.Pl.Afr. [F]; Engler (1892) [FV]; Enumeratio [F]; Fl.Afr.Cent. [F]; Fl.Descr.Mt.Nimba [FV]; Fl.Trop.Afr. [F]; Fl.Trop.E.Afr. [F]; Fl.W.Trop.Afr. [F]; Fl.Zamb. [F]; Geerling (1982) [F]; Hamilton (1982) [OCV]; Hedberg (1951) [V]; Hedberg (1957) [F]; Hedberg and Hedberg (1968) [CV]; Killick (1983) [OCFV]; Letouzey (1969-1972) [OF]; Lind and Morrison (1974) [V]; Schnell (1976, 1977) [VO]; Werger (1978) [OCV]; White (1981) [OC]; Wild and Barbosa (1967, 1968) [V].

Americas:
Prance and Elias (1977) [C].

Caribbean:
Beard (1949) [V]; Cabrera and Willink (1980) [O]; Fl.Less.Ant. [F]; Fl.Neotrop. [F]; Fournet (1976) [F]; Graham (1973) [V]; Lugo *et al.* (1981) [CV]; Sachet and Fosberg (1955, 1971) [O]; Stehlé (1945) [V]; Toledo (1985) [O]; Verdoorn (1945) [OV].

Middle America:
Cabrera and Willink (1980) [O]; Fl.Mesoam. [F]; Fl.Neotrop. [F]; Fuller and Swift (1984) [O]; Gentry (1978) [OV]; Graham (1973) [V]; Org.Est.Amer. (1967) [C]; Toledo (1985) [O]; Verdoorn (1945) [OV].

North America:
Org.Est.Amer. (1967) [C].

South America:

Cabrera and Willink (1980) [O]; Fl.Neotrop. [F]; Fuller and Swift (1984) [O]; Gentry (1978) [OV]; Graham (1973) [V]; Org.Est.Amer. (1967) [C]; Toledo (1985) [O]; Unesco (1981) [V]; Verdoorn (1945) [OV].

Asia, Continental:

Beddome (1892) [F]; Fl.Brit.Ind. [F]; Fl.Camb.Lao.Viet. [F]; Fl.E.Him. [FV]; Fl.Gén.Ind.-Chin. [F]; Fl.Iran. [F]; Fl.Males.Bull. [OC]; Grubov (1963-) [FO]; Holloway (1976) [CV]; IUCN (1980) [CF]; Lall and Moddie (1981) [OCFV]; Lebrun (1977-1979) [F]; Merrill and Walker (1938) [O]; Nayar and Kaur (1972) [F]; Norlindh (1949) [FO]; Parris (1985) [O]; Pételot (1955) [O]; Polunin and Stainton (1984) [FV]; Ranjitsinh (1979) [VC]; Reed (1969) [O]; Smithsonian Institution (1969) [O]; Unesco (1974) [VC]; Vidal (1972) [O]; Whitmore (1975a) [CV]; Whitmore (1975b) [VCF].

Atlantic Ocean incl. Macaronesia:

Fl.Mac.Check. [F]; IUCN (1980) [CF]; Knapp (1973) [V]; Lebrun (1960) [O]; Lebrun (1976) [O]; Sachet and Fosberg (1955, 1971) [O]; Threatened Plants Unit (1983) [CF]; Verdoorn (1945) [OV]; White (1983) [V].

Australasia:

Bramwell (1979) [OC]; Clark and Dingwall (in prep.) [COV]; Costin and Groves (1973) [CV]; Dahl (1980) [COV]; Douglas (1969) [OV]; Fl.Males. [O]; Fl.Males.Bull. [OC]; Given (1983) [COV]; Pac.Pl.Areas [FO]; Parris (1985) [O]; Whitmore (1975b) [VCF].

Europe:

Anon. (1984) [COV]; Añon. (1985) [CO]; Baumann and Künkele (1982) [F]; Derrick and Scheepen (1984) [O]; Fl.Eur. [F]; Goodwillie (1980) [VC]; Grey-Wilson (1979) [F]; Halliday and Beadle (1983) [F]; Heywood (1971) [C]; Jalas and Suominen (1972-) [F]; Jalas and Suominen (1984) [O]; Kent and Brummitt (1966-1971) [O]; Kornas (1976) [C]; Krüssmann (1979) [F]; Meusel (1965-) [F]; Ozenda *et al.* (1979) [V]; Polunin (1969) [F]; Polunin (1976) [F]; Polunin and Walters (1985) [V]; Priszter (1983) [O]; Schacht (1976) [F]; Synge (1980) [C]; Thonner (1980) [F]; Threatened Plants Unit (1983) [CF]; Walters (1971) [C]; Walters (1975) [O]; Walters (1976) [C]; Webb (1978) [O]; Wolkinger and Plank (1981) [VC]; Yon and Tendron (1981) [VC].

Eastern Europe:

Castroviejo (1979) [O]; CNRS (1975) [OFV]; Gómez-Campo (1979) [C]; Gómez-Campo (1985) [CV]; Götz (1975) [F]; Hamann and Wagenitz (1977) [O]; Horvat *et al.* (1974) [V]; Jávorka and Csapody (1979) [O]; Lack (1984) [O]; Med-Checklist [F]; Oberdorfer (1983) [F]; OPTIMA-Projekt [F]; Pichi-Sermolli (1979) [O]; Polunin (1980) [FV]; Polunin and Huxley (1978) [F]; Prod.Fl.Pen.Balc. [F]; Thirgood (1981) [C]; Unesco/FAO (1970) [V]; Walters (1979) [C].

Scandinavia:

Fitter *et al.* (1974) [F]; Hultén (1971) [F]; Koester (1980) [CO]; Lindman (1964) [F]; Ovesen *et al.* (1978) [C]; Ovesen *et al.* (1982) [C]; Poore and Gryn-Ambroes (1980) [C].

Western Europe:

Brosse (1977) [F]; Castroviejo (1979) [O]; CNRS (1975) [OFV]; Ellenberg (1978) [V]; Favarger (1956, 1958) [V]; Fitter *et al.* (1974) [F]; Gómez-Campo (1979) [C]; Gómez-Campo (1985) [CV]; Götz (1975) [F]; Hamann and Wagenitz (1977) [O]; Hartmann and Jahn (1967-) [V]; Hegi *et al.* (1977) [FC]; Ill.Fl.Mitteleur. [F]; Jávorka and Csapody (1979) [O]; Lack (1979) [O]; Landolt (1984) [FV]; Lebrun (1960) [O]; Med-

Checklist [F]; Oberdorfer (1983) [F]; OPTIMA-Projekt [F]; Pichi-Sermolli (1979) [O]; Polunin and Huxley (1978) [F]; Polunin and Smythies (1973) [F]; Poore and Gryn-Ambroes (1980) [C]; Thirgood (1981) [C]; Unesco/FAO (1970) [V]; Walters (1979) [C].

European Russia:
Baumann and Künkele (1982) [F]; Derrick and Scheepen (1984) [O]; Fl.Eur. [F]; Halliday and Beadle (1983) [F]; Holloway (1976) [CV]; Hultén (1971) [F]; Jalas and Suominen (1972-) [F]; Jalas and Suominen (1984) [O]; Jávorka and Csapody (1979) [O]; Kent and Brummitt (1966-1971) [O]; Krüssmann (1979) [F]; Meusel (1965-) [F]; Pichi-Sermolli (1979) [O]; Polunin (1969) [F]; Priszter (1983) [O]; Schacht (1976) [F]; Thonner (1980) [F];Threatened Plants Unit (1983) [CF]; Walters (1971) [C]; Walters (1975) [O]; Webb (1978) [O].

Indian Ocean:
Ayensu (1984) [CO]; Baker (1877) [F]; Beddome (1892) [F]; Bramwell (1979) [OC]; Cabanis *et al.* (1969-1970) [V]; Fl.Brit.Ind. [F]; Fl.Masc. [F]; Fosberg and Sachet (1972) [OCV]; Hedberg and Hedberg (1968) [CV]; Killick (1983) [OCFV]; Kunkel (1979) [OC]; Lebrun (1960) [O]; Lebrun (1976) [O]; Melville (1970) [C]; Merxmüller (1971) [OFV]; Miège and Stork (1975, 1976) [OCFV]; Peters and Lionnet (1973) [O]; Sachet and Fosberg (1955, 1971) [O]; White (1983) [V].

Malesia:
Ambio (1984) [COV]; Beddome (1892) [F]; Dahl (1980) [COV]; Fl.Brit.Ind. [F]; Fl.Males. [FOV]; Fl.Males.Bull. [OC]; Given (1983) [COV]; Jacobs (1982) [VC]; Merrill (1945) [OV]; Nayar and Kaur (1972) [F]; Pac.Pl.Areas [FO]; Parris (1985) [O]; Ranjitsinh (1979) [VC]; Reed (1969) [O]; Smithsonian Institution (1969) [O]; Steenis (1934-1936) [F]; Steenis (1958) [V]; Unesco (1974) [VC]; Whitmore (1975a) [CV]; Whitmore (1975b) [VCF]; Whitmore (1984) [VC].

Mediterranean:
Anon. (1984) [COV]; Anon. (1985) [CO]; Ayensu (1984) [CO]; Baumann and Künkele (1982) [F]; Brosse (1977) [F]; Castroviejo (1979) [O]; CNRS (1975) [OFV]; Derrick and Scheepen (1984) [O]; Eig *et al.* (1931) [F]; Fl.Afr.N. [F]; Fl.Eur. [F]; Fl.Palaes. [F]; Fl.Sahara [FOV]; Frankenberg and Klaus (1980) [F]; Gómez-Campo (1979) [C]; Gómez-Campo (1985) [CV]; Goodwillie (1980) [VC]; Götz (1975) [F]; Grey-Wilson (1979) [F]; Halliday and Beadle (1983) [F]; Hamann and Wagenitz (1977) [O]; Heywood (1971) [C]; Holloway (1976) [CV]; Horvat *et al.* (1974) [V]; IUCN (1980) [CF]; Jalas and Suominen (1972-) [F]; Jalas and Suominen (1984) [O]; Jávorka and Csapody (1979) [O]; Kent and Brummitt (1966-1971) [O]; Knapp (1973) [V]; Kornas (1976) [C]; Krüssmann (1979) [F]; Kunkel (1979) [OC]; Lack (1979) [O]; Landolt (1984) [FV]; Lebrun (1960) [O]; Lebrun (1976) [O]; Lebrun (1977-1979) [F]; Lebrun and Stork (1978) [F]; Med-Checklist [F]; Merxmüller (1971) [OFV]; Meusel (1965-) [F]; Miège and Stork (1975, 1976) [OCFV]; OPTIMA-Projekt [F]; Ozenda *et al.* (1979) [V]; Pichi-Sermolli (1979) [O]; Polunin (1969) [F]; Polunin (1976) [F]; Polunin (1980) [FV]; Polunin and Huxley (1978) [F]; Polunin and Smythies (1973) [F]; Polunin and Walters (1985) [V]; Poore and Gryn-Ambroes (1980) [C]; Post (1932) [F]; Priszter (1983) [O]; Prod.Fl.Pen.Balc. [F]; Quézel (1978) [O]; Schacht (1976) [F]; Stork and Lebrun (1981) [O]; Synge (1980) [C]; Thirgood (1981) [C]; Thonner (1980) [F]; Threatened Plants Unit (1983) [CF]; Unesco/FAO (1970) [V]; Walters (1971) [C]; Walters (1975) [O]; Walters (1976) [C]; Walters (1979) [C]; Webb (1978) [O]; White (1983) [V]; Wolkinger and Plank (1981) [VC]; Yon and Tendron (1981) [VC]; Zohary (1973) [V].

Middle East:

Castroviejo (1979) [O]; CNRS (1975) [OFV]; Eig *et al.* (1931) [F]; Fl.Iran. [F];
Fl.Palaes. [F]; Gómez-Campo (1979) [C]; Gómez-Campo (1985) [CV]; Holloway
(1976) [CV]; IUCN (1980) [CF]; Lack (1979) [O]; Lebrun (1977-1979) [F]; Med-
Checklist [F]; Pichi-Sermolli (1979) [O]; Polunin and Huxley (1978) [F]; Post
(1932) [F]; Synge (1980) [C]; Threatened Plants Unit (1983) [CF]; Unesco/FAO
(1970) [V]; Zohary (1973) [V].

Pacific Ocean:

Ambio (1984) [COV]; Balgooy (1971) [F]; Bramwell (1979) [OC]; Costin and Groves
(1973) [CV]; Dahl (1980) [COV]; Douglas (1969) [OV]; F.Males. [O];
Fl.Males.Bull. [OC]; Fl.Micronesia [F]; Fl.Micronesica [FV]; Fl.Pol.Fr. [F];
Fl.SE.Pol. [FV]; Fosberg (1973) [VC]; Fosberg *et al.* (1979) [F]; Fosberg *et al.*
(1982) [F]; Given (1983) [COV]; Merrill (1924) [O]; Merrill (1945) [OV]; Merrill
(1947) [O]; Pac.Pl.Areas [FO]; Parris (1985) [O]; Radovsky *et al.* (1984) [OFV];
Ranjitsinh (1979) [VC]; Sachet and Fosberg (1955, 1971) [O]; Whitmore (1975a) [CV];
Whitmore (1975b) [VCF].

Southern Ocean/Anatarctica:

Bramwell (1979) [OC]; Clark and Dingwall (1985) [COV]; Given (1983) [COV].

Appendix 3: The Implementation of Global Conservation Conventions Relevant to Plants

	WHC	CITES		RAMSAR	
	Acceptance, Accession or Ratification	Entry into force	Signatory States not yet Ratified	Accession or Ratification	Signature without reservation as to Ratification
AFGHANISTAN	20.03.79				
ALBANIA					
ALGERIA	24.06.74	21.02.84		04.11.83	
ANGOLA					
ANTIGUA AND BARBUDA	01.11.83				
ARGENTINA	23.08.78	08.04.81			
AUSTRALIA	22.08.74	27.10.76			08.05.74
AUSTRIA		27.04.82		16.12.82	
BAHAMAS		18.09.79			
BAHRAIN					
BANGLADESH	03.08.83	18.02.82			
BELGIUM		01.01.84			
BENIN	14.06.82	28.05.84			
BHUTAN					
BOLIVIA	04.10.76	04.10.79			
BOTSWANA		12.02.78			
BRAZIL	01.09.77	04.11.75			
BRUNEI					
BULGARIA	07.03.74				24.09.75
BURKINA FASO					
BURMA					
BURUNDI	19.05.82				
CAMEROON	07.12.82	03.09.81			
CANADA	23.07.76	09.07.75		15.01.81	
CENTRAL AFRICAN REPUBLIC	22.12.80	25.11.80			
CHAD					
CHILE	20.02.80	01.07.75		27.07.81	
CHINA		08.04.81			
COLOMBIA	24.05.83	29.11.81			
CONGO		01.05.83			
COSTA RICA	23.08.77	28.09.75			
CUBA	24.03.81				
CYPRUS	14.08.75	01.07.75			
CZECHOSLOVAKIA					
DENMARK	25.07.79	24.10.77		02.09.77	
DOMINICA					
DOMINICAN REPUBLIC	12.02.85				
ECUADOR	16.06.75	01.07.75			
EGYPT	07.02.74	04.04.78			

EL SALVADOR					
EQUATORIAL GUINEA					
ETHIOPIA	06.07.77				
FIJI					
FINLAND		08.08.76		28.05.74	
FRANCE	27.06.75	09.08.78			
GABON					
GAMBIA		24.11.77			
GERMAN DEMOCRATIC REPUBLIC		07.01.76		31.07.78	
GERMANY, FEDERAL REPUBLIC OF	23.08.76	20.06.76		26.02.76	
GHANA	04.07.75	12.02.76			
GREECE	17.07.81			21.08.75	
GRENADA					
GUATEMALA	16.01.79	05.02.80			
GUINEA	18.03.79	20.12.81			
GUINEA-BISSAU					
GUYANA	20.06.77	25.08.77			
HAITI	18.01.80				
HOLY SEE	07.10.82				
HONDURAS	08.06.79	13.06.85			
HUNGARY		27.08.85		11.04.79	
ICELAND				02.12.77	
INDIA	14.11.77	18.10.76		01.10.81	
INDONESIA		28.03.79			
IRAN	26.02.75	01.11.76		23.06.75	
IRAQ	05.03.74				
IRELAND			01.11.74	15.11.84	
ISRAEL		17.03.80			
ITALY	23.06.78	31.12.79		14.12.76	
IVORY COAST	09.01.81				
JAPAN		04.11.80		17.06.80	
JAMAICA	14.06.83				
JORDAN	05.05.75	14.03.79		10.01.77	
KAMPUCHEA			07.12.73		
KENYA		13.03.79			
KIRIBATI					
KOREA (DPR)					
KOREA, REPUBLIC OF					
KUWAIT			09.04.73		
LAOS					
LEBANON	03.02.83				
LESOTHO			17.07.74		
LIBERIA		09.06.81			
LIBYA	13.10.78				
LIECHTENSTEIN		28.02.80			
LUXEMBOURG	28.09.83	12.03.84			
MADAGASCAR	19.07.83	18.11.75			
MALAWI	05.01.82	06.05.82			
MALAYSIA		18.01.78			
MALDIVES					
MALI	05.04.77				
MALTA	14.11.78				
MAURITANIA	02.03.81			22.10.82	
MAURITIUS		27.07.75			
MEXICO	23.02.84				
MONACO	07.11.78	18.07.78			
MONGOLIA					
MOROCCO	28.10.75	14.01.76			20.06.80
MOZAMBIQUE	27.11.82	23.06.81			
NAMIBIA					

NEPAL	20.06.78	16.09.75		
NETHERLANDS		18.07.84	23.05.80	
NEW ZEALAND	22.11.84			13.08.76
NICARAGUA	17.12.79	04.11.77		
NIGER	23.12.74	07.12.75		
NIGERIA	23.10.74	01.07.75		
NORWAY	12.05.77	25.10.76		09.07.74
OMAN	06.10.81			
PAKISTAN	23.07.76	19.07.76	23.07.76	
PANAMA	03.03.78	15.11.78		
PAPUA NEW GUINEA		11.03.76		
PARAGUAY		13.02.77		
PERU	24.02.82	25.09.75		
PHILIPPINES		16.11.81		
POLAND	29.06.76	08.10.73	22.11.77	
PORTUGAL	30.09.80	11.03.81	24.11.80	
QATAR	12.09.84			
ROMANIA				
RWANDA		18.01.81		
ST LUCIA		15.03.83		
ST VINCENT				
SAO TOME & PRINCIPE				
SAUDI ARABIA	07.08.78			
SENEGAL	13.02.76	03.11.77	11.07.77	
SEYCHELLES	09.04.80	09.05.77		
SIERRA LEONE				
SINGAPORE				
SOLOMON ISLANDS				
SOMALIA				
SOUTH AFRICA		13.10.75		12.03.75
SPAIN	04.05.82		04.05.82	
SRI LANKA	06.06.80	02.08.79		
SUDAN	06.06.74	24.01.83		
SURINAME		15.02.81		
SWAZILAND				
SWEDEN		01.07.75		05.12.74
SWITZERLAND	09.75	01.07.75	16.01.76	
SYRIA	13.08.75			
TANZANIA	02.08.77	27.02.80		
THAILAND		21.04.83		
TOGO		21.01.79		
TONGA				
TRINIDAD AND TOBAGO		18.04.84		
TUNISIA	10.03.75	01.07.75	24.11.80	
TURKEY	16.03.83			
TUVALU				
UGANDA				
U.S.S.R.		08.12.76	11.10.76	
UNITED ARAB EMIRATES		01.07.75		
U.K.	29.05.84	31.10.76	05.01.76	
U.S.A.	07.12.73	01.07.75		
URUGUAY		01.07.75	22.05.84	
VANUATU				
VENEZUELA		22.01.78		
VIET NAM		03.03.73		
WESTERN SAMOA				
YEMEN, ARAB REPUBLIC	25.01.84			
YEMEN, DEMOCRATIC	07.10.80			
YUGOSLAVIA	26.05.75		28.03.77	
ZAIRE	23.09.74	18.10.76		
ZAMBIA	04.06.84	22.02.81		
ZIMBABWE	16.08.82	17.08.81		

WHC: Convention concerning the Protection of the World Cultural and Natural Heritage (Paris, France; 1972). This convention provides for the designation of areas of 'outstanding universal value' as world heritage sites, with the principle aim of fostering international cooperation in safeguarding these important areas. Sites, which must be nominated by the signatory nation responsible, are evaluated for their world heritage quality before being declared by the international World Heritage Committee. The convention entered into force 17 December 1975. For each party adopting the convention since August 1975 the convention enters into force four months after the date of adoption.

CITES: Convention on International Trade in Endangered Species of Wild Fauna and Flora. CITES is an international agreement designed to prohibit the international trade in an agreed list of currently endangered species and to control and monitor the international trade in additional species that might otherwise become endangered. The Convention works by issuance of import and export licences by designated government Management Authorities, who are advised by designated Scientific Authorities.

RAMSAR: Convention on Wetlands of International Importance especially as Waterfowl Habitat (Ramsar, Iran; 1971). An international treaty providing the framework for international cooperation for the conservation of wetland habitats. The Convention places general obligations on contracting party states relating to the conservation of wetlands throughout their territory, with special obligations pertaining to those wetlands which have been designated in a 'List of Wetlands of International Importance'. The convention entered into force 21 December 1975. For each party adopting the convention since August 1975 the convention enters into force four months after the date of adoption.

Geographical Index